About IFPRI

The International Food Policy Research Institute (IFPRI), a CGIAR Research Center established in 1975, provides research-based policy solutions to sustainably reduce poverty and end hunger and malnutrition. IFPRI's strategic research aims to foster a climate-resilient and sustainable food supply; promote healthy diets and nutrition for all; build inclusive and efficient markets, trade systems, and food industries; transform agricultural and rural economies; and strengthen institutions and governance. Gender is integrated in all the Institute's work. Partnerships, communications, capacity strengthening, and data and knowledge management are essential components to translate IFPRI's research from action to impact. The Institute's regional and country programs play a critical role in responding to demand for food policy research and in delivering holistic support for country-led development. IFPRI collaborates with partners around the world.

About IFPRI's Peer Review Process

IFPRI books are policy-relevant publications based on original and innovative research conducted at IFPRI. All manuscripts submitted for publication as IFPRI books undergo an extensive review procedure that is managed by IFPRI's Publications Review Committee (PRC). Upon submission to the PRC, the manuscript is reviewed by a PRC member. Once the manuscript is considered ready for external review, the PRC submits it to at least two external reviewers who are chosen for their familiarity with the subject matter and the country setting. Upon receipt of these blind external peer reviews, the PRC provides the author with an editorial decision and, when necessary, instructions for revision based on the external reviews. The PRC reassesses the revised manuscript and makes a recommendation regarding publication to the director general of IFPRI. With the director general's approval, the manuscript enters the editorial and production phase to become an IFPRI book.

Food Systems Transformation in Kenya: Lessons from the Past and Policy Options for the Future

Edited by Clemens Breisinger, Michael Keenan, Juneweenex Mbuthia, and Jemimah Njuki

A Peer Reviewed Publication

International Food Policy Research Institute
Washington, DC

A catalogue record for this book is available from the Library of Congress.

DOI: https://doi.org/10.2499/9780896294561

ISBN: 978-0-89629-456-1

Recommended citation: Breisinger, Clemens, Michael Keenan, Juneweenex Mbuthia, and Jemimah Njuki (eds). 2023. *Food Systems Transformation in Kenya: Lessons from the Past and Policy Options for the Future.* Washington, DC: International Food Policy Research Institute.

International Food Policy Research Institute, 1201 Eye Street NW, Washington, DC 20005, USA Telephone: 202-862-5600

Cover Design: Jason Chow/IFPRI
Book Layout: Joan Stephens/JKS Design
Cover Photo Credit: Alex Kamau

Contents

Figures, Tables, and Boxes

Figures

Tables

Boxes

Acknowledgments

This work is part of the CGIAR Research Initiative on National Policies and Strategies (NPS). CGIAR launched NPS with national and international partners to build policy coherence, respond to crises, and integrate policy tools at national and subnational levels in six countries in Africa, Asia, and Latin America. The International Food Policy Research Institute (IFPRI), a CGIAR Center participating in NPS, took the lead in preparing this publication. Other CGIAR centers participating in NPS are the International Livestock Research Institute (ILRI), International Water Management Institute (IWMI), International Potato Center (CIP), The Alliance of Bioversity International and CIAT (Alliance Bioversity-CIAT), International Institute of Tropical Agriculture (IITA), and WorldFish. We would like to thank all funders who supported this research through their contributions to the CGIAR Trust Fund: https://www.cgiar.org/funders.

We would also like to thank the Ministry of Agriculture and Livestock Development of Kenya and other partners for their inputs and support, especially in the context of the co-organized 1st Kenya Food Systems Conference. In particular, we would like to thank the Hon. Mithika Linturi, Cabinet Secretary, Ministry of Agriculture and Livestock Development (MALD); Harry Kimtai, Principal Secretary for Livestock Development, MALD; Kello Harsama, Principal Secretary for Crops Development, MALD; Leonard Kubok, Coordinator, Agricultural Transformation Office; Richard Ndegwa, Research Secretary, MALD; and Joseph Maina, Director of Research, MALD. We would also like to extend our gratitude to Prof. Appolinaire Djikeng, CGIAR Senior Director, Livestock-Based Systems, and Director General, ILRI; Prof. Nancy Mungai, Director of Research and Extension, Egerton University;

Bimal Kantaria, Chairperson, Agriculture Sector Network (ASNet); Kiplimo Arap Lagat, Chairperson, County Executive Committee Member (CECM) in charge of Agriculture Caucus and CECM in charge of Agriculture, Nandi County; Kiringai Kamau, CECM in charge of Agriculture, Murang'a County; Henry Kinyua, President's Economic Transformation Office; Agatha Thuo, General Manager, ASNet; Paul Gamba, Lecturer, Department of Agricultural Economics and Agribusiness Management, Faculty of Agriculture, Egerton University; Maru Joyce, CIP; and Joseph Karugia, ILRI/CGIAR Country Convener for Kenya, for leading the organization of the 1st Food Systems Conference. We acknowledge and thank the researchers who made presentations at the 1st Kenya Food Systems conference and all those who participated in the conference. Finally, we thank two anonymous reviewers and IFPRI's Publications Review Committee for their invaluable comments and guidance, and IFPRI's editorial team for masterly steering the manuscript to production.

Foreword

The first years of the 2020s have brought a series of challenges to the Kenyan food system. COVID-19, locust infestations, droughts, the fallout of the Russia-Ukraine conflict, and global supply chain disruptions have strained agricultural production and tightened consumers' budgets. These shocks are occurring against a backdrop of ongoing demographic shifts, urbanization, and stagnating agricultural production, and together demand a reexamination of our approach to food systems.

Creating policy-relevant, empirical research for food systems transformation is more important than ever, as policy aims to address current challenges to promote food production, resilience, sustainability, nutrition, and gender equality. *Food Systems Transformation in Kenya: Lessons from the Past and Policy Options for the Future* presents empirical research from across the various components of the Kenyan food system and is an ambitious undertaking designed to paint a holistic view of where Kenya's food system has been and where it can go. Considering the diversity of challenges and opportunities within the Kenyan food system, this volume approaches a multitude of problems facing the food system using a variety of empirical methods to give insights on evidence-based policy pathways.

The contributors for this book demonstrate the spirit of collaboration for solving the 21st century's toughest problems—academics from Kenyan universities, researchers from Kenyan research institutes, International Food Policy Research Institute (IFPRI) colleagues, other CGIAR researchers, international academics, and experts from multilateral institutions came together to make this far-reaching book a reality. Through this unique combination of national and international expertise, *Food Systems Transformation in Kenya* provides comprehensive and useful information for decision-makers in Kenya. It will also spark discussions on food systems transformation among students, researchers, and policymakers to strengthen the science-policy dialogue while also serving as source of learning and inspiration for other countries.

We look forward to the Ministry of Agriculture and Livestock Development's continued collaboration with IFPRI, CGIAR, and other partners in creating research-based policy recommendations that will lead to a brighter, healthier future for all Kenyans.

Hon. Mithika Linturi

Cabinet Secretary
Ministry of Agriculture and
Livestock Development

Dr. Johan Swinnen

Director General, IFPRI
Managing Director,
Systems Transformation, CGIAR

Acronyms

1-MCP	1-Methylcyclopropene
4WT	four-wheel tractor
AAK	Agrochemical Association of Kenya
AE	adult equivalent
AEZ	agroecological zone
AFS	agrifood system
AGRA	Alliance for a Green Revolution in Africa
ALLFED	Alliance to Feed the Earth in Disasters
AnGR	animal genetic resources
ARC	African Risk Capacity
ASALs	arid and semiarid lands
ASDS	Agricultural Sector Development Strategy
ASTGS	Agricultural Sector Transformation and Growth Strategy
AU	animal unit
BR	Biennial Review
CA	conservation agriculture
CAADP	Comprehensive Africa Agriculture Development Program
CBO	community-based organization
CCPP	contagious caprine pleuropneumonia
CIMMYT	International Maize and Wheat Improvement Center
CLIMARK	Climate, Livestock, and Markets
COMESA	Common Market for Eastern and Southern Africa
DAP	diammonium phosphate

DGE	diet deprivation–growth elasticities
DGLV	dark green leafy vegetables
EAG	East African Growers
ECC	evaporative charcoal cooler
ERS	Economic Recovery Strategy for Wealth and Employment Creation
EUREPGAP	Euro-Retailer Produce Working Group for Good Agricultural Practice
FAO	Food and Agriculture Organization of the United Nations
FGD	focus group discussion
FLW	food loss and waste
FTE	full-time employment
GAP	good agricultural practice
GDP	gross domestic product
GRV	Great Rift Valley
HCD	Horticultural Crops Directorate
HIC	high-income country
HSNP	Hunger Safety Net Program
IBLI	index-based livestock insurance
ICRISAT	International Crops Research Institute for the Semi-Arid Tropics
IMV	improved maize variety
IOT	input–output table
KACE	Kenya Agricultural Commodity Exchange
KALRO	Kenya Agricultural and Livestock Research Organization
KAM	Kenya Association of Manufacturers
KAOP	Kenya Agricultural Observation Platform
KARI	Kenya Agriculture Research Institute
KBDS	Kenya Business Development Services
KEBS	Kenya Bureau of Standards
KEPHIS	Kenya Plant Health Inspectorate Service

KIHBS	Kenya Integrated Household Budget Survey
KLIP	Kenya Livestock Insurance Program
KMDB	Kenya Maize Data Base
KNBS	Kenya National Bureau of Statistics
KSC	Kenya Seed Company
LDF	livestock-derived foods
LGP	length of growing period
LIC	low-income country
LMIC	lower-middle-income country
LSF	large-scale farmer
MAP	modified atmospheric packaging
MRL	maximum residue limit
MSF	medium-scale farmer
MTIP	Medium-Term Implementation Plan
NAAIAP	National Accelerated Agricultural Input Access Programme
NAIP	National Agriculture Investment Plan
NCD	noncommunicable disease
NCPB	National Cereals and Produce Board
ND	Newcastle disease
NEMA	National Environmental Management Agency
NGO	nongovernmental organization
NPK	nitrogen, phosphorous, and potassium
NVSP	National Value Chain Support Programme
OPV	open pollinated variety
PBI	picture-based insurance
PCPB	Pest Control Products Board
PGE	poverty–growth elasticities
PPP	public–private partnership
PRA	participatory rural appraisal
PSF	peasant subsistence farmer
R&D	research and development

RCC	risk-contingent credit
RCT	randomized controlled trial
REDD	Reference Diet Deprivation
RIAPA	Rural Investment and Policy Analysis
RIMA	Resilience Index Measurement and Analysis
SAM	social accounting matrix
SDG	Sustainable Development Goal
SMEs	small and medium enterprises
SPAM	Spatial Production Allocation Model
SPI	science–policy interface
SRA	Strategy for Revitalizing Agriculture
SSCF	small-scale commercial farmer
TFP	total factor productivity
THS	tractor-hiring service
UHT	ultra-high temperature
UMIC	upper-middle-income country
USAID	United States Agency for International Development
WASH	water, sanitation, and hygiene
WHO	World Health Organization
ZBEC	zero energy brick cooler

Chapter 1

INTRODUCTION

Michael Keenan, Clemens Breisinger, Juneweenex Mbuthia, and Jemimah Njuki

The whole world has experienced a series of global and local crises since 2019, and Kenya has been no exception. Before the COVID-19 pandemic and the Russian invasion of Ukraine, poverty and food poverty rates in the country had been declining steadily, falling from 52.3 percent to 36.1 percent and from 38.3 percent to 26.7 percent, respectively, between 1997 and 2016 (KNBS 2007, 2018). Income inequality also declined in the period from 1994 to 2015/16 (KNBS 2020).[1] Estimates suggest that, since then, progress in poverty reduction has reversed, as a result of COVID-19 (Nafula et al. 2020), and that the impacts of the Ukraine and global crises have further increased poverty levels and the number of people unable to afford a healthy diet (Breisinger et al. 2022). In addition, ongoing droughts in the arid and semiarid areas of Kenya meant that an estimated 3.5 million people were in need of assistance in May 2022 (UNICEF 2022).

Even before these crises, macroeconomic trends and acute economic disruptions had put the Kenyan economy under enormous pressure in recent years. Kenya is experiencing rapid urbanization and population growth: the urban population is projected to make up nearly 50 percent of the population by

1 Income inequality remains high in Kenya, with the Gini coefficient at 0.404; however, long-term trends point toward a decrease in income inequality over time (Bigsten et al. 2016; KNBS 2020). Some concerns have been raised over economic growth being driven by infrastructure development and growth in the information and communications technology (ICT) sector (particularly in fintech), which may be contributing to increased income inequality and replicating patterns of uneven development, particularly between rural and urban areas (Bernards 2022). However, nationwide data have not validated these concerns, and evidence from cross-country studies on African inequality point toward ICT investment lessening, not increasing, income inequality (Nchake and Shuaibu 2022; Njangang et al. 2022). Further, the widespread use of mobile money (through M-Pesa) may be a source of increased financial inclusion and reduced poverty (leading ultimately to reduced inequality)—although the empirical results for such effects from mobile money are mixed (Suri and Jack 2016; van Hove and Dubus 2019).

2050, and in absolute terms this amounts to over 40 million people (UNDESA 2018). These trends are leading to increased urban food demand, longer supply chains, and greater rural inequality (IFAD 2021). How to feed and support a growing and changing population under tumultuous climatic and market conditions is perhaps the greatest challenge facing Kenya today.

Gender inequalities are contributing to the underperformance of the economy and the agriculture sector specifically. Women represent over half of the agricultural labor force in Africa south of the Sahara (Palacios-Lopez, Christiaensen, and Kilic 2017). Their substantive contribution to agriculture and their vital role in ensuring family food security have been widely documented. However, gender-based inequalities in access to and control of productive and financial resources inhibit agricultural productivity and undermine resilience and sustainability efforts. Gender productivity gaps in the region range from about 11 percent to 28 percent, with Kenya at around 18 percent (World Bank 2012; UN Women, UNDP, and UNEP 2018). Substantial gender gaps in productivity have arisen not because women are less efficient farmers but because they have inequitable access to land and to agricultural inputs (Rodgers and Akram-Lodhi 2019). Such unbalanced distribution frequently stems from and is bolstered by deeply entrenched sociocultural norms and traditional expectations of gender roles.

In order to address these challenges, the Government of Kenya has made food security one of the key pillars of its national development strategy, Vision 2030, through the Medium Term Plan III and the Big Four Agenda, with the aim of achieving 100 percent food and nutrition security by 2022. Specifically, this is to be achieved through the construction of large-scale multipurpose dams and smaller dams for irrigation projects and of food storage facilities, and the implementation of high-impact nutritional interventions (Kenya, National Treasury and Planning 2018).

Despite these commitments, and the important role that agriculture plays in the Kenyan economy and jobs, public expenditure is lacking. Agriculture contributes 22.7 percent of gross domestic product (GDP) directly to Kenya's economy, and the broader food system contributes 33.8 percent (Chapter 2). The sector employs more than 43.3 percent of the total population (Chapter 2) and more than 70 percent of Kenya's rural people (IFAD 2019). However, public expenditure on the agriculture sector remained around 2 percent in the 2021/22 fiscal year, well below the Malabo commitments of 10 percent and below the spending of its peers in Africa south of the Sahara (AUC 2014; Kenya Parliamentary Service Commission 2021; Pernechele et al. 2021).These low levels of expenditure translate into weakened policy implementation: the agriculture, water, and environment sectors together have 216 stalled government projects—more than any other sector (Kenya Parliamentary Service Commission 2021). However, increasing spending does not automatically translate into better outcomes: quality of spending matters. There is a need to prioritize investments and to target spending strategically to make it possible to achieve the desired impacts most effectively (Fan 2008).

At the international level, the UN has coined this decade the "Decade of Action," as countries work toward reaching the Sustainable Development Goals (SDGs) by 2030 (UNSDG 2020). However, longstanding structural challenges and the ongoing crises mean that, like many countries, Kenya is not on track to achieve the SDGs. The country also risks not achieving its national goal of becoming a middle-income country by 2030, laid out in its Vision 2030 (World Bank 2020; Sachs et al. 2021). Action is urgently needed.

A promising avenue for fostering change toward achieving Kenya's national development and the SDGs is to focus policies and investments on *food systems-led transformation*. Transformation in food systems is more than just economic and agricultural transformation. It involves changes to all facets of food systems, including the supply and consumption of food as well as institutional support. On the supply side, transformation involves moving labor and other productive assets from low- to high-production activities as well as improving productivity within each value chain (UN-Habitat 2016). Such changes can come through the reprioritization of value chains and through on-farm developments or improvements in upstream and downstream supply chains. On the demand side, transformation involves changing consumption behavior and improving access to and availability of food. Policies, institutions, and research should facilitate these changes with the objectives of building nutritious, productive, resilient, sustainable, and inclusive food systems (IFPRI 2021).

The next sections provide a brief overview of the evolution of economic development thought, from an initial focus on agricultural productivity to a more market-oriented value chain approach to the current focus on food systems.

A new opportunity for development: From agriculture- to food systems-led transformation

Understanding the historical context of frameworks for food policy and rural development is critical to understanding why there is a need today for a food systems approach, and how this differs from previous initiatives. While every country's food policy follows its own path, there are undeniably global trends that underlie local policy decisions. These global trends are even more important when there is external financing of large portions of public agriculture-related budgets, as is the case in Kenya. In the 1960s, the Green Revolution led to a global effort to intensify agriculture and boost productivity through the uptake of improved varieties and chemical fertilizers, particularly for staple foods. In Kenya, after some initial success, the Green Revolution for staple foods has stalled since the 1980s (De Groote et al. 2005). With the Green Revolution generating mixed success, global trends in food policy then shifted toward value chain development at the start of the 21st century, and they are now moving toward food systems approaches (as seen in the 2021 UN Food Systems Summit). The brief discussion of the evolution of global trends in policy approaches that follows sets the stage for the presentation of the food systems framework.

The traditional Green Revolution approach

The approach to understanding economic transformation has evolved over the past decades from a narrow focus on improving on-farm productivity to a broader one on transforming food systems. This evolution in economic thought is reflective of historical transformations in agriculture, growing evidence on effective and ineffective approaches to food sector development, and changing policy strategies and goals.

Previous agricultural revolutions have been driven by the development and dramatic uptake of on-farm production technologies. The British Agricultural Revolution, taking place from the 18th century to the 19th century, for example, saw the uptake of improved practices such as crop rotation, the introduction of new crops from global trade, and the use of machinery such as seed drills to make farming more efficient. To enable these transformations, the government pursued policies of land consolidation (for example, the Enclosure Acts), market development, and physical infrastructure construction. More efficient, market-oriented farms led to undeniably transformative results: labor and capital shifted from the agriculture sector to industry, leading to the Second Industrial Revolution (O'Brien 1977). This process of increasing within-sector productivity and reallocating labor and capital from less efficient to more efficient sectors is known as structural transformation, and has been identified as a key process in achieving economic development (Breisinger and Diao 2008).

By the mid-20th century, the world was facing an impending food crisis: rapidly increasing populations threatened to outstrip food supply. Scientists searched for a way to increase food production to avoid a global, Malthusian catastrophe—and the Green Revolution was born. Cereals, particularly maize, wheat, and rice, were the focus of the Green Revolution because of their ability to provide calories *en masse*. The development of chemical fertilizers, high-yielding varieties, improved irrigation, and farm machinery proliferated on farms and allowed previously low-yield areas to become important food producers (Pingali 2012). Mexico transformed from a net importer of maize to a net exporter in under two decades (Sonnenfeld 1992). India's grain deficit turned into a surplus (Rahman 2015). Numerous other countries, notably in Southeast Asia, saw similar results and were able not only to transform their agriculture sectors but also to set the stage for industrialization and structural transformation (Hazell 2009). As in the British Agricultural Revolution before, policies enabling land consolidation and market and infrastructure development helped pave the way for the Green Revolution.

Broadly speaking, most economies in Africa south of the Sahara—including Kenya—have not seen the transformative effects of the Green Revolution. Until recently, much of the research on the African food sector has been dedicated to understanding how to replicate the results from the Green Revolution in the region, in the assumption that increasing on-farm productivity will make it possible to meet the conditions for structural transformation and spur economic transformation.

However, the results of this "traditional" approach have not been sufficiently transformative. For example, in Kenya, per hectare maize yields have not increased in 30 years (FAOSTAT 2022), food insecurity is still a major issue (Sachs et al. 2021), uptake of inputs and machinery is low (Chapter 9), and the agriculture sector still absorbs most of the labor force (World Bank 2021). Subsistence agriculture, as opposed to market-oriented production, still plays a large role in Kenyan agriculture and hinders the ability of farmers to enhance production through input use and land consolidation (Muthini, Nzuma, and Qaim 2020; Gollin, Kirchberger, and Lagakos 2021). As such, Kenya is in the process of transitioning from traditional to modern rural systems (IFAD 2021).

Market development and the value chain approach

Because market development has played a crucial enabling role in past agricultural transformations, policymakers have increasingly turned toward inclusive value chain development to transform African food sectors (Barrett et al. 2020). Value chains encapsulate the process of creating a product and delivering it to end consumers (Reardon and Timmer 2012). In Kenya, inefficiencies and uneven distribution of profits (or of the value created) often characterize agricultural value chains, and market access for farmers remains poor in many areas (Chamberlin and Jayne 2013). As such, the development of value chains to enhance the production of food and streamline the process of distributing food to end consumers has been a much-needed policy focus. Undergoing these value chain transformations in an inclusive manner has also been critical. This ensures that small or marginalized farmers (for example, women) and value chain actors (for example, processors, transporters, and retailers) can benefit from the value chain development, and existing inequalities are not exacerbated (Donovan, Stoian, and Lundy 2016). Such an approach builds on the failings of the Green Revolution, which placed little importance on inclusivity and often pushed marginalized actors out of the sector altogether (Hazell 2009). However, this approach to transforming the food sector still often remains supply-centric, and does not address the environmental issues associated with Green Revolution technologies, which are becoming increasingly important in the context of climate change.

Toward new transformative frameworks

In order to address the weaknesses of previous approaches, frameworks must not focus only on improving food supply. Demand-side dynamics and institutional arrangements play vital roles in the food sector; to drive transformation, it will be key to address constraints in these areas alongside those facing farmers. Second, frameworks must recognize that on-farm productivity is only one outcome, and it cannot be pursued at the expense of other desirable outcomes, such as sustainability. During the Green Revolution, overuse of fertilizers, pesticides, and other chemical inputs led to land degradation, the pollution of water sources, and other detrimental environmental outcomes (Hazell 2009). In the face of a growing climate crisis, it is critical to pursue the objectives of sustainability and resilience jointly with increasing productivity. Frameworks that explicitly address these issues are needed. Such frameworks should not ignore the importance of improving productivity: yields in Kenya and other countries in Africa south of the Sahara are still low, and stagnating production is a major issue currently facing the Kenyan food system that rightfully deserves policy attention (Chapter 7). However, this should not come at the price of creating further problems down the road.

The following section provides an overview of food systems frameworks and then proposes a framework for food system transformation in Kenya.

The food systems framework for Kenya

The food systems[2] framework has been proposed as a holistic framework to guide policy to meet the various needs of a modern food sector. More generally, systems thinking "involves exploring the characteristics of components within a system . . . and how they interconnect to improve understanding of how outcomes emerge from these interactions" (McNab et al. 2020). In the context of food systems, this requires identifying the proper components, or actors, and understanding how such actors interact with one another to either promote or hinder desired food system outcomes. Once an understanding of the various actors, interactions, opportunities, and constraints of a food system is attained, the policy approach to facilitating food system transformation flows naturally.

Defining food systems, understanding what the goals of transformation should be, and identifying the key levers by means of which transformation can occur are precursors to creating an enabling and coordinated policy for the food sector. "A food system gathers all the elements (environment, people, inputs, processes, infrastructures, institutions, etc.) and activities that relate to the production, processing, distribution, preparation and consumption of food, and the outputs of these activities, including socio-economic and environmental outcomes" (HLPE 2017). This complex web of actors and activities can be divided into five broad components: food supply chains, the food environment, consumer behavior, external drivers, and outcomes (shown in Figure 1.1).

Food supply chains concern the production and distribution of food and are conceptually similar to value chains, discussed above. These supply chains include upstream actors (input providers, agricultural services, and extension), farmers, and downstream actors (aggregators, processors, distributors, wholesalers, and retailers). Parallel streams, such as financial services providers and research and development, also play important roles in food supply chains (Reardon et al. 2019). Development of supply chains has been the focus of much agricultural policy in recent years, as inclusive value chain development has become an increasingly popular policy tool (Barrett et al. 2020).

The **food environment** is the context in which consumers obtain, prepare, and consume food (HLPE 2017). Physical, economic, political, and sociocultural factors affect this context—respectively, consumers' physical proximity

2 The food system is also sometimes referred to as the agrifood system, and the terms are often used interchangeably. To ensure consistency with the food systems framework laid out in this book, we use the term "food system" throughout. This choice of wording also relates to the fact that the nonfood part is of minor relevance for the topics covered in this book. The only exception is Chapter 2, which lays out an accounting framework that explicitly captures food and nonfood items across all value chains.

FIGURE 1.1 The food system framework

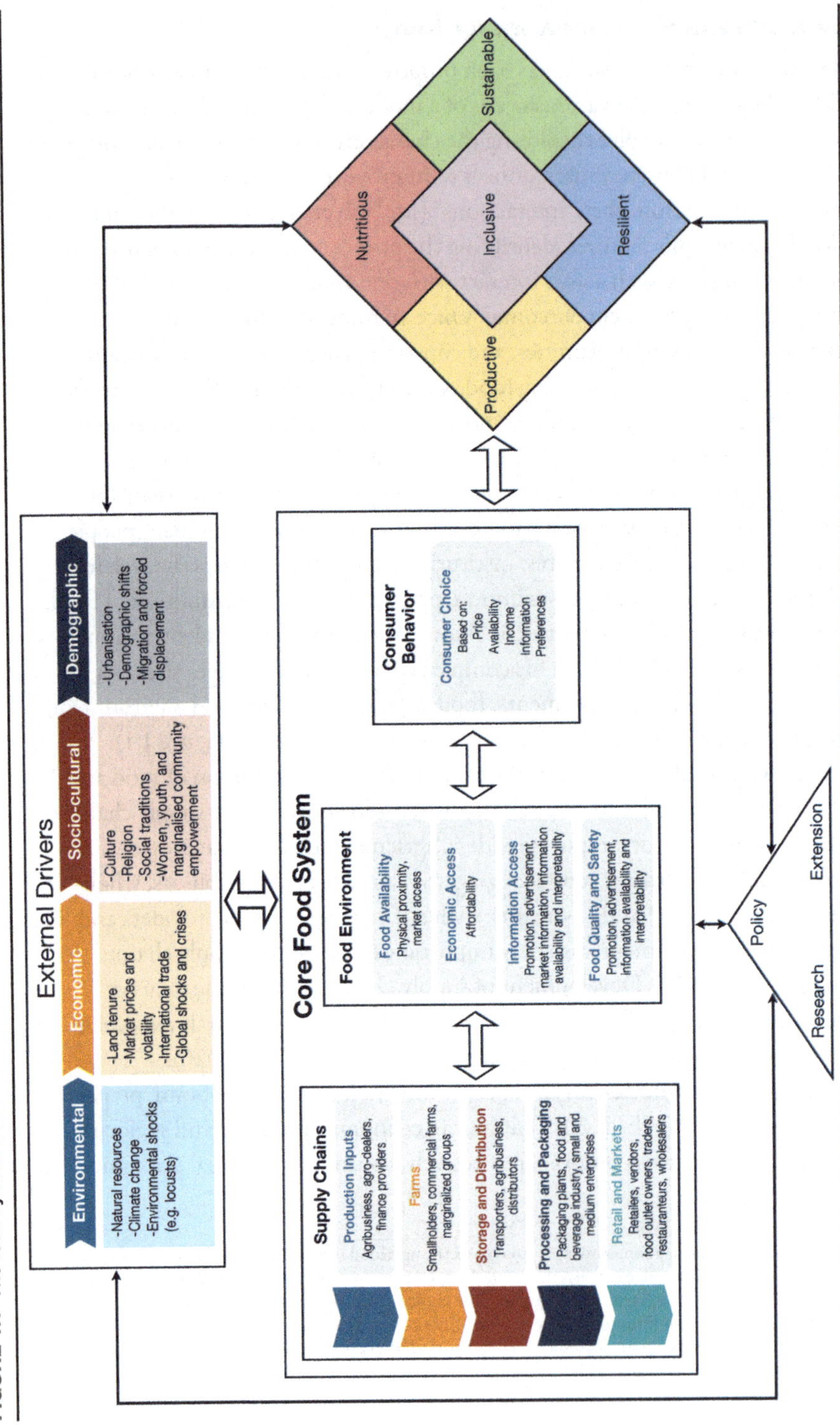

Source: Adapted from HLPE (2017); de Brauw et al. (2019); Njuki et al. (2021).

to markets, food affordability, regulations, and cultural beliefs about food have impacts on the food environment. Further, how food is marketed and promoted to consumers makes up a key component of the food environment.

Consumer behavior is the set of choices consumers make about what food to obtain, prepare, and eat (HLPE 2017). Such behavior occurs at both the household level and the individual level and encapsulates intrahousehold decisions on food allocation. Preferences, income, location, cultural beliefs, and the food environment can all shape how consumers make decisions. Placing consumers as a central component of the food system builds upon previous policy and research approaches that have placed a much larger emphasis on the supply side.

The three central components—the food supply chain, the food environment, and consumer behavior—are all shaped by external factors. Such factors include broader economic conditions, the natural environment, socio-cultural traits, demographic characteristics, and policy. Each central component responds to these factors to affect the outcomes. In the food systems framework, policy is depicted as a component distinct from other external factors because governments and other institutions can directly create policy to affect the rest of the food system. The next section discusses how to enable food system transformation through effective policy.

The food system produces more than just food: it has environmental, health, economic, and cultural outcomes. Further, overall food system results affect food supply, the food environment, and consumer behavior, creating a positive feedback loop. Early work on food systems frameworks placed a greater emphasis on nutrition and environmental outcomes (HLPE 2017; de Brauw et al. 2019); more recent work has stressed the role of gender in every aspect of food systems (Njuki et al. 2021).

The results of food systems can be categorized along five dimensions: productivity, resilience, sustainability, health, and inclusivity (IFPRI 2021). Productivity refers to the ability of food systems to efficiently produce enough food to feed a growing population. However, food supply is subject to environmental factors (for example, the increasingly volatile rainfall patterns across East Africa and the proliferation of pests such as locusts) and market risk (for example, increasing international fertilizer and fuel prices). Food systems must be designed to withstand such shocks to make it possible to avoid acute food crises. The role of food production in environmental degradation is an issue that is moving to the forefront of international policy discussions, and food systems must be able to function sustainably without ruining the most important asset for food production—the environment—for future generations. Further, food systems need to ensure that the food produced meets the nutritional needs of consumers. Safe, balanced diets with sufficient macro- and micronutrients are needed to promote health. Finally, these systems should achieve productivity, resilience, sustainability, and health for all communities and groups of people. Inclusivity is needed to prevent further marginalization, to create more equal societies, and to leverage the vast knowledge and experiences of diverse populations to transform food systems. Inclusivity is often discussed in terms of empowering female and youth populations but it also includes economically and geographically marginalized communities, such as residents of urban slums or remote rural areas.

The interactions of the food supply chain, the food environment, consumer behavior, external drivers, and food system outcomes are what make the food system run. The entire system is an interrelated web of actors and activities. Changes in one area of the food system will have ripple effects that extend beyond the immediate outcomes. For example, a policy promoting the growth of a more suitable crop in semiarid areas will immediately affect the farmers and more broadly the food supply chain. This in turn will affect the food environment, consumer behavior, and eventually food system outcomes. These outcomes are then fed back through the system, creating a cycle of change. Having this broad view of the food system can help researchers and policymakers understand the full effects of interventions, economic and environmental conditions, and policies. Such learning can translate into effective policy.

Research-based, policy-driven food system transformation

This concept of food system transformation has received much attention internationally as well as in many countries, including Kenya. The year 2021 marked an important milestone in international food policy: the UN Food Systems Summit brought together delegates from around the world to discuss a new approach to economic development. Over 100 countries—including Kenya—developed national pathways to transform their food systems. The Food Systems Summit identified five action tracks, analogous to the five key food system outcomes, for governments to focus on: ensuring safe and nutritious food for all; shifting to sustainable consumption patterns; boosting nature-positive production; advancing equitable livelihoods; and building resilience to vulnerabilities, shocks, and stress. Further, the Summit outlined four levers of change that can cut across the action tracks and drive food systems transformation: gender, human rights, finance, and innovation (Herrero et al. 2021).

However, although the Food Systems Summit placed food systems thinking and its objectives front and center for the international community and national governments, it provided less practical guidance on how policy could transform food systems. The action tracks help in policy goal setting and the levers identify areas on which governments should focus to drive change. However, setting goals is not enough. Effective policy needs research-based development and implementation to meet policy goals (Cerna 2013).

An economic lens is particularly useful in understanding the complex nature of food systems, for several reasons, according to Fan and colleagues (2021). First, economics is concerned with scarcity and choice, and resources in food systems are often scarce: producers are constrained in their ability to engage in intensive farming systems, consumers face tight budgets, and institutional actors are limited in their capacity to enable the system. Economic analysis is able to give insights across these different levels of actors. Second, economics is able to analyze the trade-offs between different food system outcomes—productivity, resilience, sustainability, health, and inclusivity. Sometimes, meeting objectives along one of these dimensions may come at the cost of another, and evidence is needed to understand these costs and benefits (de Brauw et al. 2019). Third, economics can provide analysis for individual actors (microeconomics), markets (mesoeconomics), and the economy as a whole (macroeconomics). Therefore, economic analysis can sufficiently address both the feasibility and the results of both bottom-up and top-down interventions. A fourth aspect of economic analysis is also crucial for food system analysis: the quantitative, data-driven nature of modern economics. Rigorous economic analysis relies on large sample sizes and state of the art statistical methods that can meet the objectives of data-driven policy (AUC 2014).

An overview of this book

This book takes a critical look at the Kenyan food system, where it has been, and where it can go. It brings together a mix of Kenyan and international experts from CGIAR, academia, and nongovernmental organizations, resulting in a collaborative work that offers an in-depth discussion of food systems and presents various case studies across different value chains and areas of the food system. Bringing together a diverse range of rigorous research with both a bird's-eye view of Kenyan food systems and in-depth analyses of specific components of food systems allows for a unique perspective into the policies needed to enable transformation. Reflecting this approach, the book recommends broad policy themes and specific policy recommendations that can position Kenya as a global leader in tackling the challenges of the Decade of Action.

This book presents economic studies across the food system using a variety of research methods. Some chapters present reviews and syntheses of scientific evidence on certain topics, whereas others present original, individual studies. The analysis is carried out at different levels, with some chapters looking at broader food system dynamics and other chapters zooming in on individual areas and value chains within the Kenyan food system. Chapters rely on both qualitative and quantitative methods using cutting-edge methodologies, such as randomized controlled trials, simulation modeling, and nationally representative household analysis. Taken as a whole, the varied methodological approach highlights the importance of data-driven learning, the use of mixed methods to more comprehensively understand topics, and the scale and scope of research needed to understand Kenya's food system and assess food policy. It further suggests that there is no cookie-cutter approach to rigorous research: different approaches to research are needed for different questions. In other words, as far as research methodology goes, "The gold standard is the best method for the question at hand" (Ravallion 2018).

The book is divided into six parts. The first part gives an overview of food and livestock systems in Kenya. This is followed by five parts that correspond to the five food system outcomes: health, productivity, resilience, inclusivity, and sustainability. Part 2 (health) presents research on Kenyan diets and food safety in the country. Part 3 contains four chapters related to food production, which analyze, respectively, agricultural productivity trends in general, the role of inputs in agriculture, maize intensification, and mechanization. Part 4 relates to food system resilience and presents research on overall production resilience, climate insurance, and risk contingent credit. Part 5 discusses the inclusivity of the food system with respect to gender, youth, and small farmers. This is followed by Part 6, which presents research related to the environmental and financial sustainability of transformation in the context of postharvest management and digital solutions, respectively. The book concludes with a discussion on overall policy approaches and specific policy recommendations to drive food system transformation. While the chapters are categorized based on the most relevant food system outcome, many of their insights cut across outcomes. The components, actors, and outcomes of food systems do not exist in isolation; consequently, the corresponding research must acknowledge this interconnectivity. This book demonstrates the breadth and depth needed in food systems research to inform transformative policy.

References

AUC (African Union Commission). 2014. *Malabo Declaration on Accelerated Agricultural Growth and Transformation for Shared Prosperity and Improved Livelihoods.*

Barrett, C.B., T. Reardon, J. Swinnen, and D. Zilberman. 2020. "Agri-food Value Chain Revolutions in Low-and Middle-Income Countries." *Journal of Economic Literature* 58: 1–67.

Bernards, N. 2022. "Colonial Financial Infrastructures and Kenya's Uneven Fintech Boom." *Antipode* 54 (3): 708–728.

Bigsten, A., D.K. Manda, G. Mwabu, and A. Wambugu. 2016. "Incomes, Inequality, and Poverty in Kenya: A Long-Term Perspective." In *Growth and Poverty in Sub-Saharan Africa,* eds. C. Arndt, A. McKay, and F. Tarp, 343–369. Oxford, UK: Oxford University Press.

Breisinger, C., and X. Diao. 2008. "Economic Transformation in Theory and Practice: What Are the Messages for Africa?" IFPRI Discussion Paper 797. International Food Policy Research Institute (IFPRI), Washington, DC.

Breisinger, C., X. Diao, P. Dorosh, J. Mbuthia, L. Omune, E.O. Oseko, A. Pradesha, J. Smart, and J. Thurlow. 2022. *Kenya: Impacts of the Ukraine and Global Crises on Poverty and Food Security.* Global Crisis Country Series Brief 1. Washington, DC: IFPRI.

Cerna, L. 2013. *The Nature of Policy Change and Implementation: A Review of Different Theoretical Approaches.* Paris: OECD.

Chamberlin, J., and T.S. Jayne. 2013. "Unpacking the Meaning of 'Market Access': Evidence from Rural Kenya." *World Development* 41 (1): 245–264.

de Brauw, A., M. van den Berg, I.D. Brouwer, H. Snoek, R. Vignola, M. Melesse, G. Lochetti, C. van Wagenberg, M. Lundy, M.E. d'Hotel, and R. Ruben. 2019. "Food System Innovations for Healthier Diets in Low and Middle-Income Countries." IFPRI Discussion Paper 1816. IFPRI, Washington, DC.

De Groote, H., G. Owuor, C. Doss, J. Ouma, L. Muhammad, and K. Danda. 2005. "The Maize Green Revolution in Kenya Revisited." *Electronic Journal of Agriculture and Development Economics* 2 (1), 32–49.

Donovan, J., D. Stoian, and M. Lundy. 2016. "Challenges and Approaches for Inclusive Value-Chain Development: Introduction." In *Innovation for Inclusive Value-Chain Development: Successes and Challenges,* eds. A. Devaux, M. Torero, J. Donovan, and D.E. Horton, 37–46. Washington, DC: IFPRI.

Fan, S. 2008. *Public Expenditures, Growth, and Poverty: Lessons from Developing Countries.* Washington, DC: IFPRI.

Fan, S., D. Headey, C. Rue, and T. Thomas. 2021. "Food Systems for Human and Planetary Health: Economic Perspectives and Challenges." *Annual Review of Resource Economics* 13: 131–156.

FAOSTAT. 2022. FAOSTAT Crops and Livestock Data Base. Food and Agriculture Organization of the United Nations (FAO). https://www.fao.org/faostat/en/

Gollin, D., M. Kirchberger, and D. Lagakos. 2021. "Do Urban Wage Premia Reflect Lower Amenities? Evidence from Africa." *Journal of Urban Economics* 121: 103301.

Hazell, P.B.R. 2009. "The Asian Green Revolution." IFPRI Discussion Paper 911. IFPRI, Washington, DC.

Herrero, M. M. Hugas, U. Lele, A. Wira, and M. Torrero. 2021. "Science and Innovations for Food Systems Transformation and Summit Actions." Paper presented at the *United Nations Food Systems Summit*, April 7.

HPLE (High Level Panel of Experts). 2017. *Nutrition and Food Systems*. Rome: FAO.

IFAD (International Fund for Agricultural Development). 2019. *Republic of Kenya: Country Strategic Opportunities Programme*. Rome.

IFAD. 2021. *Rural Development Report 2021: Transforming Food Systems for Rural Prosperity*. Rome.

IFPRI (International Food Policy Research Institute). 2021. *Global Food Policy Report: Transforming Food Systems after COVID-19*. Washington, DC: IFPRI.

KNBS (Kenya National Bureau of Statistics). 2007. *Basic Report on Well-Being in Kenya*. Based on the 2005/06 Kenya Integrated Household Budget Survey. Nairobi.

KNBS. 2018. *Basic Report on Well-Being in Kenya*. Based on the 2015/16 Kenya Integrated Household Budget Survey. Nairobi.

KNBS. 2020. *Inequality Trends and Diagnostics in Kenya*. Nairobi: Government of Kenya.

Kenya, Parliamentary Service Commission. 2021. *Unpacking the Estimates of Revenue and Expenditure for 2020/2021 and the Medium Term*. Nairobi.

Kenya, The National Treasury and Planning. 2018. *Third Medium Term Plan: 2018-2022. Transforming Lives: Advancing Socio-Economic Development through the 'Big Four'*. Nairobi.

McNab, D., J. McKay, S, Shorrock, S. Luty, and P. Bowie. 2020. "Development and Application of 'Systems Thinking' Principles for Quality Improvement." *BMJ Open Quality* 9: e000714.

Muthini, D., J. Nzuma, and M. Qaim. 2020. "Subsistence Production, Markets, and Dietary Diversity in the Kenyan Small Farm Sector." *Food Policy* 97: 101956.

Nafula, N., D. Kyalo, B. Munga, and R. Ngugi. 2020. "Poverty and Distributional Effects of COVID-19 on Households in Kenya." AERC Working Paper. African Economic Research Consortium, Nairobi.

Nchake, M.A., and M. Shuaibu. 2022. "Investment in ICT Infrastructure and Inclusive Growth in Africa." *Scientific African* 17: e01293.

Njangang, H., A. Beleck, S. Tadadjeu, and B. Kamguia. 2022. “Do ICTs Drive Wealth Inequality? Evidence from a Dynamic Panel Analysis.” *Telecommunications Policy* 46 (2): 102246.

Njuki, J., S. Eissler, H.J. Malapit, R.S. Meinzen-Dick, E. Bryan, and A.R. Quisumbing. 2021. *A Review of Evidence on Gender Equality, Women's Empowerment, and Food Systems.* Food Systems Summit Brief. New York: United Nations Food Systems Summit.

O'Brien, P.K. 1977. “Agriculture and the Industrial Revolution.” *Economic History Review* 30 (1): 166–181.

Palacios-Lopez, A., L. Christiaensen, and T. Kilic. 2017. “How Much of the Labor in African Agriculture Is Provided by Women?” *Food Policy*, 67: 52–63.

Pernechele, V., F. Fontes, R. Baborska, J.C. Nkuingoua Nana, X. Pan, and C. Tuyishime, C. 2021. *Public Expenditure on Food and Agriculture in Sub-Saharan Africa: Trends Challenges and Priorities.* Rome: FAO.

Pingali, P.L. 2012. “Green Revolution: Impacts, Limits, and the Path Ahead.” *PNAS* 109 (31): 12302–12308.

Rahman, S. 2015. “Green Revolution in India: Environmental Degradation and Impact on Livestock.” *Asian Journal of Water, Environment and Pollution* 12 (1): 75–80.

Ravallion, M. 2018. “Should the Randomistas (Continue to) Rule?” Working Paper 492. Center for Global Development, London.

Reardon, T., R. Echeverria, J. Berdegué, B. Minten, S. Liverpool-Tasie, D. Tschirley, and D. Zilberman. 2019. “Rapid Transformation of Food Systems in Developing Regions: Highlighting the Role of Agricultural Research and Innovations.” *Agricultural Systems* 172: 47–59.

Reardon, T., and C.P. Timmer. 2012. “The Economics of the Food System Revolution.” *Annual Review of Resource Economics* 4: 225–264.

Rodgers, Y., and H. Akram-Lodhi. 2019. *The Gender Gap in Agricultural Productivity in Sub-Saharan Africa: Causes, Costs and Solutions.* Policy Brief No. 11. New York: UN Women.

Sachs, J., C. Kroll, G. Lafortune, G. Fuller, and F. Woelm. 2021. *Sustainable Development Report 2021.* New York: Cambridge University Press for the United Nations.

Sonnenfeld, D.A. 1992. “Mexico's 'Green Revolution,' 1940–1980: Towards an Environmental History.” *Environmental History Review* 16 (4): 28–52.

Suri, T., and W. Jack. 2016. “The Long-Run Poverty and Gender Impacts of Mobile Money.” *Science* 354 (6317): 1288–1292.

UN-Habitat (United Nations Human Settlement Programme). 2016. *Structural Transformation in Developing Countries: Cross Regional Analysis.* Nairobi.

UN Women (United Nations Entity for Gender Equality and the Empowerment of Women), UNDP (United Nations Development Programme), and UNEP (United Nations Environment Programme). 2018. *Factors Driving the Gender Gap in Agricultural Productivity: Malawi.* New York: UN Women and UNDP; Nairobi: UNEP.

UNICEF. 2022. *Kenya Drought Situation.* New York.

UNSDG (United Nations Sustainable Development Group). 2020. *Decade of Action.* New York.

UNDESA (United Nations Department of Economic and Social Affairs). 2018. *World Urbanization Prospects: The 2018 Revision.* New York.

Van Hove, L., and A. Dubus. 2019. "M-PESA and Financial Inclusion in Kenya: Of Paying Comes Saving?" *Sustainability* 11 (3): 1–26.

World Bank. 2012. *World Development Report: Gender Equality and Development.* Washington, DC.

World Bank. 2020. *Kenya Economic Update: Navigating the Pandemic COVID-19.* Washington, DC.

World Bank. 2021. "Employment in Agriculture (% of Total Employment) (Modeled ILO Estimate) - Kenya." World Bank DataBank.

PART 1

An Overview of the Kenyan Food System

Part 1

AN OVERVIEW OF THE KENYAN FOOD SYSTEM

Food systems are made up of complex webs of actors, cut across sectors, and include diverse sets of economic activities. They function differently across different geographies and change over time, making them challenging to measure and understand. Further, food systems operate within a broader economic system, and their contribution to economic growth and employment must be understood in addition to their impacts on health, productivity, resilience, inclusivity, and sustainability. Contextualizing the Kenyan food system is the first step in identifying pathways to food system transformation and understanding how specific policies and programs can drive change effectively.

Part 1 of this book gives an overview of the Kenyan food system with a focus on its contributions to gross domestic product (GDP), employment, and food supply. Chapter 1 presents an overview of the food system, showing that it is critical to the Kenyan economy, providing a third of GDP and over half of employment directly and indirectly. The chapter highlights the need for greater contributions from agro-processing and food services to successfully transform food systems and create higher-value jobs.

Using an economywide modeling framework, Chapter 2 shows that different value chains within the agrifood system [1] should be prioritized for investment depending on policy goals. Results suggest that pulses and oilseeds, livestock, coffee and tea, and fruits and nuts should be the priority value chains for achieving the combined policy goals of reducing poverty, driving economic growth, creating employment, and improving diets.

Given the importance of livestock value chains in the current food system, and their potential to create economic growth, Chapter 3 provides a deep dive into the Kenyan livestock sector. It presents results from foresight analysis and stakeholder workshops to measure the supply and demand of livestock products and identifies policy pathways to sustainable transformation of the sector. The

1 Chapter 2 is an exception; it uses the term "agrifood system" because the analysis also includes nonfood agricultural production.

foresight analysis shows there is a risk that demand for livestock products will outpace supply in the coming decade but, if good economic conditions and governance prevail, then supply could outpace demand, potentially allowing Kenya to become a net exporter of livestock products. The chapter highlights the need for improved feed standards, more accessible marketing information systems, better animal health through improved veterinary services, and breed improvement.

The broad challenges and recommendations these chapters put forth are revisited in later chapters using specific case studies. For example, Chapters 15 and 17 focus specifically on priority value chains identified in Chapter 2, while nearly all chapters relate to strengthening value chains. Chapter 13 discusses efforts to improve animal health—a key recommendation from Chapter 3. Taken together, Part 1 sets the stage for the rest of the book by placing the food system in the broader context of the Kenyan economy and giving an overview of the livestock sector.

Chapter 2

KENYA'S AGRIFOOD SYSTEM: OVERVIEW AND DRIVERS OF TRANSFORMATION

Xinshen Diao, Karl Pauw, Jenny Smart, and James Thurlow

The 2010s were a decade of strong economic development in Kenya. Gross domestic product (GDP)—an indicator of the economy's size—expanded by an average of 5 percent per year (KNBS 2022). This exceeded population growth and helped raise household incomes, leading to a decline in poverty rates and, more importantly, in the number of poor people, for the first time in at least three decades (World Bank 2022). Agriculture played an important role in this. The sector grew alongside the rest of the economy, despite facing many challenges, including climate variability (Ochieng et al. 2020), weak rural infrastructure (Benin and Odjo 2018), shrinking farm sizes (Jayne et al. 2016), and inaccessibility of farm inputs combined with poor agronomic management (Worku et al. 2020). Agriculture, as part of the broader food system, also contributed to growth in downstream or off-farm sectors and helped cushion the economic damage resulting from COVID-19 in 2020 (Pauw, Smart, and Thurlow 2021).

This chapter provides a detailed description of the food (or agrifood) system. In short, the agrifood system consists of a complex network of actors, connected by their differing roles in supplying, using, and governing agrifood products. Rather than examining all the components of the agrifood system, this chapter has a narrower set of objectives. First, it measures the size and structure of Kenya's agrifood system and examines how, after a decade of rapid development, this system is transforming and contributing to national growth and structural change. This assessment is done from a supply-side perspective—that is, the chapter uses national accounts and employment statistics to track value addition, employment, and productivity changes across the different economic subsectors that form part of the agrifood system.

Second, the chapter evaluates the potential of different agricultural value chains to drive faster and more inclusive agricultural transformation in the future. This part of the analysis uses forward-looking economywide modeling that brings the supply and demand sides together. Thus, the modeling framework makes it possible not only to examine linkages across different on- and off-farm sectors on the supply side of the agrifood system but also to capture the interaction between food producers and consumers, shedding light on different households' access to and use of food. The chapter measures the inclusivity of the agricultural transformation process through the extent to which productivity growth in different agricultural value chains creates jobs in the more remunerative parts of the agrifood system, reduces poverty, and improves the quality of household diets.

The literature has described the agricultural transformation process in detail. In fact, agrifood systems are *expected* to evolve as countries develop and, eventually, they come to comprise far more than just primary agriculture (Timmer 1988; Diao, Hazell, and Thurlow 2010). Subsistence farming typically dominates during the earliest stages of development but, as agricultural productivity rises, farmers start to supply surplus production to markets, creating job opportunities for workers in the off-farm economy (Haggblade, Hazell, and Dorosh 2007). Rising rural incomes generate demand for more diverse products, leading to more processing, packaging, and transport activities. In these early stages, agriculture is an engine of rural, and even national, economic growth. Eventually, however, urban populations and nonagricultural incomes come to drive development, with urban consumers creating most of the demand for agricultural outputs via long value chains connecting the countryside to cities and towns (Dorosh and Thurlow 2013).

Although this description of the agricultural transformation process generally holds true, it is somewhat stylized and overlooks the unique structure and growth trajectory of a particular country's food system. Therefore, this chapter contributes to our understanding of the specific characteristics of Kenya's agrifood system and the potential for faster and more inclusive agricultural transformation. The next section introduces new indicators of agrifood system GDP and employment, which are used to describe Kenya's agrifood system and its component value chains from a supply-side perspective. Using these supply-side metrics, the following sections compare Kenya's agrifood system with those of other countries at different stages of development, and examine how Kenya's agrifood system has changed over the past decade and to what extent it has contributed to national development. A forward-looking economywide model is then used to identify which value chains have economic linkages that are more likely to drive inclusive agricultural transformation. The chapter concludes by summarizing findings and outlining an agenda for further research.

Kenya's agrifood system

Components and indicators

As stated earlier, the first objective of this chapter is to measure the size and structure of Kenya's agrifood system from a supply-side perspective using national accounts and employment statistics. Figure 2.1 provides a simple conceptual framework with five components (A to E). The framework follows the format of national accounts data, allowing us to estimate total value added and total employment in the agrifood system—indicators that we call AgGDP+ and AgEMP+, respectively.[1] An agrifood system is essentially the sum of all on- and off-farm GDP and employment generated across all agricultural value chains within a country.

Primary agriculture (A) is the first component of the agrifood system and includes the value added generated by all agricultural subsectors, including crops, livestock, forestry, and fishing.[2] *Agro-processing* (B) is part of the broader manufacturing sector and includes the value added from producing processed

1 Thurlow and colleagues (forthcoming) provide formal definitions of AgGDP+ and AgEMP+.

2 Note that GDP or value added is equal to the value of gross output minus the cost of intermediate inputs, such as the cost of seeds and fertilizers that farmers purchase and use.

FIGURE 2.1 Components of an agrifood system

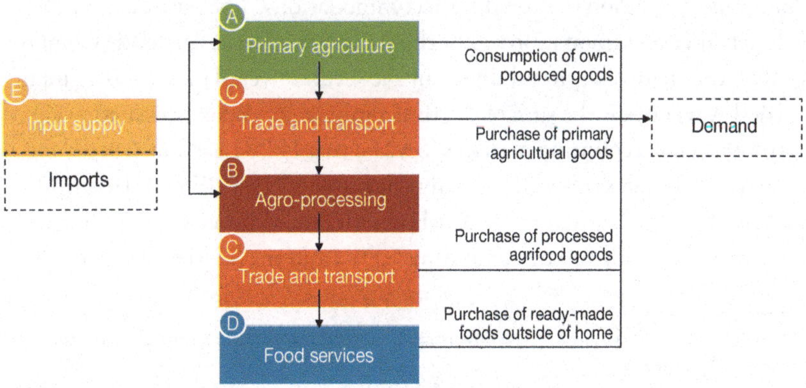

Source: Thurlow et al. (forthcoming).

foods and other agriculture-related products, such as beverages, tobacco, yarn, and timber.[3]

Input supply (E) is the GDP generated during the production of intermediate inputs that farmers and processors use directly (for example, fertilizers and financial services). Inputs that farmers and processors themselves produce are excluded to avoid double counting, since the above components (A and B) capture these. AgGDP+ includes only a portion of GDP that local input producers generate. This portion is the share of agriculture and processing's input demand in total economywide demand for that input. For example, if farmers and processors use a third of all petroleum in the economy, then a third of the petroleum sector's GDP is considered part of the agrifood system. If all petroleum is imported, then this input does not contribute to AgGDP+, because the value added occurs outside the country.

Trade and transport services (C) includes all GDP generated by the transporting, wholesaling, and retailing of agrifood products between farms, firms, and final points of sale (that is, either domestic markets or the country's border for exported goods). National accounts data do not separate the trade sector's GDP into its food and nonfood components, but it is possible to estimate this using product-level data on transaction cost margins. Transaction costs are the main source of demand for trade services, and so it is possible to attribute a portion of trade sector GDP to the agrifood system based on the total share of trade margins on agrifood products relative to the total margins on all marketed products.

Finally, *food services* (D) is the value added generated by the food services sector, plus a portion of that generated by the hotels and accommodation sector. Producers of food services (that is, meals prepared outside the home) run standalone operations (for example, restaurants or stalls), whereas hotels often operate restaurants in addition to providing accommodations. The portion of GDP in the hotel and accommodation sector that is assigned to the agrifood system is based on the share of agrifood inputs in the sector's overall intermediate inputs.[4]

Total AgGDP+ is the sum of the five components and is estimated using a series of social accounting matrices (SAMs) that IFPRI has constructed for Kenya using the latest national accounts data (see KNBS 2022). Estimating AgEMP+ follows a similar procedure but with one additional data source. Employment is estimated by combining GDP from the SAM's 90 sectors

---

3 Yarn and timber are the immediate downstream subsectors for cotton farming and the forestry sector respectively.

4 Cross-country analysis using IFPRI's Agrifood System Database indicates that this is a conservative approach that slightly underestimates the part of hotels and accommodation GDP that is associated with the agrifood system.

with labor productivity estimates (that is, GDP per worker) for 14 aggregated sectors. The employment database triangulates information from a variety of sources, including the 2009 and 2019 population censuses (KNBS 2009, 2019), household budget and labor force surveys (KNBS 2017), and international databases containing sectoral employment timeseries data (ILO 2020; de Vries et al. 2021). Our definition of employment includes all workers aged 15 years or older, assigned to sectors based on their primary sector of employment. This standard global definition allows us to compare Kenya with other countries in IFPRI's Agrifood System Database.

Current structure

Table 2.1 shows the structure of Kenya's overall economy. Primary agriculture generated almost a quarter of total GDP in 2019 and over two-fifths of total employment. Kenya exports agrifood products such as coffee and tea but generally not in primary form. For example, coffee from the farm is supplied to the manufacturing sector, where it is processed, graded, and bagged before export. Such exports appear in the table as manufactured goods. Most of Kenya's imports also are manufactured, but these are mainly nonagricultural products such as fuels, machinery, and vehicles. In total, manufacturing generates only 9 percent of GDP and 6 percent of employment. Finally, services

TABLE 2.1 Structure of Kenya's economy, 2019

	Share of total (%)			
	GDP	Employment	Exports	Imports
All sectors	100	100	100	100
Primary agriculture	22.7	43.3	20.9	2.5
Industry	18.7	12.8	30.7	81.8
Mining	1.6	0.7	0.4	0.5
Manufacturing	8.6	5.8	30.2	79.2
Utilities	2.4	0.3	0.0	0.1
Construction	6.1	5.9	0.0	2.0
Services	58.6	43.9	48.5	15.7
Trade, transport, and food services	23.1	24.3	27.6	5.9
Finance, business, and real estate	17.7	3.7	10.9	7.7
Government, health, and education	12.6	8.7	0.0	0.0
Other services	5.3	7.2	10.0	2.1

Source: Authors using IFPRI's 2019 Kenya Social Accounting Matrix.

account for most of Kenya's GDP and are as important as agriculture for employment. Within services, trade and transport are the largest and among the most labor-intensive sectors.

Table 2.2 describes the size and components of Kenya's agrifood system. Total AgGDP+ was equal to \$31 billion in 2019—well above primary agriculture, whose GDP was \$21 billion. For every \$1 of GDP generated on the farm there is an additional \$0.49 of GDP generated off the farm within the broader agrifood system. Much of this off-farm GDP is from agro-processing, although, as with manufacturing more generally, the labor intensity of agro-processing is low compared with the rest of the economy. This explains why labor productivity (annual GDP per worker) is much higher in agro-processing (\$9,928) than in primary agriculture (\$2,580). Productivity is also relatively low in agrifood trade and transport and food services, where many jobs are informal or casual.

Using the supply use table, we can decompose the agrifood system across major product groups and track how much value added is generated in each of the five agrifood system components. Table 2.3 disaggregates AgGDP+ across eight product value chains.[5] Cereals and downstream grain milling account for 16 percent of total AgGDP+ (column 1). Livestock and horticulture are also large groups of value chains within the agrifood system, although, unlike

5 The specific products included in each value chain grouping are listed in Table A.2.1 in the appendix to this chapter.

TABLE 2.2 Kenya's agrifood system, 2019

	GDP		Employment		Average GDP per worker (\$)
	Value (\$ billions)	Share of total (%)	Workers (millions)	Share of total (%)	
Total economy	92.0	100	18.7	100	4,917
Agrifood system	31.1	33.8	10.2	54.7	3,040
Primary agriculture (A)	20.9	22.7	8.1	43.3	2,580
Off-farm agrifood system	10.2	11.1	2.1	11.4	4,791
Agro-processing (B)	4.7	5.1	0.5	2.5	9,928
Trade and transport (C)	3.6	3.9	1.1	6.1	3,125
Food services (D)	0.8	0.9	0.4	2.2	1,984
Input supply (E)	1.1	1.2	0.1	0.6	10,696
Non-agrifood system	60.9	66.2	8.5	45.3	7,184

Source: Authors using IFPRI's 2019 Kenya Social Accounting Matrix.

Note: A–E correspond to the five agrifood system components in Figure 2.1.

TABLE 2.3 Decomposing AgGDP+ across value chains, 2019

	Share of total GDP (%)		
	Agrifood system	Primary agriculture	Off-farm components
Total	100	100	100
Cereals	15.6	16.3	14.1
Pulses & oilseeds	7.8	7.9	7.6
Root crops	8.9	12.2	2.2
Horticulture	17.8	21.5	10.1
Livestock	22.0	17.3	31.7
Fish	2.3	2.7	1.6
Export crops	15.5	14.8	16.9
Forestry	7.6	7.3	8.1
Unattributable	2.5	0.0	7.6

Source: Authors using IFPRI's 2019 Kenya Social Accounting Matrix.

horticulture, livestock contributes more to off-farm AgGDP+ (column 3) than to primary agriculture GDP (column 2). Finally, some products have complicated value chains that are difficult to track. This "unattributable" group includes highly processed products with fewer or more convoluted linkages back to primary agriculture (for example, baby foods and alcoholic beverages). All the value added for these products is reported as off-farm. Nevertheless, most of Kenya's agrifood system can be attributed to distinct product groups.

Table 2.4 disaggregates each value chain group across the five components in the agrifood system. Most of the value added (70 percent) in the cereals

TABLE 2.4 Decomposing AgGDP+ within value chains, 2019

	Share of value chain GDP (%)					
	Agrifood system	Primary agriculture	Agro-processing	Trade and transport	Food services	Input supply
Total	100	67.3	15.0	11.5	2.6	3.6
Cereals	100	70.4	16.3	10.8	1.1	1.4
Pulses & oilseeds	100	68.0	11.2	10.8	7.6	2.3
Root crops	100	92.0	0.9	5.7	0.4	1.0
Horticulture	100	81.3	6.3	8.8	0.5	3.1
Livestock	100	22.0	23.0	16.2	2.9	5.2
Fish	100	77.9	5.3	11.6	2.0	3.1
Export crops	100	64.3	18.1	12.8	0.8	4.1
Forestry	100	64.9	21.9	12.0	0.0	1.2

Source: Authors using IFPRI's 2019 Kenya Social Accounting Matrix.

value chain is generated by primary agriculture (that is, growing cereals on the farm), with about one-fifth (16 percent) generated in agro-processing (for example, milling grains into flour). Trade and transport margins make up a further 10 percent of GDP in the cereals value chain. Overall, despite the growing importance of purchased and processed foods in Kenya (Tschirley et al. 2015), most value chains still generate most of their value added on the farm, with cereals representative of the overall agrifood system. Root crops is the least oriented toward off-farm components in the value chain, with almost all value added generated on the farm.

In summary, Kenya's agrifood system stretches well beyond the farm. For every $1 generated on the farm, $0.49 is generated off the farm. The structural information in this section is useful for understanding the agrifood system's linkages to the rest of the economy and anticipating its contribution to national structural change. For instance, the off-farm agrifood system is far less labor-intensive than farming, which means substantial off-farm growth would be required to absorb workers exiting agriculture. However, labor productivity is, on average, higher off the farm, meaning that workers who leave agriculture and find work elsewhere in the food system should increase economywide labor productivity. One exception is food services, which not only is highly labor-intensive but also exhibits labor productivity levels even lower than in agriculture. Finally, decomposing AgGDP+ across value chains helps us anticipate how different sources of agricultural growth may affect agricultural transformation differently, by favoring either on- or off-farm growth.

The next two sections will examine past growth trends and future growth scenarios. Before this, however, we compare Kenya's current agrifood system to those in other countries at different stages of development.

Comparison with other countries

IFPRI's Agrifood System Database contains estimates of AgGDP+ and AgEMP+ for 206 countries covering 96 percent of global GDP and 94 percent of the global population (see Thurlow et al. forthcoming). Figure 2.2 shows weighted estimates across the World Bank's country income groups and compares these with Kenya (see final column in each panel). The figure therefore shows the importance and composition of agrifood systems at different stages of development, with the latter proxied by gross national income (GNI) per capita.

Panel A shows the importance of primary agriculture for developing countries, and how agriculture's share of the economy is much smaller at later

FIGURE 2.2 Decomposing agrifood systems across country income groups, 2019

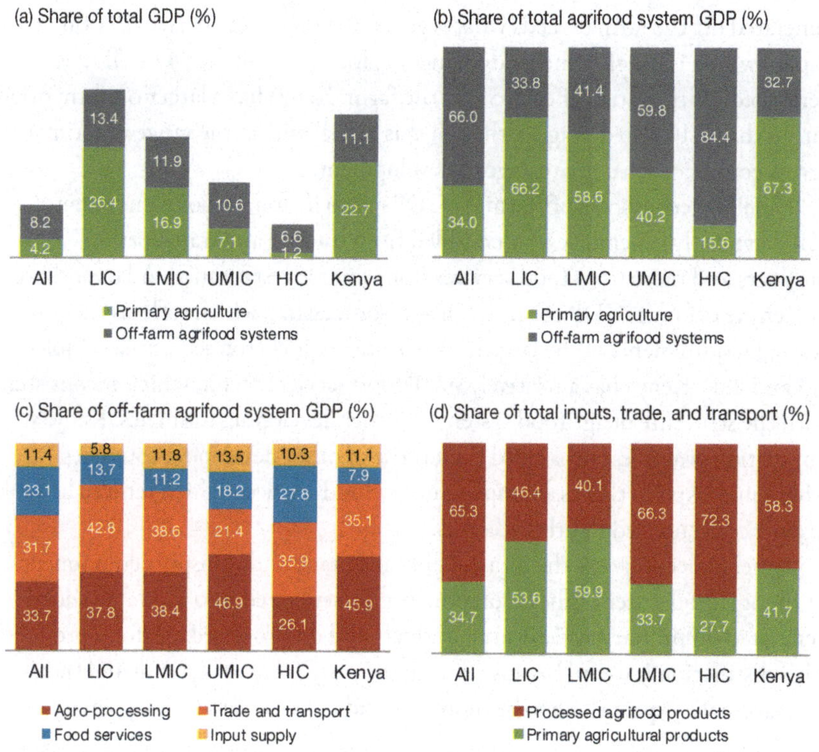

Source: IFPRI's Agrifood System Database in Thurlow et al. (forthcoming); IFPRI's 2019 Kenya Social Accounting Matrix.
Note: LIC are low-income countries; LMIC are lower-middle-income countries; UMIC are upper-middle-income countries; and HIC are high-income countries.

stages of development. On average, primary agriculture is 27 percent of total GDP for low-income countries (LICs) but only 1 percent for high-income countries (HICs). However, the agrifood system's contribution to the economy remains large, even in the most developed countries. In Hong Kong, for example, where agricultural GDP is virtually zero, the agrifood system still accounts for 8 percent of total GDP. The size of Kenya's agrifood system and its breakdown across farm and off-farm components is consistent with other countries that have recently reached lower-middle-income country (LMIC) status.

Panel B in the figure shows the contributions of farm and off-farm components to total AgGDP+ (that is, a different representation of the information in Panel A). At around $4,000 per capita, which is close to the threshold for attaining upper-middle-income country (UMIC) status, the value added generated off the farm exceeds what is generated on the farm. At $40,000 per capita, which is roughly the median income for HICs, about $4 of GDP is generated off the farm for each $1 on the farm. Kenya has a larger on-farm orientation than LICs on average, although this is well within the range of estimates across countries at a similar stage of development.

Panel C decomposes off-farm AgGDP+ into its four major components. HICs typically generate less value added from trading and transporting products and more from food services (for example restaurants). A larger share of Kenya's off-farm GDP comes from agro-processing, which reflects the processing requirements of the large livestock and export crop sectors (see Tables 2.3 and 2.4). Kenya has a relatively small food services sector, which is consistent with the structure of agrifood systems in other developing countries. Kenya's population is also less urbanized than those of other developing countries, which may explain the lesser importance of food services, which tend to be more heavily concentrated in urban centers.

Panel D decomposes the input supply and trade and transport components of the agrifood system. These components are separated into the value added generated on primary agricultural products and on processed agrifood products. This distinction is useful because during the early stages of agricultural transformation we expect there to be more value added associated with primary

FIGURE 2.3 Agrifood system share of total GDP in selected countries, 2019

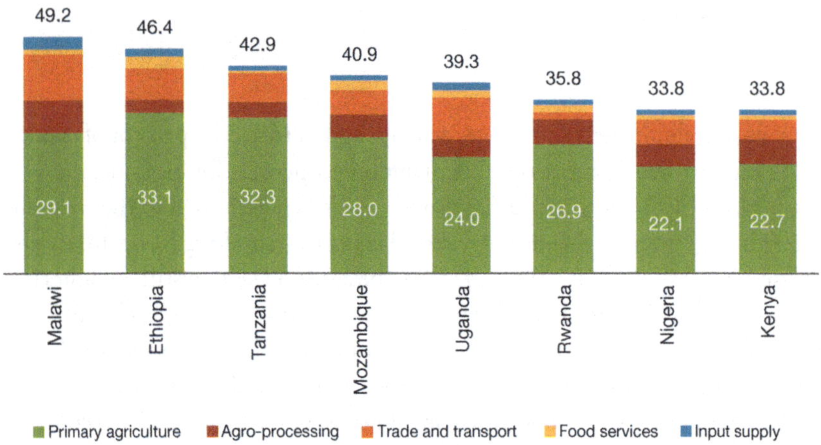

Source: IFPRI's Agrifood System Database in Thurlow et al. (forthcoming); IFPRI's 2019 Kenya Social Accounting Matrix.
Note: Figures in black are the agrifood system's percentage share of total GDP.

products than with processed products. This is supported by Figure 2.2, which shows how more value added is generated on primary products in LICs but the opposite is true for HICs. Again, Kenya has a larger-than-expected agro-processing sector, given its stage of development, and this explains why inputs and trade and transport margins on processed products generate more GDP.

Finally, Figure 2.3 compares Kenya with other low-income African countries. Most countries in eastern and southern Africa have agrifood systems that account for a larger share of their national economies than is the case in Kenya. This mainly reflects these countries' larger primary agricultural sectors, although there is some variation in the composition of off-farm agrifood systems.

Overall, Kenya's agrifood system shares many of the characteristics of other countries in the region and countries at a similar stage of development. However, not all countries have grown as rapidly as Kenya has over the past decade. This means that a closer look at Kenya's agricultural and structural transformation is warranted.

Recent growth and transformation

This section considers whether recent growth within the agrifood system has been associated with agricultural and structural transformation in Kenya. Growth between 2009 and 2019 is decomposed using data on GDP and employment across the entire agrifood system (that is, AgGDP+ and AgEMP+) estimated from an annual timeseries of Kenyan SAMs. The SAMs are constructed using a standard national accounting system and so are directly comparable. SAMs measured in current prices are deflated (that is, converted to constant prices) using Kenya's most recent GDP series (KNBS 2022).

Figure 2.4 shows how the structure of the agrifood system has changed over the past decade. Primary agriculture's share of the economy declined from 27 percent in 2009 to 23 percent in 2019. The broader agrifood system's share of total GDP has also fallen but at a slower rate than that for agriculture. As a result, the contribution of the off-farm components of the agrifood system has risen slightly, from 29 to 33 percent. As the previous section illustrated using cross-country data, countries that move toward higher levels of development tend to have faster-growing off-farm components in their agrifood systems. The changing structure of Kenya's agrifood system is consistent with a country undergoing such an agricultural transformation, which may reflect Kenyan policymakers' long-standing emphasis on market-oriented agriculture and active engagement of the private sector in the agriculture sector.

FIGURE 2.4 Changing structure and contribution of AgGDP+ in Kenya, 2009–2019

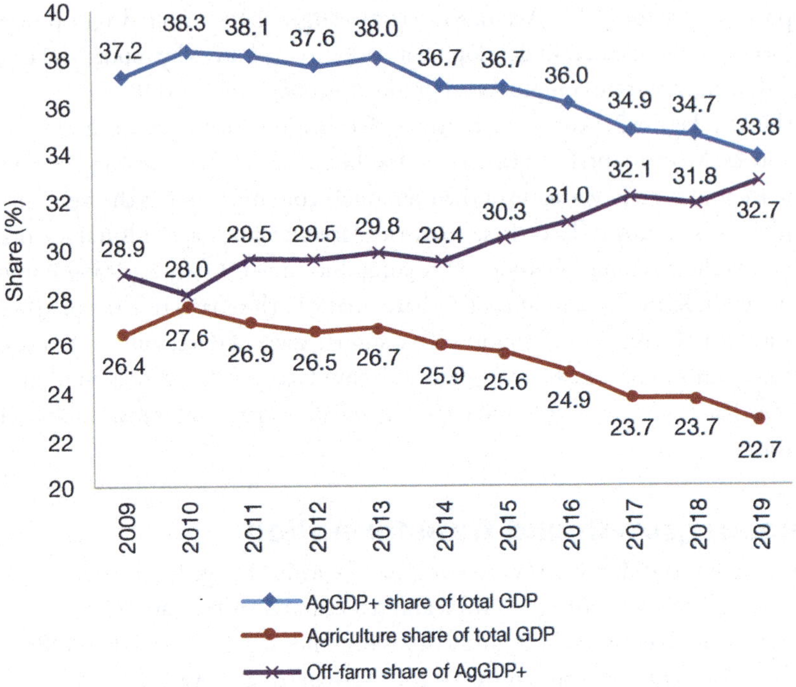

Source: Authors using IFPRI's 2009–2019 Kenya Social Accounting Matrixes.

Table 2.5 confirms that the structural changes within the agrifood system observed in Figure 2.4 coincide with a period of economic growth. Kenya's overall economy grew at 4.9 percent per year from 2009 to 2019, with agriculture growing at 3.4 percent. This explains agriculture's declining contribution to the overall economy. In contrast, the off-farm components of the agrifood system grew faster than the economy, at 5.2 percent, increasing their importance over time. The fastest growth within the agrifood system was in trade and transport services, whereas the slowest growth was in food services. However, since food services accounts for only around 1 percent of GDP, its weak performance has a minimal impact on the overall performance of the agrifood system or economy. The table also decomposes growth of inputs, trade, and transport into primary agricultural products and processed agrifood products. Growth in inputs, trade, and transport was faster for primary products—8.8 percent per year—suggesting that parts of Kenya's agrifood system are still in the

TABLE 2.5 Agrifood system GDP and employment growth decomposition, 2009–2019

	Average annual growth rate (%)			Contribution to total change (%)		Share of total GDP (%)		Share of total employment (%)	
	GDP	Employment	GDP per worker	GDP	Employment	2009	2019	2009	2019
Total economy	4.9	2.2	2.7	100	100	100	100	100	100
Agrifood system	3.9	1.5	2.3	28.3	40.5	37.2	33.8	58.1	54.7
Primary agriculture (A)	3.4	0.9	2.4	16.7	19.6	26.4	22.7	49.0	43.3
Off-farm agrifood system	5.2	4.5	0.7	11.6	20.9	10.8	11.1	9.1	11.4
Agro-processing (B)	4.7	3.4	1.2	4.9	3.8	5.2	5.1	2.2	2.5
Trade and transport (C)	6.5	4.9	1.4	4.7	12.2	3.4	3.9	4.7	6.1
Food services (D)	2.4	4.8	−2.3	0.5	4.2	1.1	0.9	1.7	2.2
Input supply (E)	6.3	2.9	3.3	1.5	0.7	1.1	1.2	0.5	0.6
Inputs, trade, and transport (C+E)	6.4	4.8	1.6	6.2	13.0	4.4	5.1	5.2	6.7
Primary products	8.8	7.4	1.3	3.2	7.0	1.5	2.1	1.6	2.7
Processed products	5.0	3.4	1.6	3.0	5.9	3.0	3.0	3.6	4.0
Non-agrifood system	5.5	2.9	2.5	69.8	59.7	62.8	66.2	41.9	45.3

Source: Authors using IFPRI's 2009–2019 Kenya Social Accounting Matrixes and employment database.

earlier stages of agricultural transformation, when the focus is on extending the linkages between primary agriculture and the immediate downstream sectors.

Total employment in the economy grew at 2.2 percent per year during the 2009–2019 period. With GDP growing faster than employment, economy-wide labor productivity, measured by GDP per worker, rose over the period. Employment growth in agriculture was much slower than in the total economy, causing the sector's share of total employment to fall from 49 percent to 43 percent over the decade. Overall, labor productivity rose in most components of the agrifood system. The only exception is food services, where rapid employment growth exceeded sectoral GDP growth, causing average GDP per worker to fall over the decade.

A decline in the share of agricultural employment is expected as a country develops. An increase in economywide labor productivity, which is strongly associated with economic development, can arise through two channels. First, productivity can increase among workers within their sectors of employment. Second, economywide productivity increases when workers migrate from less to more productive sectors. As mentioned earlier, GDP per worker in agriculture is lower than in most other parts of the agrifood system and in the rest of the economy. As such, a shift in employment patterns away from agriculture and

FIGURE 2.5 Decomposition of average annual labor productivity growth rate, 2010–2019

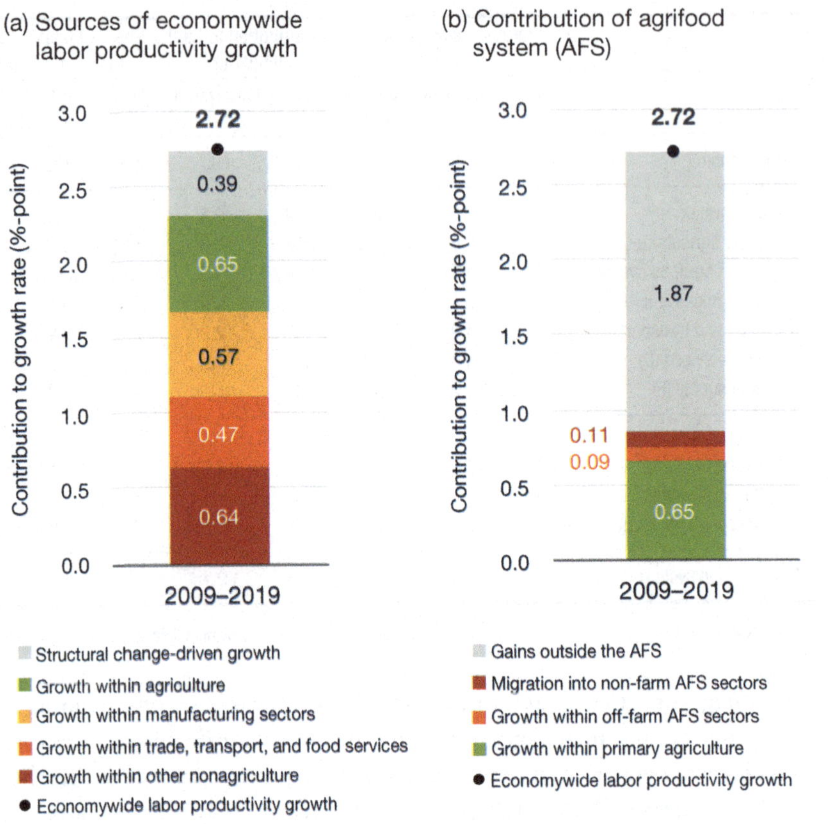

Source: Authors using IFPRI's 2009–2019 Kenya Social Accounting Matrixes and employment database.

into other sectors should increase economywide labor productivity. These two channels are usually referred to as the "within-sector" and "between-sector" (or "structural change") drivers of labor productivity growth. The individual contributions of these structural drivers of growth can be estimated using the approach McMillan, Rodrik, and Verduzco-Gallo (2014) describe.

Figure 2.5 presents the overall findings from the growth decomposition analysis. As shown in Panel A, economywide labor productivity grew at an average annual rate of 2.7 percent from 2009 to 2019. Most of this growth was driven by within-sector labor productivity gains. Growth within agriculture was one of the largest contributors, with it being a large sector (that is, 26 percent of total GDP in 2009; see Table 2.5) experiencing relatively fast labor productivity growth (that is, 2.4 percent per year). Although manufacturing employment grew even faster, its small share of the economy limited its absolute contribution to economywide growth. Overall, the figure suggests that the within-sector drivers of productivity growth were quite evenly balanced across agriculture, manufacturing, trade, and other sectors. More importantly, changes in employment patterns were consistent with workers moving out of low-productivity sectors like agriculture and into higher-productivity sectors. This contributed to an increase in economywide labor productivity—a process called structural change-led growth.

Panel B estimates the contribution of the agrifood system to total labor productivity growth. The within-sector contribution from agriculture is the same in both panels. The migration of workers into the off-farm components of the agrifood system accounted for almost a quarter of the structural change-led growth that took place in Kenya over the decade (that is, 0.09 percentage points out of a total 0.39 percentage points). This is in addition to the annual increase in labor productivity among workers who were already employed in the off-farm sectors. The agrifood system thus accounted for 30 percent of total labor productivity growth, with the remaining 70 percent originating from within-sector productivity growth and migration into sectors outside of the agrifood system.

In summary, despite rapid economic growth, Kenya's economy is undergoing a slow, but positive, process of structural change. Most labor productivity growth has been driven by rising productivity within sectors rather than a steady reorientation of employment toward more productive activities. Within this national process of growth and structural change, agricultural transformation is proceeding steadily, with faster GDP and employment growth within the off-farm components of the agrifood system. However, employment growth has been fastest in the less productive components of the agrifood system (that is, trade, transport, and food services) instead of more productive components such as agro-processing. This pattern of agrifood system growth is consistent with the earlier stages of agricultural transformation. It has, however, limited the agrifood system's contribution to structural change-led growth at the national level. Identifying sources of agricultural growth that could accelerate the growth on and beyond the farm is therefore a priority in accelerating agrifood system and broad economic transformation.

Future drivers of inclusive agricultural transformation

This section uses IFPRI's Rural Investment and Policy Analysis (RIAPA) model to compare the potential contributions of growth driven by different agricultural value chains to inclusive agricultural transformation. The latter is proxied by growth in AgGDP+ and AgEMP+, as well as reductions in the poverty headcount rate and the gap between households' consumption of major food groups and the estimated cost of a healthy reference diet. RIAPA is briefly described below; Diao and Thurlow (2012) provide a more technical description.

The Rural Investment and Policy Analysis model

RIAPA comprises a series of interlinked datasets and models that are used to assess program impacts and inform policy and investment prioritization at the country level. At the core of RIAPA is an economywide computable general equilibrium (CGE) model that simulates the functioning of a market economy, including markets for products and factors (that is, land, labor, and capital). RIAPA measures how impacts are mediated through prices and resource reallocations, and ensures that resource and macroeconomic constraints are respected, such as when inputs or foreign exchange are limited.

RIAPA divides Kenya's economy into 90 sectors. Producers in each sector maximize profits and supply output to national markets, where it may be exported and/or combined with imports depending on relative prices, with exchange rate movements affecting foreign prices. Producers combine factors and intermediate inputs using sector-specific technologies. Maize farmers, for example, use a combination of land, labor, fertilizer, and purchased seeds. Rural and urban labor markets are divided by workers' education levels, and agricultural capital is separated into crop and livestock categories. Labor and capital are in fixed supply but less educated workers are considered underemployed. The government collects taxes on products, households, and enterprises, and uses these revenues to finance public services and social transfers. Remaining revenues are added to private savings and foreign capital inflows to finance investment. RIAPA is a dynamic model, with past levels of investment determining current capital availability.

RIAPA tracks changes in household incomes and expenditures, including changes in food and nonfood consumption patterns. The 15 household groups in RIAPA are separated into rural and urban consumption quintiles, with rural households further separated into farm and nonfarm groups. Table 2.6 describes aggregate household income and consumption patterns. Kenya's population of 47.6 million people consumed, on average, $1,718 of goods and services per person in 2019 (at market exchange rates unadjusted for purchasing power parity). Consumption levels are much lower in rural areas and among the poor. Poor households spend more of their earnings on food consumption; of this, they spend a smaller share on processed foods than on primary or

TABLE 2.6 Household incomes and consumption, 2019

	National	Poor	Nonpoor	Rural farm	Rural nonfarm	Urban
Consumption per capita (US$)	1,718	514	2,521	921	1,532	2,904
Food consumption share of total consumption (%)	38.9	59.9	36.1	49.7	39.0	34.2
Food consumption share by food group (%)	100	100	100	100	100	100
Cereals & roots	28.4	43.6	25.1	39.1	33.5	20.9
Vegetables	21.8	26.3	20.8	24.0	24.2	20.0
Fruits	11.1	8.5	11.7	12.0	10.9	10.6
Milk & dairy	11.0	10.5	11.1	11.4	11.9	10.5
Meat, fish, & eggs	24.0	6.6	27.8	9.3	15.3	34.6
Added fats	3.7	4.6	3.5	4.2	4.2	3.3
Processed food share (%)	51.7	35.3	55.4	38.3	46.3	61.6
Household income share by income source (%)	100	100	100	100	100	100
Crop land returns	12.6	36.4	9.8	47.8	0.0	1.3
Labor remuneration	32.0	30.4	32.2	18.8	26.2	38.3
Less educated workers	7.9	18.9	6.6	7.5	7.9	8.1
More educated workers	24.1	11.6	25.6	11.4	18.4	30.2
Capital profits	48.8	20.0	52.2	15.4	69.7	57.8
Other sources	6.6	13.2	5.8	18.0	4.0	2.6
Population (millions)	45.4	18.1	27.2	22.7	6.5	16.2

Source: Authors using IFPRI's 2019 Kenya Social Accounting Matrix and RIAPA model.

Note: Food consumption includes meals prepared outside the household. More educated workers are those who have completed at least primary schooling. Capital income is equivalent to gross operating surplus net of taxes and other corporate transfers. Other income sources include social and foreign transfers.

unprocessed foods. Cereals and roots dominate consumption patterns, although nonpoor and urban households consume more meat, fish, and eggs. Finally, poor households are, on average, more reliant on incomes from farming and less educated labor.

RIAPA tracks poverty and dietary impacts using survey-based microsimulation analysis. Individual survey households map to the model's household groups. Estimated consumption changes in the model are applied proportionally to survey households, and post-simulation consumption values are recalculated and compared with a poverty line to determine households' poverty status, and with the cost of a healthy reference diet to estimate dietary deprivation, a measure of diet quality.

Comparing growth driven by different value chains

RIAPA is used to simulate the effects of expanding farm production within existing agricultural value chains. Total factor productivity growth in each group of agricultural products is accelerated beyond baseline growth rates, such that, in each value chain scenario, total agricultural GDP is 1 percent higher by 2025 than is expected under a business-as-usual baseline scenario. Expanding agricultural production increases supply to downstream processing activities and generates demand for agricultural trade and transport services. Agricultural subsectors differ in size; as such, to achieve the same absolute increase in *total* agricultural value added, it is necessary for smaller value chains to expand more rapidly than larger ones.

The rows in Table 2.7 show the set of value chain scenarios implemented using the RIAPA model. These are based on the same set of value chains from before (see Tables 2.3 and 2.4), except that cereals are now split into maize and rice & wheat value chains; horticulture is split into vegetables and fruits & nuts; and livestock is split into cattle & dairy and poultry & eggs. A coffee & tea scenario represents the export crops value chain, and we retain the original aggregation for pulses & oilseeds, root crops, and fish value chains. A forestry scenario is excluded.

The columns in Table 2.7 report elasticities and multipliers estimated for each value chain scenario. The poverty–growth elasticity (semi-PGE) is the

TABLE 2.7 Estimated poverty, diet deprivation, economic growth, and employment elasticities

Value chain with accelerated productivity growth	Change in indicator given 1% agricultural growth driven by productivity gains (value chain rank in parentheses)				
	National poverty–growth elasticities (semi-PGE)		Diet deprivation–growth elasticities (DGE)	AgGDP+ growth multiplier	AgEMP+ growth multiplier
	Poverty headcount	Poverty gap	ReDD index		
Maize	−0.58 (3)	−0.20 (1)	−0.27 (8)	1.49 (3)	0.03 (6)
Rice & wheat	−0.29 (7)	−0.07 (7)	−0.16 (9)	1.21 (6)	0.04 (4)
Pulses & oilseeds	−0.59 (2)	−0.19 (2)	−1.24 (3)	1.67 (2)	0.05 (3)
Root crops	−0.13 (8)	−0.05 (8)	−0.03 (10)	1.15 (8)	−0.03 (9)
Vegetables	−0.04 (10)	−0.00 (10)	−0.91 (6)	1.33 (4)	−0.03 (10)
Fruits & nuts	−0.50 (5)	−0.18 (3)	−2.09 (1)	1.03 (10)	0.14 (2)
Coffee & tea	−0.42 (6)	−0.16 (5)	−0.29 (7)	1.16 (7)	0.29 (1)
Cattle & dairy	−0.09 (9)	−0.04 (9)	−2.08 (2)	2.11 (1)	−0.01 (8)
Poultry & eggs	−0.54 (4)	−0.11 (6)	−1.09 (4)	1.11 (9)	0.04 (5)
Fish	−0.64 (1)	−0.17 (4)	−0.98 (5)	1.32 (5)	0.02 (7)

Source: Results from IFPRI's Kenya RIAPA model.

percentage point change in poverty headcount or poverty gap ratio per unit increase in primary agricultural GDP growth generated by the targeted value chain. The results in the first column indicate that all value chains are pro-poor in that productivity growth in those value chains is associated with reductions in the national poverty headcount. Among the value chains, the fish value chain has the largest (negative) elasticity, implying that growth in this value chain is the most effective at reducing poverty. The vegetables value chain is the least effective. This reflects the fact that, currently in Kenya, relatively few households near the poverty line benefit from vegetable production or lower vegetable prices. It is also because expanding vegetable production displaces other value chains that are more beneficial for poor households. This displacement effect is an important mechanism within economywide models—that is, these models account for the scarcity of land and labor resources, which may mean the expansion of one value chain comes at the expense of others.

Poverty headcount rates focus simply on the share of people who live below the poverty line, whereas poverty gaps measure how far poor people are, on average, from that poverty line. Poverty gaps therefore better reflect the depth of poverty. Policies that are effective in supporting people to move from just below the poverty line out of poverty will affect the poverty headcount rate but may be less effective at improving the lives of the poorest among the poor and therefore reducing the depth of poverty. As such, we also do not expect the same value chains that are effective at reducing poverty headcount ratios to be effective with regard to poverty gap ratios. As the second column of the table shows, fish is more effective at reducing the poverty headcount rate than it is the poverty gap; the opposite is true for maize. Productivity growth in the maize value chain will therefore benefit poor households further below the poverty line, which is not true to the same extent for the fish value chain.

The third set of results in Table 2.7 reports estimated diet deprivation–growth elasticities (DGE). This elasticity is measured using the Reference Diet Deprivation (ReDD) index (Pauw et al. 2021). ReDD is derived from food consumption gaps across six major food groups relative to reference food intake levels, which are taken from the EAT–*Lancet* healthy reference diet (Willet et al. 2019). A decline in the ReDD index indicates narrowing food consumption shortfalls and is thus associated with an improvement in diet quality. Staples, such as cereals and root crops, are already dominant food groups, and so expanding production of these crops reduces dietary diversity and shifts consumption away from other important food groups in the reference diet. This explains the low diet quality ranking of these value chains (in fact, diets deteriorate in the case of root crops). The tea & coffee value chain also has weak impacts on diets, in part because these beverages are not required in a healthy diet but also because growth in the tea & coffee value chain has weak poverty effects. Diet quality is driven both by changes in relative prices (that is, cheaper tea and coffee will not improve diet quality) and by income changes. The biggest improvements in diets are seen for the fruits & nuts, cattle & dairy, and pulses & oilseeds value chains.

The final two sets of results in Table 2.7 report the growth and employment impacts of expanding production in different value chains. The AgGDP+ growth multiplier measures the monetary change in agrifood system GDP (in US dollars) per unit increase in agricultural GDP generated in the targeted value chain (also in US dollars). Similarly, the AgEMP+ growth multiplier measures the change in number of jobs created within the agrifood system per unit of increase in agricultural GDP generated in the targeted value chain (in US dollars). Value chains with larger off-farm components typically generate larger AgGDP+ growth multiplier effects across the agrifood system. For example, as we saw previously, livestock value chains generally have larger off-farm processing components compared with other value chains (see Table 2.4). It is therefore not surprising that the cattle & dairy value chain has the strongest AgGDP+ growth multiplier effect, especially when considering that much of the supply of raw milk and slaughtered animals goes to the downstream dairy and meat processing sectors that generate downstream value added and drive agrifood system growth beyond the farm. The pulses & oilseeds and maize value chains also have large AgGDP+ growth multipliers. In contrast, most of the value addition in the fruits & nuts value chain happens on the farm, hence this value chain is least effective in driving AgGDP+ growth.

Finally, while there is some correlation between AgGDP+ and AgEMP+ elasticities, it does not necessarily follow that the value chains that are most effective at generating value added are also the most effective at creating jobs. The vegetables and root crops value chains, for example, are not as labor-intensive as other value chains, and so have smaller AgEMP+ growth multipliers. In contrast, the more labor-intensive coffee & tea and fruits & nuts sectors have large employment effects across the agrifood system.

Prioritizing value chains

Figure 2.6 compares the relative effectiveness of growth in each value chain in achieving the various development outcomes. Since we choose to include only one poverty measure in the final assessment, we arbitrarily exclude the poverty gap measure. Also, since we are comparing outcomes with different units, such as poverty–growth elasticities and employment multipliers, it is necessary to

FIGURE 2.6 **Normalized value chain scores for different outcomes**

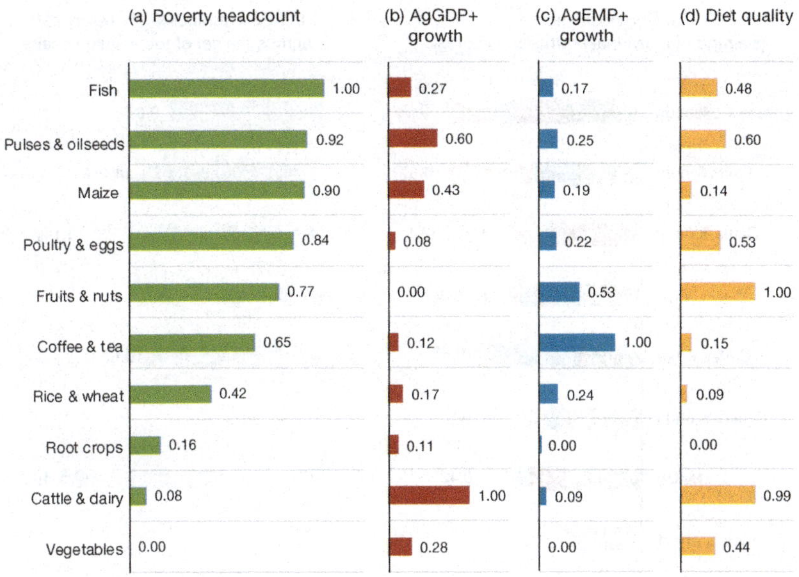

Source: Results from IFPRI's Kenya RIAPA model.

Note: Scores for each outcome category are normalized such that the value chain whose growth is most effective at improving that outcome has a score of one, while the least effective value chain has a score of zero.

normalize the individual outcome scores. The values of each indicator are scaled so that the most effective value chain is given a score of one and the least effective a score of zero. Normalization does not alter the relative effectiveness of value chains within each outcome category. For example, the fish and vegetables value chains were, respectively, the most and least effective value chains in reducing national poverty (see Table 2.7), and these rankings remain unchanged after normalization.

The value chains in the figure are ranked based on their poverty scores, with fish appearing at the top of Panel A. However, while the fish value chain is the most effective at reducing poverty, it is relatively ineffective at achieving the other outcomes. Cattle & dairy, on the other hand, is most effective at driving agrifood system growth. It is also highly effective at improving diet quality—but has weak poverty and employment impacts. This variation across outcomes shows how no single value chain is most effective at achieving every

FIGURE 2.7 Composite score and final ranking based on equally weighted outcomes

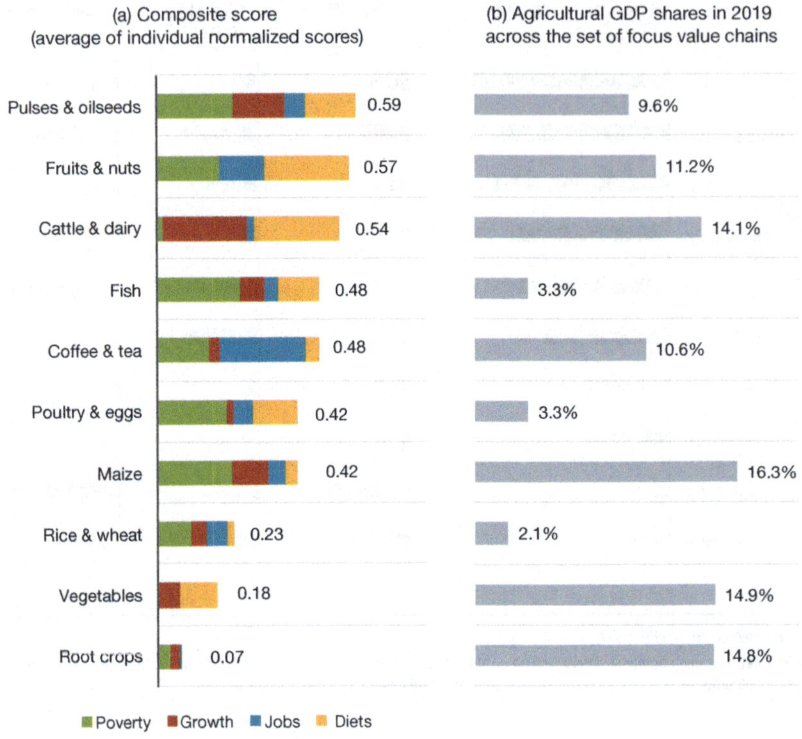

Source: Results from IFPRI's Kenya RIAPA model.

Note: Composite score is a simple average of the normalized scores for the four focus outcome indicators (see Figure 2.6).

development objective. Diversified sources of agricultural growth are therefore needed to drive inclusive agricultural transformation in Kenya.

Identifying which value chains are the most effective overall depends on the importance policymakers or society attach to each of the chosen outcome dimensions. Figure 2.7 shows composite scores estimated by assuming that each of the four outcomes is equally important. In other words, the composite score in the lefthand panel is a simple average of the four normalized scores in Figure 2.6—that is, each outcome dimension is weighted equally. The value chains are re-ranked based on the composite score, and the colors in the figure indicate the contribution of each outcome to the value chain's final score. Although the pulses & oilseeds value chain is not the highest-ranked value chain for any single development outcome, its persistently high rank means it scores highest on the composite score. Fruits & nuts, which are especially effective at improving diets and reducing poverty, and cattle & dairy are, respectively, ranked third and fourth.

The analysis suggests that, if outcomes are equally important and therefore equally weighted, then Kenya should orient its policy and investment target toward high-value products such as pulses & oilseeds, fruits & nuts, and cattle & dairy. Conversely, overinvesting in staple crops such as cereals and root crops will limit the pace of inclusiveness of agricultural transformation. Unfortunately, these crops make up a large share of the existing agriculture sector (see Panel B), which highlights the need for faster diversification.

Finally, we consider how the weighting of outcomes affects the value chain rankings. Table 2.8 ranks value chains using different weighting schemes. The first column assigns equal weights across outcomes (the same as in Figure 2.7), whereas the other columns give greater weight to each of the four outcome indicators in turn. Specifically, we arbitrarily give half the weight to a single outcome and split the rest across the other outcomes. The pulses & oilseeds value chain retains its top-three ranking irrespective of the weighting scheme adopted. The fruits & nuts value chain drops out of the top-three ranking only when the bias is in favor of the agrifood system growth outcome. The cattle & dairy value chain loses its top-three ranking when the poverty or diet quality outcomes are favored.

Although our value chain rankings are reasonably robust to different weighting schemes, our weighting schemes are limited in scope and

TABLE 2.8 Value chain rankings under different outcome weighting schemes

	Equal weights (no bias)	Outcome-biased rankings			
		Poverty bias	Growth bias	Employment bias	Diet quality bias
1	Pulses & oilseeds	Pulses & oilseeds	Cattle & dairy	Coffee & tea	Fruits & nuts
2	Fruits & nuts	Fish	Pulses & oilseeds	Fruits & nuts	Cattle & dairy
3	Cattle & dairy	Fruits & nuts	Maize	Pulses & oilseeds	Pulses & oilseeds
4	Fish	Maize	Fish	Cattle & dairy	Fish
5	Coffee & tea	Poultry & eggs	Fruits & nuts	Fish	Poultry & eggs
6	Poultry & eggs	Coffee & tea	Coffee & tea	Poultry & eggs	Coffee & tea
7	Maize	Cattle & dairy	Poultry & eggs	Maize	Maize
8	Rice & wheat	Rice & wheat	Vegetables	Rice & wheat	Vegetables
9	Vegetables	Vegetables	Rice & wheat	Vegetables	Rice & wheat
10	Root crops	Root crops	Root crops	Root crops	Root crops

Source: Results from IFPRI's Kenya RIAPA model.

Note: In the biased rankings, the focus outcome is given a 50% weight and the remaining outcomes share the remaining 50% equally.

hypothetical—that is, they may not represent the actual preferences of policymakers or society. Our results are also potentially sensitive to the definition of outcome indicators. For example, a different ranking may emerge if the poverty gap ratio is used instead of the poverty headcount ratio (see Table 2.7). Likewise, some might argue that a measure of undernourishment (that is, a lack of sufficient calories) would be more appropriate as a development outcome than our diet quality measure. And whereas we measure growth and employment effects within the agrifood system only—motivated by our interest in agrifood system transformation—others may be interested in the role of agricultural value chains in contributing to economywide growth and employment effects.

We further acknowledge that our value chain rankings offer a national perspective only. Kenya has several distinct agroecological zones, which differ in terms of the value chains that are most suited to local soil, climatic, and market conditions. The devolved political system means that county governments are mandated to design and implement their own development strategies, and this will likely include promotion of value chains depending on their suitability to local conditions and their contribution to local development goals. In an ideal setting, the analysis here would be conducted at the county or county cluster level, which would provide context-specific recommendations. This is an important area for further research in Kenya, albeit one that would require a significantly expanded data collection effort.

Lastly, the model outcomes reflect existing input–output relationships across sectors and current links between household and value chains via employment or consumption. The analysis does not account for the possibility that investments may provide access to new markets or alter input–output relationships, which may result in increased profitability or greater spillover effects between on-farm productivity growth and off-farm growth and employment. We also acknowledge the potential that policy incentives may, over time, allow certain households to engage more actively in value chains from which they are currently excluded. This could, for example, result in value chains such as vegetables or cattle & dairy having stronger poverty-reducing effects than the current data reflect. An analysis of the effects of upgrading value chains to their ideal states is beyond the scope of this study. Useful examples of such work in the context of Kenya include Davids and colleagues (2021), Delport and colleagues (2021), and Langat and colleagues (2021).

In summary, agricultural growth in Kenya is generally pro-poor, and it also contributes to broad economic development and job creation beyond the farm. However, the source or pattern of agricultural growth matters for the pace and inclusiveness of agricultural transformation, and this underscores the need to ensure a better-designed policy and investment portfolio of value chains to drive future agricultural growth.

Conclusion

Kenya's economy grew rapidly during the 2009–2019 decade, and this helped reduce the absolute number of poor people for the first time in decades. This chapter has used new economywide databases and indicators to measure the structure of Kenya's current and evolving agrifood system and its contribution to national development. It has found that the agrifood system, as measured from the supply side and using national accounts and employment statistics, stretches well beyond primary agriculture and creates jobs and income opportunities throughout the economy. These linkages will be expected to become even more important over time, as the development paths of more advanced economies demonstrate. Successful transformation in Kenya requires even larger contributions from agro-processing and food services, with more value added and jobs in the food system eventually generated off the farm. However, while Kenya's agrifood system is growing, and there is faster growth in its off-farm components, the pattern of growth is biased toward creating more value added for primary agricultural products than for processed products. This suggests that Kenya's agrifood system is, as a whole, still in an early stage of agricultural transformation, with most growth and job creation occurring close to the farm. Growth is also faster in the less productive activities of the agrifood system, which could slow the pace of agrifood system transformation.

To determine what is needed for Kenya's agrifood system to contribute more to broad development outcomes, we used RIAPA, an economywide model, to link alternative sources of agricultural growth to outcome indicators associated with inclusive agricultural transformation. The findings indicate that value chains differ considerably in their effectiveness in achieving different development goals. However, the value chains found to be the most effective in reducing poverty, generating growth and employment opportunities in the agrifood system, and improving diets are also the ones that already make up a relatively large share of Kenya's current agriculture sector. This includes value chains such as pulses & oilseeds, fruits & nuts, and cattle & dairy. That said, the value chains that are found to be least effective, including maize and root crops, often dominate agricultural landholding and account for a large share of public investments. Acceleration of structural changes within the agrifood system through the reorientation of the government's investment portfolio could enhance the potential contribution of the agrifood system to broad development outcomes.

The time series data developed for this chapter provide rich information on the changing structure of Kenya's agrifood system. The analysis here has focused on aggregate characteristics and broad trends. Further research is needed to fully exploit the database's information on detailed agricultural products and value chains, including their structure and performance in different agroecological zones. We did, for example, decompose the sources of agricultural growth across value chains, and assess their diverse contribution to national economic development. More in-depth analysis within the agriculture sector is needed to understand which sources of agricultural growth face which kinds of constraints, and which value chains are more promising for public and private investments, including at the subregional or county level. Such investments may also alter input–output relationships, allow value chain actors to access new markets, or enable new actors (such as poor households) to formally engage in value chains from which they were previously excluded.

Finally, our historical analysis of Kenya's agrifood system has focused on the supply side and decomposed changes in GDP and employment. A more comprehensive diagnosis would also require assessing changes in patterns of demand, including consumer demand (see Chapter 4 in this volume) and the role of domestic and international trade. Nevertheless, this chapter has laid the foundation for a more structured analysis of the agrifood system, including the characterization of historical patterns and drivers of change, and the identification of trade-offs and synergies between value chain investments designed to facilitate broad agricultural transformation.

Appendix 2.1 Decomposing agricultural GDP

TABLE A.2.1 Decomposing AgGDP+ across value chains, 2019

Product group	Individual products in the supply use table and their share of group's agricultural GDP (2019)
Cereals	Maize 85% I Sorghum & millet 5.3% I Rice 2.7% I Wheat & barley 6.6% I Other cereals 0.4%
Pulses & oilseeds	Pulses 90.1% I Groundnuts 7% I Other oilseeds 2.8%
Root crops	Cassava 10.2% I Irish potatoes 64.7% I Sweet potatoes 24.4% I Other roots 0.7%
Horticulture	Leafy green vegetables 27.9% I Other vegetables 22.4% I Nuts 2% I Bananas 14.6% I Plantains 0.5% I Other fruits 32.6%
Livestock	Cattle meat 27.7% I Raw milk 49.3% I Poultry meat 5.1% I Eggs 3.7% I Small ruminants 8.1% I Other livestock 6.1%
Fish	Aquaculture 9.8% I Capture fisheries 90.2%
Export crops	Sugarcane 16.8% I Tobacco 1.2% I Cotton & fibers 3% I Leaf tea 37.3% I Coffee 5.1% I Cut flowers 11.8% I Other crops & support services 24.8%
Forestry	One product line
Beverages & other foods	Cannot easily be assigned to a single agricultural activity's output

Source: Authors using IFPRI's 2019 Kenya Social Accounting Matrix

References

Benin, S., and S. Odjo. 2018. "Government Expenditures in Kenya, 1950–2014: Determinants and Agricultural Growth Effects." IFPRI Discussion Paper 1774. International Food Policy Research Institute (IFPRI), Washington, DC.

Davids, T., M. Gouse, L. Kirimi, F. Meyer, J. Thurlow, and N. Odhiambo. 2021. *Prioritising Policies for Inclusive Agricultural Transformation in Kenya: Aquaculture Value Chain Deep Dive Report.* Pretoria: Bureau for Food and Agricultural Policy.

Delport, M., S. Jackson, T. Njagi, D. Boshomane, T. Davids, F. Meyer, and J. Thurlow. 2021. *Prioritising Policies for Inclusive Agricultural Transformation in Kenya: Coffee Value Chain Deep Dive Report.* Pretoria, South Africa: Bureau for Food and Agricultural Policy.

de Vries, G., L. Arfelt, D. Drees, M. Godemann, C. Hamilton, B. Jessen-Thiessen, A. Ihsan Kaya, H. Kruse, E. Mensah, and P. Woltjer. 2021. "The Economic Transformation Database (ETD): Content, Sources, and Methods." WIDER Technical Note 2/2021. Wageningen, Netherlands: Wageningen University and Research.

Diao, X., P. Hazell, and J. Thurlow. 2010. "The Role of Agriculture in African Development." *World Development* 38 (10): 1375–1383.

Diao, X., and J. Thurlow. 2012. "A Recursive Dynamic Computable General Equilibrium Model." In *Strategies and Priorities for African Agriculture: Economywide Perspectives from Country Studies*, eds. X. Diao, J. Thurlow, S. Benin, and S. Fan, 17–50. Washington DC: IFPRI.

Dorosh, P., and J. Thurlow. 2013. "Agriculture and Small Towns in Africa." *Agricultural Economics* 44 (4–5): 435–445.

Haggblade, S., P. Hazell, and P. Dorosh. 2007. "Sectoral Growth Linkages between Agriculture and the Rural Nonfarm Economy." In *Transforming the Rural Nonfarm Economy: Opportunities and Threats in the Developing World*, eds. S. Haggblade, P. Hazell, and T. Reardon, 141–182. Washington, DC: Johns Hopkins University Press for IFPRI.

IFPRI (International Food Policy Research Institute). 2021. 2019 Social Accounting Matrix for Kenya dataset. Harvard Dataverse. Version 1. Washington, DC.

ILO (International Labor Organization). 2020. *Modeled Estimates of the Labor Market.* Geneva: ILO.

Jayne, T.S., J. Chamberlin, L. Traub, N. Sitko, M. Muyanga, F.K. Yeboah, W. Anseeuw, A. Chapoto, A. Wineman, C. Nkonde, and R. Kachule. 2016. "Africa's Changing Farm Size Distribution Patterns: The Rise of Medium-Scale Farms." *Agricultural Economics* 47 (S1): 197–214.

KNBS (Kenya National Bureau of Statistics). 2009. *2009 Population and Housing Census.* Nairobi.

KNBS. 2017. *2015-16 Kenya Integrated Household Budget Survey.* Nairobi.

KNBS. 2019. *2019 Population and Housing Census.* Nairobi.

KNBS. 2022. *Rebased GDP Series.* Nairobi.

Langat, J., J. Owande, K. Hlungwani, G. van der Burgh, T. Davids, F. Meyer, and J. Thurlow. 2021. *Prioritising Policies for Inclusive Agricultural Transformation in Kenya: Beef Value Chain Deep Dive Report.* Bureau for Food and Agricultural Policy. Pretoria, South Africa.

McMillan, M.S., D. Rodrik, and I. Verduzco-Gallo. 2014. "Globalization, Structural Change, and Productivity Growth, with an Update on Africa." *World Development* 63: 11–32.

Ochieng, J., L. Kirimi, D.O. Ochieng, T. Njagi, M. Mathenge, R. Gitau, and M. Ayieko. 2020. "Managing Climate Risk through Crop Diversification in Rural Kenya." *Climatic Change* 162: 1107–1125.

Pauw, K., O. Ecker, J. Thurlow, and A. Comstock. 2021. "Costing Healthy Diets and Measuring Deprivation: New Indicators and Modeling Approaches." IFPRI Discussion Paper 2073. IFPRI, Washington, DC.

Thurlow, J., B. Holtemeyer, K. Pauw and J. Randriamamonjy. "Measuring Agri-Food Systems: New Indicators and Estimates." IFPRI Discussion Paper. IFPRI, Washington, DC, forthcoming.

Timmer, C.P. 1988. "The Agricultural Transformation." In *Handbook of Development Economics*, eds. H. Chenery and T.N. Srinivasan, Vol. 1, 275–331. Amsterdam: North Holland.

Tschirley, D., T. Reardon, M. Dolislager, and J. Snyder. 2015. "The Rise of the Middle Class in East and Southern Africa: Implications for Food System Transformation." *Journal of International Development* 27 (5): 628–646.

Willett, W., W. Rockström, B. Loken, M. Springmann, T. Lang, S. Vermeulen, T. Garnett, et al. 2019. "Food in the Anthropocene: The EAT–Lancet Commission on Healthy Diets from Sustainable Food Systems." *The Lancet Commissions* 393 (10970): 447–492.

Worku, M., H. de Groote, B. Munyua, D. Makumbi, F. Owino, J. Crossa, Y. Beyene, S. Mugo, M. Jumbo, G. Asea, C. Mutinda, D.B. Kwemoi, V. Woyengo, M. Olsen, and B.M. Prasanna. 2020. "On-farm Performance and Farmers' Participatory Assessment of New Stress-Tolerant Maize Hybrids in Eastern Africa." *Field Crops Research* 246: 107693.

World Bank. 2022. POVCALNET Online Database. http://iresearch.worldbank.org/PovcalNet/povOnDemand.aspx

Chapter 3

LIVESTOCK SECTOR TRANSFORMATION IN KENYA: CURRENT STATE AND PROJECTIONS FOR THE FUTURE

Sirak Bahta, Francis Wanyoike, Leonard Kirui, Charles Mensah, Dolapo Enahoro, Joseph Karugia, and Isabelle Baltenweck

The livestock sector plays a major role in the Kenyan food system, contributing about 12 percent of the country's overall GDP and 40 percent of agricultural GDP, and employing about half of the agricultural labor force (Kenya Markets Trust 2019a). The livestock sector also contributes 22 percent of food system GDP. Projections from the Kenya National Bureau of Statistics (KNBS) indicate that the country's population will continue to grow, and may reach around 96 million by 2050, with nearly 50 percent residing in urban areas by then compared with 27 percent in 2019 (FAO 2019; MacMillan 2019). Demand for animal-source foods is expected to grow substantially with this population growth and as higher numbers of affluent and more urban consumers seek nutrient-rich foods and more diversified diets (FAO 2019). This change in demand could potentially drive exponential growth in the livestock sector, generating new business opportunities along various livestock value chains. Realizing such growth will require concerted investments to support increased productivity and enhanced natural resource management, including enhanced water availability and management, to ensure sustainable growth of the sector (ILRI 2019; Bosire et al. 2022).

Consistent with these expected increases in demand and supply, Kenya's Vision 2030 agenda calls for accelerated development of the livestock sector (FAO 2019). Appropriate livestock-mediated interventions in markets, institutions, and the policy space could also support the emergence of a food system with increased capacity to meet the population's growing needs for nutrition, food security, employment, and income, among other benefits. However, the choice of interventions today that can steer development toward the desired livestock sector and food system of tomorrow must consider the existing constraints, opportunities, and uncertainties. The dynamics of socioeconomic change, climate change, geopolitics, and other drivers shaping the Kenyan food system are uncertain, as are the outcomes of the interactions of these drivers.

Foresight and scenario analyses can play a key role in assessing potential alternative futures, considering uncertainties. Such analysis can also support the development of policies, processes, and programs to support sustainable, resilient food system transformation in Kenya (Wiebe et al. 2018). By incorporating foresight analysis into livestock sector planning, analysts are also better able to provide evidence on how transformative growth in the livestock sector will bring major development opportunities and challenges for the overall food system, considering strategic policy decisions investments made today that could ensure the long-term sustainability of the sector and the overall food system (FAO 2019).

This chapter provides a meta-analysis of multiple previously published scenario analyses of future demand and production of commonly consumed and/or produced livestock-derived foods in Kenya. To this end, we use the results from quantitative future scenarios to assess the demand and supply of foods of livestock origin in Kenya, adopting this as an appropriate approach for identifying policy options to guide the sustainable transformation of the livestock sector. In addition, we have synthesized the major constraints and opportunities in the livestock sector identified in a recent stakeholder workshop on value chain-specific constraints, opportunities, and improvement strategies as part of the development of a livestock master plan for Kenya.[1]

Demand for livestock products

Livestock-derived foods (meat, milk, and eggs), or LDF, comprise an essential part of people's diets in Kenya (FAO 2019). Estimates of the level of consumption are, however, subject to some uncertainty. Our review of the demand or consumption of LDF commodities in Kenya relies on data from numerous sources, including the Food and Agriculture Organization of the United Nations (FAO) and studies by various organizations and researchers, to address the challenge of inaccuracies and inconsistencies that have been reported to exist in the available information (ICPALD 2013; Kenya Market Trust 2019a). We first evaluate recent trends in the consumption levels of various livestock products over 10 years (2010–2019). We then review findings from recent demand studies for insights into how the demand for livestock food commodities is evolving. Finally, using food balance sheet data from FAO (FAOSTAT

---

1 www.ilri.org/livestock-master-plans

2022) and parameter estimates obtained from the demand studies, we derive projections of consumption levels over the next 15 years.

Trends in consumption of major livestock commodities

According to food balance sheet data (FAOSTAT 2022), the main LDF commodities consumed in Kenya are milk, beef, sheep, goat, pig, poultry, and eggs. For many commodities, there was a slight upward trend in total annual quantities consumed nationally between 2010 and 2019 (Figure 3.1). Milk consumption was highest in volume terms among all LDF consumption recorded, highlighting the importance of this livestock product for diets in Kenya (see also Chapter 4 in this book). The total annual volume of milk consumption rose from 4.43 billion liters in 2010 to 5.17 billion liters in 2019. Figure 3.2 presents the average annual rates of increase in national consumption levels of various commodities calculated based on consumption levels in 2010 and 2019. Growth in total quantities consumed varied substantially across different livestock products. For the 10 years analyzed, consumption growth was highest for poultry (13.6 percent per year), followed by pig meat (5.8 percent per year) and milk (1.7 year). The average rate of increase for beef was zero over the period

FIGURE 3.1 Annual national levels of consumption of different livestock food products in Kenya, 2010–2019

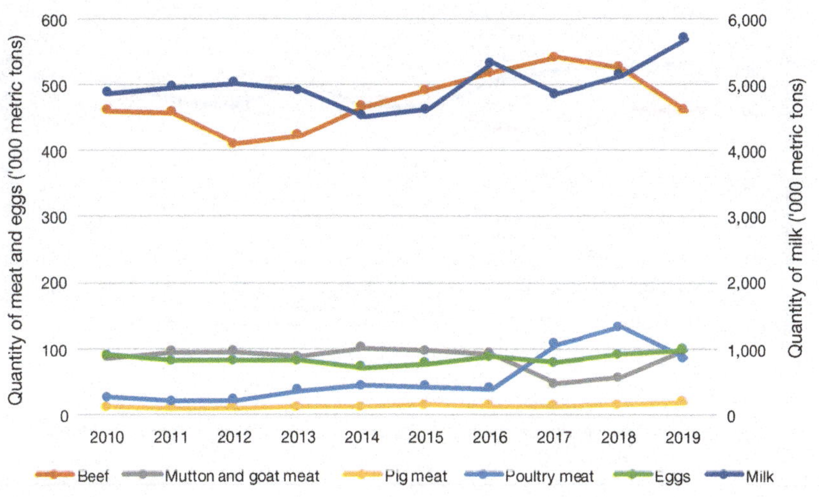

Source: FAOSTAT (2022).

FIGURE 3.2 Average annual percentage rate of change in aggregate demand for livestock food commodities, 2010–2019

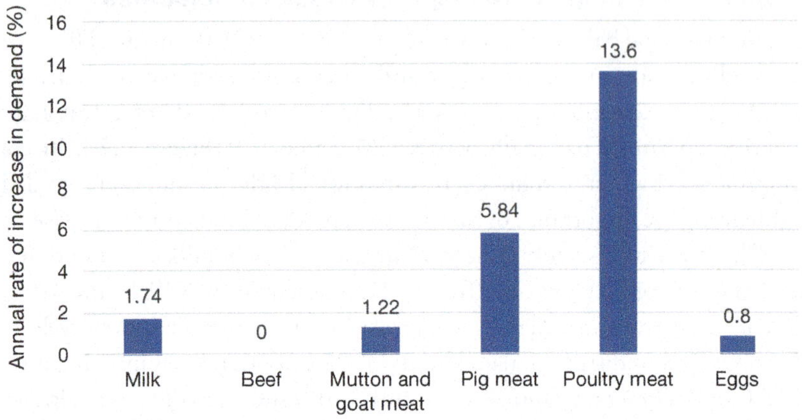

Source: FAOSTAT (2022).

FIGURE 3.3 Per capita levels of consumption of different livestock food products, 2010–2019

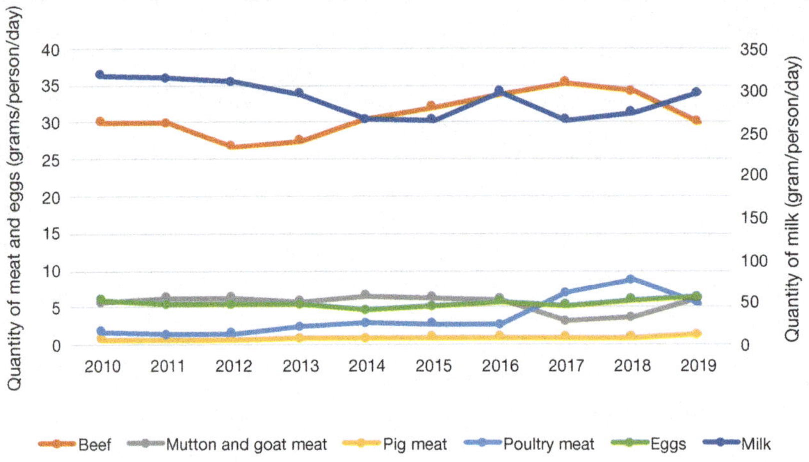

Source: FAOSTAT (2022).

(Figure 3.2), which is explained mainly by a decline in the total consumption volumes observed in 2018 and 2019 that canceled out the steady increase previously observed between 2012 and 2017 (Figure 3.1). Estimates of per capita levels of consumption based on FAO data for 2012 show that, except for poultry meat consumption, which increased by 200 percent (from 1.8 to 4.4 g/person/day), consumption per person of LDF remained more or less static in the country over the 10 years (Figure 3.3). The FAO-based estimates perhaps understate the level of consumption: a survey by Kenya Market Trust (2019a) estimated that consumption of red meat (beef, goat meat, and mutton) was 15.08 kg per person per year.

A survey by FAO (2019) presents additional details on the patterns of LDF consumption in Kenya. Results show that milk and beef constitute the main supply of LDF in diets, with approximately 86 percent and 46 percent, respectively, of households across various income groups consuming fresh milk and beef. In addition, calculations based on data from FAOSTAT (2022) show that milk contributes 6.03 grams of protein per capita per day, or 48 percent of the average daily per capita protein supply of LDF (12.62 grams), whereas meat contributes 6.17 grams, of which 4.00 grams, or 65 percent, comes from beef

Numerous studies have been conducted in recent years on the dynamics of demand for livestock products in Kenya, including Bett and colleagues

TABLE 3.1 Consumption of livestock-derived foods, 2018

LDF type	Protein consumption (grams/capita/day)	Share of average daily per capita protein supply of LDF (%)
Meat (total)	6.17	49
Beef	4	32
Chicken meat	1	8
Buffalo meat	0.09	1
Sheep and goat meat	0.43	3
Pig meat	<0.01	<0.01
Other	0.65	5
Eggs	0.42	3
Milk (total)	6.03	48
Milk, whole fresh cow	5	40
Milk, whole fresh camel	1	8
Milk, whole fresh goat and sheep	0.03	<0.01

Source: FAOSTAT (2022).

(2012), Ngula (2014), Shibia, Rahman, and Chidmi (2017), and Korir, Rizov, and Ruto (2018). The studies focus mainly on the effects of prices and income on demand. Table 3.2 presents estimates of own-price elasticities and income elasticities of demand for various livestock commodities from recent demand studies. As expected, all the studies show that price increases are likely to reduce demand for various livestock food commodities, and vice versa for price drops, authenticated by the negative own-price elasticities. Consistent with the seminal literature on the livestock revolution (Delgado et al. 2001), various studies also show that income increases positively affect demand.

TABLE 3.2 Income elasticities of demand for various livestock food commodities

Livestock product	Year data were collected	Own-price elasticity		Income elasticity	Study
		Compensated	Uncompensated		
Beef	2012	–0.1844	–0.5282	1.0010	Ngula (2014)
Shoat meat	2012	–0.2612	–0.5368	1.0652	
Chicken meat	2012	–0.26078	–0.4778	0.8519	
Pork	2012	–0.7129	–0.8764	1.1435	
Bone beef	2013	–0.5790	–0.9480	1.0257	Shibia, Rahman, and Chidmi (2017)
Boneless beef	2013	–0.7270	–0.8050	0.9894	
Goat meat/mutton	2013	–0.7490	–0.8100	0.8937	
Chicken meat	2013	–0.4950	–0.6430	1.0993	
Pork	2013	–0.4200	–0.7650	0.9594	
Indigenous chicken meat	2012	–0.1876	–0.7705	0.8537	Bett et al. (2012)
Exotic chicken meat	2012	–0.0738	–0.1089	1.5020	
Beef	2012	–0.5485	–0.6630	0.8455	
Mutton	2012	–0.5945	–0.6030	0.2547	
Goat meat	2012	–0.5150	–0.6605	1.7619	
Other meats	2012	–	–	2.6917	
Dairy products					
Rural	2015	–0.4280	–0.5096	0.8804	Korir, Rizov, and Ruto (2018)
Urban	2015	0.4202	–0.4990	0.8708	
Overall	2015	–0.4250	–0.5057	0.8770	

Source: Authors.

Note: Own-price elasticity of demand is the ratio of the percentage change in quantity demanded of a product to the percentage change in price. Income elasticity of demand is the ratio of the percentage change in quantity demanded of a product to the percentage change in income. Compensated price elasticities (also called Hicksian elasticities) ignore the income effect of a price change and comprise only the substitution effect. Uncompensated price elasticities (also called Marshallian elasticities) consider both the substitution and the income effects of a price change.

Projections of the future level of demand for livestock food commodities

Using the information on consumption levels of the various livestock commodities and additional data from the literature, projections of future levels of demand for the commodities were generated for this study following Shapiro and colleagues (2017). The projected total consumption (TLC_t) of a given livestock product in year t was calculated by multiplying the projected per capita consumption (LC_t) by the projected population (POP_t) for a given period t (Equation 1).

$$TLC_t = LC_t * POP_t \quad (1)$$

In turn, LC_t was calculated as follows:

$$LC_t = LC_0 * (1 + \eta * \gamma)^t \quad (2)$$

where LC_0 is the baseline per capita consumption of a given livestock product, η is the income elasticity of demand for the livestock product, and γ is the annual growth rate of real per capita GDP.

The results from recent demand studies in Kenya informed the values of income elasticities of demand (Table 3.2). Note that, in the results presented in Table 3.2, income elasticity estimates by Ngula (2014) and Shibia, Rahman, and Chidmi (2017) are quite close to each other but differ remarkably from those by Bett and colleagues (2012). The values of income elasticities we adopt for our projections are informed by the estimates from the studies by Shibia, Rahman, and Chidmi (2017) and Korir, Rizov, and Ruto (2018). Besides corroborating each other, the two sets of elasticities are consistent with the rates of growth in consumption available in the FAO data.

Baseline per capita consumption was estimated by dividing the national consumption values from FAOSTAT by the total human population estimate for 2019. The annual growth rate of real GDP per capita (γ) was estimated using economic survey data for the years 2016 and 2019 from KNBS (2021). To project the population (POP_t), a growth rate of 2.3 percent per year was used based on the 2009 and 2019 national census results.

Table 3.3 presents projections of annual total quantities consumed for various livestock products between 2019 and 2037. The projected quantities of various types of meat in 2037 are 108–123 percent higher than the level in the base year (2019).

The projected rates of increase in demand in this study were compared with interpolations derived from projections in 2050 by Enahoro and colleagues

TABLE 3.3 Projected levels of consumption of various livestock products between 2019 and 2037

	'000 metric tons				
Year	Milk	Beef	Mutton and goat meat	Pig meat	Poultry meat
2019	5,156.0	418.6	85.6	19.0	28.5
2020	5,370.8	437.2	89.3	19.9	29.8
2021	5,594.6	456.7	93.2	20.8	31.1
2022	5,827.7	477.0	97.3	21.8	32.5
2023	6,070.5	498.2	101.5	22.7	33.9
2024	6,323.4	520.4	106.0	23.8	35.4
2025	6,586.9	543.5	110.6	24.9	36.9
2026	6,861.3	567.7	115.4	26.0	38.6
2027	7,147.2	593.0	120.4	27.2	40.3
2028	7,445.0	619.4	125.6	28.4	42.0
2029	7,755.2	647.0	131.1	29.7	43.9
2030	8,078.3	675.8	136.8	31.1	45.8
2031	8,414.9	705.8	142.8	32.5	47.8
2032	8,765.5	737.2	149.0	34.0	49.9
2033	9,130.7	770.1	155.5	35.6	52.1
2034	9,511.1	804.3	162.3	37.2	54.4
2035	9,907.4	840.1	169.3	38.9	56.8
2036	10,320.2	877.5	176.7	40.7	59.3
2037	10,750.2	916.6	184.4	42.5	61.9

Source: KNBS (2021); FAOSTAT (2022).

TABLE 3.4 Comparison of projected increases in consumption of various livestock commodities from different studies

Livestock food commodity	Current study (%)	Increases seen in Enahoro et al. (2018) (%)				Increases seen in FAO (2019) (%)
		Moderate rates of economic growth	High rates of economic growth	Low rates of economic growth	Significant challenges to global climate change adaptation and mitigation	
Milk	108	32	26	46	32	56
Beef	119	65	67	71	64	86
Mutton and goat meat	115	65	66	71	65	–
Pig meat	123	99	117	90	97	–
Poultry meat	117	137	171	118	16	89
Eggs	–	79	90	79	80	67

Source: Enahoro et al. (2018); FAO (2019).

(2018) and FAO (2019) (Table 3.4). While the rates of increase in demand in this study compare well with those from the two other studies in the case of pig and poultry meat, they are markedly different (that is, lower than the estimates by Enahoro et al. and FAO) for milk, beef, and sheep and goat meat. These differences are likely attributable to the effects of the different methods and data used to generate the projections. Nevertheless, the results from all three studies point to some significant increase in demand for the various livestock products over the projection period.

Supply of livestock-derived food products

Historical trends of supply

The total supply of LDF in Kenya—that is, the overall quantities consumed or otherwise used by the population—consists predominantly of domestically produced items (FAOSTAT 2022). However, growth in domestic LDF production has been quite varied, with changes in the total output more pronounced in some livestock subsectors than in others. Over 2010–2019 decade, the national production of poultry meat, for example, increased by nearly 230 percent (from 27,000 metric tons in 2010 to 89,000 metric tons in 2019). Egg and milk production increased by around 6 percent, and small ruminant meat production

FIGURE 3.4 Quantities of livestock-derived foods produced domestically, 2010–2019

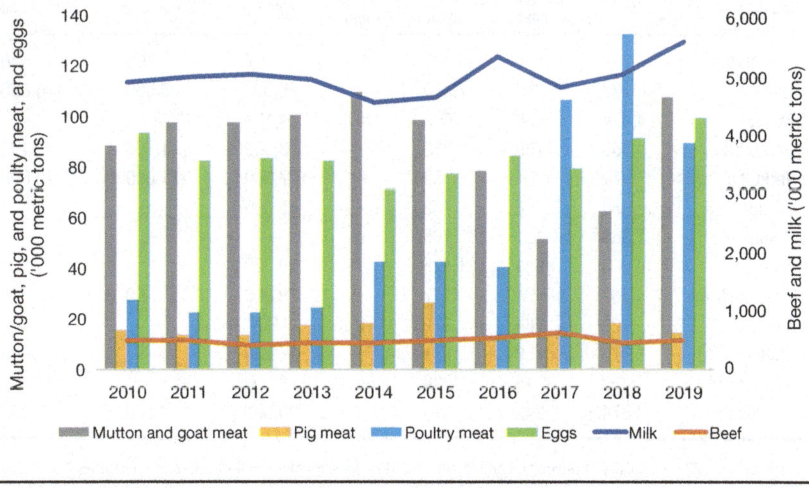

Source: Authors using FAOSTAT (2022).

by 22 percent. On the other hand, beef production was the same in 2010 and 2019, although substantial year-on-year fluctuation occurred between these years (Figure 3.4). Pig meat production declined more than 6 percent over the 10 years.

Within Kenya, dairy is produced in intensive, semi-intensive, and extensive production systems, while beef production occurs in pastoral systems, semi-intensive (agropastoralist) ranching systems, and commercial feedlots (FAO 2019). Pig and poultry production are typically not directly associated with land-based systems and may be carried out on smallholder farms or specialized poultry operations (for example, meat or egg) (Herrero et al. 2013).

Data from 2014 showed that beef production occurred mostly in arid and semiarid lands (ASALs), where 36 percent of the human population resides. Most of Kenya's beef production is recorded to have come from Marsabit, Meru, Kwale, Embu, Taita, and Laikipia counties, with two of the counties (Marsabit and Meru) together producing more than 8,400 metric tons, or 55 percent of the county-level totals recorded for that year (Table 3.5).

Cow milk production was highest in Turkana (127,000 metric tons), while Garissa accounted for the highest county-level supply of sheep and goat meat (2,660 metric tons in total). Only a few counties reported poultry and pig production data, and we have not included those in the table.

TABLE 3.5 Meat and milk production in 2014 for selected counties

	Metric tons							
County	Beef	Sheep meat	Goat meat	Chicken meat	Cow milk	Goat milk	Camel milk
Mandera	257	–	103	6	207	166	488
Garissa	137	1,380	1,380	–	28,000	8,000	436,000
Marsabit	4,474	938	663	26	4,131	–	–
Samburu	553	729	519	–	3,300	1,163	193
Turkuma	71	54	89		127,000	221,000	110,000
Embu	950	68	162	–	45,000	0.0	–
Kwale	2,368	191	105	16	–	–	–
Meru	4,000	681	1,425	–	100,320	–	–
Tharaka Nithi	551	91	4177	–	19,000	60	–
Lamu	23	84	122	8	3,700	–	–
Taita	1,077	159	31	–	17,230	–	–
West Pokot	2	15	26	–	8,000	–	–
Laikipia	874	242	134	2	99,000	2,200	2,200

Source: KNBS (2017); Ministry of Devolution and Planning (2017); Ministry of Agriculture, Livestock, Fisheries and Irrigation (2019).

Projections of future supply

While a few studies provide quantitative estimates of plausible supply of LDF in Kenya in the future, most of these studies generate the projections at the national and not subnational (for example, county) level, deriving these using aggregated statistics (for example, elasticities) of LDF demand and supply. Two key approaches to quantifying future LDF (or other agricultural/food commodity) projections will be (1) the derivation of future growth rates using knowledge of the past with probabilistic estimates of the future and (2) exploratory projections that start from the present and investigate the impacts of various drivers, trends, and interactions into the future (Wiebe et al. 2018). A relevant quantitative approach that is based on the latter derives supply growth as the endogenous output of an integrated assessment model, solving for a stated market objective (such as partial equilibrium) while accounting for biophysical realities such as Earth system changes, crop physiology, and water basin management (Robinson et al. 2015).

Projections reported by Enahoro and colleagues (2018) suggest that the production of milk, beef, small ruminant meat, pig meat, poultry meat, and eggs could increase by between 8 percent (eggs) and 69 percent (poultry meat) over 15 years. A future scenario of livestock health intervention further generates a 166 percent increase in the domestic supply of beef by 2034 following the widespread adoption of a cattle disease vaccine in areas with high mortality from diseases. However, a major practical obstacle remains that vaccine uptake rates are still very low in affected areas, hindering supply expansion (Toye et al. 2020). A different challenge to modeling the potential expansion of LDF supply has to do with weak accounting for livestock feed availability and use in available economic models (Msangi et al. 2013). Although Bosire and colleagues (2022) partially address this challenge, a key gap persists in how well the current models can represent the potential expansion of livestock production systems in Kenya, as in many developing countries.

Production-demand gaps

Historical trends in production-demand gaps

Recent trends, such as the significant increases in demand discussed earlier and other market changes, coupled perhaps with a less responsive production side, are creating scope for a larger role of other—that is, nondomestic sources of LDF supply in Kenya (FAO 2019).

FIGURE 3.5 Production and total supply in Kenya of livestock-derived foods, 2010–2019 ('000 metric tons)

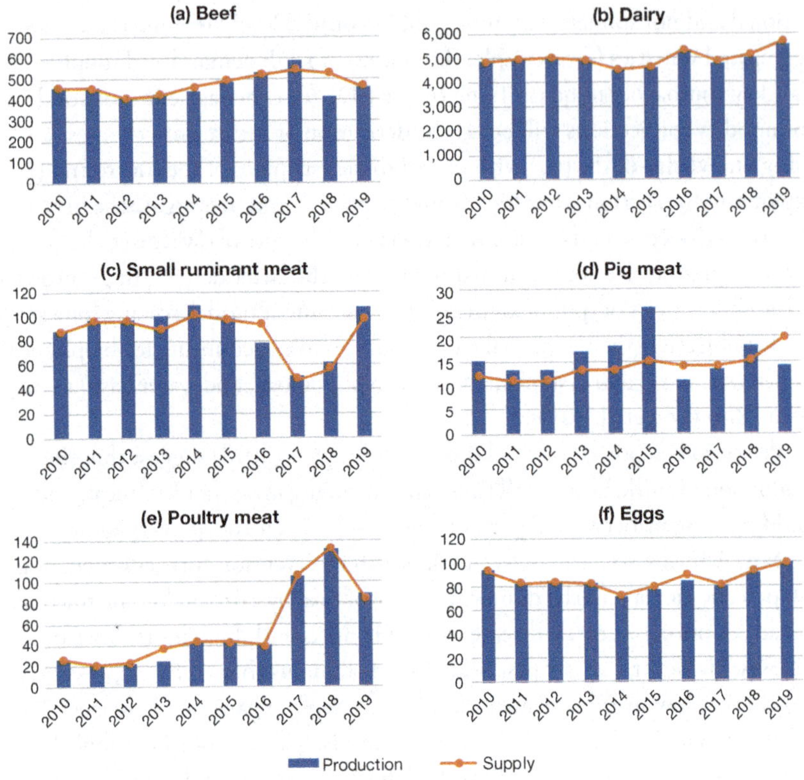

Source: Authors using FAOSTAT (2022).

Note: Production refers to the quantity of livestock products produced by the country in a calendar year; supply refers to the amount of the commodity available to consumers in a country in a calendar year.

On average, over the past decade, according to published national statistics (FAOSTAT 2022), annual national production of pig meat and small ruminant meat has generally surpassed the aggregate demand for these commodities, but this has not been the case for beef, dairy, poultry meat, or eggs (Figure 3.5).

Model projections of production–demand gaps

The projected levels of consumption of LDF were compared with projected levels of domestic production (Figures 3.6 to 3.8). The production levels were interpolated based on levels reported in the published national statistics for

2015 (FAOSTAT 2022) and projected levels in 2050 obtained from a study on the future of livestock in Kenya by FAO (2019). Our study assumes that the transformation of Kenya's livestock sector depends on several relatively well-understood factors (such as existing policies, megatrends such as population, urbanization) and some less well-understood factors (consumer behavior, government accountability, climate change, technology development). The assumption is that livestock supply chains will satisfy the demand of a growing affluent urban population in Kenya through increased uptake of technology and the nature of food-feed competition.

The FAO (2019) study generated projections of levels of production of milk, beef, and chicken meat under four scenarios intersecting macroeconomic and governance conditions, including (1) a good governance[2] good economy[3] scenario; (2) a good economy bad governance scenario; (3) a good governance bad economy scenario; and (4) a bad governance bad economy scenario. Scenario 1 represents the best context for livestock activities and Scenario 4 the worst. In Figures 3.6 to 3.8, the projected best case scenario maps the highest projected production levels whereas the projected worst case scenario maps the lowest projected levels for milk, beef, and chicken. The projected consumption level comprises two sets of data—that is, calculations using Equation 2 in this study and interpolations based on projected changes in consumption levels between 2015 and 2050 in the study by FAO (2019). Figures 3.6 to 3.8 show the highest and lowest projected consumption levels in different years from the two datasets plotted side by side. Results presented in the three figures indicate that, depending on the prevailing macroeconomic and governance situation, domestic production of milk could be higher than the highest projected level of consumption (thus leaving some surplus for export) or lower than the lowest estimated level of consumption (making the country a net importer of milk) (Figure 3.6).

For beef, the highest projected level of domestic production is much greater than the highest projected level of consumption (Figure 3.7), implying that, under the best governance and macroeconomic environment, Kenya can produce beef in excess of domestic needs.

Conversely, under the bad governance and bad economic situation, both the lowest and the highest levels of beef consumption, which are respectively

2 A good governance system is one with high levels of accountability and responsibility and strong stable institutions providing better services.

3 A good economy is described as a vibrant, thriving, and diversified economy with effective allocation of resources.

FIGURE 3.6 Projected levels of domestic consumption and production of milk, 2019–2037

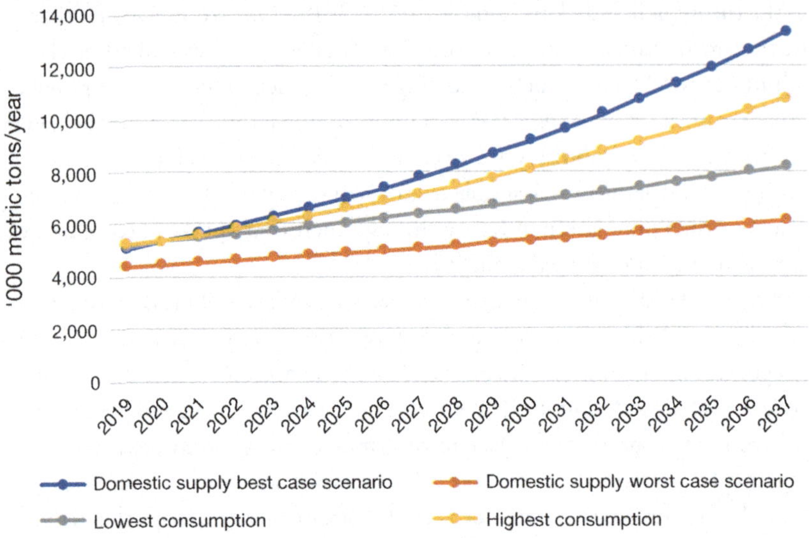

Source: Authors using data from multiple sources.

FIGURE 3.7 Projected levels of domestic consumption and production of beef, 2019–2037

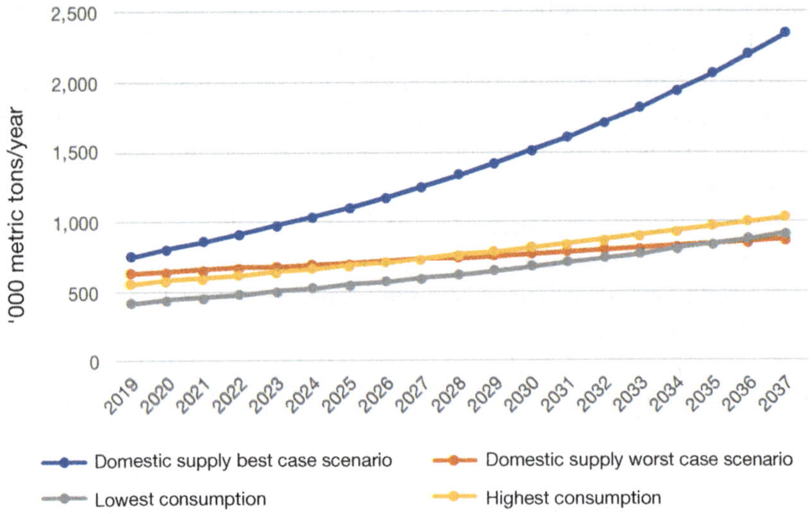

Source: Authors using data from multiple sources.

FIGURE 3.8 Projected levels of domestic consumption and production of chicken meat, 2019–2037

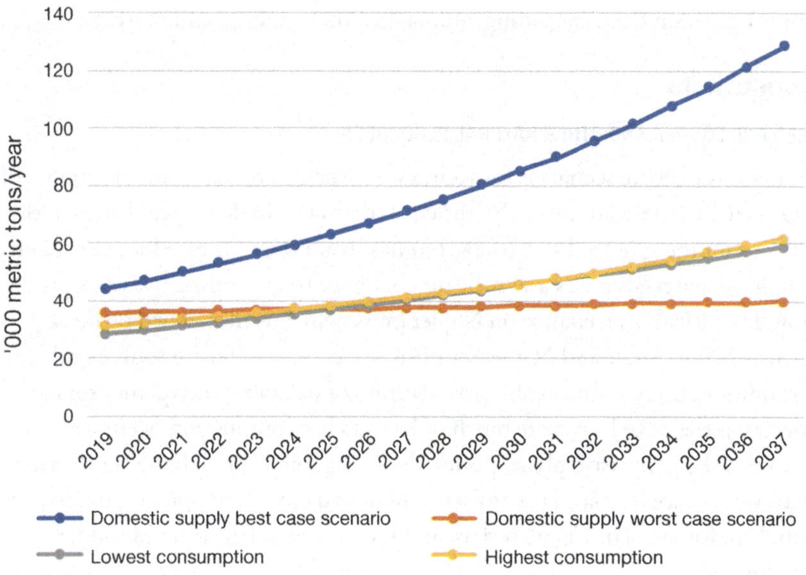

Source: Authors.

less and nearly equal to the lowest projected level of domestic production, are projected to grow progressively and rise above this level between 2029 and 2035, transitioning the country into a net importing position for beef.

In the case of chicken meat, the highest projected level of domestic production is substantially greater than the highest projected level of consumption, indicating some scope for the country to trade surplus chicken meat production (Figure 3.8).

Nevertheless, the projected levels of consumption, which are initially lower than the lowest projected level of domestic production, progressively increase and surpass production in 2025. The gap between demand and production gradually widens, and the country will be required to import chicken meat to close the gap between production and consumption.

Opportunities and constraints

Given plausible outcomes of LDF demand and production dynamics in Kenya, the future holds opportunities and constraints for transforming the livestock sector, including dairy, beef, small ruminants, pigs, and poultry, which we briefly explore in this section. Our summary is based on a Kenya livestock master plan stakeholder meeting conducted to identify value chain-specific constraints and opportunities related to feeding, animal health, breeding, and markets.[4]

Constraints

FEED AND FODDER AND OTHER NATURAL RESOURCES

Under most future scenarios for Kenya's livestock sector, additional supplies of LDF will be needed to meet the anticipated growth in demand. However, the current sourcing of feed and forage biomass from only a limited set of exploited roughage materials poses a significant challenge to expanding livestock production. Historical overreliance on Napier grass, which is threatened by diseases such as Napier smut and Napier stunting, exacerbates a feed resource problem. In addition, there is diminishing availability of palatable and quality forage species in the ASALs (where much of Kenya's beef production occurs) as a result of overgrazing, invasive plant species, declining soil health, changing climate patterns, competing land use for settlement and crop development, inadequate supply of forage planting materials, and low commercialization of fodder production.

Other challenges arise in the provision of concentrate feeds and in competition with humans using biomass as energy sources. While feed standards have been developed for most livestock species, these do not exist for some livestock categories important to Kenya (Ministry of Agriculture, Livestock, Fisheries, and Cooperatives 2020a).

ANIMAL HEALTH

Another key challenge to livestock production is poor control and management of animal diseases, fueled by, among other things, inadequate capacity among and coordination of bodies and programs responsible for sustained disease surveillance and poor enforcement of existing laws governing disease control and animal movement.

Ticks, tsetse flies, and worms are the major disease transmission agents in animals in Kenya. Cross-border movement of livestock to traditional seasonal grazing grounds and for trade is common, but coordination and collaboration with Kenya's neighbors on disease control across borders is poor. As a result, control of disease transmission poses a significant transboundary challenge, adversely affecting production and productivity in the greater region.

4 The stakeholder meeting was held on July 26–30, 2021, in Naivasha, Kenya.

BREEDING

Animal genetic resources (AnGR), embodied in live populations of animals or conserved genetic materials, are critical genetic resources, forming the biological capital for livestock development and vital to food security and sustainable rural development. In Kenya, efforts to improve livestock productivity have focused on replacing traditional/local breeds with more exotic and mixed breeds. The existing regulatory framework for the importation of germplasm has led to the erosion of indigenous AnGR. As Ruto, Garrod, and Scarpa (2007) note, while such efforts may result in short-term economic gains, the associated loss in genetic resources of the local breed is detrimental in the long run, as these local breeds possess important genetic traits that are well adapted to local conditions. Failures to design appropriate breeding programs for locally adapted breeds and to establish systems needed to facilitate sustainable use remain outstanding challenges.

MARKETS

The key issues facing the aggregation and transportation of livestock and livestock products are insecurity, poor infrastructure, seasonality of production, and inefficient distribution systems. In addition, external trade in livestock and livestock products is affected by limited capacity to meet the sanitary requirements of importing countries, inadequate volumes, and untimely deliveries.

Kenya's livestock products are currently marketed locally and internationally (primarily to neighboring countries) with limited value addition. Value addition is largely constrained by high investment costs, limited demand for value-added products by consumers, an inadequately skilled workforce, limited access to financial and business development services, poor infrastructure, and inadequate value addition technology. In addition, there is insufficient capacity for entrepreneurship among primary producers in the livestock industry, resulting in low margins for their enterprises.

Issues that affect the safety and quality of livestock and livestock products include inadequate capacity to undertake better animal husbandry practices, hygienic practices, or manufacturing practices. In addition, there is inadequate capacity to verify food quality and safety and enforce food safety regulations.

Marketing infrastructure is in a poor state or underdeveloped, and hence not favorable for efficient livestock marketing. Further, marketing information systems are poorly developed, which limits access and use.

Opportunities

The livestock sector offers opportunities in many areas, including animal-source food availability and nutrient supply, incomes and employment, and foreign exchange earnings.

ANIMAL-SOURCE FOOD AVAILABILITY AND NUTRIENT SUPPLY

There is increasing evidence that the demand for LDF will increase substantially in Africa and Asia in the coming decades. For example, Enahoro and colleagues (2018) show that demand for LDF will grow substantially by the year 2050 in eight countries—including Kenya—that are currently facing food security and nutrient supply challenges. An assessment of food demand and supply projections for the global agricultural system shows that the growth in demand for LDF will persist in Burkina Faso, Ethiopia, India, Kenya, Nicaragua, Tanzania, Uganda, and Viet Nam under a range of plausible scenarios for global economic growth and climate change. For all eight countries, the analysis shows higher total consumption of LDF in 2050 relative to 2010, reflecting expected growth in population but also some growth in income and thus in per person LDF consumption. These projections suggest strong potential for improving the livelihoods and nutrition of the rural poor through livestock sector-based transformation strategies. The challenge, therefore, remains in effectively transitioning the domestic livestock sector to enable it to meet this growing demand for livestock products.

INCOME AND EMPLOYMENT

Livestock production could be a significant source of quality employment if the sector is effectively modernized. Generally, there are not enough employment opportunities in Kenya, especially in rural areas, resulting in both underemployment and unemployment. One of the possible solutions is to grow value chains around livestock. A simulation of the employment potential of the livestock sector in two countries (Burkina Faso and Tanzania) in Africa by Frija and Enahoro (2018) found that well-targeted investments could yield annual growth in employment of 10 percent a year, with especially positive returns for women. The rural-based nature of livestock activities makes livestock-keeping a suitable enterprise to improve household incomes and contribute to employment creation.

The use of digital tools to build out value chains and improve employment via extension agents, artificial insemination agents, and veterinarians is also gaining traction in Kenya. One of the known digital tools for livestock development is iCow, an e-extension tool offered by GreenDreams Tech for farmers in Kenya. iCow serves to provide technical information and link actors across the value chain. iCow also offers a virtual marketplace (iCow Soko), where subscribers buy inputs and sell livestock and livestock products. Moreover, there is a function that enables livestock-keepers to locate veterinary and artificial insemination officers in their area. Users can also contact a resident veterinarian (Dr. iCow) through an SMS, to which Dr. iCow responds with a direct telephone call (Daum et al. 2022).

FOREIGN EXCHANGE EARNINGS

Livestock could become a major foreign exchange earner for Kenya. The livestock resource base is estimated at 17.5 million cattle, 27.7 million goats, and 17.1 million sheep and camels, making Kenya the country with the third-largest endowment of livestock in Africa, behind only Ethiopia and Botswana (Kenya Market Trust 2019b). According to a study by Kenya Markets Trust (2019a), the livestock sector could add 10 percent to the country's GDP if it reaches its latent potential. There is huge potential for meat exports from the country. For example, Namibia, a smaller country, has recently accessed US markets after making a substantial investment in the livestock sector over the past 20 years. Kenya's meat export market once thrived in the euro zone, which has been the largest importer of Kenyan meat. According to Kenya Market Trust (2019a), approximately 400,000 metric tons of processed beef were exported annually to the euro zone. However, in order to regain its former prominence in the meat export market, Kenya must make strategic investments in livestock traceability, overhaul its livestock finishing and fattening enterprises, and foster a conducive market environment. This approach, as suggested by Horizon East Africa (2021), has potential to restore Kenya's previous position in the meat export industry.

Contribution of livestock systems to income, employment, gender equality, and child nutrition

Livestock is an important contributor to the Kenyan economy and the livelihoods of people. At the farm level in Kenya, livestock production generates income through the sale of LDF, as well as the sale of products such as wool, hides, and skins. Besides income generation, livestock is used to accumulate and store wealth and, in pastoral communities, is often the only major asset (Abay and Jensen 2020). Small and large animals constitute a "walking savings account" used to purchase agricultural inputs, to invest in other income-generating activities, or to pay for expenses—planned (education and weddings) and unplanned (medical bills or funeral costs) (Baltenweck et al. 2020).

Livestock production contributes to economic development at the household and community level through this income and saving function. In ASALs, livestock contributes approximately 90 percent of the livelihood of households and accounts for nearly 95 percent of family income (Ministry of Agriculture, Livestock and Rural Development, 2008).

In terms of employment, while there are no recent detailed estimates of this in the livestock sector in Kenya, the economic benefits of the livestock sector are witnessed beyond the farm. This includes jobs for individuals engaged in the many associated industries throughout the livestock value chain. These associated industries relate to agricultural inputs and services such as feed, animal health and breeding services, equipment and machinery, and banks and insurance companies that service farmers. Focusing on the dairy sector, Staal, Nin-Pratt, and Jabbar (2008) record the employment of 77 persons per 1,000 liters of milk produced (farmers, casuals, and long-term laborers) as well as 13 and 18 persons per 1,000 liters of milk handled by the formal and informal sector, respectively, considering both the direct and the indirect effects. Other figures show that the Kenyan dairy sector provides jobs for about 3 million people, corresponding to 15 percent of the labor force (Baltenweck et al. 2020).

Livestock ownership is an essential means of enhancing gender equity in livestock systems and contributing to gender equality (Galie et al. 2019). Gender inequality is estimated to be responsible for a loss of 11 percent of Africa's total wealth, and livestock plays a pivotal role in rural women's lives (Woden and de la Brière 2018). Given the gender norms prevailing in many low- and middle-income countries, livestock is one of the few assets women can own and is a key tool for women's empowerment. Livestock provides a mechanism for women to improve their income, access information, leverage social networks, and provide nutritious food to their families. Livestock is also a key asset that women can own and transfer, unlike land and other physical assets, which might need a title deed (Baltenweck et al. 2020).

Livestock and livestock products are the key entry point to enhance the nutrition of the poor, particularly during the first 1,000 days of life, given that livestock provides nutrient-rich foods, such as milk or meat, shown to improve growth and cognitive functioning, respectively (Grace et al. 2018). LDF are energy-dense and excellent sources of protein and minerals. One of the largest randomized controlled trials conducted to date showed that toddlers in rural areas of Kenya whose diets had been supplemented with cow's milk grew taller than children consuming the usual diet or a diet supplemented with beef (Neumann 2013). Adequate quantity and quality of nutrients is important for the period from conception through pregnancy and up to two years of age for individuals to not only survive but thrive throughout life (McDonald and Thorne-Lyman 2017).

Policy considerations

Transforming the livestock system in Kenya to meet growing demand for animal-source foods will require investments in priority areas, policies to support investments, and institutions to support implementation. Two main policy documents have recently been completed. The National Assembly has adopted a national livestock policy as Sessional Paper 3 of 2020 on the Livestock Policy (Ministry of Agriculture, Livestock, Fisheries, and Cooperatives 2020a). Similarly, the National Assembly adopted a veterinary policy as Sessional Paper 2 of 2020 (Ministry of Agriculture, Livestock, Fisheries, and Cooperatives 2020b). These broad policy documents are comprehensive and consistent with the Constitution of Kenya and government strategies, including Vision 2030 and its medium-term plans, the Big Four Agenda, and sectorwide development strategies.

As acknowledged in the Livestock Policy, several policy issues are worth considering. Here we highlight a few that stand out and need immediate attention based on the constraints and opportunities identified in this chapter.

- While **feed standards** exist for most livestock species, the standardization of feeds for some categories of livestock is not complete (Ministry of Agriculture, Livestock, Fisheries, and Cooperatives 2020a). In addition, feed ingredients are not fully standardized, especially new feed innovations that arise from new research. It is important to prepare the necessary regulations on feed quality and market regulations for some livestock categories.

- To improve **animal health**, the government needs to strengthen the veterinary laboratory system to provide technical support for disease surveillance, diagnosis, and quality control. The State Department of Livestock and the Zoonotic Disease Unit under the Ministry of Health need to engage jointly in controlling zoonotic diseases within the "One Health" concept.[5] It is also important to establish mechanisms for public and private partnership for the control of cross-county and transboundary infectious diseases, and to coordinate with the Kenya Wildlife Service for the control of diseases at the livestock–wildlife interface.

5 One Health is an approach that recognizes that the health of people is closely connected to the health of animals and our shared environment (www.cdc.gov/onehealth/basics/index.html).

- To facilitate effective **breed improvement**, a uniform national system for animal identification, performance recording, and corresponding genetic evaluation is needed, but is currently missing in Kenya. Additionally, the country has very limited initiatives for in situ and ex situ conservation of AnGR.

- Unstructured **marketing** systems have a negative impact on the industry, leading to its underperformance. Producers and marketing groups need to strengthen their capacities in producing, processing, and storing livestock products. As the Livestock Policy states, it is important to facilitate the dissemination of livestock marketing information to all value chain actors and to establish mechanisms for strengthening and harmonizing market information systems and developing linkages with local and international markets.

The Livestock Policy covers a wide range of policy issues and outlines many interventions, practices, processes, guidelines, and proposals to address the challenges in the sector. However, no clear implementation plan exists, including to prioritize interventions and the resources needed to realize them. An implementation plan and a prioritization exercise should narrow down to the four main challenges highlighted in this chapter (feed and forage, animal health, breeding, and marketing) and be tailored to specific livestock species. The design of such an implementation plan and the priority-setting exercise would usually precede resource allocation and planning decisions.

References

Abay, K.A., and N.D. Jensen. 2020. "Access to Markets, Weather Risk, and Livestock Production Decisions: Evidence from Ethiopia." *Agricultural Economics* 51 (4): 577–593.

Baltenweck, I., D. Enahoro, A. Frija, and S. Tarawali. 2020. "Why Is Production of Animal Source Foods Important for Economic Development in Africa and Asia?" *Animal Frontiers* 10 (4): 22–29.

Bett H.K., M.P. Musyoka, K.J. Peters, and W. Bokelmann. 2012. "Demand for Meat in the Rural and Urban Areas of Kenya: A Focus on the Indigenous Chicken." *Economics Research International* 2012: 401472.

Bosire, C.K., N. Mtimet, D. Enahoro, J.O. Ogutu, M.S. Krol, M.S., J. de Leeuw, N. Ndiwa, and A.Y. Hoekstra. 2022. "Livestock Water and Land Productivity in Kenya and Their Implications for Future Resource Use." *Heliyon* 8 (3): e09006.

Daum, T., T. Ravichandran, J. Kariuki, M. Chagunda, and R. Birner. 2022. "Connected Cows and Cyber Chickens? Stocktaking and Case Studies of Digital Livestock Tools in Kenya and India." *Agricultural Systems* 196: 103353.

Delgado, C., M. Rosegrant, H. Steinfeld, S. Ehui, and C. Courbois. 2001. "Livestock to 2020: The Next Food Revolution." *Outlook on Agriculture* 30 (1): 27–29.

Enahoro D., M. Lannerstad, C. Pfeifer, and P. Dominguez-Salas. 2018. "Contributions of Livestock-Derived Foods to Nutrient Supply under Changing Demand in Low- and Middle-Income Countries." *Global Food Security* 19: 1–10.

FAO (Food and Agriculture Organization of the United Nations). 2019. *The Future of Livestock in Kenya: Opportunities and Challenges in the Face of Uncertainty*. Rome.

FAOSTAT. FAOSTAT database. Accessed December 4, 2022. https://www.fao.org/faostat/en/

Frija, A., and D. Enahoro. 2018. *Elaboration of Livestock R&D Scenarios and Assessment of the Impact of Alternative Investment Scenarios on Employment Generation: A Methodological Framework*. Montpellier, France: CGIAR.

Galiè, A., N. Teufel, L. Korir, I. Baltenweck, A. Webb Girard, P. Dominguez-Salas, and K.M. Yount. 2019. "The Women's Empowerment in Livestock Index." *Social Indicators Research* 142 (2): 799–825.

Grace, D., P. Domínguez Salas, S. Alonso, M. Lannerstad, E.M. Muunda, N.M. Ngwili, , et al.. 2018. *The Influence of Livestock-Derived Foods on Nutrition during the First 1,000 Days of Life*. ILRI Research Report 44. Nairobi: International Livestock Research Institute.

Herrero, M., P. Havlík, H. Valin, A. Notenbaert, M. Rufino, P. Thornton, M. Blümmel, F. Weiss, D. Grace, and M. Obersteiner. 2013. "Biomass Use, Production, Feed Efficiencies, and Greenhouse Gas Emissions from Global Livestock Systems." *PNAS* 110 (52): 20888–20893.

Horizon East Africa 2021. *Horizon Kenya: Exploring Trends Set to Shape the Future in Kenya.* Nairobi: Msingi East Africa and Kenya Markets Trust; London: Gatsby Africa.

ICPALD (IGAD Center for Pastoral Areas and Livestock Development). 2013. *The Contribution of Livestock to the Kenyan Economy.* ICPALD Policy Brief No. 4/CLE/8/2013. Nairobi: ICPALD.

ILRI (International Livestock Research Institute). 2019. "Options for the Livestock Sector in Developing and Emerging Economies to 2030 and Beyond." World Economic Forum White Paper. World Economic Forum, Geneva.

KNBS (Kenya National Bureau of Statistics). 2015. Various *County Statistical Abstract.* Nairobi.

KNBS. 2017a. *County Summaries for Livestock Products for Year 2016.* Nairobi.

KNBS. 2017b. *Economic Survey 2017.* Nairobi.

KNBS. 2021. *Economic Survey 2021.* Nairobi.

Kenya Markets Trust. 2019a. *A Study on Meat End Market Trends in Kenya.* Nairobi.

Kenya Markets Trust. 2019b. *A Study on Political Economy Analysis of the Livestock Sector.* Nairobi.

Korir, L., M. Rizov, and E. Ruto. 2018. "Analysis of Household Food Demand and Its Implications on Food Security in Kenya: An Application of QUAIDS Model." Paper presented at the 92nd Agricultural Economics Society Annual Conference, Warwick, UK, April 16–18.

McDonald, C.M., and A.L. Thorne-Lyman. 2017. "The Importance of the First 1,000 Days: An Epidemiological Perspective." In *The Biology of the First 1,000 Days,* eds. C.D. Karakochuk, K.C. Whitfield, T.J. Green, and K. Kraemer, 3–14. Boca Raton: CRC Press.

MacMillan, S. 2019. "Kenyan Livestock Sector to Grow 'Exponentially'—Kenya National Bureau of Statistics." *ILRI News Clipping,* July 31.

Ministry of Agriculture. 2012. *Sector Medium-Term Plan 2008-2012.* Nairobi.

Ministry of Agriculture, Livestock, Fisheries and Co-operatives. 2020. *Sessional Paper Number 3 of 2020 on The National Livestock Policy.* Nairobi.

Ministry of Agriculture, Livestock, Fisheries and Irrigation. 2019. *National Livestock Policy.* Nairobi.

Ministry of Agriculture, Livestock, Fisheries, and Rural Development. 2008. *Sector Medium-Term Plan 2008–2012.* Nairobi.

Msangi, S., D. Enahoro, M. Herrero, N. Magnan, P. Havlik, A. Notenbaert, and S. Nelgen. 2014. "Integrating Livestock Feeds and Production Systems into Agricultural Multi-market Models: The Example of IMPACT." *Food Policy* 49 (2): 365–377.

Neumann, C.G., N.O. Bwibo, L. Jiang, and R.E. Weiss. 2013. "School Snacks Decrease Morbidity in Kenyan Schoolchildren: A Cluster Randomized, Controlled Feeding Intervention Trial." *Public Health Nutrition* 16 (9): 1593–1604.

Ngula M. 2014. "An Assessment of the Demand for Meat in Rural and Peri-urban Areas of Central Kenya." Master's thesis, Department of Agricultural Economics, University of Nairobi, Kenya.

Robinson, S., D. Mason-D'Croz, S. Islam, T.B. Sulser, R.D. Robertson, T. Zhu, A. Gueneau, G. Pitois, and M.W. Rosegrant. 2015. "The International Model for Policy Analysis of Agricultural Commodities and Trade (IMPACT) Model Description for Version 3." IFPRI Discussion Paper 1483. IFPRI, Washington, DC.

Ruto, E., G. Garrod, and R. Scarpa. 2007. "Valuing Animal Genetic Resources: A Choice Modeling Application to Indigenous Cattle in Kenya." *Agricultural Economics* 38 (1): 89–98.

Shapiro, B.I., G. Gebru, S. Desta, A. Negassa, K. Nigussie, G. Aboset, and H. Mechale. 2017. *Ethiopia Livestock Sector Analysis.* ILRI Project Report. Nairobi: International Livestock Research Institute (ILRI).

Shibia, M., S. Rahman and B Chidmi. 2017. "Consumer Demand for Meat in Kenya: An Examination of the Linear Approximate Almost Ideal Demand System." Paper prepared for presentation at the 2017 Southern Agricultural Economics Association (SAEA) Annual Meetings in Mobile, AL, February 4–7.

Staal, S.J., A. Nin-Pratt, and M. Jabbar. 2008. "Dairy Development for the Resource Poor. Part 2: Kenya and Ethiopia Dairy. Development Case Studies." PPLPI Working Paper No. 44-2. FAO, Rome; ILRI, Nairobi.

Toye, P., H. Kiara, O. ole-MoiYoi, D. Enahoro, and K. Rich. 2020. "The Management and Economics of East Coast Fever." In *Impact of the International Livestock Research Institute*, eds. J. McIntire and D. Grace, 239–273. Nairobi: ILRI; Wallingford, UK: Centre for Agriculture and Bioscience International (CABI).

Wiebe, K., M. Zurek, S. Lord, N. Brzezina, G. Gabrielyan, J. Libertini, A. Loch, R. Thapa-Parajuli, J. Vervoort, and H. Westhoek. 2018. "Scenario Development and Foresight Analysis: Exploring Options to Inform Choices." *Annual Review of Environment and Resources* 43: 545–570.

Wodon, Q., and B. De la Brière. 2018. "Unrealized Potential: The High Cost of Gender Inequality in Earnings." Washington, DC: World Bank.

PART 2

Toward Healthier Food Systems

Part 2

TOWARD HEALTHIER FOOD SYSTEMS

Improving health is one of the five key goals in transforming food systems. Part 2 of this book presents a discussion on diets and food safety in Kenya. What food systems produce, how they deliver food to consumers, and the food choices offered to consumers have profound impacts on health through dietary quality and food safety. As Kenya faces a dual nutrition problem—with undernutrition in many rural areas and incipient overnutrition in some urban areas—understanding the role of food availability, affordability, preferences, and safety is important to designing policy that leads to healthier lives for all Kenyans. These aspects of the food system are increasingly important against the backdrop of rapid food price inflation, supply chain disruptions, and sometimes low local food production levels.

Chapter 4 presents an analysis of Kenyan diets, food affordability, and food preferences in rural, peri-urban, and urban areas. Overall, there is underconsumption of nutritious foods (for example, vegetables) and overconsumption of calorie-rich foods (for example, staples). Most Kenyan households cannot afford a healthy, diverse diet, as the cost of healthier foods is much higher than the cost of staple foods and foods with added sugars. Further, food preferences are often preventing people from consuming affordable, healthy alternatives such as pulses and nuts. The combined issues of affordability and preferences point to a need for policy that targets poverty reduction and nutrition education together.

Foodborne diseases can undermine health even if diverse, nutritious foods are made affordable and available. Chapter 5 presents an overview of food safety in Kenya. Such hazards can take various forms (for example, microbiological or chemical) and appear in any value chain. The prevalence of foodborne diseases in Kenya is high, and there are a number of approaches that can be used to improve safety. For example, harmonizing the fragmented food safety regulatory landscape and improving the enforcement of food safety regulations can lead to an improved balance between food safety and food security. Further, policy must recognize the importance of informal markets and the need for regulatory frameworks to simultaneously address the needs of formal and informal markets. Providing water, sanitation, and hygiene (WASH) infrastructure; conducting monitoring of water sources; building the capacity of value chain actors; and leveraging co-regulatory approaches with the private sector are all promising policy avenues to improve food safety.

In summary, Part 2 places health at the forefront of food system outcomes through its discussions of diets and food safety. Improving food security, nutrition, and safety for Kenyans is a primary concern in food system transformation. After all, food is ultimately produced for consumption, so any policy related to its production must keep in mind the health and well-being of the ultimate user of agricultural and livestock output—the consumer.

Chapter 4

KENYAN DIETS: QUALITY, AFFORDABILITY, AND PREFERENCES

Olivier Ecker, Andrew R. Comstock, and Karl Pauw

Dysfunctions in food systems in developing countries prevent many people from consuming a healthy diet (FAO et al. 2021), and Kenya is no exception. Globally, poor-quality diets are the leading cause of all forms of malnutrition (Afshin et al. 2019; Willet et al. 2019). In Kenya in 2020, an estimated 19 percent of children under five years of age were stunted (UNICEF, WHO, and World Bank 2021); in 2014, 33 percent of women aged 15–49 years were overweight or obese (KNBS et al. 2015), while recent regional trends in adults' body mass index suggest a rapid increase in the prevalence of overweight and obesity (Abarca-Gomez et al. 2017). The number of deaths resulting from noncommunicable diseases (NCDs), such as coronary disease, type 2 diabetes, and cancer, is projected to surpass malaria and tuberculosis by 2030 (Mkuu et al. 2021). Malnutrition and NCDs can have lifelong health consequences and high social and economic costs for individuals and societies alike, including from impaired human capital formation, reduced labor productivity, and high healthcare costs (Popkin et al. 2006; Shekar, Heaver, and Lee 2006; Victora et al. 2008; Black et al. 2013).

Public policies in developing countries have often failed to address malnutrition. The challenges have become more complex in recent decades because of the simultaneous occurrence of both under and overconsumption of food and nutrients, most notably among the poor, and often within the same population strata or even the same households (Popkin, Corvalan, and Grummer-Strawn 2020). While there is broad consensus that food system transformation is urgently needed to achieve healthier diets (HLPE 2017; Webb et al. 2020; FAO et al. 2021), there is less clarity on what are promising entry points for policy and technology and which policy levers would most effectively bring about change.

Policy considerations in this regard are mostly country- and context-specific, and thus require thorough analysis.

Analysis of current dietary patterns and the gaps between consumption levels and healthy reference intakes for nutritious food groups is an important starting point in understanding how to improve the nutritional quality of diets. The next step is to grasp the constraints that prevent consumers from obtaining high-quality diets. For poor populations, the relatively high cost of a diverse and nutritionally adequate basket of foods is a major constraint, as is consumer knowledge of the types and quantities of food required for healthy living given unique physiological needs. Finally, it is important to understand people's food consumption behavior, and specifically their food preferences and consumption responses when food prices and real incomes change. A range of policy instruments and technological innovations can target such economic variables. For example, investments in agricultural production and food value chains, consumer subsidies and food assistance, or cash transfer and employment programs can all affect relative prices of foods or household disposable incomes, which, subject to preferences, may result in dietary change.

This chapter examines the nutritional quality of Kenyan diets, the affordability of healthy diets in the country, and the food preferences of consumers. Because of rural–urban differences in food consumption patterns and rapid urbanization in Kenya, we separate our analysis into rural areas, peri-urban areas, and urban centers, with the latter two forming urban agglomerations. With more than 70 percent of the population living in rural areas, Kenya is still one of the least urbanized countries in sub-Saharan Africa, but its urbanization rate, like that of neighboring East African countries, is high (DESA 2019). The data for our empirical analysis are taken from the 2015/16 Kenya Integrated Household Budget Survey (KIHBS), a large representative survey of 12,318 rural households, 2,541 peri-urban households, and 5,353 urban households (after data cleaning).[1] Our analysis includes foods consumed at home, as

1 A detailed description of the 2015/16 KIHBS is available from the Kenya National Bureau of Statistics (KNBS 2018). We cleaned the released data for obvious reporting errors in the main variables of our analysis observation by observation. We dropped entire households from the sample if these did not complete the survey interview or did not report food consumption (at home). We also dropped households with implausible calorie consumption amounts that we derived from the survey's seven-day food consumption recall. Households were defined as having implausible calorie consumption amounts if their consumption per adult equivalent was below 600 kcal/day or above 6,000 kcal/day. Lastly, we dropped households if they reported implausibly large consumption quantities or expenditures for any of the 15 food groups used in the food demand system estimations (presented in the fourth section). Based on the two latter criteria, we dropped 7.1 percent of households that reported any food consumption.

information on their quantities and nutritional qualities can be assessed easily,[2] but excludes (prepared) foods consumed away from home, for which detailed information is unavailable.

Dietary patterns and quality

In its influential 2019 report on healthy diets from sustainable food systems, the EAT–*Lancet* Commission on Food, Planet, Health proposed a global reference diet that meets nutritional requirements, reduces the incidence of NCDs and mortality, and considers environmental sustainability of food production (Willet et al. 2019). This "healthy reference diet" provides quantitative dietary guidelines by food group. In addition to optimal food intakes in grams, it defines possible food intake ranges (except for added sugars) and specifies caloric intakes by food group that are derived from the optimal food intakes. In the absence of quantitative food-based dietary guidelines for Kenya, we use these optimal caloric intakes as the reference intakes for our dietary analysis. The global healthy reference diet of the EAT-*Lancet* Commission was complemented with four common, nutritionally balanced and predominantly plant-based, dietary patterns—namely, for flexitarian, pescatarian, vegetarian, and vegan consumers (Springmann et al. 2021).[3] Our analysis also uses the reference intakes for the flexitarian diet as an alternative set of dietary guidelines, because most diets in Kenya are mainly plant-based but often contain meat

---

2 The food consumption recall includes 196 food item categories (excluding bottled water and calorie-free stimulants such as coffee and tea) for at-home consumption. For converting reported food item consumption quantities to calorie consumption amounts, we used the National Nutrient Database of the United States Department of Agriculture (USDA 2016) and, for East Africa-specific food items, the most recent Tanzania Food Composition Tables (Lukmanji et al. 2008).

3 A modeling analysis was performed to construct the four common dietary patterns (Springmann et al. 2018, 2021). They are calorie-balanced variants of the healthy dietary guidelines as defined by the EAT–*Lancet* Commission (Willet et al. 2019), and hence their reference intakes are within the possible intake ranges of the healthy reference diet. On a calorie basis, the flexitarian diet has a larger amount of starchy staples than the optimal intake of the healthy reference diet (that accounts for about 47 percent of total caloric intake, compared with 34 percent); a small red meat amount that is still larger than the optimal healthy reference intake (equivalent to one serving per week); a modest amount of other animal-source foods (including poultry, fish, and dairy) that is smaller than the sum of the optimal healthy reference intakes (by about one-third); a generous amount of plant-based foods (including fruits, vegetables of different colors, and pulses and nuts) that is still somewhat lower than the sum of the optimal healthy reference intakes; a lower amount of oils and fats than the sum of the optimal healthy reference intakes (while the considered vegetable oils, however, are higher in saturated fat); and a lower maximum amount of added sugars compared with what is allowed under the healthy reference diet. The more specialized diets were constructed from the flexitarian diet by replacing meat with two-thirds fish and seafood and one-third fruits and vegetables—for the pescatarian diet; by replacing meat with two-thirds pulses and one-third fruits and vegetables—for the vegetarian diet; and by replacing all animal-source foods with two-thirds pulses and one-third fruits and vegetables—for the vegan diet (Springmann et al. 2018).

or other animal-sourced foods. (For simplicity, we refer to the healthy reference diet and flexitarian diet together as the "EAT–*Lancet* diets.")

While the reference intakes should not be interpreted as strict caloric thresholds that individual consumers must achieve, they provide useful benchmarks for an average diet that yields sufficient calories from a diverse set of food groups that are likely to also provide adequate amounts of essential macro and micronutrients for most people. The major food groups of the EAT–*Lancet* diets are starchy staples (separated into cereals and starchy roots and tubers), vegetables, fruits, protein sources, dairy foods, added fats and oils, and added sugars. In our analysis, we separate the protein sources into animal-source proteins (that is, meat, fish, and eggs) and plant-based proteins (that is, pulses and nuts); combine the starchy staples into one group (because starchy root and tuber consumption is very low across Kenya); and add a "discretionary foods" group. Discretionary foods include snacks, sweets, and beverages that all provide calories but that the EAT–*Lancet* Commission considers nutritionally non-essential and partly even unhealthy foods (Willet et al. 2019). All reference diets are scaled for a total daily caloric intake of 2,500 kcal—the healthy intake of a moderately active, average-size adult. Households' food consumption in our analysis is therefore expressed based on calories per adult equivalent (AE).[4]

Figure 4.1 shows average calorie consumption amounts for rural, peri-urban, and urban areas in Kenya and relates them to the reference intakes of the healthy reference and flexitarian diets. Apart from the fact that the average person obtains less than the requisite quantity of daily calories, average dietary patterns differ significantly from dietary guidelines, which is suggestive of poor dietary quality overall. The average person overconsumes starchy staples and underconsumes nutritious foods such as vegetables, fruits, and both animal-source and plant-based protein foods. Around 60 percent of total calories consumed in rural and peri-urban areas and 55 percent in urban areas are obtained from starchy staples. According to the healthy reference diet, only one-third of total calories should come from staples, while the flexitarian diet allows for just less than half of total calories to come from staples. Over 90 percent of staple calories come from cereals. These are consumed primarily in refined form, predominantly as maize meal but also as wheat flour or polished rice. The high consumption of refined grains is concerning because removal of the bran during

4 An AE expresses an individual household member as a fraction of an adult person—here, in terms of daily calorie requirements. We calculated household-specific AE values from detailed dietary energy requirements for individuals (provided by FAO, WHO, and UNU 2004). These calculations account for household compositions by sex and age and the dietary energy needs of breastfeeding mothers. In our sample, the average household member corresponds to about 0.95 AE.

the milling process results in loss of fiber and much-needed micronutrients. Moreover, the EAT–*Lancet* Commission emphasizes the importance of whole grain consumption as this is associated with reduced risk of coronary disease, type 2 diabetes, and mortality (Willet et al. 2019).

Figure 4.1 also shows that the average diets in Kenya are lacking vegetables, fruits, pulses or nuts, and meat, fish, or eggs, in both relative and absolute quantities. For meat, fish, and eggs, the average shares of consumed calories in the reference intake of the flexitarian diet range from 26 percent in rural areas to 47 percent in urban areas. The respective average calorie consumption shares

FIGURE 4.1 **Mean calorie consumption amounts per AE and reference intakes of the EAT–*Lancet* diets by major food group**

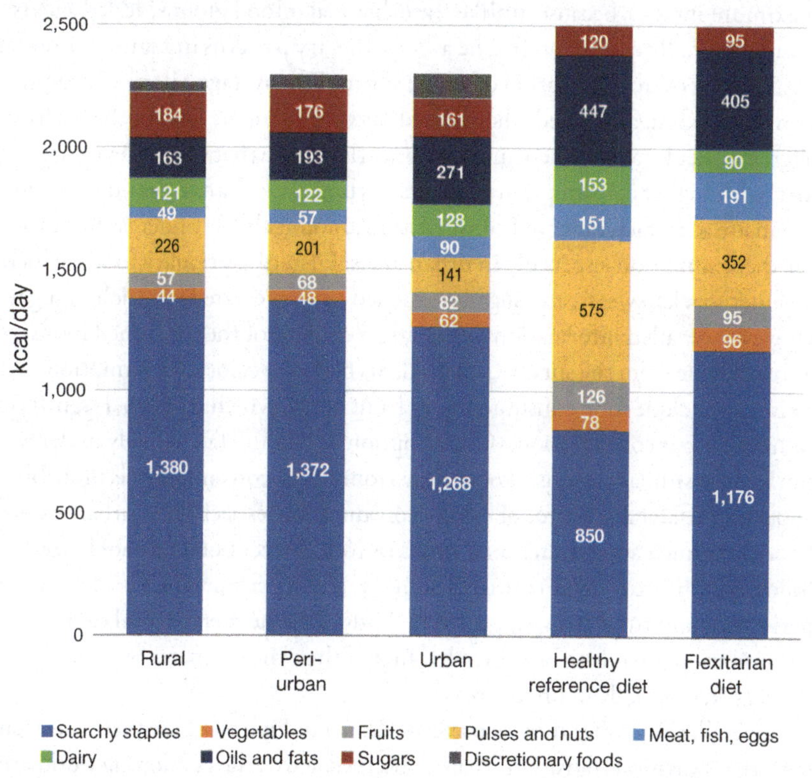

Source: Authors' estimates using 2015/16 KIHBS data.

Note: Consumption estimates refer only to foods consumed at home. Starchy staples include cereals, starchy roots/tubers, and plantains. Discretionary foods include snacks, sweets, and beverages and are considered as non-required foods according to the EAT–*Lancet* Commission. AE = adult equivalient.

account for 46–65 percent for vegetables and 60–87 percent for fruits. For pulses and nuts, the average calorie consumption shares are larger in rural aeras than in peri-urban and urban areas, accounting for 64 percent and 40 percent of the flexitarian diet's reference intake in rural and urban areas, respectively. When relating the average calorie consumption amounts to the intakes of the healthy reference diet, these shares are somewhat higher for meat, fish, and eggs and for vegetables, and lower for fruits and for pulses and nuts. The average calorie consumption amounts obtained from dairy foods, an essential source of key micronutrients for children in particular, meet the reference intake of the flexitarian diet and equate to around 80 percent of the reference intake of the healthy reference diet. This highlights the importance of dairy in the diet of many Kenyans.

For added sugars, the EAT–*Lancet* Commission defined acceptable maximum intakes, because, unlike the other major food groups, added sugars are nutritionally not essential. The average dietary patterns in Figure 4.1 reveal an alarming overconsumption of added sugars. The average calorie consumption of added sugars exceeds the allowed maximum intake level of the healthy reference diet by about one-third and is nearly double that of the flexitarian diet, with higher consumption amounts in rural areas than in peri-urban and urban areas. In fact, observed consumption amounts likely underestimate total sugar consumption, especially in urban areas. First, discretionary foods, which includes snacks, sweets, and sugar-sweetened beverages, are often rich in sugar. However, detailed information on the sugar content of the consumed foods is unavailable from the survey data used. Second, our calorie consumption estimates exclude food consumed away from home, which is another significant source of sugar consumption. (Consumption of oils and fats is likely underestimated for similar reasons.) Food-away-from-home consumption is probably another considerable source of calorie consumption, especially in urban areas. Food consumed away from home amounts to 10 percent of households' total food expenditure in urban areas, on average, and 5 percent and 3 percent in peri-urban and rural areas, respectively. Thus, the true average total calorie consumption is likely to be somewhat higher than the estimates presented in Figure 4.1, especially in urban areas.[5]

Overall, the average dietary patterns shown in Figure 4.1 are very consistent with the "nutrition transition" that is observed across the developing world and

5 It should also be noted that people's calorie expenditures tend to be higher in rural areas than in peri-urban and urban areas because of higher physical activity levels in common economic activities such as farming.

that is closely associated with urbanization and other food systems dynamics (Popkin 1999, 2006, 2017).

Our examination of the distribution of food consumption provides first evidence of large inequalities in household access to healthy diets across the Kenyan population (which will be further explored in the next subsection). While large proportions of the population overconsume starchy staples and calorie-rich non-required foods, underconsumption of nutritious food groups is even more prevalent. According to our estimates, 28 percent of all households obtain more than two-thirds of their total calories from starchy staple food consumption. This proportion is considerably larger in rural areas (34 percent) and peri-urban areas (29 percent) than in urban areas (18 percent). In contrast, Figure 4.2 shows that at least an estimated 80 percent of rural households, 75 percent of peri-urban households, and 70 percent of urban households have calorie consumption amounts from vegetables, fruits, or animal-source or plant-based protein foods that are lower than the reference intakes of the healthy reference and flexitarian diets, and the consumption by most of these households falls short in multiple nutritious food groups. Dairy consumption is insufficient in more than 50 percent of all households relative to the reference intake of the flexitarian diet, and more than 70 percent relative to that of the healthy reference diet. Yet even the consumption of starchy staples is below the flexitarian diet's reference intake among more than 20 percent of all households, and nearly twice that percentage have calorie consumption amounts for starchy staples below the staple reference intake of the healthy reference diet. In each residential area, almost all households have consumption amounts that are lower than the reference intakes of both the healthy reference diet and the flexitarian diet for at least one of the six major nutritious food groups of the EAT–*Lancet* diets, suggesting widespread overall poor diet quality.

Figure 4.2 also shows the average gaps in calorie consumption amounts by major nutritious food group. The calorie consumption gaps are calculated as the differences between the reference intakes and households' food consumption amounts that are averaged across all households, while households that have consumption amounts above the reference intakes enter the calculation with a zero deficit. The gaps in rural, peri-urban, and urban areas are largest for the two protein food groups, varying between 50 percent and 80 percent. Thus, closing the gaps for protein foods requires more than doubling the current average consumption. The next largest consumption gaps are found for fruits, vegetables, and dairy foods, in that order. The average consumption gaps for starchy staples are relatively small in all three residential areas, accounting for less than 10 percent of the reference intake of the healthy reference diet and less

FIGURE 4.2 **Proportions of population with calorie consumption amounts below the reference intakes and mean calorie consumption gaps by major nutritious food group of the EAT–*Lancet* diets**

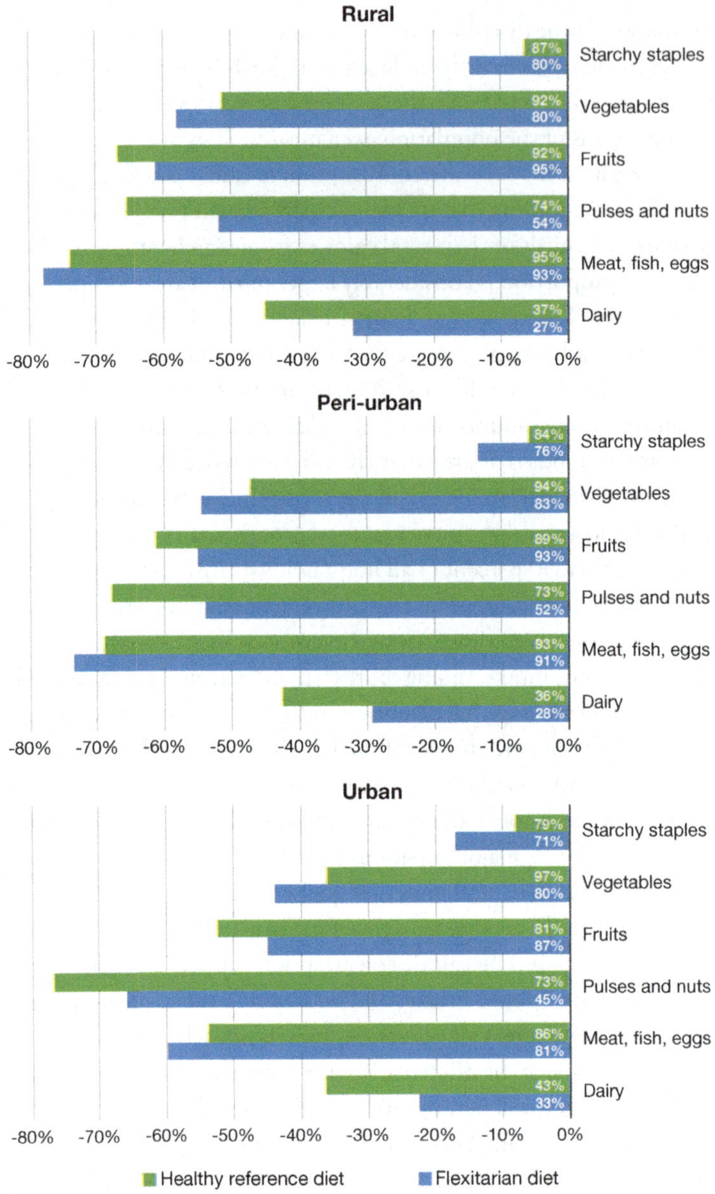

Source: Authors' estimates using 2015/16 KIHBS data.

Note: The bars indicate the average size of the calorie gap for each food group. The white percentages at the inside base of the bars indicate the proportions of the population with calorie consumption amounts below the reference intakes of the EAT–*Lancet* diets. Households with calorie consumption amounts above the reference intakes enter the calculation of the mean calorie consumption gaps with a zero deficit. Consumption estimates refer only to foods consumed at home. Starchy staples include cereals, starchy roots/tubers, and plantains.

than 20 percent of the flexitarian diet's reference intake. The consumption gaps for all major nutritious food groups are considerably smaller in urban areas than in peri-urban and rural areas, with the important exception of pulses and nuts.

Diet costs and affordability

Large consumption gaps for nutritious foods observed in Kenya reflect, at least in part, the high costs of a healthy diet relative to household incomes. Figure 4.3 shows the median prices per 100 kcal for the major food groups of the EAT–*Lancet* diets in Kenya's rural, peri-urban, and urban areas. Prices are derived from the food expenditures reported in the 2015/16 KIHBS.[6] The meat, fish, and egg group is by far the most expensive food group in all three residential areas, as Figure 4.3 demonstrates. On a calorie basis, the median price of meat, fish, and eggs is about 14–16 times the price of oils and fats; 12–15 times the price of starchy staples; and 10–11 times the price of sugars and the price of pulses and nuts. While the nutritional value of meat, fish, and eggs is certainly more than as a source of calories, this comparison demonstrates that the problem of food affordability when satisfying dietary energy needs is a primary motivation in food consumption decisions. The comparison of the prices of meat, fish, and eggs and of pulses and nuts clearly suggests that plant-based protein foods are the much cheaper source of high-quality protein and essential micronutrients, beyond the provision of calories. This prompts the question why plant-based proteins are vastly underconsumed in Kenya—a question that this chapter will address later. Meat, fish, and eggs are also two to three times more expensive than dairy products, which is another animal-source food group rich in essential amino acids and key micronutrients, as well as calories.

The second most expensive food group per calorie in all three residential areas is vegetables. Observed vegetable prices per weight may not be high but many vegetables (and especially green leafy vegetables) are low in calories, which yields high per calorie prices. Again, the nutritional properties of vegetables

6 In the absence of local market prices for the variety of foods consumed in Kenya, we estimated the median food group prices from reported food item-specific unit values. We calculated the food item unit values from 2015/16 KIHBS data on households' expenditures on purchased foods. As we obtained the food item unit values through a stepwise average calculation procedure starting with sub-counties as the lowest level of aggregation, the median food group prices shown in Figure 4.3 captures consumer market price differences. They also account for differences between households' food group compositions, consisting of combinations of food items yielding a total of 100 kcal. Our food group price approximation procedure uses medians for averaging across households (instead of means) to select food group compositions whose estimated item prices are not inflated by non-nutrition quality aspects of foods and outlier observations owing to reporting errors. Thus, the interquartile ranges presented in Figure 4.3 and Figure 4.4, which include household-level estimates from the 25th to the 75th percentile, have interpretational value.

are other than a source of calories but this is a main reason why food-insecure households find it difficult to choose such micronutrient-dense foods. Compared with vegetables and meat, fish, and eggs, fruits were found to be surprisingly cheap per calorie. However, the median price of this food group is driven primarily by the high share of bananas in fruits, which, compared with other fruits, are rich in calories but lack important vitamins such as vitamins A and C. In all three residential areas, the interquartile range, proportional to the median price level, is also largest for fruits, indicating large heterogeneity in this food group. In contrast, the interquartile ranges, in relative terms, are smallest for sugars; oils and fats; and pulses and nuts. Differences between the median prices in rural and urban areas are largest for dairy products, followed by fruits and starchy staples. There are virtually no rural–urban differences between the median prices of sugars and of oils and fats. Moreover, Figure 4.3 illustrates the inexpensiveness of added sugars and added oils and fats as calorie sources—roughly equivalent to starchy staples.

Figure 4.4 presents the cost structures of the healthy reference diet and the flexitarian diet in Kenya's rural, peri-urban, and urban areas. Depending on the place of residence, the median costs for consuming the reference intake

FIGURE 4.3 Median food group prices per 100 kcal (in 2015 KSh)

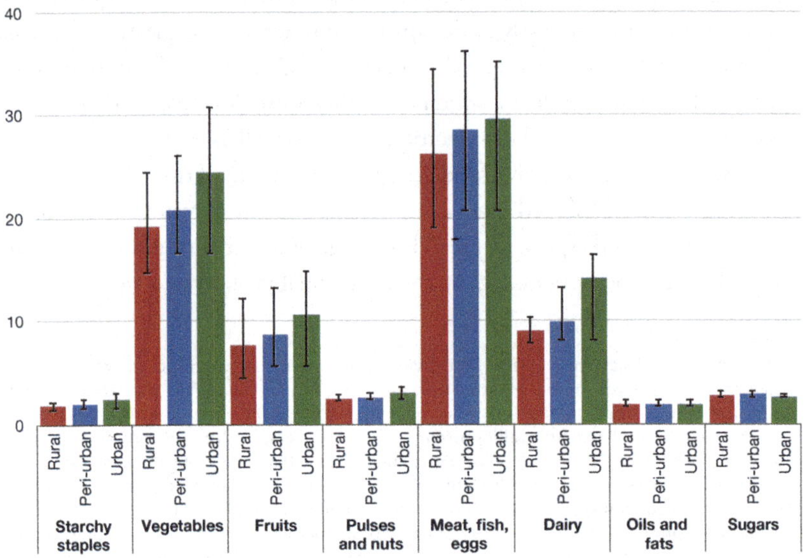

Source: Authors' estimates using 2015/16 KIHBS data.
Note: The error bars indicate the interquartile ranges.

Figure 4.4 Median daily costs of the EAT–*Lancet* reference intakes per AE (in 2015 KSh)

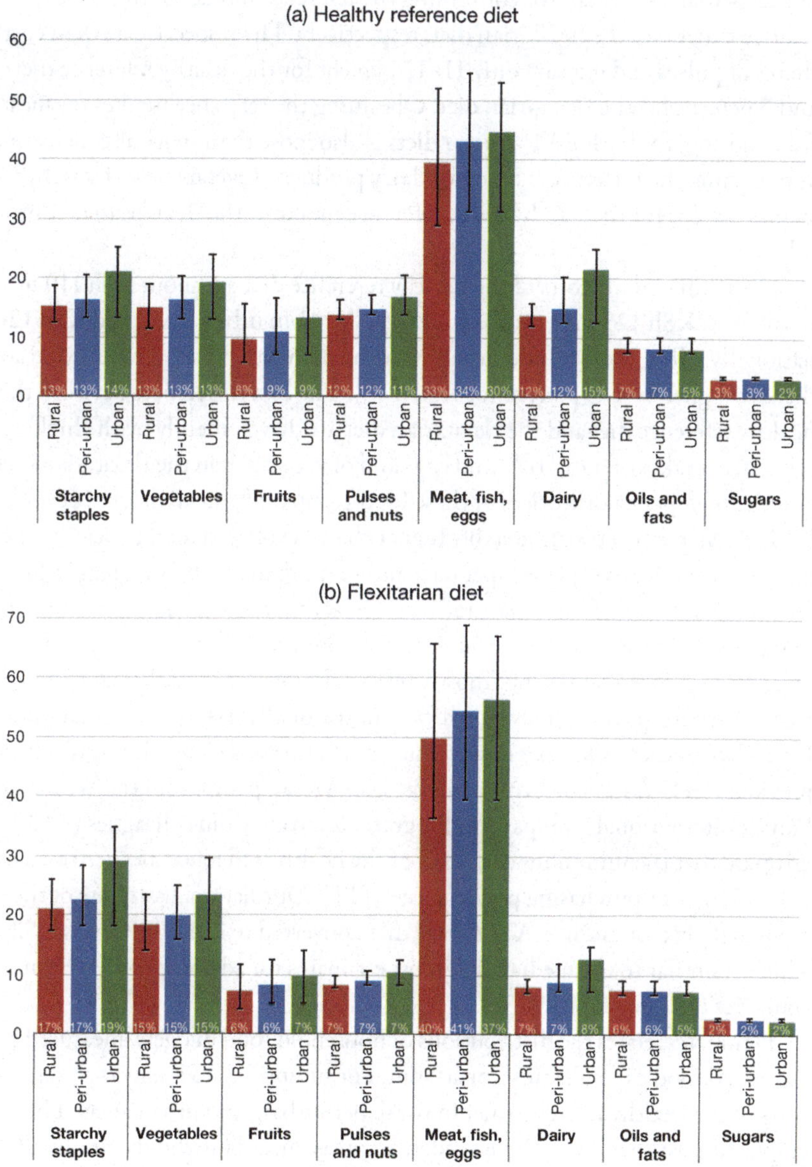

Source: Authors' estimates using 2015/16 KIHBS data.

Note: The error bars indicate the interquartile ranges. The percentages are the shares of the median food group costs in the totals of all food groups' median costs.

of meat, fish, and eggs amount to between 30–34 percent and 37–41 percent of the median cost totals for consuming the reference intakes of the healthy reference diet and the flexitarian diet, respectively. The respective median cost shares of pulses and nuts are only 11–12 percent for the healthy reference diet and 7 percent for the flexitarian diet. Obtaining the reference intakes for meat, fish, and eggs for both EAT–*Lancet* diets is also more than twice as expensive as obtaining the reference intakes for dairy products, vegetables, and starchy staples (except for the starchy staple reference intake of the flexitarian diet in urban areas).

The median daily costs of the healthy reference diet are about KSh 110 in rural areas, KSh 123 in peri-urban areas, KSh 141 in urban areas, and KSh 120 nationally (all at 2015/16 price levels). The median daily costs of the flexitarian diet are higher by 4–5 percent (or KSh 4–6). The cost differences between the healthy reference diet and the flexitarian diet are driven mainly by the higher reference intakes of meat, fish, and eggs and of vegetables in the flexitarian diet and the high per calorie prices of these food groups. The median costs of both EAT–*Lancet* diets are considerably higher than the official food poverty lines in rural areas (KSh 69 per AE per day) and peri-urban and urban areas (KSh 84 per AE per day). This in part reflects the fact that Kenya's poverty lines are calculated for a total calorie consumption amount of 2,250 kcal/day per AE as opposed to 2,500 kcal/day but mostly reflect the lower cost of the basic food basket used for poverty analysis, which is nutritionally less balanced than the EAT–*Lancet* diets. Our diet cost estimates are also consistent with results from previous EAT–*Lancet* diet cost analysis. Using retail prices from the World Bank's International Comparison Program, Hirvonen and colleagues (2020) estimate that the minimum daily cost of the healthy reference diet across Kenya is $2.17 in 2011 purchasing power parity (PPP). Our national estimate of the median daily cost of this EAT–*Lancet* diet converted to 2011 PPP levels is $2.26, which is similar to the median daily cost estimates for adjacent East African countries (Headey et al. 2023).

Figure 4.5 relates the distributions of household food budgets, measured by estimated food expenditures (for at-home consumption), to the median daily costs of the healthy reference diet in rural, peri-urban, and urban areas. The graphs suggest that, in all three residential areas, most households cannot afford the healthy reference diet. According to our data, 75 percent of rural households, 74 percent of peri-urban households, and 65 percent of urban households have food expenditures below the median costs of the healthy reference diet. These percentages are more than twice the official proportions of food-poor

Figure 4.5 Distributions of household food expenditures and median costs of the healthy reference diet

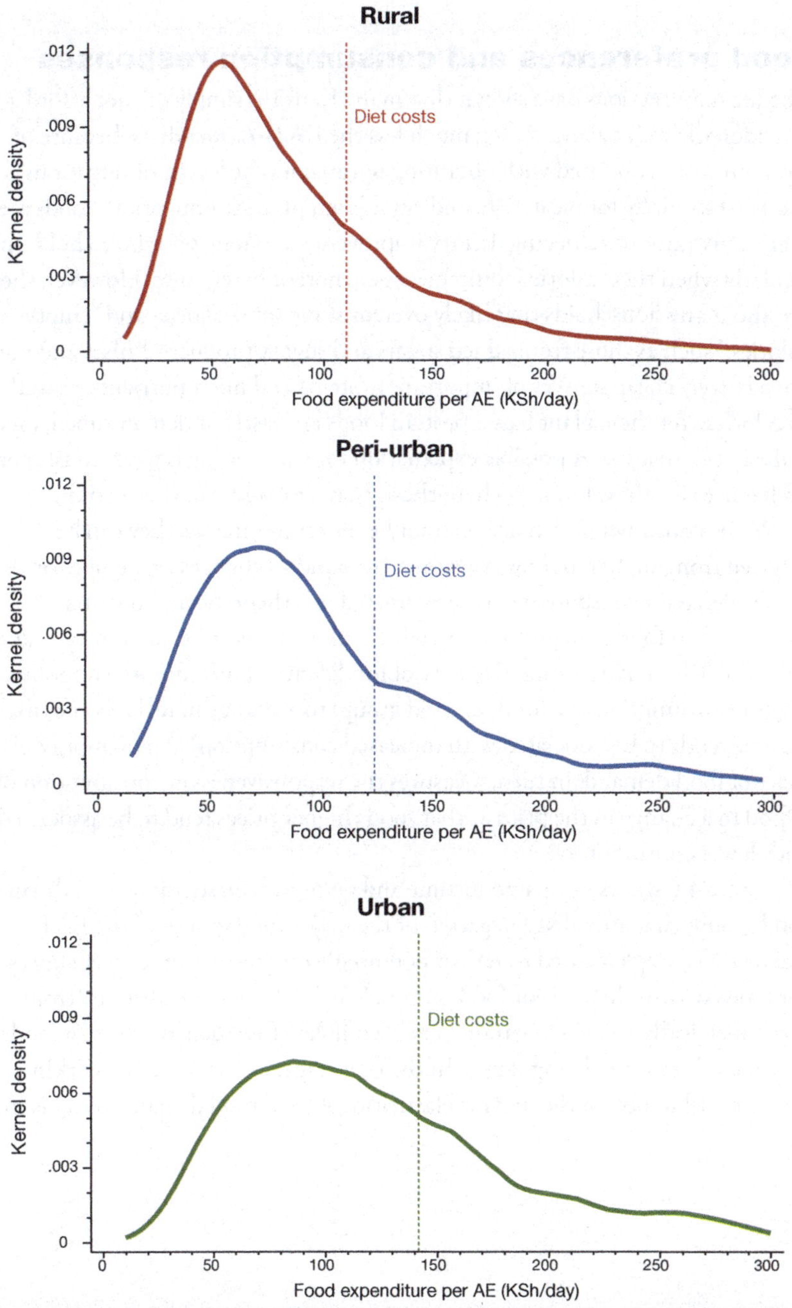

Source: Authors' estimates using 2015/16 KIHBS data.

Note: Food expenditure distributions are truncated at 300 KSh/day. AE = adult equivalent.

households, estimated at 28 percent in rural areas, 22 percent in peri-urban areas, and 18 percent in urban areas (KNBS, World Bank, and UNICEF 2018).

Food preferences and consumption responses

The previous sections have shown that many Kenyans simply cannot afford a nutritionally well-balanced diet, much less the EAT–*Lancet* diets, because of the high costs associated with obtaining optimal intake levels for nutritious foods—especially for meat, fish, and eggs. High prices for nutritious foods are a main constraint to achieving dietary improvement among poor households, particularly when their calorie requirements are not, or barely, met. However, there are also many households that likely overconsume total calories and "empty calories," such as those from added sugars and sugary products. Pulses and nuts are relatively cheap sources of important proteins and micronutrients, as well as calories. Yet these plant-based protein foods are vastly underconsumed, especially in urban areas. A possible explanation for this finding is weak consumer preferences for these foods—a hypothesis that we explore in this section.

While consumer preferences cannot be observed directly, they can be inferred from analysis of how consumers respond to changes in incomes or prices. Behavioral responses can be estimated in a theoretically consistent manner from food demand systems, which allow the derivation of income and price elasticities. An income elasticity of food demand measures the responsiveness of consumption of a food (or food group) to a change in real income (higher income tends to be associated with increased consumption). An own-price elasticity of food demand, in turn, measures the responsiveness of consumption of a food to a change in the price of that food (higher prices tend to be associated with lower consumption).

Figure 4.6 shows estimated income and own-price elasticities for 15 distinct food groups that are a disaggregation of the major food groups of the EAT–*Lancet* diets, implemented to reflect potentially diverse consumer preferences for foods within these major food groups. We derived these elasticities from econometrically estimated parameters of complete food demand system models that include two modeling stages. In the first stage, we estimated a Working-Leser model to obtain the income elasticities of total food demand vis-à-vis the aggregate demand for nonfood consumption (Working 1943; Leser 1963).[7] We then, in the second stage, modeled within-food budget allocations, allowing for full substitutability between all food groups, conditional on the available food budget. To estimate the demand for different food groups, we used a Quadratic Almost Ideal Demand System that accounts for censoring of food consumption observations (Banks, Blundell, and Lewbel 1997; Shonkwiler and Yen 1999). The estimation methodology is documented elsewhere in detail (see Ecker and Comstock 2021). We separately estimated the models for rural areas and for urban areas (combining urban centers and peri-urban areas) to allow for structurally different food demand curves between these areas, which may exist because of a greater share of farm households, and a larger dependence on subsistence-oriented agriculture, in rural areas. Figure 4.6 presents mean elasticity estimates for the lowest-, middle-, and highest-income quintiles within each residential area, which together span a range in which the individual elasticity estimates of most households fall.

The income elasticities of demand for most food groups are smaller in urban areas than in rural areas, as Figure 4.6 shows. This reflects higher incomes among urban households. Engel's Law states that, as household incomes rise, the percentage of income spent on food declines (Engel 1857). This translates into lower income elasticities for food among wealthier households. According to our estimates, the mean income elasticities of total food across all households are 0.707 in rural areas and 0.565 in urban areas. The income elasticities are largest for the two animal-source protein food groups (meat and fish; and eggs) in rural areas, with elasticity estimates for the middle-income quintile of around 1.5. The differences between the income elasticities of the lowest and highest income quintiles in rural areas are also very large. These results suggest that income growth in rural areas, and particularly among the poor, is likely to substantially increase people's consumption of animal-source foods and thus contribute to narrowing the large consumption gap found for this food group. In rural areas, the income elasticities for pulses and nuts, vegetables other than green leafy vegetables, and cereals other than maize (which includes mostly wheat-based products and rice) are also above the income elasticity of total food demand,

7 Food-away-from-home consumption is here considered as part of nonfood consumption, and related consumption behavior is therefore captured in the first modeling stage. This allocation is carried out to account for the fact that most of the costs of meals and drinks consumed in restaurants and bars are likely payments for food preparation and services (rather than for the market value of the raw meal ingredients and beverages) and to accommodate model requirements. Food demand system estimations require price information for each considered food category, which, however, are unavailable in the used household survey data for food-away-from-home consumption.

while the elasticities of the lowest income quintile for these food groups exceed unity. With growing household incomes, the consumption of these food groups can hence be expected to increase faster than the consumption of other food groups.

This holds true in urban areas for other cereals and beverages, which both include high proportions of highly processed foods, as well as sugary foods in the case of the latter. The observed tendencies for rapidly increasing consumption of highly processed foods are consistent with findings from previous studies on the positive association between the growing spread of modern food retailers (most notably supermarkets) and increasing prevalence of overweight, obesity, and related NCDs in Kenya's urban areas (for example, Rischke et al. 2015; Demmler, Ecker, and Qaim 2018; Khonje, Ecker, and Qaim 2020). Moreover, our income elasticity estimates suggest that, in urban areas, income growth likely leads to faster increases in the consumption of meat, dairy, and fruits (both bananas and other fruits) than for total food consumption, and slower increases in the consumption of pulses and nuts, dark green leafy vegetables, other vegetables, and fish and eggs. This provides support for our hypothesis of weak preferences among urban consumers for plant-based protein foods.

In addition to high prices per calorie, weak preferences for vegetables in general may explain the large consumption gaps found for vegetables in urban centers and peri-urban areas. Dairy is the only nutritious food group that shows lower (and statistically significant) income elasticity estimates for rural areas than for urban areas, which may be largely explained by better access to dairy products as a result of widespread livestock husbandry in rural areas. Maize consumption in urban areas is least elastic to income changes. As maize is the primary staple food, this result reflects Bennett's Law. This states that, as economies grow and per capita incomes rise, the share of calories from staple foods declines (Bennett 1941). In rural areas, and particularly for lower income quintiles, the income elasticities for maize are similar to the income elasticities of total food demand. The sensitivity of maize consumption to income changes within this population stratum confirms that food insecurity is a key factor of rural consumers' food choices.

Figure 4.6 shows that, compared with the estimated income elasticities, the estimated own-price elasticities of demand show much less differentiation between the estimates for the lowest income quintiles and the highest income quintiles in both rural and urban areas, suggesting similar consumption responses to relative price changes. Also, the elasticity patterns for prices across the food groups are less clear, as the price elasticity estimates for most food

FIGURE 4.6 Income and price elasticities of total food demand and the demand for 15 main food groups

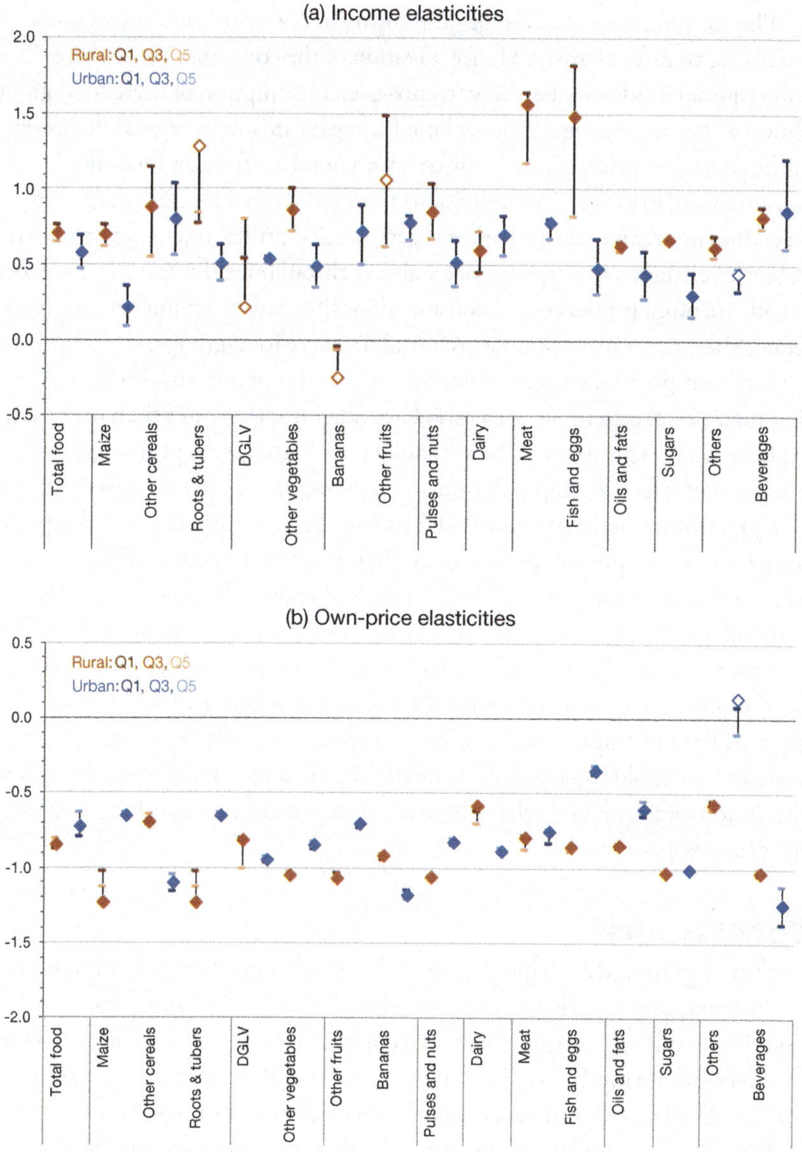

Source: Authors' estimates using 2015/16 KIHBS data.

Note: Q1 = lowest income quintile, Q3 = middle income quintile, Q5 = highest income quintile. DGLV = dark green leafy vegetables. Other cereals are mostly wheat, wheat products, and rice. The food group of bananas includes plantains. Other foods include sweets, snacks, and condiments. Beverages include sugar-sweetened beverages. Hollow diamond-shaped markers indicate that the elasticities are statistically insignificant at the 5 percent level as per bootstrapped standard errors.

groups vary relatively closely around the elasticity estimates for the total food average. Nevertheless, there are some notable tendencies.

The consumption of beverages and sugars in both rural and urban areas is sensitive to price changes. Hence, taxation of the consumption of these non-required foods may be a way to curb overconsumption of these food groups. However, because the prices per calorie for sugars are low, marginal increases in the consumer prices are unlikely to have a notable effect on household consumption. Price-sensitive nutritious food groups include dark green leafy vegetables in urban areas and other vegetables and pulses and nuts in rural areas. Public investments to support price stability throughout the year for these foods (which are often produced and consumed locally) may be an important policy to reduce the large consumption gaps found for these food groups.

The own-price elasticities of starchy roots and tubers in rural areas and non-maize cereals in urban areas also have magnitudes greater than unity. An explanation for this result is the availability of alternative staple foods and substitution effects in staple consumption in response to relative price changes.[8] Dairy consumption in rural areas and fish and egg consumption in urban areas seem to be less responsive to own-price changes than the consumption of all other nutritious food groups. The former result may reflect the fact that milk and milk products are often obtained from own livestock and own processing in rural areas. The low price elasticity for fish and eggs in urban areas should be interpreted in conjunction with the low income elasticity for this food group. The consumption of both fish and eggs is generally very low, and its share in household food budgets is small. Marginal income or price changes are therefore associated with relatively small average consumption changes across the urban population.

Conclusions

The findings from this chapter's analysis have four important policy implications.

First and foremost, Kenya's diet problem—the underconsumption of nutritious foods and high consumption of calorie-rich foods with increasing amounts of empty-calorie foods—is primarily a poverty problem, as most Kenyans simply cannot afford a healthy diet. Low household incomes are a significant constraint to dietary improvement, and poverty reduction measures are therefore likely to have nutritional benefits as well.

8 This explanation may also hold for the found own-price elasticities for the "bananas" food group, which includes plantains. Plantains are consumed as a staple food in some parts of Kenya.

Second, our diet costing exercise has shown large differences between the costs of meeting dietary guidelines for highly nutritious foods and the costs of obtaining adequate amounts of staple foods and maximum amounts of calories from nutritionally non-required foods such as added sugars and sugary foods. These cost discrepancies have particularly strong effects on household diets because the food choices of many Kenyans are essentially driven by food insecurity and geared toward satisfaction of calorie requirements. This finding points to a relative food price problem, which interacts with the poverty problem. The relative food price problem is most apparent for animal-source protein foods and somewhat less so for vegetables. Observed consumption gaps for both food groups are large across Kenya's rural, peri-urban, and urban populations, and their prices per calorie are high. Although the reference intakes of the healthy reference diet and flexitarian diet for meat, fish, and eggs are low compared with the global average consumption (Willet et al. 2019), the costs for obtaining the reference intakes amount to at least one-third of the total costs of these EAT–*Lancet* diets in Kenya. Thus, policy interventions and technological innovations that address this relative food price problem can help narrow the consumption gaps for some nutritious food groups. While most households' consumption of nutritious food groups is moderately or highly responsive to price signals (as our estimated price elasticities of food demand suggest), more detailed analysis is necessary to assess the dietary effects of specific policies and innovations.

Third, our analysis of revealed food preferences highlights that Kenya's nutrition challenge goes beyond economic issues. This is most obvious for plant-based protein foods, including pulses and nuts, which are also important calorie sources. The per calorie price of this food group is low, and the costs of obtaining the reference intakes of either EAT–*Lancet* diet are similar to, or even lower than, the costs of obtaining the starchy staple reference intakes. However, compared with other foods groups, the observed consumption gaps for pulses and nuts are the largest in urban areas and among the largest in rural and peri-urban areas. Our income elasticity estimates reveal that current consumer preferences for pulses and nuts are relatively weak, indicating that these highly nutritious foods are less desired than others. Hence, income growth is also unlikely to lead to large increases in their consumption. This is different for meat, for instance, where consumption can be expected to increase faster than total food consumption in urban areas and in rural areas—even faster than household incomes grow. Moreover, the estimated own-price elasticities for pulses and nuts are larger in magnitude than those for total food demand in urban areas, rising above unity in rural areas. This means that pulse and nut consumption is also more sensitive to price changes than the rest of the food basket among urban households and is price-elastic in rural areas. Thus, seasonal price fluctuations or price shocks from poor harvests, for example, are likely to have significant impacts on pulse and nut consumption. Weak consumer preferences for pulses and nuts, as well as high price sensitivity, may be an indication of a lack of consumer knowledge of the nutritional value of these foods and their importance for healthy diets. Nutrition education, for example, in schools and through public information campaigns, may aid in changing consumer behavior. Our estimation results also suggest that the weak preference problem is common across the entire country—and largely independent of household wealth.

Fourth and finally, our analysis calls for a strategic focus of food systems transformation policy on consumers and their dietary needs. Transforming Kenya's current food systems for better nutrition and health will require a paradigm shift that puts consumer diets at the center of policymaking. As agriculture is by far the dominant sector in Kenya's food systems, such a shift will entail striking a balance between traditional objectives like agricultural productivity growth, export stimulation, and farmer support, on the one hand, and the new responsibility for better nutrition and health for all Kenyans, on the other hand.

References

Abarca-Gómez, L., Z.A. Abdeen, Z.A. Hamid, N.M. Abu-Rmeileh, B. Acosta-Cazares, C. Acuin, R.J. Adams, et al. 2017. "Worldwide Trends in Body-mass Index, Underweight, Overweight, and Obesity from 1975 to 2016: A Pooled Analysis of 2416 Population-based Measurement Studies in 128.9 Million Children, Adolescents, and Adults." *Lancet* 390 (10113): 2627–2642.

Afshin, A., P.J. Sur, K.A. Fay, L. Cornaby, G. Ferrara, J.S. Salama, E.C. Mullany, et al. 2019. "Health Effects of Dietary Risks in 195 Countries, 1990–2017: A Systematic Analysis for the Global Burden of Disease Study 2017." *Lancet* 393 (10184): 1958–1972.

Banks, J., R. Blundell, and A. Lewbel. 1997. "Quadratic Engel Curves and Consumer Demand." *Review of Economics and Statistics* 79 (4): 527–539.

Bennett, M.K. 1941. "Wheat in National Diets." *Wheat Studies* 18 (2): 37–76.

Black, R.E., C.G. Victora, S.P. Walker, Z.A. Bhutta, P. Christian, M. de Onis, M. Ezzati, et al. 2013. "Maternal and Child Undernutrition and Overweight in Low-Income and Middle-Income Countries." *Lancet* 382 (9890): 427–451.

Demmler, K.M., O. Ecker, and M. Qaim. 2018. "Supermarket Shopping and Nutritional Outcomes: A Panel Data Analysis for Urban Kenya." *World Development* 102: 292–303.

Ecker, O., and A.C. Comstock. 2021. *Income and Price Elasticities of Food Demand (E-FooD) Dataset: Documentation of Estimation Methodology*. Washington, DC: International Food Policy Research Institute.

Engel, E. 1857. "Die Vorherrschenden Gewerbszweige in Den Gerichtsämtern mit Beziehung auf Die Productions- und Consumtionsverhältnisse Des Königlichen Sachen." *Zeitschrift des Statistischen Bureaus des Königlich Sächsischen Ministerium des Inneren* 3 (8–9): 153–182.

FAO (Food and Agriculture Organization of the United Nations), WHO (World Health Organization), and UNU (United Nations University). 2004. *Human Energy Requirements: Report of a Joint FAO/WHO/UNU Expert Consultation*. Rome: FAO.

FAO, IFAD (International Fund for Agricultural Development), UNICEF, WFP (World Food Programme), and WHO. 2021. *The State of Food Security and Nutrition in the World 2021: Transforming Food Systems for Food Security, Improved Nutrition and Affordable Healthy Diets for All*. Rome: FAO.

Headey, D.D., O. Ecker, A.R. Comstock, and M.T. Ruel. "Poverty, Price and Preference Barriers to Improving Diets in Sub-Saharan Africa." *Global Food Security* 36: 100664.

Hirvonen, K., Y. Bai, D. Headey, and W.A. Masters. 2020. "Affordability of the EAT-*Lancet* Reference Diet: A Global Analysis." *Lancet Global Health* 8 (1): E59–E66.

HLPE (High Level Panel of Experts on Food Security and Nutrition). 2017. *Nutrition and Food Systems*. Rome: FAO.

Khonje, M.G., O. Ecker, and M. Qaim. 2020. "Effects of Modern Food Retailers on Adult and Child Diets and Nutrition." *Nutrients* 12 (6): 1714.

KNBS (Kenya National Bureau of Statistics). 2018. *Basic Report Based on 2015/16 Kenya Integrated Household Budget Survey*. Nairobi.

KNBS, Kenya Ministry of Health, Kenya National AIDS Control Council, Kenya Medical Research Institute, Kenya National Council for Population and Development, and ICF International. 2015. *Kenya Demographic and Health Survey 2014*. Rockville, MD: ICF International.

KNBS, World Bank, and UNICEF. 2018. *Basic Report on Well-Being in Kenya: Based on 2015/16 Kenya Integrated Household Budget Survey*. Nairobi: KNBS.

Leser, C.E. 1963. "Forms of Engel Functions." *Econometrica* 55 (4): 694–703.

Lukmanji, Z., E. Hertzmark, N. Mlingi, V. Assey, G. Ndossi, and W. Fawzi. 2008. *Tanzania Food Composition Tables*. Dar es Salam: Muhimbili Universty College of Health and Allied Sciences; Boston, MA: Harvard School of Public Health.

Mkuu, R., A. Barry, G. Yonga, et al. 2021. "Prevalence and Factors Associated with Overweight and Obesity in Kenya." *Preventive Medicine Reports* 22: 101340.

Popkin, B.M. 1999. "Urbanization, Lifestyle Changes and the Nutrition Transition." *World Development* 27 (11): 1905–1916.

Popkin, B.M. 2006. "Global Nutrition Dynamics: The World is Shifting Rapidly toward a Diet Linked with Noncommunicable Diseases." *American Journal of Clinical Nutrition* 84 (2): 289–298.

Popkin, B.M. 2017. "Relationship between Shifts in Food System Dynamics and Acceleration of the Global Nutrition Transition." *Nutrition Reviews* 75 (2): 73–82.

Popkin, B.M., C. Corvalan, and L.M. Grummer-Strawn. 2020. "Dynamics of the Double Burden of Malnutrition and the Changing Nutrition Reality." *Lancet* 395 (10217): 65–74.

Popkin, B.M., S. Kim, E.R. Rusev, S. Du, and C. Zizza. 2006. "Measuring the Full Economic Costs of Diet, Physical Activity and Obesity-related Chronic Diseases." *Obesity Reviews* 7 (3): 271–293.

Rischke, R., S.C. Kimenju, S. Klasen, and M. Qaim. 2015. "Supermarkets and Food Consumption Patterns: The Case of Small Towns in Kenya." *Food Policy* 52: 9–21.

Shekar, M., R. Heaver, and Y.K. Lee. 2006. *Repositioning Nutrition as Central to Development: A Strategy for Large Scale Action*. Washington, DC: World Bank.

Shonkwiler, J.S., and S.T. Yen. 1999. "Two-Step Estimation of a Censored System of Equations." *American Journal of Agricultural Economics* 81 (4): 972–982.

Springmann, M., M.A. Clark, M. Rayner, P. Scarborough, and P. Webb. 2021. "The Global and Regional Costs of Healthy and Sustainable Dietary Patterns: A Modelling Study." *Lancet Planetary Health* 5 (11): e797–e807.

Springmann, M., K. Wiebe, D. Mason-D'Croz, T.B. Sulser, M. Rayner, and P. Scarborough. 2018. "Health and Nutritional Aspects of Sustainable Diet Strategies and their Association with Environmental Impacts: A Global Modelling Analysis with Country-level Detail." *Lancet Planetary Health* 2 (10): e451–e461.

UN-DESA (United Nations Department of Economic and Social Affairs). 2019. World Urbanization Prospects: Urban Agglomerations datasets. Accessed February 17, 2022. https://population.un.org/wup

UNICEF, WHO (World Health Organization), and World Bank. 2021. *Levels and Trends in Child Malnutrition: Key Findings of the 2021 Edition of the Joint Child Malnutrition Estimates.* Geneva: WHO.

USDA (United States Department of Agriculture). 2016. USDA National Nutrient Database for Standard Reference, Release 28. Accessed October 5, 2021.

Victora, C.G., L. Adair, C. Fall, P.C. Hallal, R. Martorell, L. Richter, H.S. Sachdev, et al. 2008. "Maternal and Child Undernutrition: Consequences for Adult Health and Human Capital." *Lancet* 371 (9609): 340–357.

Webb, P., T.G. Benton, J. Beddington, D. Flynn, N.M. Kelly, and S.M. Thomas. 2020. "The Urgency of Food System Transformation is Now Irrefutable." *Nature Food* 1 (10): 584–585.

Willett, W., J. Rockström, B. Loken, M. Springmann, T. Lang, S. Vermeulen, T. Garnett, et al. 2019. "Food in the Anthropocene: The EAT–Lancet Commission on Healthy Diets from Sustainable Food Systems." *Lancet* 393 (10170): 447–492.

Working, H. 1943. "Statistical Laws of Family Expenditure." *Journal of the American Statistical Association* 38 (221): 43–56.

Chapter 5

FOOD SAFETY IN KENYA: STATUS, CHALLENGES, AND PROPOSED SOLUTIONS

Vivian Hoffmann, Silvia Alonso, and Erastus Kang'ethe

Foodborne disease—that is, disease caused by consuming foods contaminated with biological or chemical hazards—is an important and often underrecognized public health concern in low- and middle-income countries around the world, including Kenya. While comprehensive national statistics on the foodborne disease burden are not available, the best available evidence from the region indicates that diseases transmitted via contaminated food have a greater impact on public health in Africa than either tuberculosis or diabetes (Havelaar et al. 2015; GBDCN 2016). Beyond its contribution to illness and death, foodborne disease also plays a role in child stunting, which remains a persistent problem in Kenya, affecting 16 percent of children (KNBS and ICF 2023). One study based on 20 years of data from 5 countries attributed 25 percent of stunting to repeated diarrheal episodes (Checkley et al. 2008), many of which can be traced to microbial contamination of food.

Kenya's food system is in the midst of various transitions, which imply both new challenges and opportunities to ensure the safety of the food supply. First, we are seeing a shift away from the dominance of starchy staples in diets, and toward more fruits, vegetables, animal-source foods, and ready-to-eat foods. Second, the scale at which food is produced, processed, and distributed is increasing, and the distances over which it is transported to rapidly urbanizing population centers are growing. As observed by Jaffee and colleagues (2018), both of these trends increase the probability that food is contaminated with microbial hazards. Further, the intensification of food production may lead to increased use of potentially hazardous chemicals such as pesticides and antibiotics. On the other hand, the growing importance of larger-scale food production and processing firms could enable better food safety monitoring and regulatory enforcement in some food value chains.

While the market share of larger-scale processors is growing, small-scale, often informal, food businesses remain critical to Kenya's food supply and the food security of low-income populations (Owuor 2020). Application of traditional punitive regulatory enforcement approaches to this more fragmented sector presents enormous challenges, risks harming livelihoods and limiting food access, and fails to address the underlying problem of low compliance capacity. Developing context-appropriate systems through which the food that Kenya's multitude of farmers, butchers, milk traders, and roadside fruit and vegetable vendors grow, process, and sell can be made safer is critical to reducing the foodborne disease burden its population shoulders.

This chapter reviews recent evidence on the major food safety risks affecting Kenya and discusses strategies for addressing these. We begin by introducing the major categories of foodborne hazards and discussing evidence from Kenya on the prevalence and sources of these hazards in the key affected value chains. We then describe the current regulatory structure and challenges to effective governance. Finally, we identify opportunities for public action to reduce Kenya's foodborne disease burden, including through engagement of the small-scale, informal businesses that dominate Kenya's food supply but to which standard regulatory tools are ill-suited. We close the chapter with a summary of recommendations.

Food safety problems and affected value chains in Kenya

Microbiological hazards

Microbiological hazards, which include diarrheal and other disease agents as well as parasitic organisms, are by far the most important type of food safety hazard in terms of health impact, both at the global level and within Africa (Havelaar et al. 2015; GBD 2016). Perishable foods, including animal-source foods and fresh fruits and vegetables, are most prone to microbiological contamination.

Since most microorganisms are inactivated by heat treatment, contamination of perishable foods consumed raw (for example, fruits and salads) constitutes the greatest risk to health. However, even cooked foods may not be heated to a sufficient temperature, or for a sufficient time, to render all pathogens harmless. Also, certain microorganisms, such as Staphylococcus aureus, produce toxins that resist heat treatment. Moreover, microbes present on these foods may be transferred to previously cooked foods, or to those eaten raw, during handling (Knechtges 2011).

Many organisms carried in the feces of livestock (for example, Salmonella spp, Campylobacter spp, and shiga toxin producing E. coli), are important foodborne pathogens. These pathogens may lead to on-farm contamination of animal products—milk and meat—as well as crops via contaminated irrigation water or manure applied as fertilizer, if this is not well composted (Jiang, Chen, and Dharmasena 2015). We next turn to the evidence regarding microbial contamination in vegetable, meat, and milk value chains.

VEGETABLES

Water used to irrigate vegetables on small-scale commercial farms near Nairobi and Machakos has been shown to exceed allowable standards for microbial contamination by several orders of magnitude (Kitulu et al. 2020; Kutto et al. 2011). Despite the illegality of using wastewater for irrigation in Kenya, this practice is common in and near urban centers as a result of water scarcity (Kaluli et al. 2011). Water used to wash vegetables in informal wet markets around Nairobi has been found to be similarly contaminated with high levels of bacteria, including Salmonella (Kutto et al. 2011). Many roadside vendors offer shredded kale and cabbage; cutting the vegetables in this way provides bacteria with additional surfaces on which to grow.

Personal hygiene of food handlers is an important determinant of microbiological food safety throughout the value chain for any food product, in particular perishables such as fruits, vegetables, and animal-source foods. While Kenyan Standard 1758-II (Sections 13.1.6A, 13.12.5A, and 13.13.3) requires that all food business operators, from farm to retail, have a clean on-site toilet and handwashing facilities, a study of the tomato value chain in Laikipia found that in practice these were often absent (Gatere et al. 2020).

MEAT

The safety of animal-source foods likewise begins on the farm. A study of small-scale poultry producers (defined as those with a minimum of 2 and a maximum of 800 birds) in Dagoretti and Kibera detected the pathogen Campylobacter in 33–44 percent of indigenous and broiler chicken farms in these two communities. This pathogen is one of the most frequently isolated from children suffering from diarrhea in Kenya (Carron et al. 2018). Unrestricted access to the flock (allowed at 88 percent of these farms) was associated with a higher likelihood of Campylobacter detection, while disinfection of enclosures (practiced by only 15 percent of farmers) reduced this risk.

Slaughter of animals is a particularly high-risk node for pathogen contamination of meat, primarily because of contact of meat with pathogens present in the digestive tract and feces of animals. Poor hygiene of personnel and the environment in abattoirs also facilitates the spread of pathogens between carcasses, amplifying the problem (Wambui et al. 2017). Data collected in 2012 by CGIAR researchers revealed that only 60 percent of slaughter facilities in western Kenya had a toilet, and 20 percent had handwashing facilities (Cook et al. 2017).

Food hygiene and temperature control practices are critical to meat safety beyond the slaughterhouse. Practices associated with Campylobacter contamination at small-scale chicken retailers in Nairobi included display of meat on a difficult-to-clean surface such as wood or cardboard, and selling defrosted meat (Carron et al. 2018).

MILK

Smallholder dairy farmers in Kenya often have low levels of knowledge about hygiene and biosecurity (Nyokabi et al. 2021a). Common practices among these producers, including unrestricted access to livestock, insufficient cleaning of enclosures, and unsanitary handling practices, lead to high levels of microbial contamination in milk (Nyokabi et al. 2021b).

A recent paper showed that, of all the foods consumed by infants in low-income neighborhoods of Kisumu, milk was the most likely to contain a pathogen (Tsai et al. 2019). This finding confirms analysis of nationally representative household data showing that children under 24 months who had consumed milk or fresh fruit were more likely to have suffered an episode of diarrhea within the previous two weeks, compared with those who had been exclusively breastfed (Hoffmann and Baral 2019). As these foods have high nutritional value for young children, it is critical that they are also safe, so that their benefits for child health can be fully harnessed and potential harms are minimized.

A study of infant food safety conducted among households in low-income neighborhoods of Kisumu found that over 90 percent of caregivers used ultra-high temperature (UHT) treated milk to feed their infants (Hoffmann et al. 2022). As contamination of UHT milk was low at point of purchase, most microbial contamination observed in infant food was introduced via household handling. However, contamination present at purchase also contributed to infants' exposure to diarrheal pathogens. This was especially true for milk that had been purchased raw, in which the rate of contamination with pathogens was high: 14 of 34 raw milk samples analyzed contained the pathogen Salmonella enterica, and 20 of 34 tested positive for Shigella sonnei. The same bacterial species present in a typical sample of raw milk at purchase were detected in 33 percent of infant food samples prepared with that milk, indicating that milk heating practices were not sufficient to eliminate microbiological risk.

While formally packaged milk was far less likely to contain pathogens—2.2 percent of UHT pasteurized milk samples and 4.8 percent of conventionally pasteurized samples contained either S. enterica or S. sonnei—the probability of detecting the same organism in both these milk types at purchase and in the linked infant food sample was far higher (65 percent for UHT milk). This suggests that caregiver handling practices may differ by milk type, and raises the concern that, as food systems modernize, consumer expectations regarding the safety of food may shift, leading them to rely more on pre-market controls.

Chemical hazards

HEAVY METALS

Heavy metals, which can increase cancer risk and lead to lifelong cognitive impairment, are the most important class of chemical foodborne hazards, in terms of disability-adjusted life years lost (Havelaar et al. 2015; Gibb et al. 2019). Heavy metals may enter the environment through industrial effluent, disposal of household waste, disintegration of lead-based paint, and current or previous use of pesticides or fertilizer containing heavy metals in growing sites (ATSDR 2012; NASEM 2017). They then enter crops through irrigation water and agricultural soils (Makokha et al. 2008; Inoti et al. 2012). While regulations exist in Kenya limiting the use of lead (for example, in paint), there are concerns with compliance and enforcement (WHO 2018). A 2017 study found that 71 percent of paints sold in Kenya exceeded this standard, with 50 percent over 7 times above the limit, and 1 almost 18,000 times over the limit (CEJAD and IPEN 2017).

While systematic data are lacking, the available evidence suggests that heavy metal contamination of vegetables in Kenya could be significant. For example, a CGIAR study of contamination in tomatoes grown in Laikipia county, in an area not known for high levels of pollution, found that 52 percent of samples exceeded the World Health Organization (WHO) standard for lead, and 24 percent exceeded that for cadmium (Gatere et al. 2020). This is notable as the rate of uptake of heavy metals into tomato fruit is relatively low, and the sites from which samples were taken were not known to be high in risk. Other studies that have sampled crops from areas known to be heavily polluted have found average levels of heavy metals up to 125 times over the WHO standard (Makokha et al. 2008; Karanja et al. 2010; Gallaher et al. 2013).

AFLATOXIN

Kenya is a global hotspot for aflatoxin, a fungal toxin (mycotoxin) that is highly carcinogenic and associated with child stunting (Liu et al. 2012; Ismail et al. 2021). Aflatoxin contamination arises in crops, especially maize and groundnuts, as a result of the presence of molds that occur naturally in soils, and is exacerbated by poor postharvest practices.

Fatal outbreaks of aflatoxin poisoning resulting from consumption of highly contaminated maize caused an average of nine deaths per year in Kenya during the period from 1981 to 2014, and over twice this rate from 2001 to 2014 (Hoffmann and Jones 2021, Table D1). In recent years, government food safety surveillance efforts, media attention, and international funding for both food safety research and capacity building in Kenya have been focused on this hazard (Mutua, Grace, and Watts 2021). While aflatoxin represents a relatively minor portion of the total regional food safety burden as quantified by WHO (Havelaar et al. 2015),[1] much of the research on firm and consumer responses to food safety risk in Kenya relates to aflatoxin.

Heavy reliance on maize, combined with low dietary diversity among low-income segments of the population, implies high levels of aflatoxin exposure among Kenyans (Leroy, Wang, and Jones 2015; Mutegi, Cotty, and Bandyopadhyay 2018). Indeed, a comparison of blood aflatoxin levels across six African countries showed that levels in Kenya during an outbreak year were an order of magnitude higher than those observed in any other country (Xu, Gong, and Routledge 2018). As the potential harms resulting from aflatoxin, in particular its suspected effect on children's development, are not fully understood, a cautious approach to control of this hazard is warranted.

However, if we consider only the health consequences of aflatoxin exposure for which the scientific case is clear (that is, deaths as a result of cancer and acute aflatoxin poisoning), the impact of this contaminant is dwarfed by those of other, primarily biological, foodborne hazards. CGIAR researchers estimated, based on observed aflatoxin levels in food during a typical year, that the resulting level of exposure would lead to 67 liver cancer deaths annually in eastern Kenya, the region with the highest aflatoxin levels in crops and also the highest human exposure based on blood analysis (Hoffmann and Jones

1 The WHO calculations do not include the potential impact of aflatoxin on child growth or immune function, as the authors did not consider these sufficiently well established. A randomized controlled trial by CGIAR researchers testing the impact of removing aflatoxin from children's diets in eastern Kenya showed ambiguous results, with an impact on child growth detected at 11–19 months after enrollment but not on the primary study outcome of growth at 24 months (Hoffmann, Jones, and Leroy 2018).

2021).[2] Scaling the cancer risk resulting from aflatoxin exposure using regionally representative data from the analysis of blood samples (Yard et al. 2013), and combining this with reported cases of fatal aflatoxin poisoning, the total number of deaths owing to this toxin nationally can be estimated at 155 per year, 146 from cancer and 9 from aflatoxicosis. To put this in context, the estimated number of child deaths as a result of foodborne diarrheal disease in Kenya is 8.5 times greater, at approximately 1,328 annually (Hoffmann and Baral 2020), than the number of deaths due to aflatoxin.

PESTICIDES

Most of the available data on pesticide contamination are based on crops destined for export and thus of limited use for estimating risk in the domestic food supply. One study found that none of 61 samples of tomatoes grown in Laikipia in 2019 for the local market were found by the Kenya Plant Health Inspectorate Service to contain pesticides above levels allowed under EU law (Gatere et al. 2020). Negative health effects of pesticide exposure have, however, been documented among Kenyan vegetable farm workers (Macharia 2015).

Challenges and potential solutions to improve food safety in Kenya

Kenya faces some of the same challenges to the effective control of food safety risks as governments all over the world, including regulatory fragmentation across line ministries and levels of government, and limited public resources. In addition, inappropriate regulatory standards and a large informal food sector are challenges that are common to Kenya and other low- and middle-income countries.

Regulatory fragmentation

Food safety governance in Kenya is highly fragmented, across levels of government (national versus county), commodities (horticulture, dairy, meat, fish, grains), and locations (environment, farm, factory, retail). Food safety policies and laws are generally enacted at the national level with participation of county governments but responsibility for enforcing laws and standards often falls to counties.

---

2 Reported levels of aflatoxin in maize grown in eastern Kenya regularly include observations in the thousands of parts per billion, while maximum values in other parts of the country tend to be in the hundreds of parts per billion. The regulatory limit for aflatoxin in Kenya is 10 parts per billion (Mutegi, Cotty, and Bandyopadhyay 2018).

Harmonization of food safety governance across these levels of government is achieved through alignment of county policies and laws with those adopted at the national level. For example, following the Kenya National Nutrition Action Plan, the Murang'a County Nutrition Action Plan includes several food safety-specific actions (Murang'a, Department of Health Services 2020). However, during the years covered by the Murang'a Plan (2020/21–2024/25), none of these activities were allocated any budget. Budget constraints have generally been the bottleneck in implementing county functions across sectors, as services are devolved but the funds transferred by the national government are insufficient to implement them. Analysis conducted by CGIAR researchers based on budget tracking for Murang'a, Laikipia, Nakuru, Nyandarua, and Nairobi counties found that public spending on food safety at both the national and the county level was low compared with the estimated cost of foodborne illness (Hoffmann and Baral 2019; Guthiga, Kirui, and Karugia 2020).

Beyond the challenges of implementing policies in a devolved governance structure, multiple line ministries and other public agencies share responsibility for food safety. Taking as an example the value chain for maize flour, the Ministry of Agriculture, Livestock, Fisheries, and Cooperatives has jurisdiction over practices applied on farm, while the crop is grown and stored prior to sale. The Agriculture and Food Authority manages the system, allowing traceability from the miller back to the farm. The Kenya Bureau of Standards is responsible for the inspection and registration of the maize flour brand. And finally, the Ministry of Health monitors safety of this product once it is available for purchase. While the National Food Safety Coordinating Committee constitutes an important channel for communication across these bodies, the demarcation of specific responsibilities is often unclear, leading to overlap of mandates and wastage of resources. It is not uncommon to find several different agencies inspecting the same factory at different times.

This patchwork of jurisdictions is not unique to Kenya. The multisectoral nature of this issue means that many countries have food safety responsibilities spread across disparate government agencies, a fact that has often been criticized as leading to inefficiencies and system failures (Heinzerling 2015). Other member states of the East African Community have created food and drug authorities but these have not always reduced overlaps and wastage as hoped, as the mandates of existing agencies involved in food safety have never been adjusted (Kang'ethe et al. 2021). Tanzania is a case in point: the functions of the Tanzanian Food and Drug Authority overlapped with those of the Tanzanian Bureau of Standards, leading to the former's disbandment and the redistribution of functions after several years. Kenya should take this as a cautionary tale when implementing any institutional solution to integrate food safety activities across agencies.

Inappropriate regulatory standards

Like many low- and middle-income countries, Kenya has to a large extent adopted food safety standards developed by high-income countries and adopted by international bodies (Sirma et al. 2018). While harmonization with such standards is critical for gaining access to export markets, compliance may be infeasible in the context of inadequate water, sanitation, and cold chain infrastructure.

Significant rates of noncompliance with standards are observed in the formal and informal sectors alike, including in leafy greens (Kutto et al. 2011), milk (Wanjala et al. 2017), and maize flour (Hoffmann and Moser 2017; Hoffmann, Moser, and Herrman 2021). The development of locally appropriate standards, in line with Kenya's food safety priorities and current hazard prevalence, combined with investment in the food safety capacity of food business operators and the creation of incentives for compliance, may offer a way to progressively register and formalize informal businesses (Blackmore, Alonso, and Grace 2015).

Heavy-handed enforcement of standards should be avoided as this can lead to tragic unintended consequences. For example, forcible relocation of butchers to an upgraded but inconveniently located market led to riots and deaths in Nigeria (Grace, Dipeolu, and Alonso 2019). CGIAR authors have also noted that destruction of foodstuffs that exceed Kenya's aflatoxin standards would have a devastating impact on food security, far worse than the public health benefit to be gained by such enforcement (Sirma et al. 2018).

Importance of informal food markets

Although informal markets are critical to the supply of animal-source foods and fruits and vegetables in Kenya, this sector is not adequately addressed in current food safety policies or practice (Kang'ethe et al. 2021). This means that much of the Kenyan food supply is outside the reach of public systems and private incentive structures that could be used to improve its safety. Unless effort is directed to addressing the food safety challenges of informal food markets that supply most animal-source foods, fruits, and vegetables to the majority of Kenyans, the foodborne disease burden will remain high.

Even in value chains where part of the food chain is formalized, typically from the point of processing, primary production is often still informal. This implies challenges to implementing the recommended farm-to-fork management approach of food safety. For example, the government regulates large-scale maize processors. Some of these pay a significant premium for maize that meets food safety and other quality standards in order to ensure their compliance with food safety regulations (Hoffmann and Moser 2017). However, as traders typically do not have traceability systems in place, and procure grain from a large number of small-scale farmers, this premium does not flow back to farmers, whose actions are critical to aflatoxin control. The logistical complexity and cost of testing farmers' grain prior to purchase, as well as the option to sell maize found to be noncompliant on the informal bulk grain market, means traders have little incentive to impose quality requirements on farmers.

The coexistence of formal and informal food systems also poses regulatory challenges when the two operate in parallel. While more comprehensive surveillance and stringent application of standards in maize is likely to improve compliance within the formal sector, this would imply that a greater share of maize sold as whole grain, or processed by small-scale, informal hammer mills known locally as *posho* mills, would previously have been rejected by formal millers. Formally processed maize flour is already significantly less contaminated on average than the lower-cost flour processed by *posho* mills, which is primarily consumed by lower-income Kenyans (Kariuki and Hoffmann 2021). Stricter enforcement in the formal sector would be expected to widen this inequitable food safety gap. A more effective approach in this particular case would be to inform consumers of the relative safety of formally marketed grain.[3]

To address the more general challenge of improving food safety in the informal market, we propose a progressive regulatory approach that combines training food business operators on practices to improve food safety, offering voluntary certification based on observed food safety practices and/or outcomes, and building consumer awareness of and demand for food safety. Such an approach meets vendors where they are and encourages incremental improvements in practices rather than expecting immediate compliance with standards that are often infeasible in the context of limited access to water and cold chain infrastructure. Thailand's Clean Food Good Taste project, jointly implemented by Thailand's Department of Health, the Tourism Authority, and local governments through the Ministry of the Interior since 1989, is a successful example of a voluntary public food safety certification program (Kongchuntuk 2002).

3 The lower aflatoxin contamination level in formally marketed maize flour appears to owe primarily to the fact that the milling process removes most of the germ and bran, where aflatoxin is concentrated. While this removes valuable fiber and nutrients, the mandated addition of micronutrient premix to formally marketed flour more than compensates for the loss of nutrients (though not fiber).

Limited public resources

Kenya's food inspectorate service lacks the budget and capacity to fully address the country's food safety problems. In addition to increasing the public resources available for this purpose, the capacity of private firms to monitor their own food safety outcomes could be leveraged. Since the early 1990s, food safety regulation in high-income countries has moved away from relying on command-and-control policies toward increased involvement of regulated firms in determining how they will meet standards (Henson and Caswell 1999; Rouvière and Caswell 2012). This approach, variously referred to as "coregulation" or "enforced self-regulation," shifts responsibility for routine monitoring of food safety practices and outcomes to food businesses. An important distinction between coregulation and industry self-regulation is that the regulator provides oversight (for example, through duplicate testing of samples), and retains the right to sanction firms for noncompliance. The organization Aflatoxin Proficiency Testing and Control in Africa has worked with Kenyan maize milling firms and the Meru county government to promote a coregulatory approach to aflatoxin management.[4]

Interventions to improve food safety

This section outlines recommendations for specific public actions to improve the safety of foods in Kenya's markets and reduce the foodborne disease burden. We discuss these in sequence from production to consumption, beginning with the control of environmental contaminants that may become foodborne.

Enforce regulations regarding environmental pollutants

Regulations that limit the use of lead and other harmful heavy metals in industrial and household products should be enforced, as these contaminants make their way into food supplies via contaminated irrigation water and soil. While the existing code of practice for production of fruits and vegetables, KS 1758-II, specifies that irrigation water should be tested annually, few farmers follow this guideline in practice (Gatere et al. 2020). Making information available on the microbiological and heavy metal contamination of water bodies, and on recommended mitigation strategies, would enable farmers to manage the risks these hazards pose. For example, crops with lower uptake of heavy metals could be grown in polluted areas, and preharvest watering intervals could be observed to reduce microbial contamination risk.

4 https://apteca.tamu.edu/

Build food safety capacity throughout value chains

Research by CGIAR and others shows that, across a wide range of contexts, both food safety practices and hazard prevalence can be significantly improved by building the capacity of farmers and food handlers through the provision of training and tools.

In eastern Kenyan communities, where training and postharvest technologies including drying sheets, a mobile drying service, and hermetic storage bags were made available and partially subsidized, aflatoxin levels reduced by half (Pretari, Hoffmann, and Tian 2019).[5] In Ghana, training on postharvest handling recommendations for aflatoxin control led to significant improvements in farmers' practices. Effects were stronger when tools to implement recommended practices were also provided (Magnan et al. 2021). Farmers' food safety capacity can be built by including this topic within agricultural extension programs and rural development projects that aim to boost smallholder productivity and incomes. An example of this approach is the East African Dairy Development Project, which includes training on milking hygiene and its benefits.[6]

Improving firms' capacity for food safety testing can have impacts even among formal sector firms (Herrman et al. 2020). This can have significant effects on population health if the market share of these firms is large, as it is for large-scale maize processors in Kenya.

Building the capacity of small-scale, often informal, retailers is relatively costly per unit of food affected, but reaching these vendors is critical, as most Kenyan consumers rely on such firms for fresh produce, meat, and milk—the riskiest foods in terms of microbial contamination. CGIAR has pioneered the "three-legged stool" approach to improving food safety in informal markets, which relies on three pillars that underlie food safety interventions: capacity building of business operators, incentives to support behavior change, and promotion of an enabling policy environment (Alonso Alvarez, Grace, and Nguyen-Viet 2021). Several intervention studies based on this model have shown that significant improvements in food safety practices and outcomes can be achieved by providing food vendors with training and simple materials

5 Many of the studies conducted in Kenya have focused on aflatoxin control, as detection of this contaminant is relatively easy and low in cost. High levels of public concern and regulatory attention to this hazard create the conditions for strong behavioral responses to information and the promotion of control technologies. This allows researchers to detect impacts of interventions on the responses of consumers and food business operators relatively easily. However, it is important to bear in mind that responses to similar interventions addressing other food safety hazards may vary.

6 www.heifer.org/our-work/flagship-projects/east-africa-dairy-development-project.html

or equipment, such as easy-to-clean cutting boards (Lindahl et al. 2018; Chea et al. 2021).

Two caveats are in order regarding the importance of capacity building. First, in the absence of access to appropriate infrastructure and equipment, training alone may be insufficient to change practices or improve food safety. This was the case for Kenyan abattoir workers, who failed to adopt the hygienic practices on which they were trained (Mwai 2011). Access to sufficient water and the means to heat it, as well as supplies such as disinfectants and soap, remained barriers. The fact that most workers were paid on a piece rate basis implied an incentive to work fast but not necessarily carefully (Mwai 2011). Second, as the effect of capacity building on food safety knowledge, practices, and outcomes can fade over time, repeated engagement with businesses is important to sustaining the impact of such interventions (Kinyua et al. 2021).

Support the development of incentives for safer food

Incentives for food business operators and food handlers to invest in food safety are critical. Potential motivators for the adoption of better food safety practices include reduced risk of spoilage, higher prices or sales volumes, and avoidance of regulatory enforcement action or a negative consumer response should a food safety problem become known.

A study conducted in Meru and Tharaka-Nithi counties showed that maize growers producing solely for home consumption used better postharvest practices than did farmers who sold a portion of their maize. Providing incentive payments to farmers for maize that met aflatoxin standards narrowed this gap (Hoffmann and Jones 2021). A separate trial in the same region found that most farmers who had purchased the aflatoxin control product Aflasafe had done so to ensure the safety of their own food supply. Introducing a small market incentive for safe grain more than doubled the amount of the product purchased (Hoffmann et al. 2022).

While this evidence points to the potential for market incentives to improve practices, the fragmented nature of Kenyan value chains can make this difficult. Examples of value chains in which requirements for farm-to-fork food safety practices have been successfully implemented include the horticultural export sector and a handful of fruit and vegetable distributors supplying premium markets.[7]

7 An example of a fruit and vegetable distributor that monitors food safety is Instaveg (COLEAD 2021).

There may be opportunities to incentivize better hygiene practices in milk value chains through quality-based payment systems. Lower microbial load is an important quality criterion used by cooling plants and dairy processors as it both reduces spoilage and allows for value addition and processing into dairy products, while also implying better food safety. Premium payments based on quality parameters including hygiene have therefore been adopted by milk processors in Indonesia, Brazil, and India, and piloted in Kenya (Ndambi, Dido, and Gülzari 2020; Treurniet 2021). Even outside of formalized quality payment systems, qualitative research shows that milk vendors perceive economic benefits to employing hygienic practices, through lower spoilage, the ability to charge higher prices, and customer retention (Alonso et al. 2018).

Other incentives for participation in trainings and uptake of improved practices could include facilitation of a license to operate legally, improved business skills, and earning esteem within peer networks (Blackmore, Alonso, and Grace 2015).

Invest in water, sanitation, and hygiene infrastructure

Public provision of water, sanitation, and hygiene (WASH) infrastructure and promotion of good hand and food hygiene practices at points through which large volumes of food pass could potentially be a cost-effective way to improve food safety. Such high-volume nodes include abattoirs, designated markets, and other areas where vendors congregate. Handwashing stations deployed to markets as part of the COVID-19 response are likely to reduce transmission of other infectious agents including foodborne diseases and should continue to be maintained.

While provision of modern infrastructure alone is not sufficient to address food safety challenges in informal markets (Grace, Dipeolu, and Alonso 2018), without access to adequate infrastructure even well-trained and motivated business operators will struggle to maintain appropriate food and hand hygiene. As informal businesses often have limited access to appropriate equipment, such as regulation-compliant food containers, cutting boards, or fridges, facilitating access to credit, especially among women, and making basic equipment available can allow food business operators to overcome financial constraints to food safety upgrading.

Build consumer demand

Pairing training and certification programs with interventions to build consumer demand for safer food could increase their effectiveness by strengthening market incentives. A recent experimental study found that communicating both risks and recommendations for less risky products had an impact on consumer food choices (Kariuki and Hoffmann 2021). Through this study, conducted in Meru town, some consumers were told which maize flour brands were most likely to comply with Kenya's aflatoxin regulation; others were given this information plus test results for the maize flour they were currently consuming. Nine weeks later, those given test results were more likely to be consuming one of the safer brands, whereas those given only the safer brands recommendation had not significantly changed their maize choices. The results were driven, unsurprisingly, by those whose maize had tested above the regulatory aflatoxin limit, indicating that change is much more likely for those made aware of a problem. While household-level testing for food safety hazards is not feasible, systematic surveillance of high-risk foods and provision of relative risk information by certification status, vendor, or product type could be used to steer consumers toward safer choices.

Educate caregivers of infants and young children on food hygiene practices

As infants and young children face the greatest health risk from foodborne infectious disease, improving household handling of foods consumed by this group is critical. Despite the higher levels of contamination found in food relative to water (Lanata 2003; Kung'u et al. 2009), food safety has been a relatively neglected aspect of infant and young child health and feeding programming globally. High-quality evidence shows that training caregivers of young children on safe food handling practices can improve practices and reduce microbial contamination (Islam et al. 2013; Touré et al. 2013; Gautam et al. 2017).

A study involving over 4,000 infants living in informal settlements within Nairobi found that only 2 percent were exclusively breastfed up to six months, and the mean age at which complementary foods were introduced was one month (Kimani-Murage 2011). Information on safe handling of infant food should thus be provided as part of standard postnatal care and support, while promotion of the WHO recommendation of exclusive breastfeeding up to six months should be continued.

Summary of recommendations

Conduct monitoring of water sources used for irrigation, and remediate problems. Compliance with the requirement for testing of irrigation water under the horticultural code of practice for fruits and vegetables is low. Public testing of commonly used water bodies could be used to identify where agricultural practices should be adapted to mitigate risks and inform efforts to address pollution sources.

Provide WASH infrastructure at markets and abattoirs. Informal food markets and processing facilities often lack adequate infrastructure for WASH. Access to safe water, toilets, and handwashing facilities is a prerequisite for food handlers to maintain appropriate hand and food hygiene, and public provision of these necessities is likely to be a cost-effective way to reduce foodborne disease.

Build capacity and incentivize food safety among small-scale, informal businesses. Small-scale businesses, from farm to fork, are critical to Kenya's food supply but often lack the capacity to adhere to food safety standards. Improving the capacity of food business operators, especially in the area of hand and food hygiene, and creating mechanisms to incentivize improvements in food safety can lead to meaningful gains in public health. In this context, rewarding improved food safety performance is likely to be more effective than enforcing compliance with standards that may not be attainable.

Implement regular and comprehensive surveillance of high-risk foods. Surveillance of high-risk foods should include collection and analysis of a representative set of samples, including from small-scale, informal businesses. Information on the relative hazard rates observed across food and business types could be used to steer consumers toward safer choices.

Leverage private sector capacity for self-monitoring under a coregulatory approach. Regular and representative food safety inspections require significant resources. Putting responsibility for routine monitoring of food safety compliance in the hands of businesses, with oversight provided by the regulator, can stretch budgets further and increase the efficiency of food safety surveillance.

Include food safety in infant and young child feeding recommendations for caregivers. Information on child health and nutrition given to caregivers of infants and young children through perinatal and child health clinics and via community health volunteers should including recommendations on food hygiene and heat treatment, while continuing to promote exclusive breastfeeding up to six months of age.

References

Alonso Alvarez, A., D. Grace, and H. Nguyen-Viet. 2021. "Informal Food Markets: What It Takes to Make Them Safer." *The Conversation,* June 3, 2021.

Alonso, S., E. Muunda, S. Ahlberg, E. Blackmore, and D. Grace. 2018. "Beyond Food Safety: Socio-economic Effects of Training Informal Dairy Vendors in Kenya." *Global Food Security* 18: 86–92.

ATSDR (Agency for Toxic Substances and Disease Registry). 2012. *Toxicological Profile for Cadmium*. Washington, DC: US Department of Human and Health Services.

Blackmore, E., S. Alonso, and D. Grace. 2015. *Legitimising Informal Markets: A Case Study of the Dairy Sector in Kenya.* IIED Briefing Papers. London: International Institute for Environment and Development; Nairobi: International Livestock Research Institute.

Carron, M., Y.M. Chang, K. Momanyi, J. Akoko, J. Kiiru, J. Bettridge, G. Chaloner, J. Rushton, S. O'Brien, N. Williams, and E.M. Fevre. 2018. "*Campylobacter*, a Zoonotic Pathogen of Global Importance: Prevalence and Risk Factors in the Fast-Evolving Chicken Meat System of Nairobi, Kenya." *PLoS Neglected Tropical Diseases* 12 (8): e0006658.

CEJAD (Centre for Environment Justice and Development) and IPEN (International Pollutants Elimination Network). 2017. *Lead in Solvent-Based Paints for Home Use in Kenya.* Nairobi.

Chea, R., S. Tum, H. Nguyen-Viet, S. Dang-Xuan, D. Grade, J.F. Lindahl, C. Ty, S. Koam, T. Heng, O. Phirum, S. Boqvist, and F Unger. "Good Hygiene Practices Intervention for Safer Pork in Traditional Markets in Cambodia: Preliminary Results." Abstract in Safe Food, Fair Food for Cambodia Project Final Workshop Proceedings, June 21–22, 2021.

Checkley, W., G. Buckley, R.H. Gilman, A.M. Assis, R.L. Guerrant, S.S. Morris, K. Mølbak, P. Valentiner-Branth, C.F. Lanata, R.E. Black, et al. 2008. "Multi-country Analysis of the Effects of Diarrhoea on Childhood Stunting." *International Journal of Epidemiology* 37 (4): 816–830.

COLEAD (Committee Linking Entrepreneurship–Agriculture–Development). 2021. "Access to Finance: Instaveg Benefits from Agrifi Challenge Fund after COLEACP Support." January 19. https://news.coleacp.org/en/instaveg-towards-new-horizons/

Cook, E.A.J., W.A. de Glanville, L.F. Thomas, S. Kariuki, B.M. de Clare Bronsvoort, and E.M. Fèvre. 2017. "Working Conditions and Public Health Risks in Slaughterhouses in Western Kenya." *BMC Public Health* 17 (1): 14.

Gallaher, C.M., D. Mwaniki, M. Njenga, N.K. Karanja, and A.M. WinklerPrins. 2013. "Real or Perceived: The Environmental Health Risks of Urban Sack Gardening in Kibera Slums of Nairobi, Kenya." *EcoHealth* 10 (1): 9–20.

Gatere, L., V. Hoffmann, M. Murphy, and P. Scollard. 2020. *Food Safety in Tomatoes Produced in Laikipia County.* Voice for Change Partnership Policy Brief. Washington, DC: International Food Policy Research Institute (IFPRI).

Gautam, O. P., W.P. Schmidt, S. Cairncross, S. Cavill, and V. Curtis. 2017. "Trial of a Novel Intervention to Improve Multiple Food Hygiene Behaviors in Nepal." *American Journal of Tropical Medicine and Hygiene* 96 (6): 1415–1426.

Gibb, H.J., A. Barchowsky, D. Bellinger, P.M. Bolger, C. Carrington, A.H. Havelaar, S. Oberoi, Y. Zang, K. O'Leary, and B. Devleesschauwer. 2019. "Estimates of the 2015 Global and Regional Disease Burden from Four Foodborne Metals–Arsenic, Cadmium, Lead and Methylmercury." *Environmental Research* 174: 188–194.

GBDCN (Global Burden of Disease Collaborative Network). 2016. Global Burden of Disease Study 2010 (GBD 2010) Results. Seattle: Institute for Health Metrics and Evaluation (IHME). http://ghdx.healthdata.org/

Grace, D., M. Dipeolu, and S. Alonso. 2019. "Improving Food Safety in the Informal Sector: Nine Years Later." *Infection Ecology & Epidemiology* 9 (1): 1579613.

Guthiga, P., L. Kirui, and J. Karugia. 2020. *Tracking Government Food Safety Budgets in the Dairy and Horticulture Sectors in Kenya: A Methodological Brief.* Prepared for the Voice for Change Partnership (V4CP). Nairobi: ILRI.

Havelaar, A.H., M.D. Kirk, P.R. Torgerson, H.J. Gibb, T. Hald, R.J. Lake, N. Praet, D.C. Bellinger, N.R. De Silva, N. Gargouri, and N. Speybroeck. 2015. "World Health Organization Global Estimates and Regional Comparisons of the Burden of Foodborne Disease in 2010." *PLoS Medicine* 12 (12): e1001923.

Heinzerling, L. 2015. "Divide and Confound: The Strange Allocation of US Regulatory Authority Over Food." In *Food and Drug Regulation in an Era of Globalized Markets*, ed. S.F. Halabi, 125–134. London: Elsevier.

Henson, S., and J. Caswell. 1999. "Food Safety Regulation: An Overview of Contemporary Issues." *Food Policy* 24 (6): 589–603.

Herrman, T., V. Hoffmann, A. Muiruri, and C. McCormick. 2020. "Aflatoxin Proficiency Testing and Control in Kenya." *Journal of Food Protection* 83 (1): 142–146.

Hoffmann, V., and S. Baral. 2019. *Foodborne Disease in Kenya: County-Level Cost Estimates and the Case for Greater Public Investment.* Washington, DC: IFPRI.

Hoffmann, V., and K.M. Jones. 2021. "Improving Food Safety on the Farm: Experimental Evidence from Kenya on Agricultural Incentives and Subsidies as Public Health Investments." IFPRI Discussion Paper 1746. IFPRI, Washington, DC.

Hoffmann, V., K. Jones, and J.L. Leroy. 2018. "The Impact of Reducing Dietary Aflatoxin Exposure on Child Linear Growth: A Cluster-Randomized Controlled Trial in Kenya." *BMJ Global Health* 3 (6): e000983.

Hoffmann, V., S. Kariuki, J. Pieters, and M. Treurniet. 2022. "Upside Risk, Consumption Value, and Market Returns to Food Safety." *American Journal of Agricultural Economics* 105 (3): 914–939.

Hoffmann, V., and C. Moser. 2017. "You Get What You Pay For: The Link between Price and Food Safety in Kenya." *Agricultural Economics* 48 (4): 449–458.

Hoffmann, V., M. Moser, and T. Herrman. 2021. "Demand for Aflatoxin-Safe Maize in Kenya: Dynamic Response to Price and Advertising." *American Journal of Agricultural Economics* 103 (1): 275–295.

Inoti, K.J., K. Fanuel, O. George, and O. Paul, O. 2012. "Assessment of Heavy Metal Concentrations in Urban Grown Vegetables in Thika Town, Kenya." *African Journal of Food Science* 6 (3): 41–46.

Ismail, A., I. Naeem, Y.Y. Gong, M.N. Routledge, S. Akhtar, M. Riaz, L.N.Z. Ramalho, C.A.F. de Oliveira, and Z. Ismail. 2021. "Early Life Exposure to Dietary Aflatoxins, Health Impact and Control Perspectives: A Review." *Trends in Food Science & Technology* 112: 212–224.

Islam, M.S., Z.H. Mahmud, P.S. Gope, R.U. Zaman, Z. Hossain, M.S. Islam, D. Mondal, M.A.Y. Sharker, K. Islam, H. Jahan, and A. Bhuiya. 2013. "Hygiene Intervention Reduces Contamination of Weaning Food in Bangladesh." *Tropical Medicine & International Health* 18 (3): 250–258.

Jaffee, S., S. Henson, L. Unnevehr, D. Grace, and E. Cassou. 2018. *The Safe Food Imperative: Accelerating Progress in Low- and Middle-Income Countries*. Washington, DC: World Bank.

Jiang, X., Z. Chen, and M. Dharmasena. 2015. "The Role of Animal Manure in the Contamination of Fresh Food." In *Advances in Microbial Food Safety*, ed. J. Sofos, 312–350. Sawston, UK: Woodhead Publishing.

Kaluli, J.W., C. Githuku, P. Home, and B.M. Mwangi. 2011. "Towards a National Policy on Wastewater Reuse in Kenya." *Journal of Agriculture, Science and Technology* 13 (1): 116–125.

Kang'ethe, E., F. Mutua, K. Roesel, M. Ntawubizi, C. Kankya, S. Niragira, B. Kilima, P. Gathura, K.K. Kuorwel, A. Kiri, and D. Grace. 2021. "A Review of the Food Safety Architecture in the East African Community: Animal-Source Foods, Fruits and Vegetables." ILRI Discussion Paper 41. ILRI, Nairobi.

Karanja, N., M. Njenga, G. Prain, E. Kang'ethe, G. Kironchi, C. Githuku, P. Kinyari, and G.K. Mutua. 2010. "Assessment of Environmental and Public Health Hazards in Wastewater Used for Urban Agriculture in Nairobi, Kenya." *Tropical and Subtropical Agroecosystems* 12 (1): 85–97.

Kariuki, S.W. and V. Hoffmann. 2021. "Can Information Drive Demand for Safer Food? Impact of Brand-Specific Recommendations and Test Results on Product Choice." *Agricultural Economics* 53 (3): 454–467.

Kimani-Murage, E.W., N.J. Madise, J.C. Fotso, C. Kyobutungi, M.K. Mutua, T.M. Gitau, and N. Yatich, 2011. "Patterns and Determinants of Breastfeeding and Complementary Feeding Practices in Urban Informal Settlements, Nairobi Kenya." *BMC Public Health* 11 (1): 1–11.

Kinyua, C., S. Alonso, E. Blackmore, and V. Thebe. 2021. "Interventions that Fail to Sustain: Evidence from a Training and Certification Intervention in Tanzania." Poster presented at the virtual Agriculture, Nutrition and Health (ANH) Academy Week 2021, June 29–July 1. Nairobi: International Livestock Research Institute (ILRI).

Kitulu, L., J.K. Nzeve, F. Waswa, E. Kitur, and D. Shitanda. 2020. "Irrigation Water Quality Analysis of Mitheu Stream in Machakos Municipality, Kenya." *African Journal of Environmental Science and Technology* 14 (9): 241–249.

KNBS (Kenya National Bureau of Statistics) and ICF. 2023. *Demographic and Health Survey 2022. Key Indicators Report*, Nairobi, Kenya, and Rockville, Maryland, USA: KNBS and ICF

Knechtges, P.L. 2011. *Food Safety: Theory and Practice*. Jones and Bartlett Publishers.

Kongchuntuk, H. 2002. "Thailand's Food Safety Project on Restaurants and Street Vendors: The 'Clean Food Good Taste Project'." Paper presented at the FAO/WHO Global Forum of Food Safety Regulators, Marrakech.

Kung'u, J.K., K.J. Boor, S.M. Ame, N.S. Ali, A.E. Jackson, and R.J. Stoltzfus. 2009. "Bacterial Populations in Complementary Foods and Drinking-Water in Households with Children Aged 10–15 Months in Zanzibar, Tanzania." *Journal of Health, Population, and Nutrition* 27 (1):41–52.

Kutto, E.K., M.W. Ngigi, N. Karanja, E. Kangethe, L.C. Bebora, C.J.L. Lagerkvist, P.G. Muthia, L.W. Njagi, and J.J. Okello. 2011. "Bacterial Contamination of Kale (*Brassica oleracea Acephala*) along the Supply Chain in Nairobi and Its Environment." *East African Medical Journal* 88 (2): 46–53.

Lanata C.F. 2003. "Studies of Food Hygiene and Diarrhoeal Disease." *International Journal of Environmental Health Research* 13 (Suppl 1): S175–183.

Leroy, J.L., J.S. Wang, and K. Jones. 2015. "Serum Aflatoxin B1-lysine Adduct Level in Adult Women from Eastern Province in Kenya Depends on Household Socio-economic Status: A Cross Sectional Study." *Social Science & Medicine* 146: 104–110.

Lindahl, J.F., R.P. Deka, R. Asse, L. Lapar, and D. Grace. 2018. "Hygiene Knowledge, Attitudes and Practices among Dairy Value Chain Actors in Assam, North-East India and the Impact of a Training Intervention." *Infection Ecology & Epidemiology* 8 (1): 1555444.

Liu, Y., C.C.H Chang, G.M. Marsh, and F. Wu. 2012. "Population Attributable Risk of Aflatoxin-related Liver Cancer: Systematic Review and Meta-analysis." *European Journal of Cancer* 48 (14): 2125–2136.

Macharia, I. 2015. "Pesticides and Health in Vegetable Production in Kenya." *BioMed Research International* 2015: 241516.

Magnan, N., V. Hoffmann, N. Opoku, G.G. Garrido, and D.A. Kanyam. 2021. "Information, Technology, and Market Rewards: Incentivizing Aflatoxin Control in Ghana." *Journal of Development Economics* 151: 102620.

Makokha, A.O., L.R. Mghweno, H.S. Magoha, A. Nakajugo, and J.M. Wekesa. 2008. "Environmental Lead Pollution and Contamination in Food around Lake Victoria, Kisumu, Kenya." *African Journal of Environmental Science and Technology* 2 (10): 349–353.

Murang'a, Department of Health Services. 2020. *County Nutrition Action Plan 2020/21–2024/25*. Murang'a, Kenya.

Mutua, F., D. Grace, and C. Watts. 2021. *Feed the Future Innovation Lab for Food Safety: Food Safety Investments in East Africa*. Washington, DC: USAID; Nairobi: ILRI.

Mutegi, C.K., P.J. Cotty, and R. Bandyopadhyay. 2018. "Prevalence and Mitigation of Aflatoxins in Kenya (1960-to Date)." *World Mycotoxin Journal* 11 (3): 341–357.

Mwai, C.W. 2011. "Risk of Contamination of Beef Carcasses with *Escherichia coli* O157:H7 from Slaughterhouses in Nairobi, Kenya." Master's thesis, University of Nairobi, Kenya.

NASEM (National Academies of Sciences, Engineering, and Medicine). 2017. "Sources of Lead Contamination." In *Investigative Strategies for Lead-Source Attribution at Superfund Sites Associated with Mining Activities*. Washington, DC: National Academies Press.

Ndambi, A., S. Dido, and S. Ö. Gülzari. 2020. *Milk Quality Assurance in Smallholder Dominated Dairy Chains: Lessons from Uganda and Kenya*. Wageningen Livestock Research Report. Wageningen, Netherlands: Wageningen University & Research.

Nyokabi, S., P.A. Luning, I.J. de Boer, L. Korir, E. Muunda, B.O. Bebe, J. Lindahl, B. Bett, and S.J. Oosting. 2021. "Milk Quality and Hygiene: Knowledge, Attitudes and Practices of Smallholder Dairy Farmers in Central Kenya." *Food Control* 130: 108303.

Nyokabi, S.N., I.J. de Boer, P.A. Luning, L. Korir, J. Lindahl, B. Bett, and S.J. Oosting. 2021b. "Milk Quality along Dairy Farming Systems and Associated Value Chains in Kenya: An Analysis of Composition, Contamination and Adulteration." *Food Control* 119: 107482.

Owuor, S. 2020. *Inclusive Growth and Informal Food Vending in Nairobi, Kenya*. Hungry Cities Partnership Report No. 21. Cape Town, South Africa and Waterloo, Canada: Hungry Cities Partnership.

Pretari, A., V. Hoffmann, and L. Tian. 2019. "Post-harvest Practices for Aflatoxin Control: Evidence from Kenya." *Journal of Stored Products Research* 82: 31–39.

Rouvière, E. and J.A. Caswell. 2012. "From Punishment to Prevention: A French Case Study of the Introduction of Co-regulation in Enforcing Food Safety." *Food Policy* 37 (3): 246–254.

Sirma, A.J., J.F. Lindahl, K. Makita, D. Senerwa, N. Mtimet, E.K. Kang'ethe, and D. Grace. 2018. "The Impacts of Aflatoxin Standards on Health and Nutrition in Sub-Saharan Africa: The Case of Kenya." *Global Food Security* 18: 57–61.

Treurniet, M. 2021. "The Potency of Quality Incentives: Evidence from the Indonesian Dairy Value Chain." *American Journal of Agricultural Economics* 103 (5): 1661–1678.

Touré, O., S. Coulibaly, A. Arby, F. Maiga, and S. Cairncross. 2013. "Piloting an Intervention to Improve Microbiological Food Safety in Peri-Urban Mali." *International Journal of Hygiene and Environmental Health* 216 (2): 138–145.

Tsai, K., S. Simiyu, J. Mumma, R.E. Aseyo, O. Cumming, R. Dreibelbis, and K.K. Baker. 2019. "Enteric Pathogen Diversity in Infant Foods in Low-Income Neighborhoods of Kisumu, Kenya." *International Journal of Environmental Research and Public Health* 16 (3): 506.

Yard, E.E., J.H. Daniel, L.S. Lewis, M.E. Rybak, E.M. Paliakov, A.A. Kim, J.M. Montgomery, R. Bunnell, M.U. Abudo, W. Akhwale, and R.F. Breiman. 2013. "Human Aflatoxin Exposure in Kenya, 2007: A cross-sectional study." *Food Additives & Contaminants: Part A* 30 (7): 1322–1331.

Wambui, J., E. Karuri, P. Lamuka, and J. Matofari. 2017. "Good Hygiene Practices among Meat Handlers in Small and Medium Enterprise Slaughterhouses in Kenya." *Food Control* 81: 34–39.

Wanjala, G.W., F.M. Mathooko, P.M. Kutima, and J.M. Mathara. 2017. "Microbiological Quality and Safety of Raw and Pasteurized Milk Marketed in and around Nairobi Region." *African Journal of Food, Agriculture, Nutrition and Development* 17 (1): 11518–11532.

WHO (World Health Organization). 2018. *Global Alliance to Eliminate Lead Paint/Lead Paint Alliance Advisory Council Meeting*. Geneva.

Xu, Y., Y.Y. Gong, and M.N. Routledge. 2018. "Aflatoxin Exposure Assessed by Aflatoxin Albumin Adduct Biomarker in Populations from Six African Countries." *World Mycotoxin Journal* 11 (3): 411–419.

PART 3

Toward More Productive Food Systems

Part 3

TOWARD MORE PRODUCTIVE FOOD SYSTEMS

Productivity is key to ensuring the food system can provide a healthy diet for all. In past approaches to agricultural and rural development, much of the focus has been on improving food supply through the uptake of on-farm technologies. While the food systems approach takes a broader perspective, improving food production is still fundamental to driving transformation. Part 3 looks at agricultural productivity in Kenya, its evolution, and policies to improve it.

Agricultural production in Kenya is not uniform: the diversity of the country's agroecological zones means that productivity drivers and trends can differ by location, as Chapters 6 and 7 show. Chapter 6 analyzes the total factor productivity of Kenyan farms, and the question of whether to promote smallholder-dominated agricultural systems or systems based on larger farms. On the one hand, evidence suggests smaller farms are more efficient; on the other hand, small farmers face more barriers to market access, which can erode their efficiency gains. Chapter 7 shows that maize productivity, in particular, has stagnated in the past few decades, pointing toward a tapering-off of the Green Revolution. To reverse these trends, the chapter recommends boosting on-farm technology by including farmers in variety evaluations and by improving extension services. Further, market-based interventions, such as promoting private sector competition with the parastatal Kenya Seed Company, could boost adoption and lead to higher maize productivity.

Chapters 8 and 9 focus specifically on agricultural inputs and mechanization to drive productivity. Chapter 8 discusses trends, policies, challenges, and lessons learned from fertilizer, seed, pesticide, and knowledge usage. It highlights the need to improve market access to inputs through better transportation infrastructure, reduced regulations and levies, and promotion of the private sector. In particular, the use of input subsidies may crowd out private sector involvement in the input sector. Chapter 9 shows that mechanization faces slightly different challenges, related to the high fixed costs involved in adopting mechanized production techniques. Mechanization rates in Kenya remain low, partly because of the large portion of smallholder farmers in Kenya and the country's diverse agroecological conditions. In terms of policy, successful mechanization efforts in other countries have relied on the private sector, with the government playing a facilitating role. Promoting hiring services may help overcome the challenges involved in smallholder adoption of mechanization by removing the high fixed costs in accessing machinery. Further, policy can support the local manufacturing of machinery parts that are optimal for Kenyan farms. This could also be an important steppingstone toward full-scale machinery production in Kenya.

In summary, Part 3 looks at the many challenges to increasing agricultural productivity in Kenya. Solutions here must address the diverse agroecological landscape of Kenya and include a coherent plan on whether and how to promote a smallholder-driven transformation or a transformation focused on large-scale, commercialized farmers—or potentially both in parallel. Further, the role of market access in stimulating productivity is not to be understated; the book will return to this issue in Chapter 18, which discusses digital technologies.

Chapter 6

AGRICULTURAL PRODUCTIVITY IN KENYA, 2000–2020

Alejandro Nin-Pratt

Agriculture is key to economic growth and poverty reduction in Kenya as it plays a pivotal role in employment creation, food security, exports, and sustainable development. In 2019, it directly contributed 22.7 percent of GDP, accounted for 20.9 percent of total exports, and generated 43.3 percent of employment (Chapter 2). The sector is thus not only an important driver of Kenya's economy but also the means of livelihood for many Kenyan people.

Given the economic and social importance of agriculture in Kenya, policies have revolved around the main goal of increasing productivity and incomes, especially for smallholders, to enhance food security and equity, with an emphasis on production intensification, commercialization, and environmental sustainability (Alila and Atieno 2006). In this context, the declining performance of the sector measured in terms of its growth has been a major concern for policymakers.

This chapter looks at the performance of agriculture in Kenya—the first link in the food system chain of activities—in the context of the agricultural development strategies that the Government of Kenya has implemented since the year 2000. This entails the analysis of trends and of the evolution of output and total factor productivity (TFP) at the country level and a comparison of TFP and the technical efficiency of sub-counties with a focus on the Central Highlands, Rift Valley, and Western agroecologies. These zones contribute more than two-thirds of Kenya's agricultural output. Productivity is measured using a TFP index calculated using secondary data from the Kenya National Bureau of Statistics (KNBS). The regional analysis employs a decomposition of TFP into different measures of efficiency.

The chapter is organized as follows. The next section reviews the Kenyan agricultural policy context and the consequent productivity from independence to the present day. The chapter then lays out the methodology and data used to calculate TFP and to analyze production efficiency, followed by an analysis of output, input, and productivity trends in Kenya's agriculture sector between 2000 and 2020. After this, an efficiency analysis of the agricultural production of sub-counties in the Western, Central Highland, and Rift Valley zones is presented. The last section discusses alternative interpretations of the results obtained, with implications for Kenya's food security and agricultural policies.

An overview of agricultural policy and productivity in Kenya

Three distinct periods of policy and rural development can be observed between independence, achieved in 1963, and the year 2000, according to Kirori (2003) and Mwega and Ndung'u (2008). The 1960s saw enhanced flows of foreign direct investment, supported by import substitution policies for industrialization that were adopted before independence and deepened during this period. Simultaneously, the opening of the White Highlands in 1961, an area in the central uplands of Kenya that had officially been reserved for the exclusive use of Europeans, indirectly increased access to land. These two developments combined to enhance the productive capacity of agriculture, delivering relatively high economic growth but at the expense of increased regional inequality (Mwega and Ndung'u 2008). Between 1961 and 1972, Kenya's GDP per person grew at 3.0 percent on average as a result of the high GDP growth (7.2 percent); the average annual agricultural growth rate was 6.9 percent during the same period (World Bank 2022).

The second period extends from the late 1970s and to the mid-1990s. It opened with President Moi's election and the introduction of policies to address regional disparities (Mwega and Ndung'u 2008). This period was preceded by a series of exogenous shocks that eroded the performance of Kenya's economy. These included the oil crisis of the 1970s and the consequent world recession, high external interest rates and a decline in capital inflows, severe droughts, and the collapse of the East African Community in 1977, which soured the market for Kenya's nontraditional exports (Onjala 2002). Under Moi's regime, weaker budget management and the introduction of policies to control inflationary pressures that later created distortions in the economy were associated with a lower growth trajectory.

This situation lingered on until the early 2000s, slowing production expansion in both firms and smallholder farms, and overall economic growth (Mwega and Ndung'u 2008). Average annual GDP and agricultural GDP growth between 1973 and 1995 dropped to 4.0 and 3.2 percent, respectively. Slower growth together with fast population expansion resulted in an annual growth rate of GDP per person of only 0.5 percent.

The third period corresponds to the 1990s. It was characterized by economic reforms to aid markets to work better. During this period, Kenya shifted its economic development strategy from import substitution to export promotion and trade openness, significantly reducing restrictions on international trade and actively engaging in regional and continental integration initiatives. These reforms were instituted to reactivate the economy in response to declining growth. They included price decontrols, the removal of tariff and nontariff barriers, and the adoption of export promotion initiatives, including manufacturing in export processing zones, investment incentives, and increasing export market access through regional integration and bilateral trade agreements (Kenya, Ministry of Agriculture 2010; Wamalwa and Were 2021).

By the new millennium, most markets were fully liberalized, but these measures were often subject to reversal. Moreover, the impact of liberalization was not immediate (Kimenyi, Mwega, and Ndung'u 2015). Between 1995 and 2002, years of economic transformation in Kenya, GDP and agricultural GDP increased at approximately 2.0 percent annually, whereas GDP per person shrank at an average rate of −0.7 percent (World Bank 2022).In addition, between 1993 and 2000, the export sector performed poorly even for commodity exports such as tea, coffee, and horticulture, in which Kenya has a comparative advantage.

A study by Nyoro and Jayne (1999) looked at the changes undergone by the agriculture sector in Kenya after the implementation of structural adjustment and sectoral reform programs in the 1990s. It found a decline in labor productivity of about 20 percent between 1970/74 and 1990/94, while land productivity had increased up to about 1990 and fallen in the last years of the analyzed period. The authors also found that large increases in land and labor productivity in the most productive areas (the then-Central province) reflected changes in the crop mix, with a significant expansion of the area allocated to crops like coffee, tea, maize, and wheat, together with the introduction of new varieties of maize and wheat. In low-potential areas, on the other hand, crop yields had declined because of a lack of enhanced technologies adapted to those areas.

In the year 2000, Kenya's economy recorded an all-time low growth rate of 0.6 percent. However, following a peaceful change of government in December 2002 from the Kenya African National Union, which had ruled the country since independence, to the National Rainbow Coalition under Mwai Kibaki, the growth rate accelerated (Kimenyi, Mwega, and Ndung'u 2015).

In 2003, the new government launched the Economic Recovery Strategy for Wealth and Employment Creation (ERS; 2003–2007) as the blueprint for setting the country back on a growth path. The ERS focused on agriculture, trade and industry, and tourism as the key sectors to drive the economic recovery and contribute to improving food security and reducing rural poverty. The Strategy for Revitalizing Agriculture (SRA) followed in 2004, to "transform Kenya's agriculture into a profitable, commercially oriented and internationally and regionally competitive economic activity," providing an enabling environment for increasing agricultural productivity, promoting investment, and encouraging private sector involvement in agriculture. The overall aim was to refocus the government on the provision of key public goods, such as research and extension services, roads, and irrigation infrastructure (Poulton and Kanyinga 2014).

Significant progress was made during the ERS period. The economy recovered from low growth of 2.0 percent in 1995–2000 to growth of 5.5 percent between 2003 and 2007 (World Bank 2022). As a result, real per capita income increased at an annual average rate of 2.4 percent. In theory, these changes ought to have benefited agricultural producers, but this did not happen, at least at the aggregated sectoral level, despite the priority the government attached to agricultural recovery and the support the SRA received from Kenya's international development partners. Average annual agricultural growth of 0.9 percent between 2003 and 2007 was well below overall GDP growth (World Bank 2022).

There were, however, improvements in some subsectors. Kibaara and colleagues (2008), analyzing trends in crop yields using household panel survey data from eight agro-regional zones between 1996/97 and 2006/07, found consistent growth in maize productivity across most zones. This was driven by an increased percentage of households using fertilizer and high-yielding crop varieties, coinciding with an increased density of fertilizer retail outlets and a decline in the distances between farmers and sellers of agricultural inputs. Their findings also showed impressive growth of the dairy subsector as a result of increased production of fodder crops and the adoption of improved cattle breeds.

Meanwhile, Kibaara and colleagues found that tea yields had grown slightly, driven by increased fertilizer use, while the productivity of sugarcane and coffee had declined during the decade. These results on the productivity of export crops confirm findings showing that, while growth of nominal exports increased from an average rate of 4.1 percent in 1990–1999 to 11 percent in

2000–2009, during a period of fast-growing commodity prices, the actual growth of export quantities was only 1.6 percent in the latter period, compared with 2.8 percent growth in imports. This widened the trade gap and was a drag on economic growth, which was heavily dependent on the domestic market (Wamalwa and Were 2021).

The elections held in 2007 marked a new phase in Kenya's policy and development context. They were followed by a serious outbreak of ethnic violence, drought, and the global financial crisis, which eroded the achievements of the previous half-decade. There was significant disruption to the economy, which grew only 0.23 percent in 2008 (Kimenyi, Mwega, and Ndung'u 2015). After a year-long political crisis, and with the ERS set to expire, in June 2008 the newly elected government launched the Kenya Vision 2030 as the new long-term development blueprint for the country, with a vision of transforming Kenya into "a globally competitive and prosperous country with a high quality of life by 2030" (Kenya, Ministry of State Planning 2007). The SRA was revised to capture new developments and to strategically position the agriculture sector as a key driver in delivering the 10 percent annual economic growth rate envisaged under the economic pillar of Vision 2030.

The Agricultural Sector Development Strategy (ASDS) 2010–2020 aimed to deliver the Millennium Development Goal targets, with the main objective of achieving a food-secure and prosperous nation by 2020 through the transformation of smallholder agriculture from subsistence to commercially oriented and modern approaches. The new strategy identified four major challenges to Kenyan agriculture: persistent low productivity; suboptimal land use, mainly related to the growth of the population; inefficient markets owing to insufficient storage capacity and poor access; and low levels of value addition and largely informal value chains. The ASDS provided the basis for the implementation of the Comprehensive Africa Agriculture Development Program (CAADP) Compact and the formulation of the Medium-Term Implementation Plan (MTIP) 2010–2015.

The Government of Kenya developed its latest strategy, the Agricultural Sector Transformation and Growth Strategy (ASTGS, Kenya, Ministry of Agriculture 2018), and the National Agricultural Investment Plan (NAIP) after 2015, faced with suboptimal performance of agriculture in terms of production, value addition, food security, and nutrition. The new strategy was anchored in the belief that food security requires a vibrant, commercial, and modern agriculture sector that sustainably supports economic development, national priorities, and commitments to the Malabo Declaration under CAADP and the UN Sustainable Development Goals. The ASTGS divides the country into seven distinct agroecological zones based on soil type and rainfall and uses this division as the basis for value chain and intervention selections, to ensure the latter are sensitive to the needs of farmers in these areas.

Decades of policy changes were finally reflected in the country's economic performance after 2015. Data from the World Bank (2022) show that Kenya's economy achieved broad-based growth between 2015 and 2019, with GDP growth averaging 4.9 percent per year. Poverty declined significantly, falling to an estimated 34.4 percent at the $1.90/day line in 2019 (World Bank 2023). In 2020, the COVID-19 pandemic hit the economy hard, disrupting international trade and transport, tourism, and urban services activity. Fortunately, the agriculture sector, a cornerstone of the economy, remained resilient, helping limit the pandemic-driven contraction in GDP to only 0.3 percent (World Bank 2023).

Despite progress made, there are still signs that Kenya's agriculture sector is not yet on a sustainable path of fast growth driven by TFP. For example, De Groote (2022), using 2022 data from the Food and Agriculture Organization of the United Nations (FAO), on the area, production, and yield of maize between 1961 and 2022, shows that Kenya has not been able to sustain growth in maize productivity. While maize yields almost doubled between 1961 and the mid-1980s, they have remained stagnant for the past 30 years, with output growth driven by area expansion. If this is the case for most agricultural subsectors, it will be important to look again at Kenya's agricultural productivity growth to analyze the impact of policy changes after the economic transformation of the 1990s and the new policies and strategies implemented between 2000 and 2020.

In a review of the literature on agricultural productivity in Kenya, Birch (2018) identified some of the principal barriers to agricultural productivity growth, clustered in different areas:

- Land and population pressures mean that average farm size is falling and land is becoming more concentrated. Ever-smaller farm sizes may undermine the capacity of households to generate a surplus and the economic incentive to invest to improve productivity.
- Government interventions in markets distort input and cereal markets; and institutional barriers, high transaction costs, and limited access to credit further hamper markets.
- Low adoption of sustainable land management practices leads to increasing land degradation, and changes in temperature and variability of rainfall pose a growing threat to agricultural production (see, for example, De Groote and Omondi 2021; Jena et al. 2021).

- Public expenditure on agriculture is low, and spending on agricultural research in particular has fallen steadily over the past decade, by 2016 declining to one-third of its value in 2006 (Beintema et al. 2018). Low spending has also resulted in insufficient qualified personnel in extension services, with a ratio of national extension staff to farmers at 1:1,000, compared with the recommended 1:400 (Wanyama et al. 2016, 23).

Approach

The approach to calculating TFP and the data used are described in this section. The calculation of TFP used at the aggregated country level and in the regional comparison at the sub-county level follows O'Donnell (2012) in that TFP is expressed as the ratio of an index of total output and an index of total input based on a simple linear aggregation of inputs and outputs. O'Donnell refers to this index as the Lowe index of TFP, one of the indexes in a class of TFP indexes that are particularly suited to intertemporal and cross-sectional comparisons of production units (farms, counties, countries). Starting with the analysis of aggregated agricultural productivity across time, calculation of the TFP index to measure productivity changes between 2000 and 2020 involves the calculation of total output and input quantity indexes and the total output–input ratio for each year. For example, the change in agricultural TFP in Kenya between 2000 and 2020 can be expressed as $TFP_{2000\text{-}2020}=(QI_{2000\text{-}2020}/XI_{2000\text{-}2020})$, where $QI_{2000\text{-}2020}$ and $XI_{2000\text{-}2020}$ are the changes in aggregated output and total aggregated input between 2000 and 2020, respectively. More formally, the TFP index between period t and a reference period s is expressed as the ratio of an output index (QI) and an input index (XI):

$$MFP_{st} = \frac{QI_{st}}{XI_{st}} = \frac{Q(q_t)}{Q(q_s)} \div \frac{X(x_t)}{X(x_s)} \qquad 1$$

where $Q(q_t)$ is the weighted sum of m outputs produced in year t: $Q(q_t)=\Sigma_m p_m q_{mt}$ and $X(x_t)=\Sigma_n w_n x_{nt}$ is the weighted sum of n inputs used in production in the same year, with p_m and w_n being predetermined time-invariant reference prices of outputs and inputs, respectively. In the case of sub-county comparisons in the same year, the index is calculated in the same way but, instead of comparing TFP between periods s and t, the comparison is between the sub-county of interest A and the reference sub-county h:

$$MFP_{Ah} = \frac{QI_{Ah}}{XI_{Ah}} = \frac{Q(q_A)}{Q(q_h)} \div \frac{X(x_h)}{X(x_A)} \qquad 2$$

These indexes are ratios of the values of baskets of outputs and inputs from sub-counties *A* and *h* evaluated at the same set of reference prices. Note that the same prices are used to build the output and input indexes in all years for the country-level analysis. Similarly, for the cross-sectional analysis, the same prices are used to calculate output and input indexes for each sub-county. O'Donnell (2012) recommends the use of price vectors that are representative of the price vectors facing all production units that are to be compared.

The spatial analysis of the performance of agricultural production is conducted by clustering sub-counties into groups with the same or a similar agroecology and production environment, and comparing all sub-counties against the sub-county with the highest TFP value in each group. The performance of agriculture at the sub-county level is analyzed by measuring TFP efficiency (*TFPE*) for each sub-county. A measure of *TFPE* for sub-county *A* is defined as the ratio of *A*'s *TFP* and the *TFP* of the most productive sub-county in its group (*TFP**).

$$TFPE_A = TFP_A/TFP^* \quad (3)$$

The maximum value of *TFPE* is 1, only obtained if *A* is the sub-county with the highest *TFP* (*TFPA=TFP**). Using the definition of efficiency, unit *A*'s *TFP* can then be expressed as the product of the maximum observed *TFP* and *TFP* efficiency:

$$TFP_A = TFP^* \times TFPE_A \quad (4)$$

Differences between sub-counties and the most productive sub-county in each group in a particular year are the result of inefficiency in the use of inputs.[1] For this purpose, the *TFP* index in Equation 2 comparing sub-counties *A* and *h* can be exhaustively decomposed into different measures of efficiency. As O'Donnell (2012) shows, total *TFPE* of a production unit can be decomposed into different measures of efficiency by changing the reference efficient production unit used in the comparison. The intuition of this decomposition and the different efficiency measures follows. A more formal approach to this decomposition can be found in Appendix 6.1.

1 If comparisons of performance were conducted across sub-counties and years, then the maximum value of TFP in each group could potentially change between periods as the result of technical change. In this case, differences in TFP levels result from differences in efficiency and in the level of technology used. Inefficient sub-counties could increase TFP as the result of technical change (the shift of the technological frontier expressed as a change in the maximum value of TFP) and by increasing efficiency (reducing the difference between their own TFP and the maximum TFP). In this chapter, comparisons between sub-counties are conducted for the year 2019, so differences in TFP between sub-counties are explained by differences in efficiency in the use of the available technology in that particular year.

One of the possible decompositions of *TFPE* proposed by O'Donnell (2012) is:

$$TFPE_A = TE_A \times ME_A \times RSE_A \qquad 5$$

where *TFPE* is *TFP* efficiency of production unit (sub-county) *A*, as defined in Equation 3; *TE* is "pure" technical efficiency; *ME* is mix efficiency; and *RSE* is residual scale economy efficiency. TE_A is obtained by comparing production unit *A* with production units using the same combination of inputs to produce the same combination and amount of outputs as *A*. TE_A is referred to as "pure" technical efficiency because differences between *A* and the reference production unit are not related to differences in the output or input mix, nor to the scale of production, but only to the differences in management of the same combination of inputs and outputs. A value of $TE_A<1$ means that output produced by *A* can be obtained with the same input mix and a smaller quantity of aggregated input than the one used by *A*.

The measure of mix efficiency (ME_A) is obtained by comparing production unit *A* with units producing the same quantity of aggregated output as *A* but with different input mixes than the one used by *A*. A value of $ME_A<1$ indicates that it is feasible to produce the same quantity of aggregated output as A using less inputs by using a different input mix.

Notice that, to improve *ME*, unit *A* can change the mix of inputs to further reduce the aggregated level of inputs used to produce the original level and mix of aggregated outputs (Q_A). However, it is still possible to further increase *TFPE* if *A* is allowed to change the level of aggregated output. For example, a higher or a lower level of aggregated output with a different output mix than the one used by *A* could result in higher *TFP* than the one obtained with the level of output produced by *A*. This is captured by the last term in Equation 5, reflecting differences in scale between unit *A* and the unit with the highest *TFP*. However, *RSE* is not a measure of "pure" scale efficiency because it is calculated as a residual and reflects differences in scale and in the mix of inputs between *A* and the most efficient production unit.

O'Donnell (2012) also defines an alternative decomposition of *TFPE* that includes a measure of pure scale efficiency (*SE*) instead of *ME*. In this decomposition, the residual term is a residual mix efficiency (*RME*) and the *TE* is the same as in Equation 5:

$$TFPE_A = TE_A \times SE_A \times RME_A \qquad 6$$

To obtain *SE*, production unit *A* is compared with production units obtaining the same level of aggregated output but using different levels of aggregated input with the same input mix as *A*. If *SEA<1*, production unit *A* can increase productivity by proportionally increasing (or reducing) the total level of input. This is "pure scale" efficiency because no differences in the mix of inputs are involved. Unlike *ME*, *RME* is not a measure of "pure" mix efficiency because it is calculated as a residual and results from differences in the input mix and in the level of output between *A* and the most efficient production unit.

Data

Agricultural TFP at the country level was calculated using data on total output, materials, agricultural land, irrigated land, labor, animal stock, and machinery. Definitions and sources of the variables used are as follows:

Output: An agricultural output and input series in millions of current and constant Kenyan shillings for the period 2000–2020 was obtained from several issues of the KNBS Economic Survey, published annually. An index of output prices was built using Kenya's producer prices from FAO (2022) in current Kenyan shillings for the analyzed period. Prices of individual commodities were aggregated into a price index using average quantities of each commodity for the period.

Materials: Values of materials and input services used in production and their respective price indexes were also obtained from the KNBS Economic Survey. Materials include fertilizers, other chemicals, livestock drugs and medicines, fuel, power, spares and maintenance of machinery, bags, manufactured feeds, seeds, and others. Input services include artificial insemination, aerial spraying, tractor services, private veterinary services, and government veterinary inoculation services. Values were converted into quantities and aggregated into a single input (materials) that includes input services.

Land: Cultivated land for the year 2019 was obtained from the 2019 Kenya Population and Housing Census: Volume IV. Using 2019 as the reference, a timeseries for the period 2000–2020 was built using data on cropland from FAO (2022). The price of land used as weight to include land in the aggregate input index is from the Kenya Integrated Household Budget Survey (KIHBS) 2005/06. The KIHBS collected data on the cost of land in various regions and used this information to compute the median sale price of an acre of farmland and the cost of renting or leasing land parcels over the 12 months preceding the survey at the county level. The median price for renting/leasing an acre of land in rural areas for a year in Kenya was KSh 2,000, with the lowest value of KSh 333 in Mwingi and a high of KSh 9,600 in Isiolo.

Labor: Employment in Kenya is categorized into three sectors—namely, formal (modern), informal, and small-scale agriculture or subsistence farming and pastoralist activities. The KNBS Economic Survey keeps track of employment in the first two categories but no information was found on small-scale agriculture and pastoralists. To build the labor series, information on the number of households farming and on the total number of people employed by sub-county working from the 2019 Census was used. The total number of workers in each sub-county was allocated to agriculture proportionally to the number of households farming. Annual data for 2000–2020 were built using share of total employment in agriculture and total employment from the World Bank (2022). Labor prices used in the aggregate input index are wage earning per employee per year (in Kenyan shillings) in formal agriculture from the KNBS Economic Survey (several issues).

Animal stock: A similar approach to the one used for land was used to build the total animal stock series. Detailed information at the sub-county level on the number of heads of beef and dairy cattle, sheep, goats, camels, chickens, and pigs was obtained from the 2019 Census and used as a reference to build the time series for the period 2000–2020 using data on animal stock from FAO (2022). Average prices per ton of live weight in current Kenyan shillings from FAO (2022) were used to calculate the total value of the animal stock, and a real interest rate of 7 percent was used to determine the contribution of animal stock to total input.

Machinery: No data on mechanization are available from the Government of Kenya, including in the 2009 and 2019 Censuses. Information used on trends in tractor use was obtained from De Groote, Marangu, and Gitona (2020), who used four household surveys conducted between 1992 and 2012 to analyze the evolution of agricultural mechanization in Kenya. Number of tractors in World Bank (2023) was used to build the trend in machinery use after 2012. The flow of services from tractor use was calculated using the market price of an average tractor in Kenya to determine the total capital in tractors, and then a depreci ation rate of 10 percent and an interest rate of 7 percent to establish the input from machinery to production.

Two major sources of information were used to build the dataset for the regional cross-sectional analysis at the sub-county level. The 2019 Kenya Population and Housing Census was the source of cultivated area, labor, and animal stock at the sub-county level by species. In the case of the animal stock, the number of heads of each species was converted to animal units (AUs) based on animal weights from FAO (2022). The total cultivated area by sub-county is also drawn from the 2019 Census. The number of workers in agriculture was calculated using the proportion of farming households and the total number of working persons per sub-county.

The second source of information was the Spatial Production Allocation Model (SPAM) (IFPRI 2020). This accesses a variety of information sources to generate plausible, disaggregated estimates of crop distribution, which are useful for understanding production and land use patterns. SPAM uses a variety of inputs together with a cross-entropy approach to make plausible estimates of crop distribution, moving the data from coarser units such as countries and subnational provinces to finer units such as grid cells at 10×10 km resolution, to create a grid for 42 crops and 2 production systems within disaggregated units. Grid cell data on harvested area, production, the value of production, and yields were aggregated to the sub-county level. The portion of rainfed and irrigated crops produced using "high inputs" was also calculated at the sub-county level and used to allocate total crop materials calculated at the country level.

Production values from SPAM were used to calculate the output shares of each sub-county in the total value of crop output so that the aggregate of the total value of production of sub-counties added up to the value of crop production from KNBS. Yields were then recalculated with the adjusted production values. Animal stocks by sub-county from the 2019 Census were used to allocate livestock production across sub-counties. The proportion of exotic dairy, beef, and poultry AUs in each sub-county was used as an indicator of "high inputs" use to allocate livestock materials across sub-counties. Livestock output was allocated based on total number of AUs in each sub-county adjusted by the proportion of exotic AUs, assuming that the larger the proportion of exotic animals, the larger the production per animal in the sub-county.

Production and productivity trends

Policy changes since the year 2000 are reflected in the evolution and growth rates of agricultural output and its components, total input, and TFP (Figures 6.1 and 6.2). Figure 6.1 shows that the performance of agriculture between 2000 and 2007 was still poor after the country saw its lowest GDP growth in 2000, and policy changes of the 1990s and the SRA introduced under President Kibaki did not show an immediate impact. Output increased at an average annual rate of 1.2 percent, driven by input growth of 2.7 percent, resulting in negative growth of −1.5 percent (Figure 6.2). The poor performance of agriculture is explained in part by adverse climatic conditions during the period that negatively affected agricultural incomes and investments in rural areas (Balié et al. 2018). Mutsotso, Sichangi, and Makokha (2018) characterized the period between 1998 and 2001 as one of prolonged and moderate drought, and 2005 and 2006 as years of mild drought.

FIGURE 6.1 Evolution of agricultural production and output decomposition, 2000–2020

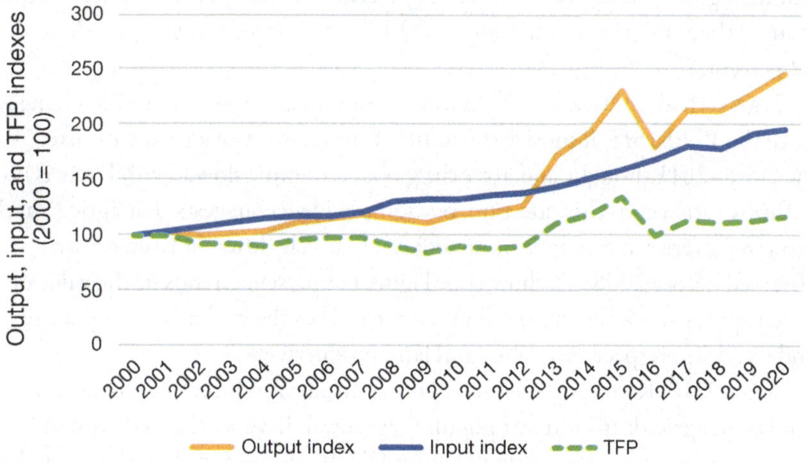

Source: Elaborated by authors based on KNBS (various years; 2006; and 2019), FAO (2022), and World Bank (2022).

FIGURE 6.2 Average growth rates of output, input, and TFP in different periods

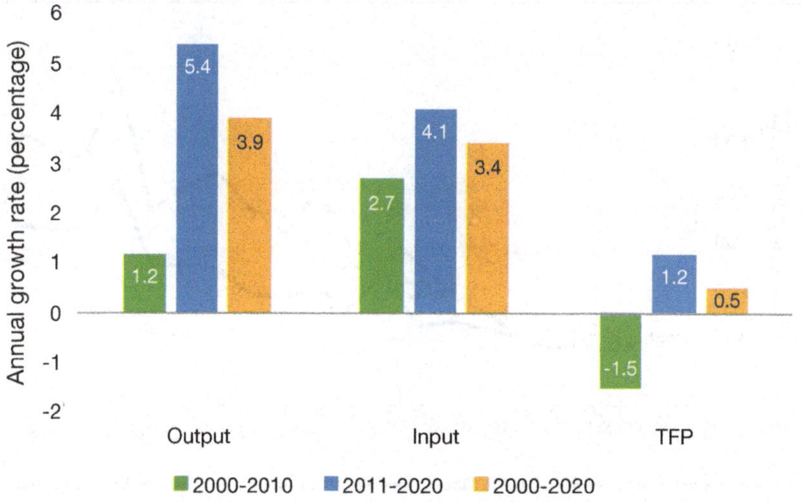

Source: Elaborated by authors based on KNBS (various years; 2006; and 2019), FAO (2022), and World Bank (2022).

The implementation of the ASDS after 2008 and the launch of the first MTIP 2008–2013 had to deal with the consequences of a prolonged severe drought between 2008 and 2011 (Mutsotso, Sichangi, and Makokha 2018), further delaying the expected positive effects of policy changes of previous years. Figure 6.2 shows that agricultural production started its recovery after 2010, and then grew at an average rate of 5.4 percent until 2020, with input growing at more than 4.0 percent annually and TFP reaching an average growth of 1.2 percent.

Is growth after 2012 an indication of a significant impact of policy changes since the 1990s, or is it mostly the result of the recovery of agriculture after the 2008–2011 drought and after the global economic slowdown? There is no definitive answer to this question, but some evidence suggests that agricultural growth patterns in recent years are not qualitatively different from growth observed before 2008. To show this, Figure 6.3 presents trends in the value of food exports and imports, and Figures 6.4 and 6.5 the evolution of input use and trends in, respectively, labor and land productivity.

Since 2010, Kenya has no longer been a surplus producer of food, as slow growth in agriculture and fast population growth have accelerated growth in imports of agricultural products (Figure 6.3). With the population projected to double between 2020 and 2050, food production and agricultural exports will need not only to sustain fast growth in the next 30 years but also to diversify

FIGURE 6.3 Evolution of the value and volume of agricultural imports and exports since 2000

Source: Authors using data from FAO (2022).

and to increase value added in the case of agricultural exports. As Wamalwa and Were (2021) stress, Kenya's major agricultural exports like tea, coffee, and animal products, with low income elasticities, yield lower and volatile foreign earnings compared with higher-value agricultural exports and manufactures. Wamalwa and Were argue that the prevalence of primary commodities, low productivity externalities, and stiff competition from cheap exports from developing and emerging economies have contributed to the decline in competitiveness of Kenya's merchandise exports, as made evident by shrinking net merchandise exports to Africa, the Common Market for Eastern and Southern Africa (COMESA), and the East African Community since 2011. Kenya's loss of export competitiveness applies not only to African countries but also to the rest of the world.

The pattern of agricultural intensification observed in the past 10 years shows that increased production and productivity have been the result of intensive use of labor and materials per hectare of cropland. Figure 6.4 shows that land productivity more than doubled between 2012 and 2020 as a result of

FIGURE 6.4 **Trends in land productivity and use of inputs per hectare of cultivated land, 2000–2020**

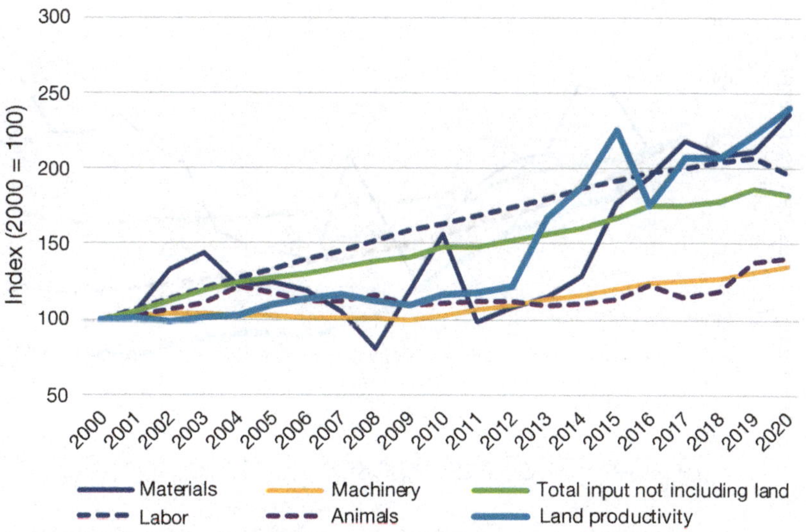

Source: Elaborated by authors based on KNBS (Various years; 2006; and 2019), FAO (2022), and World Bank (2022).

Note: Land productivity is measured as the ratio of agricultural output to land used in agriculture. Materials, labor, machinery, and animals refer to the ratio of each individual input to land used in agriculture. Total input not including land is calculated as the index of total input in the section on Approach and Data above, the difference being that land is not included.

sustained growth in the number of workers per hectare and a very large increase in the use of materials per hectare of cropland. Fertilizer was one of the major drivers of the observed growth in materials, explained in part by the introduction of a fertilizer subsidy program in 2006. As Jayne and colleagues (2018) point out, however, even though fertilizer subsidies can quickly raise national food production and grain yields at least in the short term, the overall production and welfare effects of subsidy programs tend to be smaller than expected. According to Jayne and colleagues, two characteristics of these programs consistently mitigate their intended effects. The first is that subsidy programs partially crowd out commercial fertilizer demand owing to difficulties associated with targeting and sale of inputs by program implementers. The second is that the crop yield response to fertilizer is lower than expected. Jayne and colleagues conclude that improved seed and fertilizer are not sufficient to achieve profitable and sustainable farming systems in most parts of Africa.

Increasing land productivity by relying mostly on increased fertilizer per worker and intensive use of labor per hectare, as observed in Kenya in recent years, has yet to yield benefits in terms of labor productivity. Despite the rapid

FIGURE 6.5 Trends in labor productivity and use of inputs per worker, 2000–2010

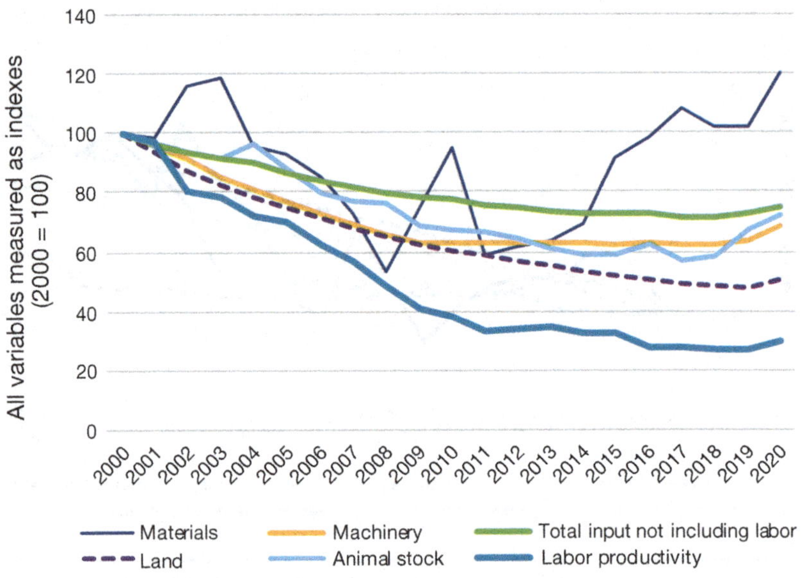

Source: Elaborated by authors based on KNBS (Various years; 2006; and 2019), FAO (2022), and World Bank (2022).

increase in the use of materials per worker, labor productivity in agriculture reached its lowest historical level after 2012 and has remained unchanged since that year (Figure 6.5). A study by De Groote, Marangu, and Gitonga (2020) on mechanization in Kenya using data for the period 1992–2012 shows persistently low levels of mechanization. According to this study, in 2012 most farm households still used only hand tools; from 1992 to 2012, the percentage of farmers with oxen increased from 17 to 33 percent but those with tractors decreased from 5 percent to 2 percent. No data on mechanization are available for recent years but De Groote, Marangu, and Gitonga conclude that mechanization in Kenya is likely to continue to depend on animal traction as it is not linked to farm size, complements labor, helps reduce fertilizer use, increases commercial maize production, and has room to grow—particularly in the highlands.

Policy changes in the past two decades have had some success in the medium run, improving the performance of agriculture and increasing the use of materials per hectare, land productivity, and TFP. However, in the long term, the key determinant of agricultural growth will be the growth in the stocks of productive capital and knowledge (agricultural research) affecting the productivity of land and labor in the production of agricultural goods. For this reason, the low public expenditure in agriculture that Birch (2018) notes—with spending on agricultural research falling steadily over the past decade—should be a concern regarding future growth in agriculture.

Figure 6.6 shows the evolution of public R&D investment in Kenya and the knowledge stock that agricultural research generates.[2] Note that the evolution of research spending is correlated with the policy changes discussed above. The peak of government spending in agricultural R&D occurred in 1994; it dropped sharply after that year in concert with policy changes favoring less government intervention, export promotion, and trade openness and remained stagnate after 2000. As a result, knowledge stock has not grown since 2012 and, because of the lagged effect of research (it takes several years for an investment to have an impact on productivity), even if Kenya increases R&D spending in the coming years it could take a decade or more for this to be reflected in faster productivity growth.

2 The knowledge stock can be thought of as the total knowledge accumulated as a result of past research. A measure of this knowledge in a particular year is obtained by adding up all R&D spent before that year (in this case going back 30 years). The contribution of investments on the path to the knowledge stock depends on how long ago the investment was made as knowledge generated in the past could become obsolete.

FIGURE 6.6 Evolution of public R&D investment in agriculture and of the knowledge stock from agricultural research, 1981–2019

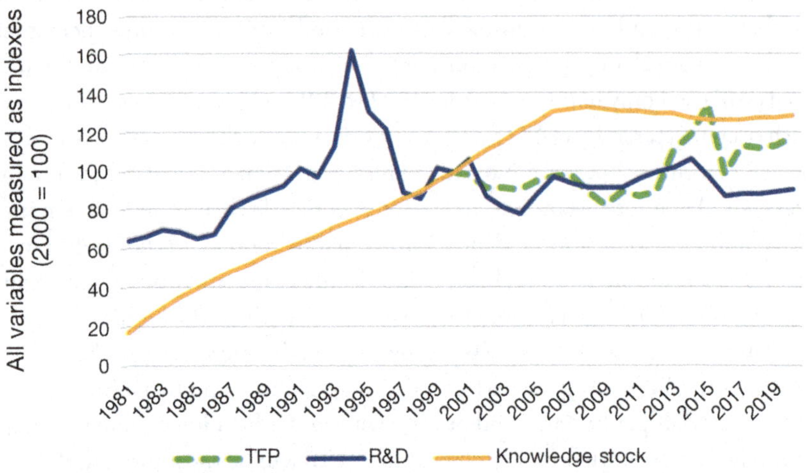

Source: Elaborated by authors based on ASTI (2021).

Regional analysis of productivity and efficiency

The starting point for the regional analysis is the division into seven distinct agroecological zones based on soil type and rainfall as used in the ASTGS (Kenya, Ministry of Agriculture 2018) as the basis for value chain and intervention selections. Table 6.1 gives a brief description of the seven zones and Appendix 6.2 presents a general characterization of the seven zones, including information on Mombasa–Nairobi and total country values for comparisons.

The most important agricultural zones in Kenya for their production, their contribution to total output, and the number of people working in the sector are the Western and Central Highlands zones. More than 30 percent of Kenya's agricultural output, almost half of the people working in agriculture, 40 percent of total materials used, and about 30 percent of cropland are in the Western zone, the agroecology with the highest agricultural potential. As Table 6.1 shows, this region produces cereals and root crops (23 percent of regional output), fruits and vegetables (30 percent), and coffee and tea (15 percent). Dairy production (8 percent) is the main livestock activity. Pulses, oil crops, sugarcane, and other livestock products are also produced. The Central Highlands zone

TABLE 6.1 Characterization of agroecological zones

	Northern ASALs	Central ASALs	Semiarid Uplands	Coast	Rift Valley	Central Highlands	Western
Agricultural potential							
High potential land (%)	0	10	8	10	39	52	55
Poor potential land (%)	99	73	37	65	41	12	7
Length of growing period	106	200	183	181	253	227	286
Zone's share in total input use (%)							
Materials	10	10	13	3	10	14	40
Animal units	29	18	8	2	11	6	25
Cropland	12	10	19	6	14	10	29
Labor in agriculture	4	6	12	5	10	16	47
Zone's share in total output (%)							
Agriculture	6	10	13	7	7	23	33
Crops	2	7	13	9	5	28	37
Livestock	21	20	11	3	15	10	20
Output composition in each zone (%)							
Cereals	2	5	16	5	20	9	16
Roots and tubers	0	3	4	4	15	20	7
Pulses	1	3	27	1	6	9	9
Oil crops and sugarcane	1	1	0	3	0	0	8
Coffee and tea	0	0	2	0	3	18	15
Fruits and vegetables	15	39	28	76	4	34	30

Source: Elaborated by authors based on Kenya, Ministry of Agriculture (2018).

Note: High and poor potential land refers to the proportion of land that is classified as of high and poor potential within each zone, respectively. Input and output shares refer to the share of each zone in Kenya's total use of inputs and in total output. Output composition refers to the output mix produced by each zone.

also has high potential; it produces 23 percent of Kenya's total agricultural output, mostly cash crops like coffee and tea, roots and tubers (Irish potatoes), fruits and vegetables (French beans, bananas, tomatoes), and livestock.

Showing lower agricultural potential, the Rift Valley and Coastal zones contribute 7 percent each to total output. In the case of Rift Valley, the share of livestock production in total output is 50 percent. This zone also produces mixed staples (35 percent of regional output) and a smaller share of cash crops and fruits and vegetables. The Coastal zone, on the other hand, produces fruits and vegetables (76 percent of regional output), mixed staples (10 percent), and livestock (10 percent).

Finally, the arid and semiarid lands (ASALs) are the zones with the lowest potential for agriculture. Here, farming households are mostly pastoralists, raising beef cattle, goats, sheep, and camels, with occasional maize cultivation on raised plateaus. The northern ASALs have poorer infrastructure and are more remote from major markets compared with the central ASALs (see Appendix 6.2).

To analyze agricultural performance at the regional level, this section focuses on the main producing zones: Western, Central Highlands, and Rift Valley. As Appendix 6.2 shows, these three zones together generate 52 percent of the country's GDP, and concentrate more than 60 percent of the population, including 1.85 million farming households representing more than 70 percent of total households in agriculture. These are also the zones with the highest population density (between 138 persons per km^2 in the Rift Valley and 478 persons per km^2 in the Western zone) and the best connectivity measured in terms of the proportion of the population with access to a cell phone (between 42 percent and 57 percent) and travel time to towns of 20,000 to 100,000 people. The production unit used for the analysis is the sub-county, a decentralized unit through which the 47 county governments provide functions and services. There are in total 345 sub-counties in Kenya. Using information on the length of the growing period, sub-counties of the three zones were classified into two distinct groups. The first includes sub-counties in and around the Central Highlands and the second centers on the Western zone.[3] For the purposes of the analysis, these two groups are referred to as the Central and Western zones, respectively.

To compare and analyze performance, sub-counties in the two zones were ranked separately using the calculated values of *TFPE*. The sub-counties in the top 30 percent of this ranking were defined as "efficient performers." Sub-counties within the bottom 30 percent of the *TFPE* ranking were grouped as "inefficient performers." The remaining 40 percent were "average performers." Comparisons of different measures of output, input, productivity, input and output combination, and environmental factors were made between the groups of efficient and inefficient sub-counties.

Figures 6.7 and 6.8 present the decomposition of output into total input, TFP, and different measures of efficiency. Figure 6.7 shows values of production, input, and productivity for an average household in the efficient group of

3 As information is at the sub-county level, county limits are not followed. This means, for example, that sub-counties in neighboring counties that are not in the Central Highlands, Rift Valley, or Western zone may be included if their length of growing period (LGP) is closer to the average LGP of these zones than to the average of their county's zone. LGP is the period (in days) during a year when precipitation exceeds half the potential evapotranspiration (FAO 1978).

FIGURE 6.7 **Aggregate output, input, and productivity of efficient sub-counties measured relative to inefficient sub-counties**

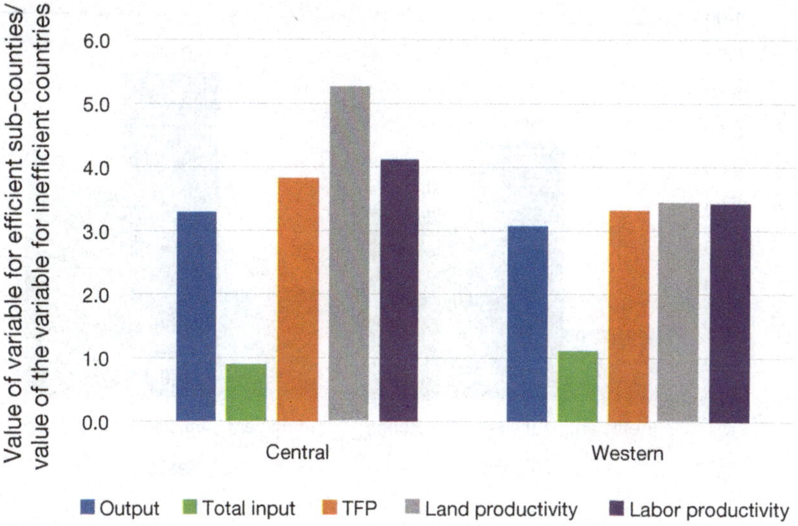

Source: Elaborated by author based on KNBS (Various years; 2006; 2019), FAO (2022), and IFPRI (2020).

Note: Values are calculated as the ratio value for efficient sub-counties divided by the value of the variable for inefficient sub-counties.

sub-counties presented as indexes, where 1 is the value of the indicator in the group of inefficient sub-counties. The average household in the efficient group in both the Central and the Western zones produces about three times more output than the average household in the inefficient group. Most of the differences in output between groups of performance in the two zones are explained by differences in *TFP*, given that differences in the total level of input used by efficient and inefficient sub-counties are small in all cases (all values are close to 1). Note that differences in land productivity in the Central zone are higher than differences in TFP and labor productivity, indicating that efficient sub-counties use more input per hectare (including labor) than inefficient sub-counties. This is not the case in the Western zone, where differences in land and labor between efficient and inefficient counties are the same as differences in TFP.

Figure 6.8 shows differences in *TFPE* for each zone. The *TFPE* of inefficient sub-counties in the Central zone is below 0.2, compared with almost 0.7 in efficient sub-counties. These differences are even larger in the Western zone, where *TFPE* of efficient sub-counties is 0.85 compared with 0.29 in inefficient sub-counties. To explain the observed differences in efficiency,

FIGURE 6.8 TFP efficiency and efficiency differences between efficient and inefficient sub-counties by agroecological zone

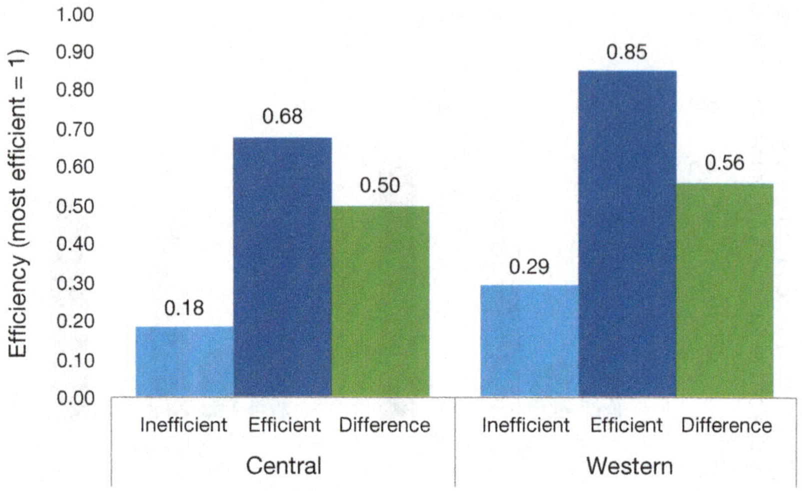

Source: Elaborated by author based on KNBS (Various years; 2006; 2019), FAO (2022), and IFPRI (2020).

Figure 6.9 displays two decompositions of *TFPE*. The first (Figure 6.9 panel A) decomposes *TFPE* into technical efficiency (*TE*), pure mix efficiency (*ME*), and residual scale efficiency (*RSE*). In Figure 6.9 panel B, *TFPE* is decomposed into the same technical efficiency as in Figure 6.9 panel A (*TE*) but the second component is now a measure of pure scale efficiency (*SE*) and the last term is a residual mix efficiency term (*RME*).

Results of the efficiency decomposition show small differences in *TE*, indicating that most sub-counties produce close to the technological frontier in their respective agroecological zones. The fact that differences in pure scale efficiency between efficient and inefficient groups are close to zero in the Central zone while showing a value of only 0.13 in the Western zone (Figure 6.9 panel B) seems to indicate that differences in scale efficiency have a small impact on overall *TFPE*. The large differences in *TFPE* observed between performance groups are primarily the result of differences in output composition and input mix and are significant only when associated with differences in the output and input mix.

To understand the effect of the use of an efficient input or output mix on overall efficiency in the two zones, Table 6.2 presents indicators of input use of efficient and inefficient sub-counties. It shows that the intensity in the use of materials per worker and hectare of cultivated land is a major factor determining

FIGURE 6.9 Decomposition of TFP efficiency into technical, mix, scale, and residual mix-scale efficiencies and differences between efficient and inefficient sub-counties by agroecological zone

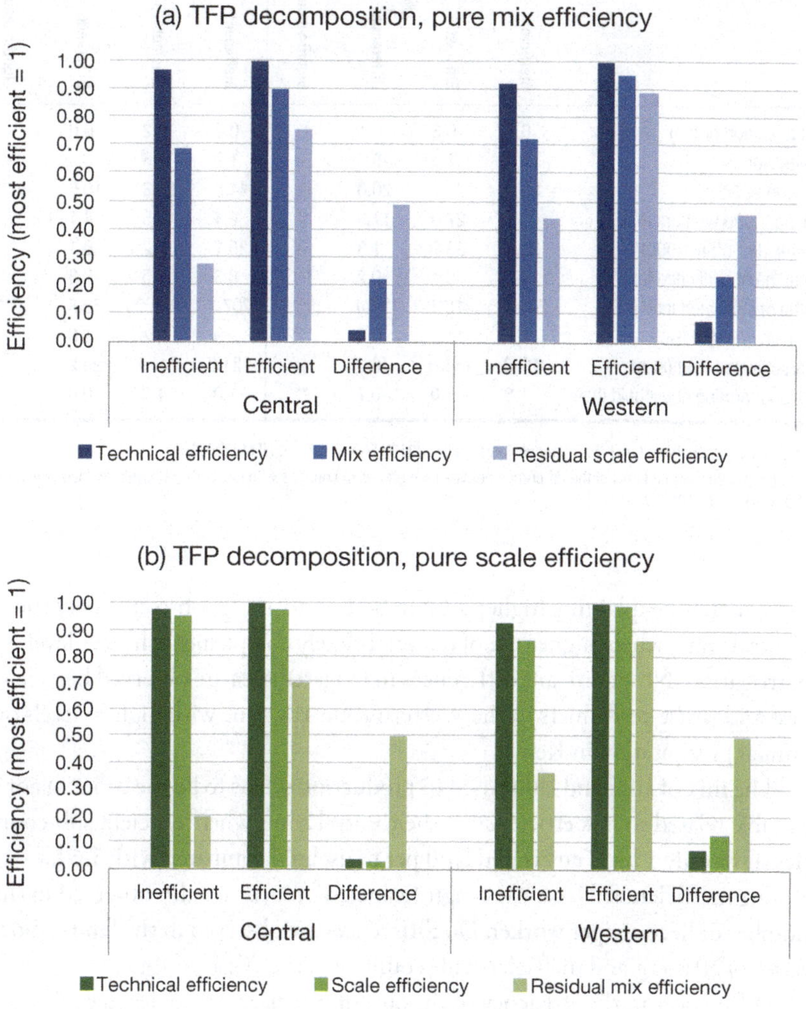

Source: Elaborated by author based on KNBS (Various years; 2006; 2019), FAO (2022), and IFPRI (2020).

performance in the two zones. Without major differences in the use of land and labor between efficient and inefficient sub-counties within each zone, a more intensive use of materials increases outputs more than proportionally across all inputs, resulting in higher TFP. A larger proportion of irrigated area is also

TABLE 6.2 Input use in agricultural production and differences between efficient and inefficient sub-counties by agroecological zone

	Central				Western			
	Inefficient	Efficient	Difference	Significance	Inefficient	Efficient	Difference	Significance
Cultivated land/worker (ha)	0.4	0.3	−0.1	**	0.2	0.2	0.0	–
Animal units/worker	2.2	1.6	−0.6	–	1.7	1.6	−0.1	–
Materials/worker (KSh)	56.6	76.7	20.0	*	44.8	61.3	16.5	***
Irrigated land/1,000 workers (ha)	9.1	27.0	17.9	**	2.4	10.5	8.1	–
Total input/worker (KSh 1,000)	32.6	33.9	1.3	–	25.1	31.2	6.1	**
Animal units/ha of cultivated land	6.5	6.4	−0.2	–	8.3	7.5	−0.8	–
Materials/ha of cultivated land (KSh)	200.7	318.1	117.4	**	207.1	285.9	78.8	***
Cultivated land irrigated (%)	2.7	11.2	8.6	***	1.9	3.7	1.8	–
Total input/cultivated land (KSh)	107.8	154.0	46.3	–	121.0	141.2	20.2	–
Cultivated land/farming household (ha)	1.6	1.0	−0.7	**	1.0	1.0	0.0	–

Source: Elaborated by authors based on KNBS (Various years; 2006; 2019), FAO (2022), and IFPRI (2020).
Note: * refers to the p-value from a t-test of the difference between the groups of counties with low and high levels of poverty and food insecurity: *** p<0.01, ** p<0.05, * p<0.1.

a major factor explaining higher TFP in best-performing sub-counties of the Central zone, where higher use of materials likely complements higher levels of irrigation. No significant differences in irrigated area are observed between best and worst performers in the Western zone, the zone with highest levels of annual precipitation in Kenya.

The mix of land and labor used in production seems to be the other major variable related to mix efficiency in the Central zone, where efficient sub-counties show only 1 ha of cultivated land per household compared with 1.6 ha among inefficient sub-counties. Significant differences are also observed in the number of hectares per worker. No differences are observed in the land–labor ratios of efficient and inefficient sub-counties in the Western zone.

Differences in the efficiency of the output mix used are analyzed by comparing land allocation with different crop activities by efficient and inefficient sub-counties. Figure 6.10 shows that efficient sub-counties in the Central zone allocate more area to cash crops and fruits and vegetables and less to staple crops than inefficient sub-counties. Inefficient sub-counties allocate on average almost 70 percent of cultivated area to staple crops and 5 percent to fruits and vegetables, and tea and coffee. Efficient sub-counties, on the other hand, allocate only 44 percent of their area to mixed staples and 17 percent to high-value crops.

FIGURE 6.10 Allocation of cultivated land in efficient and inefficient sub-counties by agroecology

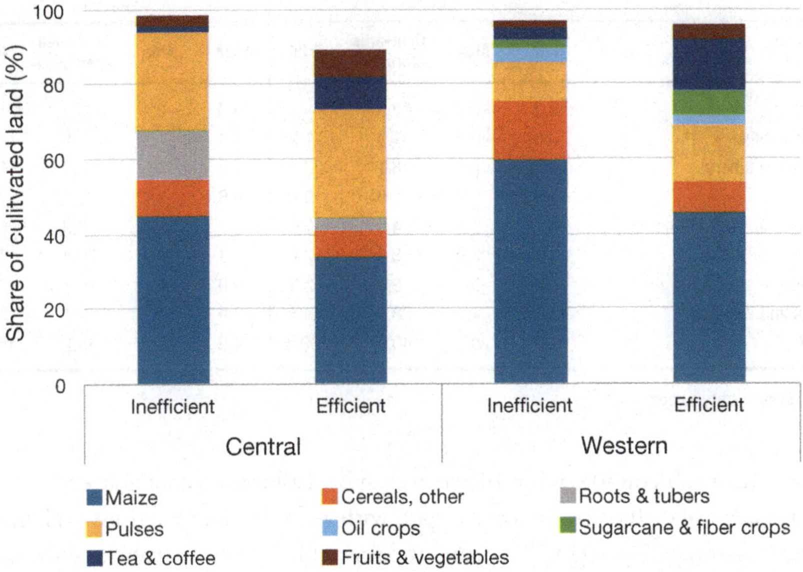

Source: Elaborated by authors based on KNBS (various years; 2006; 2019), FAO (2022), and IFPRI (2020).

In the Western zone, the importance of maize production is reflected in the 46 percent of the cultivated area allocated to this crop by the most efficient sub-counties compared with almost 60 percent in the group of inefficient sub-counties. Overall, inefficient sub-counties in the Western zone allocate 75 percent of their land to cereals and 10 percent to cash and high-value crops (tea; fruits and vegetables; sugarcane, fiber, and oil crops) compared with 55 and 28 percent allocated to cereals, and high-value and cash crops, respectively, by efficient sub-counties.

Differences in the area allocated to different crops and differences in the yields of these crops result in higher agricultural production by efficient sub-counties, explaining most of the differences in mix efficiency observed in the *TFPE* decomposition. Table 6.3 shows that output differences between efficient and inefficient sub-counties in the Central zone are mostly the result of higher yields (52 percent) and the interaction effect between yields and differences in land allocation (40 percent). Most of this difference is explained by fruits and vegetables, roots and tubers, and tea. In the Western zone, yields explain 45 percent of the difference in crop production between efficient and

TABLE 6.3 Contribution of yield and area of different crops to differences in total crop output between efficient and inefficient sub-counties by agroecology (%)

	Central				Western			
	Yield	Area	Yield–area interaction	Total	Yield	Area	Yield–area interaction	Total
Maize	8.2	−0.7	−2.0	–	14.4	−2.4	−3.3	8.7
Cereals, other	3.0	−0.4	−1.3	1.2	4.4	−1.2	−1.5	1.7
Roots and tubers	12.5	2.8	8.1	23.4	5.5	1.5	3.2	10.2
Pulses	7.9	0.2	0.9	9.0	4.9	1.4	2.4	8.7
Sugarcane, oil, and fiber crops	4.7	−0.9	−4.4	−0.6	3.7	6.6	3.9	14.1
Tea	1.5	2.9	8.3	12.7	7.7	4.2	23.4	35.3
Coffee	0.8	0.3	2.2	3.3	0.0	0.1	0.1	0.3
Fruits and vegetables	13.3	4.0	28.1	45.4	3.9	10.1	7.1	21.1
Total	51.8	8.2	40.0	100.0	44.5	20.3	35.2	100.0

Source: Elaborated by authors.

inefficient sub-counties while differences in land allocation contribute to 20 percent of differences in total output, with yield–land interactions explaining the remaining 35 percent. Tea, fruits and vegetables, roots and tubers, and cash crops (oil and fiber crops and sugarcane) explain most production differences between efficient and inefficient sub-counties in this zone.

Finally, Table 6.4 compares the value of different indicators related to the production environment of efficient and inefficient sub-counties. In the case of the Central zone, efficient sub-counties show a more diversified economy, where agriculture represents less than 50 percent of country GDP and where greater access to mobile phones is expected to be related to better infrastructure and a more diversified economy. Shorter travel time to towns of 20,000 is used as an indicator of better access to local markets. No differences are observed in access to the internet, travel time to larger markets, access to credit, employment in agriculture, or education.

There are no major differences between inefficient and efficient sub-counties in the Western zone. Efficient sub-counties appear to have better access to larger markets (towns of 250,000) but the differences are small and significant only at the 10 percent level. Also, small differences are observed in the percentage of the population with access to credit, where access is higher in inefficient sub-counties. A possible explanation for this is that households use access to credit to diversify income into nonagricultural activities, which could explain the importance of maize and staple crops and the less commercial orientation of production systems in inefficient sub-counties.

TABLE 6.4 Differences in the value of indicators of production environment in efficient and inefficient sub-counties by agroecology

	Central				Western			
	Inefficient	Efficient	Difference	Significance	Inefficient	Efficient	Difference	Significance
Length of growing period	242	230	–12	–	298	298	0.6	–
Access to mobile phone (%)	46.8	55.2	8.4	**	42.7	42.3	–0.4	–
Access to internet (%)	21.1	23.4	2.3	–	16.1	16.4	0.3	–
Travel time to towns of 20,000	1.5	1.1	–0.5	**	0.7	0.8	0.1	–
Travel time to towns of 250,000	2.1	2.2	0.1	–	3.8	3.4	–0.4	*
Access to credit (% of population)	28.5	27.6	–0.9	–	42.2	34.2	–8.0	*
Share of agriculture in county GDP (%)	62.6	48.9	–13.7	***	48.5	51.6	3.1	–
Employment in agriculture (%)	69.2	73.8	4.6	–	84.0	84.3	0.3	–
Population with primary education (%)	40.7	43.5	2.7	–	44.6	44.9	0.3	–
Population with secondary education (%)	19.7	21.7	2.0	–	18.9	18.5	–0.4	–

Source: Elaborated by authors based on KNBS (various years; 2006; 2019), FAO (2022), and IFPRI (2020).

Note: * refers to the p-value from a t-test of the difference between the groups of counties with low and high levels of poverty and food insecurity: *** p<0.01, ** p<0.05, * p<0.1.

In sum, efficient sub-counties in the Central and Western zones allocate a larger share of harvested area to export and high-value crops and a lower share of this area to maize and other staple crops. Efficient sub-counties in the two zones show a positive correlation between higher intensity in the use of materials per worker and hectares of cultivated land, with efficiency in land allocation. Efficient sub-counties in the Central zone also show a larger proportion of irrigated area than do inefficient sub-counties. As a result of these differences, the average household in the efficient group in both the Central and Western zones produces about three times more output than the average household in the inefficient group, while *TFP*, land, and labor productivity in efficient sub-counties are at least three times bigger than in inefficient sub-counties. Output differences between efficient and inefficient sub-counties in the Central zone are mostly the result of higher yields of fruits and vegetables, roots and tubers, and maize and the interaction between higher yields and differences in land allocation. In the Western zone, output differences result from larger areas allocated to fruits and vegetables, tea, and other cash crops, and higher yields in almost all crops but especially in maize and tea. There is some evidence showing that efficient sub-counties in the Central zone are part of counties with a more diversified economy, better infrastructure, and better access to local markets.

No major differences were observed in the economic environment of sub-counties in the Western zone; however, better access of the population to credit and longer travel time to larger markets in inefficient sub-counties could be related to the less commercial orientation of production systems in these sub-counties.

Two interpretations of results, and policy implications

Policy changes in the past two decades have had some success in the medium run, improving the performance of Kenya's agriculture as shown in Figures 6.1 and 6.2. After poor performance between 2000 and 2007, explained in part by adverse climatic conditions, followed by a prolonged severe drought between 2008 and 2011, agricultural production started its recovery in 2010, with farmers increasing the use of materials per hectare, land productivity, and TFP. These improvements did not yield benefits in terms of labor productivity, however (Figure 6.3). With the population projected to double by 2050 with respect to its level in 2020, food production and agricultural exports will need to sustain fast growth and increase their value added in the next 30 years to avoid further deterioration of the country's trade food balance.

The results of the regional analysis give some insights on the challenges Kenya faces in sustaining fast agricultural growth in the future. They show that the more productive sub-counties in and around the Central and Western zones are more market-oriented; use more inputs per worker and hectare; allocate more land to fruits and vegetables, tea and coffee, and other cash crops; and obtain much higher yields from these activities and from maize and other staples than do inefficient sub-counties. Efficient sub-counties in the Central zone also show a smaller average farm area, a higher proportion of irrigated area, and a smaller cultivated area than do inefficient sub-counties.

At least two possible interpretations of these results can be made, with very different policy implications. The first interpretation could use results showing that sub-counties with smaller farm areas are more efficient and productive than sub-counties with more land available as a confirmation of the inverse relationship between farm size and productivity frequently observed in Africa. Assuming higher efficiency of smallholders, policies under this interpretation would target smaller, efficient farms (those cultivating 1 ha or less as shown in Table 6.2) with extensive interventions in markets and support services—extension, subsidies to inputs, investment in irrigation schemes, roads—to favor the use of improved crop varieties and the intensive use of fertilizer and labor per hectare (Collier and Dercon 2014). In the case of Kenya, the goal of these policies would be to close the productivity gap between efficient and inefficient sub-counties. This interpretation has been supported by donors and adopted by several African governments, which have spent large shares of their budgets on subsidies for technology adoption. However, little evidence of widespread progress in technology adoption and productivity growth has been observed so far (Wise 2020), particularly in Kenya.

A second interpretation of the results is based on significant evidence from the most recent literature. Under this interpretation, the inverse relationship between area and productivity is flawed because of the aggregation used and the small variability between sub-counties in average farm size. Muyanga and Jayne (2019) tested the inverse relationship hypothesis on a much wider range of farm sizes than in most studies and found that farms between 20 and 70 ha were substantially more productive than farms under 5 ha. Results like these are a major challenge to the hypothesis of efficient smallholders as agents of change and of the transformation of agriculture by facilitating farmers' access to new technologies. Rather than an indication of efficiency, the small size of farms in Kenya could be part of a poverty trap whereby frictions in land markets prevent households from exiting agriculture to the extent that would be efficient (see, for example, Chen 2017; Gottlieb and Grobov 2019; discussion in Gollin 2021). Spatial frictions that alter crop choice, affect input use, and prevent local specialization could also be behind the differences observed between the Central and Western zones, as the Western zone lags the Central zone in income per capita, specialization in cash crops, and diversification of its economy, with a higher share of agriculture in county GDP and labor markets.

In this context, an explanation of low agricultural productivity in Africa that has received considerable attention in recent years is that there are simply too many farmers (Gollin 2021). With more than half of the adults in Kenya earning their living from agriculture, it seems plausible that not all are equally capable. Gollin (2021) argues that, with well-functioning markets, the least effective farmers would be expected to move out of agriculture into other occupations, either selling or renting their land to farmers who are more skillful. That this is not happening could imply that Kenya may have institutional frictions or rigidities that prevent unproductive farmers from exiting the market (see discussion and references in Gollin 2021). The outcome is aggregate inefficiency resulting from the misallocation of labor, capital, and managerial effort that creates a consequential drag on aggregate productivity. Studies by Chen (2017), Restuccia and Santaeulalia-Llopis (2017), and Gollin and Udry (2021) point in this same direction, finding that misallocation could not only lead to substantial losses in aggregate efficiency and sizable reductions in overall agricultural output but also prevent efficient allocation of resources across sectors as a result of frictions in land markets. Muraoka, Jin, and Jayne (2015) show that, in Kenya, land rental markets are the most important means available to land-constrained rural households to access additional land for cultivation even when rental markets perform below their potential. Muraoka and colleagues conclude that there appears to be untapped potential for land rental markets to play a positive role in promoting agricultural production and food security in rural Kenya in the future.

The implications for investment and technical change of the "land market frictions" interpretation is that land constraints mean there is little incentive to invest the careful and timely attention to agronomic management needed for the efficient use of fertilizer. Instead, most vulnerable households sell labor and land and diversify income to off-farm sources to minimize risks. Smallholders following this strategy are unlikely to intensify their production, which limits their ability to contribute to their own, or national, food self-sufficiency. There are also few incentives for intensification where land is more abundant. Particularly if animal traction is available, households are predisposed to increase their production by cultivating more land, through extensification, rather than through increasing yields, which has happened in Kenya (as De Groote, Marangu, and Gitonga 2020 have shown). Better-endowed households, on the other hand, have tended to diversify and acquire land that has enabled them to adapt to and benefit from the major changes observed in external drivers. This could have happened more often in efficient sub-counties in the Central zone but no information is available on the distribution of land by farm size.

Under the "land market frictions" interpretation of the results, the implementation of policies and institutions that support a better allocation of resources in the agriculture sector is critical to allow farms to grow and become economically *and* agronomically viable while keeping the urban population well fed. The efficient reallocation of factors and the increase in productivity that could result from these policies would encourage productive farmers to invest and grow by using modern inputs (mechanization, chemical seeds, and other intermediate inputs) and by investing in better farm management practices, triggering a profound process of structural transformation (Adamopoulos and Restuccia 2014; Chen et al. 2021a).

Further analysis is needed to define specific policies that could deliver the transformation of Kenya's agriculture under the second interpretation of the productivity results obtained here, but some policy areas seem to be relevant given the evidence so far. First, to facilitate the reallocation of land and other factors of production to more productive uses, Chen, Restuccia, and

Santaeulalia-Llopis (2021b) point to the need for well-defined property rights over land and the development of well-functioning land and complementary markets. In this regard, Chen and colleagues present the examples of a property rights reform associated with digitization of land titles in Pakistan (Beg 2022), a rural land contracting law in China that formalizes leasing rights (Chari et al. 2021), and a land certification reform in Ethiopia (Chen et al. 2021a). In all these cases, the reform induces more land rental activity that improves resource allocation and productivity in the agriculture sector. Reallocation of inputs also results in reductions in agricultural income inequality and poverty because the poorest agricultural households happen to be the least productive and hence benefit the most from secure property rights and the rental income associated with an efficient allocation (Chen et al. 2021b).

A second set of policies includes measures to support the emerging commercial farmers who are expected to foster labor productivity growth, wage labor income, and integration in retail value chains toward domestic and export markets. Also relevant are policies and investments to shape the development of the industrial structure of the food and agriculture sector and the links at different levels of the value chain (Neven et al. 2009 for horticulture and supermarkets in Kenya; Lowder, Skoet, and Raney 2016).

A third relevant policy area includes policies and institutions to facilitate the movement of labor out of agriculture and into nonagricultural sectors in this process. This further requires the creation of rural and urban jobs in industry and services and other forms of social protection in the form of social safety nets.

Finally, it seems to be particularly relevant for Kenya to obtain a better understanding of the drivers of agricultural growth, and the role of the domestic market, agricultural exports, and the agro-processing industry. As stated by Gollin (2021), one lesson that emerges from the literature is that there is substantial heterogeneity in agriculture's role in structural transformation, across both geographical contexts and time, and depending critically on the nature of demand for agricultural goods. Could agricultural exports of tea, coffee, fruits, and vegetables—the most dynamic activities in agriculture—drive growth and transformation of the Kenya's economy? Is there a role for the shrinking manufacturing sector to play? Could a more productive agriculture sector trigger growth and transformation of the food industry sector? Chapter 2 of this book provides some initial estimates but more research is needed to answer these questions, to better understand the reasons behind slow productivity growth, and to identify the most appropriate policies and pathways for the transformation of Kenya's agriculture.

Appendix 6.1 Productivity index

The Lowe index used to calculate TFP satisfies several properties of index numbers needed to make sound comparisons of production units across time and/or space. The property of most interest for the analysis in this study is the transitivity property. An index is transitive if the index number that directly compares the TFP of a sub-county (or a year) with the TFP of a reference sub-county (or reference year) is identical to the index number computed when the comparison is made through an intermediate sub-county (or year). This means that, if an TFP index is transitive, *then* if $TFP_A > TFP_B$ and $TFP_B > TFP_C$ then $TFP_A > TFP_C$. Indexes like the Fisher, Tornqvist, and Malmquist do not satisfy the transitivity property, so it is possible to have $MPF_A < TFP_C$ *in the comparison even when* $TFP_A > TFP_B$ and $TFP_B > TFP_C$.

Explaining differences in TFP between sub-counties in Kenya involves estimating measures of differences in technology and in efficiency. For this purpose, the TFP index in Equation 2 in the main text, comparing sub-counties A and h, can be exhaustively decomposed into different measures of efficiency. In what follows, the efficiency decomposition of TFP is presented using Figure A6.1 as the reference.[4] The figure depicts values of aggregated input (horizontal axis) and aggregated output (vertical axis) while point *A* represents a production unit (for example, a sub-county) producing total output Q_A using aggregated input X_A. The definition of TFP as the ratio of an aggregated output and an aggregated input can be used here to determine TFP_A as the slope of the line going from the origin to point *A*. The steeper the slope of the line *OA*, the higher the TFP_A.

Production unit *A* produces under an available technology defined as all the points to the right and below the full curve passing through point *E*, the production frontier. This frontier envelopes all aggregate input–output feasible combinations that can be produced with the available technology. No production unit can produce to the left or above this frontier. The point of maximum feasible TFP, given the observed technology, is determined by the line with the biggest slope passing through the origin and a technically feasible point. In Figure A6.1, the maximum value of TFP is given by $TFP^* = Q^*/X^* = slope\ OE$. $TFPE_A$ is then defined as the ratio of *A*'s *TFP* and the maximum *TFP* given the available technology:

$$MFPE_A = \frac{MFP_A}{MFP^*} = \frac{Q_A}{X_A} \times \frac{X^*}{Q^*} = \frac{slope\ OA}{slope\ OE} \qquad \text{A1}$$

4 To simplify notation, in what follows $Q_A = Q_{hA} = Q(q_A)/Q(q_h)$; $X_A = X_{hA} = X(x_A)/X(x_h)$, $MFP_A = MFP_{hA}$.

To proceed with the full decomposition of differences in *TFPE*, it is necessary to identify the reference points with which unit *A* is to be compared to obtain measures of pure technical efficiency (TE_A), mix efficiency (ME_A), and the residual scale efficiency (RSE_A). The first of these points is production unit B in Figure A6.1. *B* produces the same quantity and combination of outputs as unit *A* but uses less input ($X_B<X_A$), which means it is more technically efficient than A. The "pure" technical efficiency component of $TFPE_A$ is then calculated as:

$$MFPE_A = \frac{MFP_A}{MFP^*} = \frac{Q_A}{X_A} \times \frac{X^*}{Q^*} = \frac{slope\ OA}{slope\ OE} \qquad \text{A2}$$

FIGURE A6.1 Measures of efficiency in aggregate output–input space

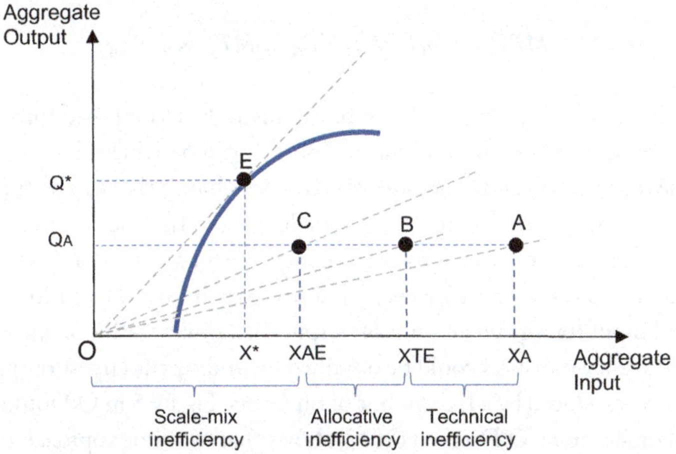

Source: Adapted from O'Donnell (2012).

Comparing production unit A against a unit producing the same output but using a different combination of inputs, as is the case with production unit C in Figure A6.1, the result is a measure of allocation efficiency of inputs (XAE). In other words, XAE in Figure A6.1 is the minimum value of aggregated input needed to produce Q_A quantities of output. *A* can produce Q_A using quantities XAE of aggregated input only if it employs the mix of inputs used by C. The "pure" mix efficiency of *A* in Figure A6.1 is calculated as:

$$ME_A = \frac{Q_A}{XTE} \times \frac{XAE}{Q_A} = \frac{XAE}{XTE} = \frac{slope\ OB}{slope\ OC} \qquad \text{A3}$$

Finally, the difference between *A*'s *TFP* and the maximum *TFP* after accounting for pure technical change and mix efficiency is the residual scale efficiency, which requires changes in X_A together with changes in the mix of outputs and inputs:

$$RSE = \frac{Q_A}{XAE} \times \frac{X^*}{Q^*} = \frac{slope\ OC}{slope\ OE} \qquad \text{A4}$$

Putting all together, *Equation 4* can now be expressed in terms of the full *TFP* decomposition:

$$MFP_A = MFP^* \times \frac{XTE}{X_A} \times \frac{XAE}{XTE} \times \frac{Q_A}{XAE} \times \frac{X^*}{Q^*} = MFP^* \times \frac{Q_A}{X_A} \times \frac{X^*}{Q^*} = MFP^* \times MFPE_A \qquad \text{A5}$$

Or, equivalently:

$$MFP_A = MFP^* \times TE_A \times ME_A \times RSE_A \qquad \text{A6}$$

Equation A6 shows that TFP can be exhaustively decomposed into a measure of technology (the maximum TFP that can be achieved with the available technology) and a measure of efficiency change that can be further decomposed into technical, mix, and a residual scale efficiency terms.

The efficiency measures derived from Figure A6.1 are what the literature refers to as "*input-oriented efficiency*," as it involves finding the minimum potential input for a given amount of output. In a similar fashion, output-oriented efficiency measures could be obtained by finding the largest output set that can be produced by a fix amount of input (see Figure 5 in O'Donnell 2012). Furthermore, O'Donnell (2012) shows that the same approach can be used to decompose the efficiency change component into any number of meaningful output- or input-oriented measures. For example, a TFP index can be decomposed into measures of technology, pure technical efficiency, pure scale efficiency, and a residual mix efficiency, instead of the pure mix efficiency and residual scale efficiency derived here. For the analysis in this study, a geometric mean of the output- and an input-oriented versions of Equation A6 is used in the reported results.

Appendix 6.2 Kenya's agroecologies

TABLE A6.1 Kenyan agroecological zones as defined by the Agricultural Sector Transformation and Growth Strategy

Western	Moderate to deep red soils of medium–high fertility and two seasons of medium rains Mixed staples and cash crops including maize, French beans, sugar cane, groundnuts, sweet potatoes, Irish potatoes, dairy, poultry, and a variety of fish species
Rift Valley	Mixed shallow/low with deep/highly fertile soils and one season of moderate rainfall Mixed staples, cash crops, and livestock, including maize, wheat, sorghum, Irish potatoes, honey, goats, sheep, chicken, and dairy cattle
Central Highlands	Deep red highly fertile soils and two seasons of high rainfall Cash crops, including coffee, tea, Irish potatoes, French beans, bananas, tomatoes, and other staples, including dairy, cattle, and poultry
Semiarid uplands	Red, acidic, low to moderately fertile soils, with one season of low rains Dryland crops such as sorghum and pigeon peas, and beef cattle
Northern ASALs	Sandy, saline, shallow, low-fertility soil with one season of rain at best Livestock pastoralism, including camels, goats, and sheep, with occasional maize cultivation on raised plateaus
Central ASALs	Saline, low-fertility soils, with one season of rain at best Livestock pastoralism, including beef, cattle, goats, and sheep, with occasional maize cultivation on raised plateaus
Coast	Mix of sandy, deep, low, and highly fertile soil and two seasons of moderate rainfall Mixed staples and cash crops, including maize, sorghum, millet, cashew nuts, mangoes, marine fish, crustaceans and mollusks, and poultry

Source: Kenya, Ministry of Agriculture (2018).

TABLE A6.2 **Characterization of Kenya's agroecologies and major urban centers**

	North ASALS	Central ASALS	Semi-arid Uplands	Coast	Rift Valley	Central Highlands	Western	Mombasa-Nairobi	KENYA
Structural									
GCP per capita (2019 KSh)	37,103	50,962	80,510	56,035	91,502	97,963	71,339	234,274	93,471
Share in GCP	3	3	8	3	8	17	28	30	100
Agriculture (% of GCP)	45	29	20	26	41	35	36	0	24
Agricultural labor (% employment)	50	60	61	58	61	54	76	0	71
Infrastructure									
Persons with cell phone (%)	23	38	49	38	42	57	42	68	47
Travel time to towns of 20K people	6.3	3.4	2.3	2.8	1.9	1.0	0.9	–	1.9
Travel time to towns of 100K people	10.5	4.8	3.3	3.4	2.7	2.3	1.8	–	3.1
Travel time to towns of 500K people	14.2	6.8	3.6	3.4	4.5	2.8	5.4	–	5.5
Education									
Primary education (%)	13	26	42	41	38	41	44	30	38
Secondary education (%)	4	11	20	11	18	24	19	28	19
Population									
Rural population (%)	77	78	73	77	69	68	86	0	70
Total population ('000)	3,035	2,900	4,373	2,465	3,987	7,421	16,781	5,605	46,568
Persons per square km.	17	31	110	81	138	368	478	5,871	82
Share of total population	7	6	9	5	9	16	36	12	100
Number of households ('000)	496	627	1,233	510	1,000	2,281	3,919	1,885	12,144
Farming households (%)	18	23	23	23	23	22	29	1	21
Household size	6.1	4.6	3.5	4.8	4.0	3.3	4.3	3.0	3.8

Source: Elaborated by authors based on Kenya, Ministry of Agriculture (2018) and KNBS (various years).

Note: GCP is gross county product.

References

Adamopoulos, T., and D. Restuccia. 2014. "The Size Distribution of Farms and International Productivity Differences." *American Economic Review* 104 (6): 1667–1697.

Alila, P.O., and R. Atieno. 2006. "Agricultural Policy in Kenya: Issues and Processes." Paper prepared for the Future Agricultures Consortium Workshop, Institute of Development Studies, March 20–22.

ASTI (Agricultural Science and Technology Indicators). 2021. ASTI database. Accessed February, 2021. https://asti.cgiar.org/

Balié, J., P. Boulanger, A. Mainar Causapé, H. Dudu, E. Ferrari, and L. Battaglia. 2018. *Policy Options to Support the Agriculture Sector Growth and Transformation Strategy in Kenya: A CGE Analysis.* Luxembourg: European Commission, Joint Research Centre.

Beg, S. 2022. "Digitization and Development: Property Rights Security, and Land and Labor Markets." *Journal of the European Economic Association* 20 (1): 395–429.

Beintema, N., L. Mose, T. Kibet, R. Emongor, F. Murithi, I. Kimani, V. Ndungu, and P. Mwangi. 2018. *Kenya: Agricultural R&D Indicators Factsheet Update.* Washington, DC: International Food Policy Research Institute (IFPRI); Nairobi: Kenya Agricultural and Livestock Research Organization (KALRO).

Birch, I. 2018. *Agricultural Productivity in Kenya: Barriers and Opportunities.* K4D Helpdesk Report. Brighton, UK: Institute of Development Studies.

Chari, A., E.M. Liu, S.Y. Wang, and Y. Wang. 2021. "Property Rights, Land Misallocation, and Agricultural Efficiency in China." *Review of Economic Studies* 88 (4): 1831–1862.

Chen, C. 2017. "Untitled Land, Occupational Choice, and Agricultural Productivity." *American Economic Journal: Macroeconomics* 9 (4): 91–121.

Chen, C., D. Restuccia, and R. Santaeulalia-Llopis. 2021a. "The Effects of Land Markets on Resource Allocation and Agricultural Productivity." *Review of Economic Dynamics* 45: 41–54.

Chen, C., D. Restuccia, and R.Santaeulalia-Llopis. 2021b. "Land Misallocation and Productivity." NBER Working Paper 23128. National Bureau of Economic Research, Cambridge, MA.

Collier, P., and S. Dercon. 2014. "African Agriculture in 50 Years: Smallholders in a Rapidly Changing World?" *World Development* 63: 92–101.

De Groote, H. 2022. "Intensification of the Maize Systems in Kenya with Improved Varieties and Fertilizer: What Happened to the Maize Green Revolution?" Unpublished paper.

De Groote, H., C. Marangu, and Z.M. Gitonga. 2020. "Evolution of Agricultural Mechanization in Kenya." In *An Evolving Paradigm of Agricultural Mechanization Development: How Much Can Africa Learn from Asia?* Eds. X. Diao, H. Takeshima. and X. Zhang, 401–422. Washington, DC: IFPRI.

De Groote, H., and L.B. Omondi. 2021. "Adoption of Improved Maize Varieties, Varietal Turn-over and Their Effect on Yield and Food Security-Evidence from Four Household Surveys Over 20 Years in Kenya." Paper presented at the 31st Virtual International Conference of Agricultural Economists (ICAE), August 17–31.

FAO (Food and Agriculture Organization of the United Nations). 1978. *Report on the Agro-ecological Zones Project, Vol. 1: Results for Africa.* World Soil Resources Report 48/1. Rome: FAO.

FAO. 2022. FAOSTAT database. Accessed February 8, 2022. http://faostat.fao.org.

Gollin, D. 2021. *Agricultural Productivity and Structural Transformation: Evidence and Questions for African Development.* STEG Pathfinding Paper. London: Foreign, Commonwealth and Development Office (FCDO).

Gollin, D., and C. Udry. 2021. "Heterogeneity, Measurement Error, and Misallocation: Evidence from African Agriculture." *Journal of Political Economy* 129 (1): 1–80.

Gottlieb, C., and J. Grobov. 2019. "Communal Land and Agricultural Productivity." *Journal of Development Economic* 138: 135–152.

IFPRI (International Food Policy Research Institute). 2020. "Spatially-Disaggregated Crop Production Statistics Data in Africa South of the Sahara for 2017." Harvard Dataverse, V3.

Jayne, T. S., N.M. Mason, W.J. Burke, and J. Ariga. 2018. "Taking Stock of Africa's Second-Generation Agricultural Input Subsidy Programs." *Food Policy* 75: 1–14.

Jena, P.R., H. De Groote, B.P. Nayak, A. Hittmeyer. 2021. "Evolution of Fertiliser Use and Its Impact on Maize Productivity in Kenya: Evidence from Multiple Surveys." *Food Security* 31: 95–111.

Kenya, Ministry of Agriculture. 2010. *Agricultural Sector Development Strategy 2010–2020.* Nairobi.

Kenya, Ministry of Agriculture. 2018. *Agricultural Transformation and Growth Strategy: Towards Sustainable Agricultural Transformation and Food Security in Kenya.* Nairobi.

Kenya, Ministry of State Planning. 2007. *Kenya Vision 2030: The Popular Version.* Nairobi.

Kibaara, B., J. Ariga, J. Olwande and T.S. Jayne. 2008. Trends in Kenyan Agricultural Productivity: 1997–2007. WPS 31/2008, Tegemeo Institute of Agricultural Policy and Development, Nairobi.

Kimenyi, M., F.M. Mwega, and N. Ndung'u. 2015. "The African lions: Kenya Country Case Study." UNU-WIDER Working Paper No. 2015/134. UNU-WIDER, Helsinki.

Kirori, G.N. 2003. "Rural Development Policies in Kenya: A Critical Analysis." Master's thesis, Department of Economics, University of Nairobi.

KNBS (Kenya National Bureau of Statistics). Various years. *Economic Survey.* Nairobi.

KNBS. 2019. *2019 Kenya Population and Housing Census, Volume IV: Distribution of Population by Socio-economic Characteristics.* Nairobi.

KNBS. 2006. *Kenya Integrated Household Budget Survey 2005/06*. Basic report. Nairobi.

KNBS. 2020. *Children, Youth, Women, Men & the Elderly: From National to County Level.* Comprehensive Poverty Report. Nairobi.

Lowder, S.K, J. Skoet, and T. Raney. 2016. "The Number, Size, and Distribution of Farms, Smallholder Farms, and Family Farms Worldwide." *World Development* 87: 16–29.

Muraoka, R., S. Jin, and T.S. Jayne. 2015. "Land Access, Land Rental and Food Security: Evidence from Kenya." Paper presented at the International Conference of Agricultural Economists (ICAE), Milan, Italy, August 8–14.

Mutsotso, R.B., A.W. Sichangi and G.O. Makokha. 2018. "Spatio-Temporal Drought Characterization in Kenya from 1987 to 2016." *Advances in Remote Sensing* 7: 125–143.

Muyanga, M., and T.S. Jayne. 2019. "Revisiting the Farm Size-Productivity Relationship Based on a Relatively Wide Range of Farm Sizes: Evidence from Kenya." *American Journal of Agricultural Economics* 101 (4): 1140–1163.

Mwega, F.M., and N.S. Ndung'u. 2008. "Explaining African Economic Growth Performance: The Case of Kenya." In *The Political Economy of Economic Growth in Africa, 1960–2000. Volume 2: Country Case Studies*, eds. B.J. Ndulu, S.A. O'Connell, J.-P. Azam, R.H. Bates, A.K. Fosu, J.W. Gunning, and D. Njinkeu. Cambridge: Cambridge University Press.

Neven, D., M.M. Odera; T. Reardon, and H. Wang. 2009. "Kenyan Supermarkets, Emerging Middle-Class Horticultural Farmers, and Employment Impacts on the Rural Poor." *World Development* 37 (11): 1802–1811

Nyoro, J., and T.S. Jayne. 1999. "Trends in Regional Agricultural Productivity in Kenya." Paper presented at the Kenya Agricultural Monitoring and Policy Analysis Project Conference Proceedings, July 1999. Tegemeo Institute, Nairobi.

O'Donnell, C.J. 2012. "Nonparametric Estimates of the Components of Productivity and Profitability Change in U.S. Agriculture." *American Journal of Agricultural Economics* 94 (4): 873–890.

Onjala, J.O. 2002. *Total Factor Productivity in Kenya: The Links with Trade Policy*. AERC Research Paper 118. Nairobi : African Economic Research Consortium.

Poulton, C., and K. Kanyinga. 2014. "The Politics of Revitalising Agriculture in Kenya." *Development Policy Review* 2 (32): 151–172.

Restuccia, D., and R. Santaeulalia-Llopis. 2017. "Land Misallocation and Productivity." Technical Report. Cambridge, MA: National Bureau of Economic Research.

Wamalwa, P.S., and M. Were. 2021. "Is it Export- or Import-Led Growth? The Case of Kenya." *Journal of African Trade* 8 (1): 33–50.

Wanyama, R., M.W.K. Mathenge, and Z.S. Mbaka. 2016. Agricultural Information Sources and Their Effect on Farm Productivity in Kenya." Tegemeo Institute of Agricultural Policy and Development, WPS 62/2016.

Wise, T.A. 2020. "Failing Africa's Farmers: An Impact Assessment of the Alliance for a Green Revolution in Africa." Working Paper 20-01. Global Development and Environment Institute, Medford, MA.

World Bank. 2019. *Unbundling the Slack in Private Sector Investment: Transforming Agriculture Sector Productivity and Linkages to Poverty Reduction*. Kenya Economic Update, Edition No. 19. Washington, DC.

World Bank. 2022. World Development Indicators. Accessed February 2022. https://data.worldbank.org/country/KE

World Bank. 2023. The World Bank in Kenya. Last updated March 31, 2023. https://www.worldbank.org/en/country/kenya/overview

Chapter 7

INTENSIFICATION OF MAIZE-BASED FARMING: WHAT HAPPENED TO THE MAIZE GREEN REVOLUTION?

Hugo De Groote

Maize is the major food crop in eastern and southern Africa, including Kenya. Maize-based farming systems make up the largest proportion of agricultural land, and maize is central to the food system, in both rural and urban areas. Because of its importance, maize has received wide attention from the government, including in policy and research. As a result, Kenya has been at the forefront of the "maize green revolution" in Africa (Hassan and Karanja 1997; Hassan, Njoroge et al. 1998c). It was one of the first countries in Africa (with South Africa and Zimbabwe) to develop its own maize hybrids and combine them with fertilizer in demonstration trials, demonstrations, and dissemination (Hassan and Karanja 1997). In both South Africa and Zimbabwe, the settler communities continued to dominate commercial maize production (Eicher 1995), but in Kenya, indigenous African farmers took over most of the maize production right after independence.

The new government supported research and dissemination of improved maize varieties, initially focusing on large-scale farmers in the highlands but quickly expanding to small-scale farmers and to other maize areas (Harrison 1970; Gerhart 1975). The government also continued the colonial policies of controlled input and output markets, with pan-territorial prices (Wangia, Wangia, and De Groote 2004). These efforts were successful and resulted in high adoption rates for the new improved maize varieties and increased yields and production, especially in the highlands (Lynam and Hassan 1998b).

Unfortunately, since the 1980s, maize yields have stagnated (De Groote et al. 2005). A range of policies, projects, and other measures have been put in place to boost yields and production, to keep up with population growth, but with little success. The government, facing inefficient markets and high intervention costs, and under pressure from the donor community, liberalized agricultural input and output markets in the 1990s (Wangia, Wangia, and De Groote 2004). The resulting privatization of the maize sector brought new actors from the private sector, including maize seed companies, which increased the number of maize varieties and agrodealers to distribute them, with accompanying fertilizers and pesticides (Ariga and Jayne 2009). However, privatization had limited impact on maize yields (De Groote and Omondi 2023).

Other efforts were undertaken, including the United States Agency for International Development (USAID)-funded Maize Development Program (Smale et al. 2012), a second-generation input subsidy program (Jayne et al. 2018), and continuing control of maize prices by the National Cereals and Produce Board (NCPB) in combination with import and export control by the government. At the same time, maize research continued at high levels with, among other initiatives, the Stress Tolerant Maize for Africa, the Improved Maize for African Soils, and the Accelerated Genetic Gains projects, mostly focusing on the breeding of improved maize varieties.

Despite all these different efforts, statistics show an increase in maize production but not in yields, as the next section describes in detail. As production cannot keep up with population growth, substantial maize imports are required most years. This chapter first provides some historical background, needed to understand the Kenyan maize sector and the food system it dominates. Next, we study the trends in the adoption of improved maize technologies, in particular improved maize varieties and fertilizer. Finally, we analyze the effect of research, policies, and other factors on maize yields and production, identify weaknesses in current approaches, and try to formulate alternatives.

A good understanding of historic trends in the intensification of maize production systems is essential to understanding the food system in Kenya. Maize is grown by almost all rural households, which constitute two-thirds of the population. Moreover, maize constitutes the major food staple of the rural as well as the urban population, especially the poor. Therefore, it plays a central role in the food system in Kenya, as in the rest of eastern and southern Africa.

Background on maize in Kenya

Geography, climate, and demography

Maize production in Kenya is closely linked to geography and climate. Kenya's climate is largely a result of its position on the equator and its geography, dominated by the Great Rift Valley (GRV), which was shaped by tectonic drift and related volcanic activity (see map in Figure 7.1, panel A). The rims on the sides of the GRV were pushed up to form the highlands, with a valley in

FIGURE 7.1 Geography, climate, and population in Kenya

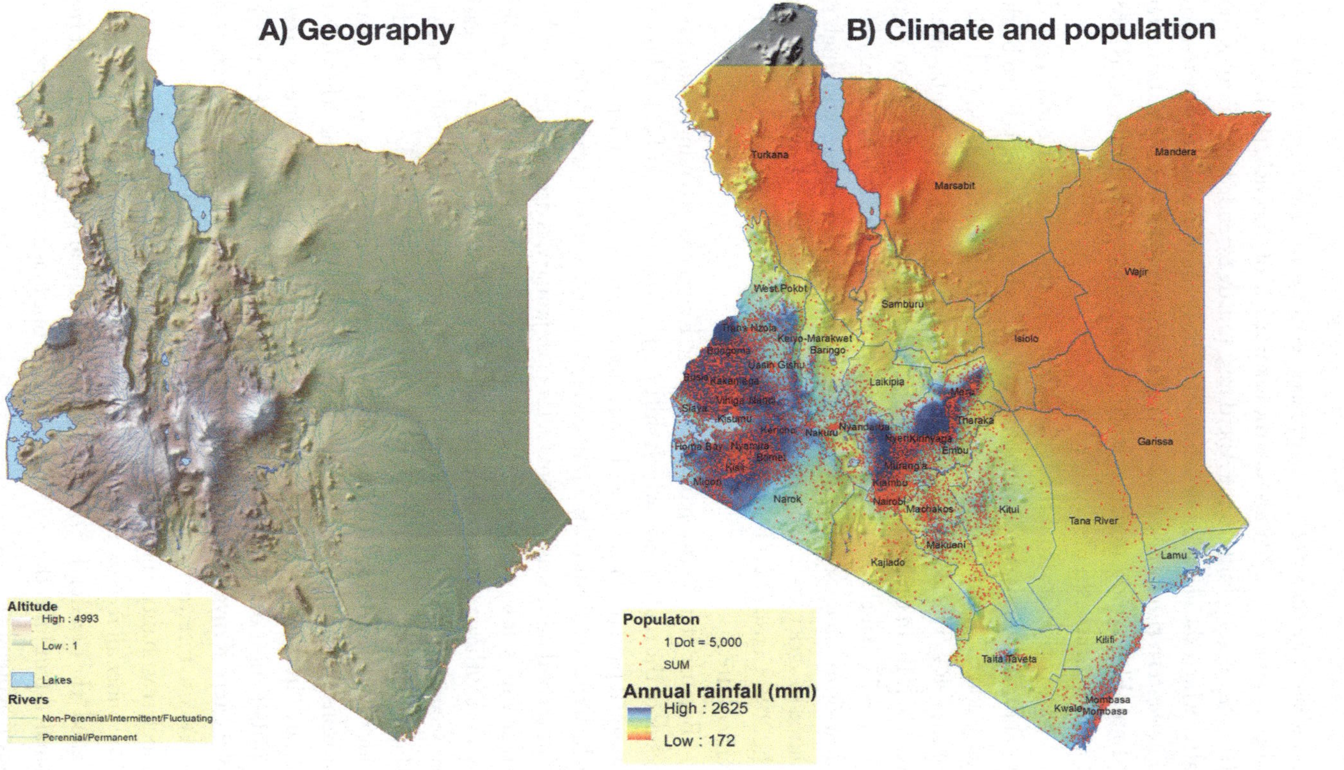

Source: https://datasets.wri.org/dataset/kenya-digital-elevation-model-dem-250m-resolution; www.worldclim.org/; www.worldpop.org/

between. The center of the valley has a relatively high elevation and is therefore also classified as highlands. On the eastern side of the GRV, the highlands range from the Ngong Hills to the Aberdares, and are extended by Mount Kenya, after which the elevation descends into the lower plains all the way to the coast. The western rim forms the Mau Mau escarpment, and the highlands extend to Mount Elgon in the north; westward, the altitude gradually descends into the Lake Victoria basin.

Rains originate from the Indian Ocean in the east and are transported by the trade winds, and orographic rain falls where moist air is lifted by the geography, in particular on the eastern side of Mount Kenya and on both rims of the GRV. The lee sides, however, form rain shadows, in particular after the coastal hills, in the lower parts of the GRV and on the shores of Lake Victoria (Figure 7.1, panel B). Most of the country does not have enough rainfall for agriculture, so most of the population is concentrated in areas with high rainfall, suited to agriculture. Most of the population is still rural—69 percent as per the 2019 Census (KNBS 2019), and their livelihood largely depends on rainfed agriculture.

Because of its location on the equator, Kenya has two rainy seasons as a result of the trade winds, which shift from northeast to southeast following a seasonal pattern. Most agricultural areas are in the northern hemisphere, to which the southeast trade winds bring the major rains between March and June. The second rainy season, from October to January, is driven by the northeast trade winds; it is usually called the minor season, although it is more important in the southeast, except for at the coast. The seasonality of rainfall can easily be seen in the patterns of average monthly rainfall in the different regions (Figure 7.2). Central Kenya, close to the equator and located on the windward side of Mount Kenya, receives ample rains, in two distinct but roughly equal seasons, and almost no rain in between. Eastern Kenya, on the other hand, is located in the low and mid-altitudes, far from the mountains, and receives little rainfall; most of what falls is in the second season (October to December). Most of western Kenya, located between Lake Victoria and the western ridges of the GRV, also receives ample rain, concentrated in the main season but without much seasonality: there is hardly a dry period and the months from June to November all receive similar amounts of rain. In the North Rift Valley, often called "the breadbasket of Kenya," there is hardly any seasonality: apart from a small dip in June, the region basically has one long rainy season, from March until September.

FIGURE 7.2 Seasonal rainfall patterns in the different regions of Kenya

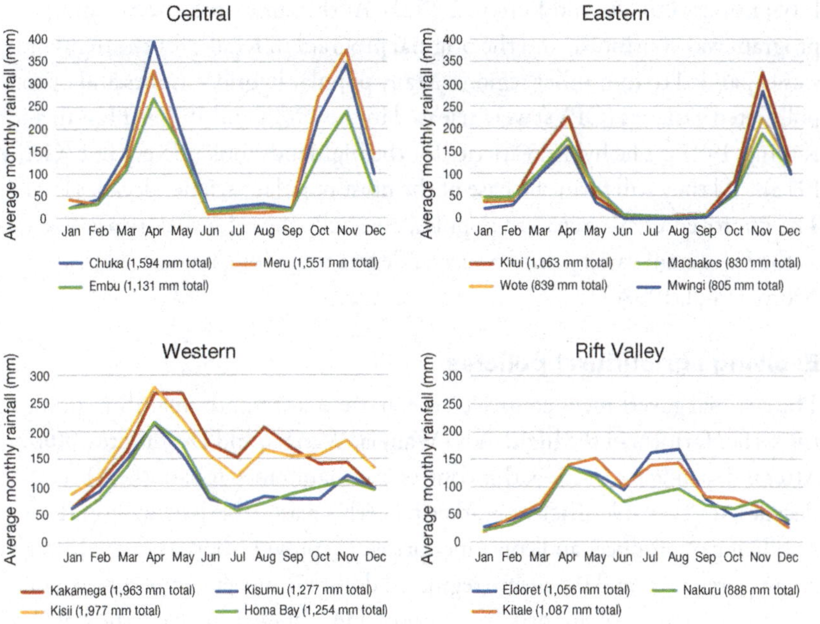

Source: https://en.climate-data.org/africa/kenya

Maize arrival and distribution

Maize is a relatively recent introduction in Africa. The first maize varieties that came to East Africa were Caribbean flints, most likely introduced by Portuguese traders at the end of the 15th century (Miracle 1965; McCann 2001). These varieties were mostly used as a garden crop, and spread only slowly. Maize became the major food staple relatively recently, with the establishment of European settlers, who imported white dent varieties from South Africa that had originated in North America (Smale and Jayne 2003). Maize gradually replaced local cereals such as millet and sorghum because of its wide adaptability, better yields, and good resistance to birds and other pests (McCann 2005). At the beginning of the 20th century, maize covered only an estimated 20 percent of Kenya's crop area; by 1960, this area had risen to 44 percent (Hassan and Karanja 1997), boosted by the interest of the European settlers in the crop, which was grown on large-scale settler farms. The colonial government strongly supported these farmers, and a successful maize breeding program was started in 1955.

After independence in 1963, land ownership was transferred to African farmers (often members of the new political elite), under the "willing buyer, willing seller" principle, with programs providing credit to the buyers (Jones 1965; Boone, Lukalo, and Joireman 2021). At the same time, the breeding program was continued, and the original program in Kitale, for the highlands, was expanded to four other regions. Many popular hybrid varieties and open pollinated varieties (OPVs) were released in the 1960s and 1970s (Hassan and Karanja 1997). The hybrid varieties for the highlands took off very fast (Gerhart 1975), and they still form the base of the most popular varieties decades later. The improved OPVs were also popular, especially those for the dry areas, such as the Katumani variety, and to a lesser degree those for the coast (Hassan, Njoroge et al. 1998c).

Evolving agricultural policies

The colonial government controlled the maize sector tightly, mainly to protect the settler farmers in the highlands (Wangia, Wangia, and De Groote 2004). After independence in 1963, this control was maintained for another three decades, to (1) enable efficient marketing with a reasonable balancing and stabilizing of producer and consumer prices, (2) provide food security through strategic reserves, and (3) ensure regulated domestic movement of maize with strict management of imports and exports (DAI 1989). Implicitly, the state had a "social contract" with the majority of citizens to ensure the supply of maize, which had now become the basic food staple, at low and stable prices (Jayne et al. 1999).

State control of the maize sector (and the wider agricultural economy) included research, seed production, extension and dissemination of seed and fertilizer, and marketing of the grain, all through different specialized parastatals. New maize varieties were developed by public research institutes, initially within the East African Community and from 1977 incorporated in the Kenya Agriculture Research Institute (KARI), with different programs for the different agroecological zones (AEZs) (Karanja 1996). The Kenya Seed Company (KSC), initially a private seed company founded by European settler farmers, was asked to produce the seed of the first improved maize varieties (IMVs), and over time parastatals acquired a majority of shares in the company. A seed unit within KARI managed quality control, and the Ministry of Agriculture was charged with the extension of new technologies. Distribution of seed, and the accompanying fertilizers and other inputs, was carried out through the retail network of the Kenyan Farmers Association (Hassan, Karanja, and Mulamula 1998a).

From the late 1960s to the early 1980s, this system was remarkably effective at producing and disseminating many popular varieties, which, in combination with improved agricultural practices such as use of chemical fertilizers, led to rapidly increasing maize yield and production (Gerhart 1975; Karanja 1996). While government intervention is essential in areas without established markets, such as hybrid seed in the 1960s, in the long run it tends to be costly and inefficient, as it does not take advantage of the flexibility and efficiency of private sector initiatives (Gisselquist and Grether 2000). Liberalization of input and output markets in developing countries has therefore been advocated since the 1980s as part of structural adjustment programs (Gisselquist, Nash, and Pray 2002), including in Kenya. The liberalization has aimed at three major changes: (1) lifting of the controls in the maize market, (2) restructuring the NCPB, and (3) market development (Lewa and Hubbard 1995). The European Community sponsored the Cereal Sector Reform Programme (CSRP) from 1988, with the main goals of decontrolling the maize grain market and restructuring the NCPB (Jayne, Robert, and James 2008). The private sector was supported through the USAID-sponsored Kenya Market Development Project, specifically targeting maize and beans (DAI 1989).

The evolution of the parastatal-based system to an open input and output market system with increased private sector participation has been a long and hard process because of the existence of many entrenched interests (Lewa and Hubbard 1995). The reforms were only firmly established in 1995, after an extended period of uncertainty. The seed sector was opened up to include the private sector, with national, regional, and multinational companies competing. New national and regional companies benefited from publicly owned varieties, while multinationals could tap into new markets with their own germplasm (Tripp and Rohrbach 2001). The Kenya Plant Health Inspectorate Service was established as a regulatory agency in 2012. However, efforts to privatize KSC, as with national seed companies in other African countries, failed and its parastatal status was re-established.

In the fertilizer market, the monopoly of distribution of the Kenyan Farmers Association was cancelled in 1985, fertilizer prices were decontrolled in 1990, and foreign exchange and import licensing controls were removed in 1993, leading to a fully liberalized fertilizer market (Mwangi, Lynam, and Hassan 1998). Fertilizer subsidies were reintroduced in 2007–2009, albeit with limited results (Mather and Jayne 2018). Deregulation in the seed and fertilizer market led to a proliferation of small agrodealers throughout the country. In the grain markets, finally, control of prices and movement was removed, and NCPB was established as a buyer of last resort, to buy maize at floor prices (Jayne, Robert, and James 2008).

The December 2002 election created a window of opportunity for issue- and performance-based politics in Kenya (Poulton and Kanyinga 2014).

In March 2004, the new government set out its Strategy for Revitalizing Agriculture (SRA). However, the government coalition began to unravel soon after attaining power, and the return to ethnically based patronage politics undermined the SRA's chances of success. After the 2008 elections, the new government launched the 2010 Agricultural Sector Development Strategy (Kenya, Government of Kenya 2010), with clearly assigned roles for the private and public sectors—divesting from all state corporations production, processing, and marketing that could be better done by the private sector while reforming and streamlining agricultural services such as in research, extension, training, and regulatory institutions, to make them more effective and efficient. Nevertheless, the KSC's status as parastatal was confirmed.

Historic trends in maize production

Maize production trends are shown in Food and Agriculture Organization of the United Nations (FAO) statistics of the past 60 years, which illustrate the importance of maize in the farming system and the initial success and later stagnation of the sector (Figure 7.3). Currently, it covers 37 percent of crop area. The success of the maize sector is shown in the substantial yield increase, from 1.2 metric tons/ha[1] to 1.5 tons/ha in the 1990s. In the 1960s and 1970s, there was also a substantial increase in maize production per person, from 110 kg/person to 180 kg/person (FAOSTAT 2022). However, since the 1990s, yields have stagnated at around 1.5 tons/ha. Maize production continues to increase but this can be attributed largely to an increase in area, which has doubled from about 1 million ha in the 1960s to about 2 million ha now. Further, the increase in production has not kept up with population growth: per capita production decreased to 90 kg/person in the 1990s and further to 70 kg/person in the 2000, where it has remained.

The 2019 Census counted a population of 45.6 million, still growing at a rate of 2.2 percent (down from 2.9 percent since the prior census) (KNBS 2019). The rural population remains very large, at 69 percent of the total, and most rural people live in maize production areas. Agriculture remains the mainstay of the economy and an important source of employment and income for Kenyans (see Chapter 2).

Currently, of a 5.7 million ha cropping area, more than half is covered by two crops: maize (37 percent) and beans (21 percent) (FAOSTAT 2022). Other

1 Tons refers to metric tons throughout this volume.

FIGURE 7.3 Maize production trends, 1961–2021

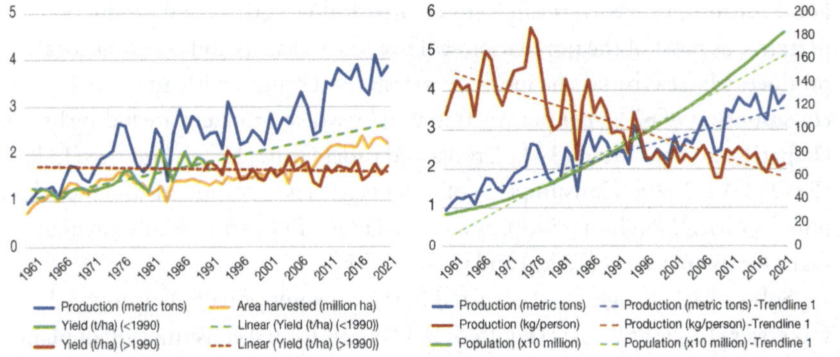

Source: FAOSTAT (2022).

cereals are much less important, including sorghum (3 percent) and millet and wheat (2 percent each). Similarly, area shares of tuber crops are particularly small, with the most popular tubers being potatoes (3 percent), cassava (2 percent), and sweet potatoes (1 percent). Maize to date retains its prime role in the economy, not only as food but also as a source of cash. Maize is not a major cash crop, except for a decreasing number of large-scale farmers, but many smallholder farmers do sell maize when there is a surplus or when there are immediate cash needs in the household (Jena et al. 2020). The Kenyan government estimates that 12.5 percent of maize production is used for animal feed, especially for poultry and dairy cattle.

Several cash crops are grown in the various AEZs: tea is most common in the upper highlands (at 1,500–2,700 meters) whereas coffee is grown in the lower highlands (at 1,400–2,000 meters) and sugarcane in the low and mid-altitudes. Several other cash crops have been introduced in the past, including cotton, sisal, pyrethrum, and others, but all have largely been abandoned because of marketing difficulties after the collapse of the specific marketing parastatals. The remaining cash crops still have marketing issues, and interested farmers typically need a certain minimum size to make their work worthwhile. Smallholder farmers therefore tend to sell vegetables and fruits, products that enjoy a steady market and entail no minimum size requirement. Finally, in many maize areas, dairy cattle have become an important source of cash. In the mid- and high altitudes, even small farms can maintain a dairy cow, with zero grazing, based on Napier grass (Odero-Waitituh 2017).

Shift in consumption patterns

Food consumption patterns in Kenya closely follow agricultural production patterns, as most of the population still live in rural areas and consume local produce. Maize is by far the most important food source, with an annual consumption of 62 kg per person, followed by wheat products (including bread, chapatis, and *mandazis*) (34 kg), roots and tuber crops (21 kg), and rice (13 kg) (FAOSTAT 2020). Consumption of other cereals is very limited, and includes pulses (4.5 kg), sorghum (3 kg), and millet (1 kg). The other main sources of plant foods are tubers (22 kg) and pulses (4.5 kg).

Substantial changes have occurred in the consumption of plant-sourced food over the past 60 years (Figure 7.4). Overall, annual consumption of plant-sourced food has declined from more than 180 kg in the 1960s to about 140 kg now, although this reduction took place mostly in the late 1970s and 1980s, with the situation remaining stable since then. Another major change over time has been the reduction of maize consumption, from more than 100 kg per

FIGURE 7.4 Trends in the consumption of food from plant sources over time

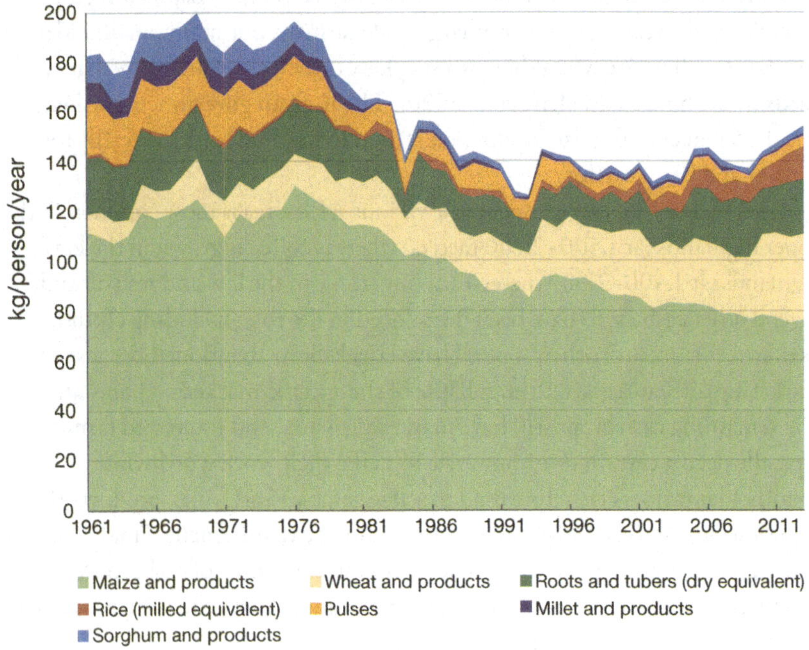

Source: FAOSTAT (2020).

person per year from the 1960s through the mid-1980s, but down to less than 70 kg now. In addition, consumption of coarse grains has fallen dramatically to the point of becoming negligible: sorghum has decreased from 10 kg to 3 kg/person/year, and millet from 9 kg to 1 kg/person/year. Similarly, consumption of pulses has decreased by three-quarters, from 21 kg per person to 4.5 kg. Finally, the decrease in the consumption of locally produced maize and coarse grains has been compensated by an uptake in the consumption of the new cereals: wheat (from 8 kg to 34 kg) and rice (from 1 to 13 kg), most of which is imported.

Household surveys on the use and impact of improved agricultural inputs

Farm household surveys, the Kenya Maize Data Base, and the establishment of maize production zones

Farm household surveys are important to gauge farmers' appreciation and adoption of new technologies, and the suitability of technologies in the pipeline. Over the years, many such surveys have been conducted in Kenya, including four nationally representative surveys by the International Maize and Wheat Improvement Centre (CIMMYT) and KARI, now the Kenya Agriculture and Livestock Research Organization (KALRO), as well as four panel surveys by the Tegemeo Institute. This chapter uses the data from the former and compares the results to the latter in the discussion.

The first national household survey in Kenya was conducted in 1992, by CIMMYT and KARI, and resulted in the Kenya Maize Data Base (KMDB) (Hassan, Lynam, and Okoth 1998b). This survey covered the adoption of maize technologies such as improved varieties and fertilizer. The households were georeferenced, and the data were used to adapt the standard agroecological classification into six major AEZs for maize production in Kenya, important to guide the maize breeding programs (Corbett 1998) (Figure 7.5). The major adjustment was the creation of a transitional zone between the mid-altitude AEZ and the highlands. When moving from east to west, the first zone is the lowland tropics at the coast; this is followed by the dry mid-altitude and the dry transitional zones around Machakos. These three zones are characterized by low yields (less than 1.5 tons/ha); although they cover 29 percent of maize area in Kenya, they produce 11 percent of the country's maize (Table 7.1). In Central and Western Kenya, we find the highland tropics, bordered to the west and east by the moist transitional zone (transitional between the mid-altitude AEZ and the highlands). These zones have high yields (more than 2.5 tons/ha)

FIGURE 7.5 Agroecological zones and the communities of different household surveys over time

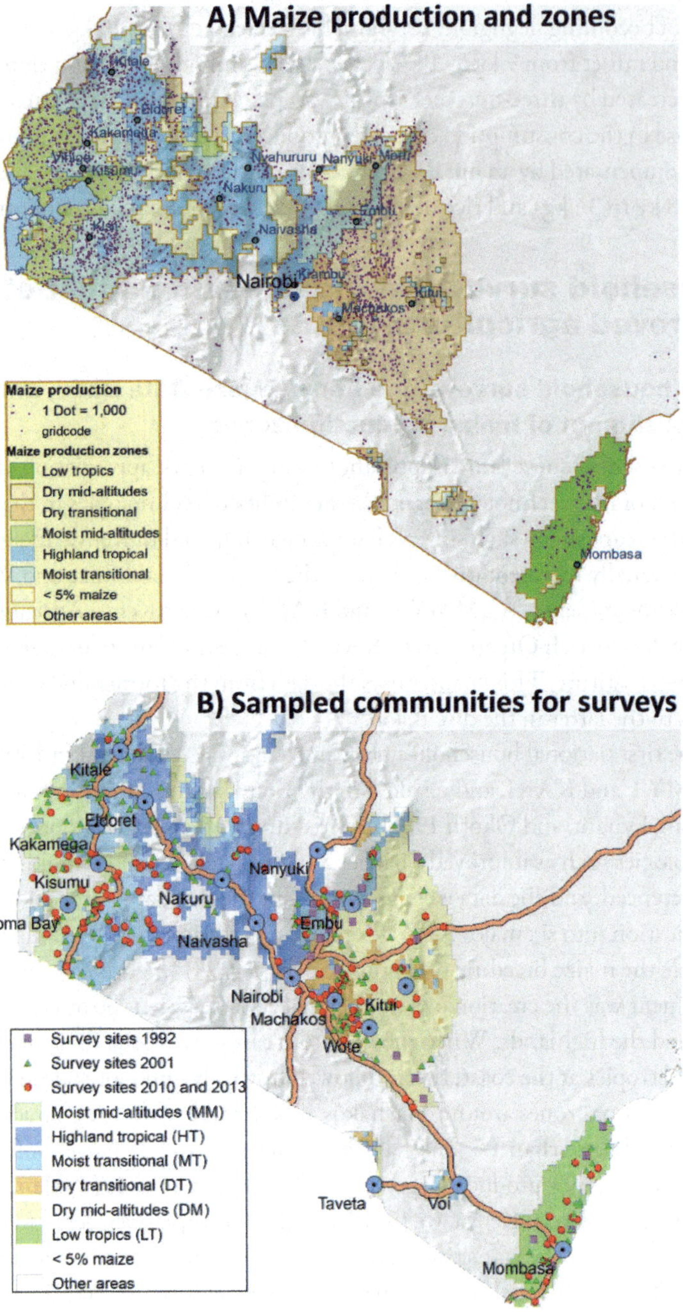

Source: www.mapspam.info/; De Groote, Marangu, and Gitonga (2018).

and produce 80 percent of the maize in Kenya on 30 percent of the area. Finally, around Lake Victoria is the moist mid-altitude zone, which has moderate yields (1.44 tons/ha): it covers 22 percent of the area and produces 9 percent of maize in the country.

While Kenya in general has two maize-growing seasons, these differ in importance between zones (Figure 7.2). In the highlands, for example, almost all of maize production takes place in the main season (March–July) (Figure 7.2, panel D). In the moist transitional zone, on the other hand, more than half of the maize is produced in the minor season (October–February) (Figure 7.2, panel B).

Consecutive national surveys by CIMMYT and KARI

Apart from the KMDB, CIMMYT, in collaboration with ARI, has conducted three nationally representative household surveys in the major maize growing areas over the past 30 years. We will use the data from these four surveys to analyze and synthesize trends in the adoption of new technologies. The data and the surveys have been described in more detail elsewhere, as the same data were used to analyze the trends in mechanization (De Groote, Marangu, and Gitonga 2018), fertilizer use (Jena et al. 2021), and varieties (De Groote and Omondi 2023).

All surveys used the same two-stage stratified design, with maize AEZs as strata, census clusters or sublocations as primary sampling units, and maize-growing households as secondary sampling units (see map in Figure 7.5, panel B). The first survey, for the KMDB, was conducted in 1992 by CIMMYT and KARI and covered 79 clusters totalling 1,397 farmers (Hassan, Lynam, and Okoth 1998). This survey also defined the six AEZs as described above, and these were subsequently used to stratify the next three surveys. The second survey was conducted in 2002 as a baseline for the Insect Resistant Maize for Africa project, and covered 185 sublocations based on the 1999 Census, with 1,652 households (De Groote et al. 2005). The third survey was conducted in 2010 as a baseline for the AflaControl project, and covered 120 sublocations with 1,341 households (De Groote et al. 2016). The fourth and last survey, in 2012, interviewed the same farmers with a replacement of 20 percent of randomly sampled households (Wainaina, Tongruksawattana, and Qaim 2016).

Micro adoption studies

CIMMYT also helped build capacity for social science research in East Africa by supporting a series of small local adoption surveys (Doss et al. 2003). Four of these studies took place in Kenya, covering two districts in Western Kenya (Salasya et al. 2007), one district in Central Kenya (Makokha et al. 2001), one district in Eastern Kenya (Ouma et al. 2002), and two districts at the coast (Wekesa et al. 2003).

Tegemeo adoption studies

The Tegemeo Institute, in collaboration with Michigan State University, has conducted panel household surveys in 5 rounds over 13 years (Mathenge, Smale, and Olwande 2014; Smale and Olwande 2014). These households represent the major maize AEZs of Kenya, although they do not fully overlap with Hassan's zones and also do not cover all maize production areas (Hassan, Lynam, and Okoth 1998). Further issues with these panel data are that no partial replacement over time was used, causing the whole panel to age over time, and that the sampling strategy of the first round (including the sampling stages and the sampling frames) is not clearly defined and the proceedings were not well recorded. We will compare our results to those of the Tegemeo surveys in the discussion.

Combining household surveys with SPAM data

Kenya does not, unfortunately, produce regional maize statistics. To estimate maize production by AEZ, we therefore used the map of the zones as developed by Hassan et al. (1998) and overlayed this with the 2017 SPAM (IFPRI 2020) and calculated the maize area and production for 2017 for the different AEZs (Table 7.1). To estimate the population in each AEZ, we used the 2015

TABLE 7.1 Maize agroecological zones in Kenya, with estimated maize area and production in 1992 and 2017

AEZ	Elevation (masl)	Maize 1992				Maize 2017			Population 2020 ('000s)
		Area ('000 ha)	Production ('000 metric tons)	% long rains	Yield (t/ha)	Area ('000 ha)	Production ('000 metric tons)	Yield (t/ha)	
Lowland tropics	0–700	33	45	0.62	1.36	58	37	0.65	2,857
Dry mid-altitude	700–1,400	118	122	0.41	1.03	401	196	0.49	3,825
Dry transitional	1,100–1,700	37	45	0.51	1.21	588	486	0.83	5,403
Moist transitional	1,200–2,000	424	1170	0.74	2.76	386	524	1.36	7,931
Highlands	1,600–2,900	307	893	0.99	2.91	248	586	2.36	1,801
Moist mid-altitude	1,110–1,500	118	170	0.61	1.44	103	109	1.06	12,137
<5%						91	119	1.30	1,858
Other						210	326	1.55	10,076
Total		1,037	2,445	0.70	2.31	2,086	3,186	1.53	45,890

Source: Author's calculations based on Hassan (1998); IFPRI (2020); Stevens et al. (2015); and Wainaina, Tongruksawattana, and Qaim (2016).

population density dataset from WorldPop (www.worldpop.org) (Stevens et al. 2015). Finally, we allocated the annual production data for each zone to the two seasons, proportionate to the distribution found in the 2013 household survey (Wainaina, Tongruksawattana, and Qaim 2016).

Results on the adoption of improved maize varieties and fertilizer, by agroecological zone

Improved maize varieties

Over time, as pictured by the four consecutive national surveys, the average adoption rate of IMVs (weighted by AEZ) did increase, but only slightly, from 72 to 79 percent (Figure 7.6). However, adoption rates over the first three surveys were nearly the same: the small increase was realized only between the last two surveys, between 2010 and 2013. Adoption rates also differed substantially between AEZs, and the graph shows a clear increase in adoption rates from low- to high-potential zones. The highest adoption rates are found

FIGURE 7.6 Trends in adoption of improved maize varieties, by survey and agroecological zone

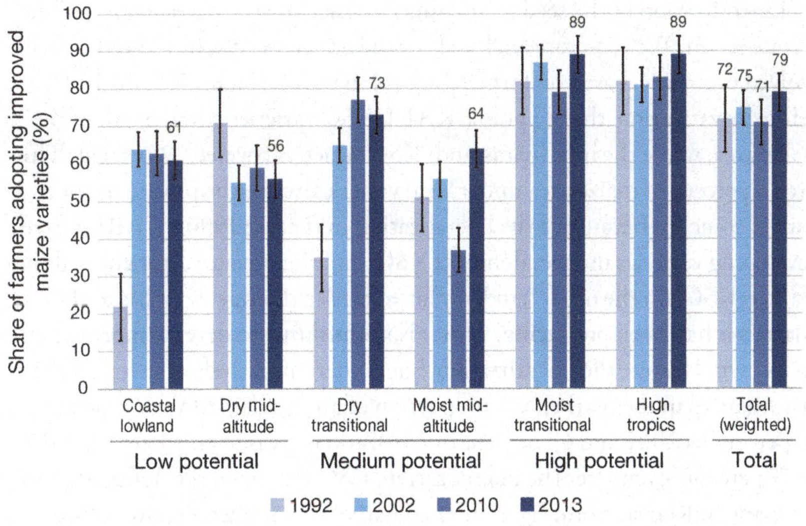

Source: De Groote and Omondi (2023).

Note: Error bars are standard errors.

in the high-potential areas (moist transitional and high tropics), where almost all farmers (89 percent) had adopted IMVs by 2013. In the medium-potential areas, the results vary between the moist mid-altitude (64 percent) and the dry transitional (73 percent) zones. In the low-potential areas, finally, adoption rates were the lowest, but IMVs had still been adopted by more than half the farmers in both the coast (61 percent) and the dry mid-altitude (56 percent) zones.

Using regression analysis (with a random effects probit model), the factors affecting the adoption of IMVs can be analyzed over the four surveys (De Groote and Omondi 2023). The household characteristics that significantly and positively affected adoption of IMVs were education of household head, household size, if the household sold maize, and access to extension services. Gender and age of household head, on the other hand, did not have a significant effect on adoption rates. A major factor was market participation: households selling maize were much more likely to grow IMVs. Market participation increased from 20 to 49 percent over the study period, although that latter level remains low, reflecting the importance of production for home consumption. Among institutional factors, access to extension services increased the adoption rate but access to credit or distance to market did not have significant effects. There were also substantial differences between AEZs, as seen in the graph, with farmers in the high-potential zones more likely to adopt IMVs than their colleagues in low-potential areas.

Liberalization of the seed sector aimed to increase the participation of the private sector. We therefore analyze how the private seed sector's market share evolved over time, based on farmers' adoption trends (Figure 7.7). In 1992, before privatization, the old public KARI/KSC varieties covered half of the maize area, while the new (parastatal) KSC varieties covered 22 percent. This left 28 percent of maize area under local varieties, while the private sector had not yet come in. Because of the liberalization in the late 1990s, KARI started developing varieties independently of KSC, and the private sector entered the market. KSC, on the other hand, obtained the legal property rights to the varieties it had been producing, while also continuing to develop more varieties on its own. Despite the liberalization, but perhaps negatively affected by the disturbances in the market, the proportion of area under IMVs dropped to 66 percent in 2002 and 62 percent in 2010, and increased only later (in 2013) to 77 percent. Even after the liberalization, KSC remained the dominant seed company in Kenya, retaining more than half of the market. Private sector participation evolved slowly: varieties owned by the private sector (mostly multinationals and companies from southern Africa) made up only 2 percent in 2002 and increased their share to 15 percent by 2013. Varieties owned by the

FIGURE 7.7 Market share of maize varieties, by source and over time

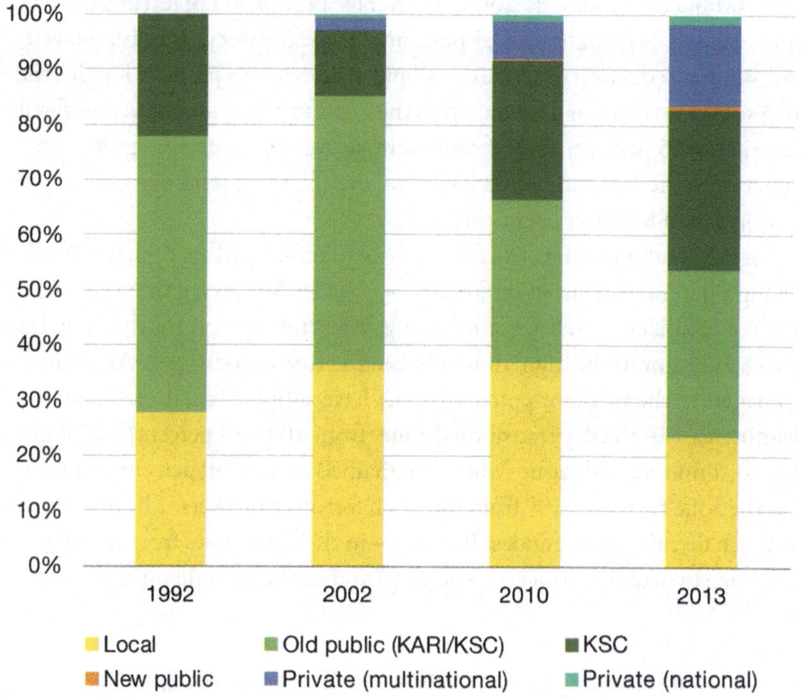

Source: De Groote and Omondi (2023).

public sector (KARI, its successor KALRO, and CIMMYT), with seed mostly produced and disseminated by local seed companies, had a share of 6 percent in 2002 but increased it to only 9 percent in 2013.

We did not include OPVs as a separate class in the analysis as they were important only in the lowlands (going from 18 percent in 1992 to 14 percent in 2013) and the dry midlands (from 49 to 8 percent); in the other regions they never covered more than 2 percent. Moreover, as KSC is phasing out its OPVs and replacing them with popular hybrids, they are also decreasing rapidly in those two areas (for details on OPVs, see Supplementary Material 3 in De Groote and Omondi 2023).

Adoption of fertilizer

The data from the four household surveys allow us to map the trends in the adoption of fertilizer over time (Figure 7.8). In 1992, the first year, 62 percent of farmers used fertilizer, but with a strong variation between the different maize production zones. By 2002, the average proportion of fertilizer users had increased only slightly, to 65 percent. The next survey, in 2010, showed a substantial reduction of fertilizer adoption rates, to 58 percent. Finally, the 2013 survey, which was a follow-up to the 2010 survey, showed average fertilizer adoption of 65 percent, a significant increase over 2010 yet not significantly higher than the 1992 and 2002 levels. So, overall, there is no significant increase in adoption of fertilizer over time.

Fertilizer adoption rates differ highly between the different maize AEZs. In the high-potential areas, most farmers—generally 80 percent or more—have adopted fertilizer, with little or no change over time (except for the jump from 57 to 89 percent in the highlands between the first two surveys). About half of farmers in the medium-potential zones have adopted fertilizer, but with a clear increase in the dry transitional zone (from 40 to 64 percent) but not in the moist mid-altitude zone (where it remained around 50 percent), which is also the zone furthest away from the main fertilizer markets. The low-potential areas, finally, also saw a modest increase—in the low tropics from no adoption to about a quarter of farmers (24 percent) and in the dry mid-altitude zone

FIGURE 7.8 Trends in the adoption of fertilizer over time, by agroecological zone

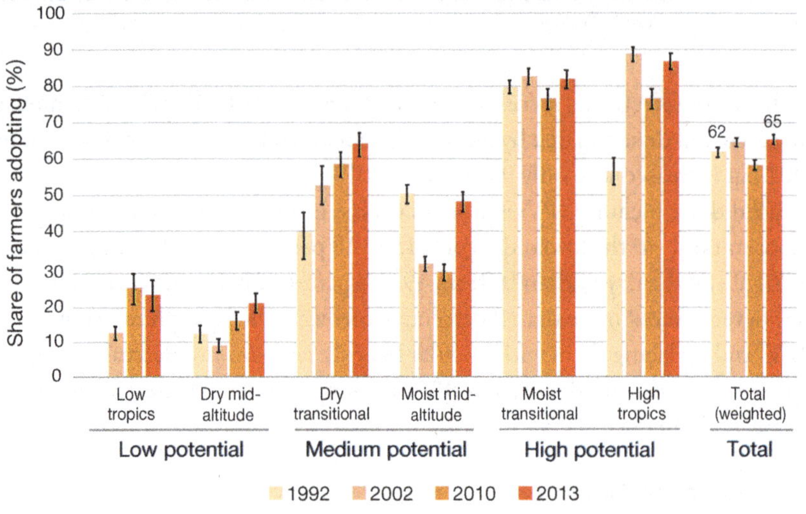

Source: De Groote and Omondi (2023).
Note: Error bars are standard errors.

from 12 to 21 percent of farmers. However, the three zones with an increase in adoption are rather small maize producers compared with the high-potential areas, so their increase in adoption did not affect the overall trend of stagnation in fertilizer adoption.

Next, we calculate the average dose of fertilizer used per hectare; again, we calculate weighted average using the maize areas in the different AEZs as weights to ensure representativeness at a national level. The seasonally weighted averages for the four years indicate a modest increase over the study period, from 82 kg/ha in 1992 to 100 kg/ha in 2002 but with a dip in 2010 (to 68 kg/ha). These doses are still substantially below the recommended dose for fertilizer application. Again, application rates vary among AEZs: the high-potential zones have high rates (between 140 and 160 kg/ha in the last year) while the low-potential zones have particularly low rates (between 8 and 30 kg/ha). The high standard deviations for the average figures also indicate a high variability in fertilizer application rates between farmers, with a standard deviation higher than the mean in all years.

Analyzing the change in application rates by zone, however, shows that the overall increase stems mostly from the increase in the high-potential areas (again

FIGURE 7.9 Trends in the fertilizer application rate by agroecological zone

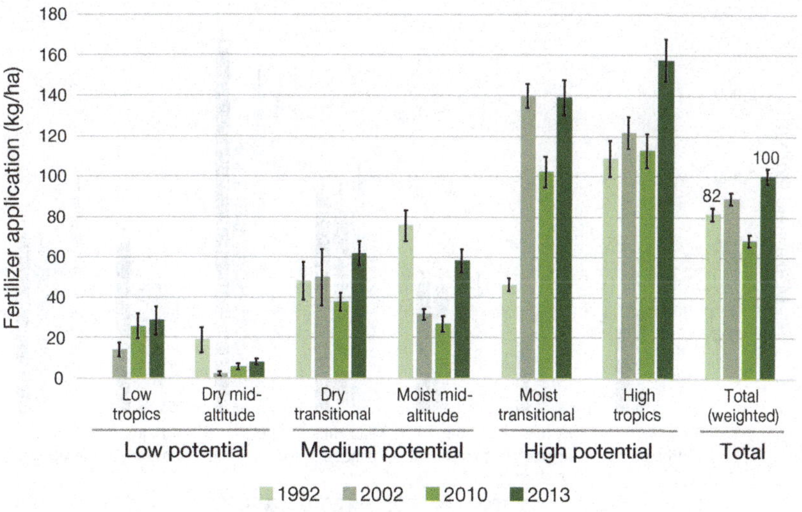

Source: Jena et al. (2020).

Note: Error bars are standard errors.

with the dip in 2010). In the low- and medium-potential zones, no increase in application rates was observed, except for in the dry transitional zone.

Effect of improved maize varieties and fertilizer on yields

Analyzing maize yield trends in Kenya over the same four surveys shows no increase, rather a slight decrease from 1,360 to 1,116 kg/ha (Figure 7.10). The yield estimates from our surveys follow the trend of the FAO statistics but are actually systematically lower (see Jena et al. 2020 for details of the comparison). Yields differ strongly between AEZs, and these differences remained high over the 30 years of the study period. Yields are understandably higher in the high-potential zones, at around 1,500 kg/ha in the moist transitional zone and 2,000 kg/ha in the high tropics. In the low-potential areas, however, they barely reach 500 kg/ha (500–1,000 kg/ha in the mid-potential zones). Trends also differ by zone but without a clear overall trend: while the high tropics and the moist mid-altitude zone show an increase, in the moist transitional zone yields decreased, and in the other zones they just stayed stagnant.

The question remains: What are the major drivers of maize yields? A previous analysis, using endogenous switching regression coefficients, showed

FIGURE 7.10 Average maize yields, by agroecological zone and year of survey

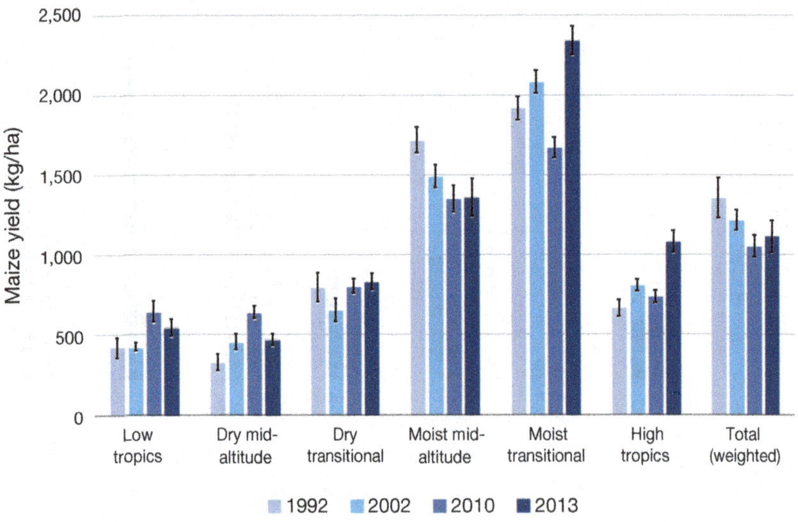

Source: Jena et al. (2020).

Note: Error bars are standard errors.

the factors that affect maize yields for fertilizer adopters and non-adopters (Jena et al. 2020). The use of hybrid maize also positively affects the yield for both groups, with an increase of 291 kg/ha for adopters and 174 kg/ha for non-adopters, indicating the synergistic effect of combining both technologies. Maize area per household, in contrast, has a negative and significant coefficient for non-adopters, indicating that a larger maize area leads to a yield reduction for this group. Among institutional variables, access to extension service has a positive and significant impact, increasing maize yield by 240 kg/ha for fertilizer adopters and 184 kg/ha for non-adopters, indicating a synergy between extension and fertilizer use. Household size has a positive impact on yield for both adopters and non-adopters. Among weather variables, an increase in minimum temperature has a positive effect on yield, while an increase in annual rainfall has a positive effect on yields for adopters. Finally, the AEZ are a major driver of yields, with the high-potential areas easily reaching 2.5 tons/ha and the low-potential areas having difficulties reaching 1.5 tons/ha.

Another analysis shows yields to increase for younger varieties, with about 4 kg/ha for each reduction of a year (De Groote and Omondi 2023). Fertilizer also increased yields, by 5 kg/ha for each 1 kg, with a significant, negative cross-effect, indicating that younger varieties are more responsive to fertilizer. At the average fertilizer rate by fertilizer users of 132 kg, the effect of a year is an additional 2.5 kg/ha.

Discussion

This analysis, based on data from four household surveys conducted over 30 years, shows a very slow process of intensification in the maize production sector. Adoption rates for IMVs show a slight increase (from 72 to 79 percent), and poor progress in varietal turnover, with a continuous domination of the parastatal KSC in the seed market. Similarly, adoption rates of fertilizer increased slightly (from 62 to 65 percent), with a modest increase in application rates (from 82 to 100 kg/ha). However, maize yields did not increase during the study period. This is confirmed by the FAO data (FAOSTAT 2022), which show a stagnation in yields since 1990, in contrast with a steady increase from the 1960s till then. Stagnating yields for a major food crop are particularly problematic given the rapidly increasing population. This was partly upset by an increase in maize area, but maize production per person has been reducing steadily over time. This shortage is being countered with increasing imports of maize, but also wheat and rice, reflecting a change in consumer preferences. However, continually increasing imports of the major food staple is likely to create political problems in the long run, especially in times of increasing food prices.

The adoption rates of IMVs in the CIMMYT surveys are similar to those in the Tegemeo panel survey: between 38 to 82 percent of hybrid adopters, depending on AEZ (Mathenge, Smale, and Olwande 2014), with an overall level of hybrid adoption by more than 80 percent of farmers (Smale and Olwande 2014). Our results on fertilizer, however, stand in contrast with those from the Tegemeo panel data, which show an increase of fertilizer use and maize yields from 1997 to 2007 (Olwande, Sikei, and Mathenge 2009; Ariga and Jayne 2011). The FAO yield data for this period also show an increase in maize yields. However, our analysis, both based on our survey and the FAOSTAT data, indicate that this short-period increase is not part of a larger trend (Jena et al. 2020). Further, our results indicated yields reduce with maize area, while other studies find a U-shaped relationship, including an older study (Carter and Wiebe 1990) and one based on the Tegemeo data (Muyanga and Jayne 2019). However, the yield increase of the latter study was only found in the 5–70 ha farm size range. Our data, based on representative household surveys, do not include enough households in that range to test that relationship. Other factors not included in our analysis were climate and economic data. Important climate variables are volatility and timing of rainfall, which would be good to include when better data become available. Also, output and output prices, in particular grain, seed, and fertilizer prices, would be good to include in future surveys; the Tegemeo data showed a strong positive effect of the grain/price ratio on adoption of hybrids (Smale and Olwande 2014).

Our results do confirm that intensification of maize production, in particular the use of fertilizers and improved varieties, has a positive effect on yield, income, and food security. This has, of course, been reported before (Mathenge, Smale, and Olwande 2014). What remains puzzling, though, is the low rate of intensification. The slow rate of intensification and the stagnation of maize yields are major concerns for a country that relies heavily on maize for its food security. Some of the factors that affect the adoption of improved maize technologies can be influenced by policy, in particular availability of affordable inputs, education, and extension. Others, in particular the commercialization of maize production, are more difficult.

On the provision of inputs, the liberalization of the maize seed sector was not particularly effective. While many private seed companies have entered the market and many varieties are now available, farmers seem to prefer their old varieties, and are risk-averse in trying new ones. This has led to a low varietal turnover and a continued domination by the parastatal KSC. The combination of public research and local seed companies does not, however, seem very successful as compared with international private seed companies and, especially, KSC.

In the fertilizer market, liberalization was initially successful, leading to the entry of the private sector and the development of an efficient distribution system. However, the liberalization was later countered by new subsidy programs, which did increase input use for some time, with, for example, a 34 percent increase in smallholder fertilizer use over 1997–2007 (Ariga and Jayne 2011). Unfortunately, government subsidies and intervention in distribution do crowd out private investment in the sector (Makau et al. 2016), and do not necessarily reach targeted households. Moreover, distribution of subsidized fertilizer by NCPB is contrary to the earlier liberalization policy and expands its role from buyer of last resort. What is particularly lacking in Kenya is systematic research on soil fertility management to counter the continuous cultivation without nutrient replenishment.

Agricultural extension was also found to affect adoption of technologies, but the interaction of the public and private sectors remains problematic. New apps hold some promise but, while our surveys confirm that most farmers now have access to phones, these are not usually smartphones, hindering access to the new developments in apps for pest management, variety selection, and soil fertility management.

Finally, the major factor not yet included in the analysis is the low level of urbanization in the country, linked to relatively low population density, low prices for agricultural land, and high levels of home consumption of maize. Less than 30 percent of the Kenyan population lives in urban areas (KNBS 2019), and more than half of the farmers produce maize only for home consumption. Kenya may therefore not have reached the conditions that will lead to the intensification of the major food crop, in particular population density (Boserup 1965), especially as related to arable land (De Groote 1999). In this case, the stagnation of the intensification may be worrisome but not much can be done to speed it up. What would be very useful, however, is a regular farm household survey, like the living standard measurement surveys (LSMS) or the household survey in Ethiopia (CSA 2014), to follow and understand the trend. It is therefore unfortunate that no new rounds for either the CIMMYT or the Tegemeo surveys have been conducted in the past 10 years.

Conclusion and policy recommendations

Based on the results of our review, several policy recommendations can made. First, it is important to create an enabling environment for the private sector to compete on a level playing field with the parastatal KSC. Second, at this stage in agricultural intensification, soil fertility is the major factor in maize yields, so it should receive the appropriate level of attention from research, extension, and policy.

Finally, to improve the adoption of IMVs, our results indicate several recommendations. More participatory variety evaluation would ensure varieties fit farmers' needs and reduce the number of varieties released that do not get adopted. Access to extension also needs to be promoted, both through the public extension service and the private sector, which now dominates seed dissemination, to ensure farmers choose the right varieties for their situation. Policies should also promote universal education, especially in rural areas. And, given its important effect on the adoption of IMVs, market participation of farmers needs to be encouraged and supported, by promoting good drying and storage practices and increasing access to markets, in particular through the provision of market information and infrastructure and a reduction in transaction costs.

References

Ariga, J., and T.S. Jayne. 2009. "Private Sector Responses to Public Investments and Policy Reforms: The Case of Fertilizer and Maize Market Development in Kenya." IFPRI Discussion Paper 921. International Food Policy Research Institute (IFPRI), Washington, DC.

Ariga, J., and T.S. Jayne. 2011. "Fertilizer in Kenya: Factors Driving the Increase in Usage by Smallholder Farmers." In *Yes Africa Can: Success Stories from a Dynamic Continent,* eds. P. Chuhan-Pole and M. Angwafo, 269–288. Washington, DC: World Bank.

Boone, C., F. Lukalo, and S.F. Joireman. 2021. "Promised Land: Settlement Schemes in Kenya, 1962 to 2016." *Political Geography* 89: 102393.

Boserup, E. 1965. *The Conditions of Agricultural Growth: The Economics of Agrarian Change under Population Pressure.* Piscataway, NJ: Transaction Publishers.

Carter, M.R., and K.D. Wiebe. 1990. "Access to Capital and Its Impact on Agrarian Structure and Productivity in Kenya." *American Journal of Agricultural Economics* 72 (5): 1146–1150.

Corbett, J.D. 1998. "Classifying Maize Production Zones." In*Maize Technology Development and Transfer. A GIS application for Research Planning in Kenya*, ed. R.M. Hassan. 1998. New York: CABI for CIMMYT.

CSA (Central Statistical Agency). 2014. *Agricultural Sample Survey 2013/2014 (2006 E.C.). Volume I -Report on Area and Production of Major Crops (Private Peasant Holdings, Meher Season).* Statistical Bulletin 532. Addis Ababa.

DAI (Development Alternatives Inc.). 1989. *Economic and Social Analyses for the Kenya Market Development Program.* Washington, DC: USAID; DAI, Institute for Development Anthropology (IDA).

De Groote, H. 1999. "Population Pressure and Agricultural Technological Change." *Quarterly Journal of International Agriculture* 38: 185–202.

De Groote, H., and L.B. Omondi. 2023. "Varietal Turn-Over and Their Effect on Yield and Food Security: Evidence from 20 Years of Household Surveys in Kenya." *Global Food Security* 36 (March): 1000676.

De Groote, H., G. Owuor, C. Doss, J. Ouma, L. Muhammad, and K. Danda. 2005. "The Maize Green Revolution in Kenya Revisited." *Electronic Journal of Agricultural and Development Economics* (eJADE) 2: 32–49.

De Groote, H., C, Narrod, S. Kimenju, C. Bett, R. Scott, and M. Tiongco. 2016. "Measuring Rural Consumers' Willingness to Pay for Quality Labels Using Experimental Auctions: The Case of Aflatoxin Free Maize in Kenya." *Agricultural Economics* 47: 33–45.

De Groote, H., C. Marangu, and Z.M. Gitonga. 2018. "Trends in Agricultural Mechanization in Kenya's Maize Production Areas from 1992-2012." *Agricultural Mechanization in Asia, Africa and Latin America* 49: 20.

Doss, C., W. Mwangi, H. Verkuijl, and H. De Groote. 2003. *Adoption of Maize and Wheat Technologies in East Africa: Synthesis of the Findings of 22 Case Studies.* Mexico City: International Maize and Wheat Improvement Center (CIMMYT).

Eicher, C.K. 1995. "Zimbabwe's Maize-Based Green Revolution: Preconditions for Replication." *World Development* 23: 805–818.

FAO (Food and Agriculture Organization of the United Nations). 2020. FAOSTAT Food Supply—Crops Primary Equivalent. Accessed June 9, 2020. http://www.fao.org/faostat/en/#data/CC

FAO. 2022. FAOSTAT Crops and Livestock Data Base. Rome.

Gerhart, J. 1975. *The Diffusion of Hybrid Maize in Western Kenya.* Mexico City: CIMMYT.

Gisselquist, D., and J.M. Grether. 2000. "An Argument for Deregulating the Transfer of Agricultural Technologies to Developing Countries." *World Bank Economic Review* 14: 111–127.

Gisselquist, D., J. Nash, and C. Pray. 2002. "Deregulating the Transfer of Agricultural Technology: Lessons from Bangladesh, India, Turkey, and Zimbabwe." *World Bank Research Observer* 17: 237–265.

Harrison, M.N. 1970. "Maize Improvement in East Africa." In *Crop Improvement in East Africa*, ed. C.L.A. Leakey, 27–36. Buckinghamshire, UK: Commonwealth Agricultural Bureaux.

Hassan, R.M., and D.D. Karanja. 1997. "Increasing Maize Production in Kenya: Technology, Institutions and Policy." In *Africa's Emerging Maize Revolution,* eds. D. Byerlee and C.K. Eicher, 81–94. Boulder, CO: Lynne Rienner Publishers.

Hassan, R.M., D.D. Karanja, and H.H.A Mulamula. 1998a. "Availability and Effectiveness of Agricultural Extension Services for Maize Farmers in Kenya." In *Maize Technology Development and Transfer: A GIS Application for Research Planning in Kenya*, ed. R.M. Hassan, 175–190. Oxon, UK: CAB International.

Hassan, R.M., J. Lynam, and P. Okoth. 1998b. "The Spatial Sampling Frame and Design for Farmer and Village Surveys." In *Maize Technology Development and Transfer: A GIS Application for Research Planning in Kenya*, ed. R.M. Hassan, 27–41. Oxon, UK: CAB International.

Hassan, R.M., K. Njoroge, M. Njore, R. Otsyula, and A. Labos. 1998c. "Adoption Patterns and Performance of Improved Maize." In *Maize Technology Development and Transfer: A GIS Application for Research Planning in Kenya*, ed. R.M. Hassan, 107–136. Wallingford, UK: CABI Publishing.

IFPRI (International Food Policy Research Institute). 2020. *Spatially-Disaggregated Crop Production Statistics Data in Africa South of the Sahara for 2017, Harvard Dataverse, V2, V1.* Washington, DC: IFPRI.

Jayne, T.S., M. Mukumbu, M. Chisvo, D. Tschirley, M.T. Weber, B. Zulu, R. Johansson, P. Santos, and D. Soroko. 1999. "Successes and Challenges of Food Market Reform: Experiences from Kenya, Mozambique, Zambia, and Zimbabwe." MSU International Development Working Paper No. 72. Michigan State University, East Lansing, MI.

Jayne, T.S., J.M. Robert, and N. James. 2008. "The Effects of NCPB Marketing Policies on Maize Market Prices in Kenya." *Agricultural Economics* 38: 313–325.

Jayne, T.S., N.M. Mason, W.J. Burke, and J. Ariga. 2018. "Taking Stock of Africa's Second-Generation Agricultural Input Subsidy Programs." *Food Policy* 75: 1–14.

Jena, P.R., H. De Groote, B.P. Nayak, and A. Hittmeyer. 2020. "Evolution of Fertiliser Use and its Impact on Maize Productivity in Kenya: Evidence from Multiple Surveys." *Food Security* 13: 95–111.

Jones, N.C. 1965. "The Decolonization of the White Highlands of Kenya." *Geographical Journal* 131 (2): 186–201.

Karanja, D.D. 1996. "An Economic and Institutional Analysis of Maize Research in Kenya." MSU International Development Working Paper No. 57. Michigan State University, East Lansing, MI.

Kenya, Government of Kenya. 2010. *Agricultural Sector Development Strategy, 2010–2020.* Nairobi.

KNBS (Kenya National Bureau of Statistics). 2019. *2019 Kenya Population and Housing Census. Volume II Distribution of Population by Administrative Units.* Nairobi.

Lewa, P.M., and M. Hubbard. 1995. "Kenya's Cereal Sector Reform Programme: Managing the Politics of Reform." *Food Policy* 20: 573–584.

Lynam, J., and R.M. Hassan. 1998. "A New Approach to Securing Sustained Growth in Kenya's Maize Sector." In *Maize Technology Development and Transfer: A GIS Application for Research Planning in Kenya*, ed. R.M. Hassan, 3–14. Oxon, UK: CAB International.

Makau, J.M., P. Irungu, R.A. Nyikal, and L.W. Kirimi. 2016. "An Assessment of the Effect of a National Fertiliser Subsidy Programme on Farmer Participation in Private Fertiliser Markets in the North Rift Region of Kenya." *African Journal of Agricultural and Resource Economics* 11: 292–304.

Makokha, S., S. Kimani, W. Mwangi, H. Verkuijl, and F. Musembi. 2001. *Determinants of Fertilizer and Manure Use for Maize Production in Kiambu District, Kenya.* Mexico City: CIMMYT.

Mathenge, M.K., M. Smale, and J. Olwande. 2014. "The Impacts of Hybrid Maize Seed on the Welfare of Farming Households in Kenya." *Food Policy* 44: 262–271.

Mather, D.L., and T.S. Jayne. 2018. "Fertilizer Subsidies and the Role of Targeting in Crowding Out: Evidence from Kenya." *Food Security* 10: 397–417.

McCann, J.C. 2001. "Maize and Grace: History, Corn, and Africa's New Landscapes, 1500–1999." *Comparative Studies in History and Society* 43 (2): 246–272.

McCann, J.C., 2005. *Maize and Grace: Africa's Encounter with a New World Crop, 1500–2000.* Cambridge, MA: Harvard University Press.

Miracle, M.P. 1965. "The Introduction and Spread of Maize in Africa." *Journal of African History* 6: 39–55.

Muyanga, M., and T.S. Jayne. 2019. "Revisiting the Farm Size-Productivity Relationship Based on a Relatively Wide Range of Farm Sizes: Evidence from Kenya." *American Journal of Agricultural Economics* 101: 1140–1163.

Mwangi, W., J. Lynam, and R.M. Hassan. 1998. "Current Challenges and Strategic Future Choices for Maize Research and Policy in Kenya: A Synthesis of the Maize Data Base Project Methods and Results." In *Maize Technology Development and Transfer. A GIS Application for Research Planning in Kenya*, ed. R.M. Hassan, 191–203. Oxon, UK: CAB International.

Odero-Waitituh, J. 2017. "Smallholder Dairy Production in Kenya: A Review." *Livestock Research for Rural Development* 29: 139.

Olwande, J., G. Sikei, and M. Mathenge. 2009. "Agricultural Technology Adoption: A Panel Analysis of Smallholder Farmers' Fertilizer Use in Kenya." CEGA Working Paper Series No. AfD0908. Center of Evaluation for Global Action, Berkely, CA.

Ouma, J.O., F.M. Murithi, W. Mwangi, H. Verkuijl, M. Gethi, and H. De Groote. 2002. *Adoption of Maize Seed and Fertiliser Technologies in Embu District, Kenya*. Mexico City: CIMMYT.

Poulton, C., and K. Kanyinga. 2014. "The Politics of Revitalising Agriculture in Kenya." *Development Policy Review* 32: s151–s172.

Salasya, B., W. Mwangi, D. Mwabu, and A. Diallo. 2007. "Factors Influencing Adoption of Stress-Tolerant Maize Hybrid (WH 502) in Western Kenya." *African Journal of Agricultural Research* 2: 544–551.

Smale, M., and T. Jayne. 2003. "Maize in Eastern and Southern Africa: 'Seeds' of Success in Retrospect." EPTD Discussion Paper No. 97. IFPRI, Washington, DC.

Smale, M., and J. Olwande. 2014. "Demand for Maize Hybrids and Hybrid Change on Smallholder Farms in Kenya." *Agricultural Economics* 45: 409–420.

Smale, M., M.K. Mathenge, T.S. Jayne, E.C. Magalhaes, J. Olwande, L. Kirimi, M.W. Kamau, and J. Githuku. 2012. "Income and Poverty Impacts of USAID-funded Programs to Promote Maize, Horticulture, and Dairy Enterprises in Kenya, 2004–2010." MSU International Development Working Paper No. 122. Michigan State University, East Lansing, MI.

Stevens, F.R., A.E. Gaughan, C. Linard, and A.J. Tatem. 2015. "Disaggregating Census Data for Population Mapping Using Random Forests with Remotely-Sensed and Ancillary Data." *PloS ONE* 10: e0107042.

Tripp, R., and D. Rohrbach. 2001. "Policies for African Seed Enterprise Development." *Food Policy* 26: 147–161.

Wainaina, P., S. Tongruksawattana, and M. Qaim. 2016. "Tradeoffs and Complementarities in the Adoption of Improved Seeds, Fertilizer, and Natural Resource Management Technologies in Kenya." *Agricultural Economics* 47: 351–362.

Wangia, C., S. Wangia, and H. De Groote. 2004. "Review of Maize Marketing in Kenya: Implementation and Impact of Liberalisation, 1989-1999." In *Integrated Approaches to Higher Maize Productivity in the New Millennium*, eds. D.K. Friesen and A.F.E. Palmer, 10–20. Mexico City: CIMMYT.

Wekesa, W.E., H. Mwangi, H. Verkuijl, K. Danda, and H. De Groote. 2003. *Adoption of Maize Production Technologies in the Coastal Lowlands of Kenya*. Mexico City: CIMMYT and Kenya Agricultural Research Institute (KARI).

Chapter 8

AGRICULTURAL INPUTS IN KENYA: DEMAND, SUPPLY, AND THE POLICY ENVIRONMENT

Lilian Kirimi, John Olwande, Jackson Langat, Timothy Njagi, Mercy Kamau, and Gideon Obare

Agricultural inputs, including fertilizers, seeds, breeding stock, crop protection chemicals, machinery, irrigation, and knowledge, are key to innovation and productivity improvement, and are the backbone of any agricultural revolution. They are an integral part of the food supply chain, which comprises the production and distribution of food, and as such a key component of the food system (HLPE 2017). The food supply chain involves various actors at different stages of the chain but this chapter focuses only on agricultural inputs, including both farm inputs and agricultural advisory services.

Agricultural inputs play a critical role in supporting the farming sector and contributing to the productivity of the food system. They complement other components of the food system to enable the efficient and sustainable production of sufficient, safe, and nutritious food for diverse groups of a growing population. Hence, they are part of a support system involving other services like storage, transport, marketing, and processing that work together synergistically toward the modernization and transformation of the agriculture sector (Tiyambe 1991; Nin-Pratt 2016; Gulati et al. 2021).

This chapter presents trends in the supply and use of agricultural inputs and the evolution of policies that have affected these trends. Kenya launched its planned economic development in 1964, immediately after independence. The associated policies were based on Sessional Paper 10 of 1965 (Kenya, Republic of Kenya 1965) and were designed to revolutionize agriculture through land consolidation, extension services and training, and the introduction of modern farming methods (Kamande 2009). In the subsequent decade, the thrust of policy was agricultural intensification practices aimed at replicating the successes of the Asian Green Revolution. The focus was on addressing technical constraints in smallholder production, which included poor access to farm inputs and credit, and lack of knowledge of agricultural production. Consequently, the government provided key services, including inputs, credit, research, and extension, to smallholders (Argwings-Kodhek 1996; Kelly, Adesina, and Gordon 2003).

In the 1980s and 1990s, structural adjustment programs were introduced to restore resource use efficiency in all sectors of the economy and hence raise the rate of economic growth. This included liberalization of input and output markets, remedying distortions in agricultural markets, and generating "efficiency" in the agriculture sector by transforming state agricultural agencies and parastatals and transferring services provision to the private sector (Kelly, Adesina, and Gordon 2003). During the 1980s, government policies focused on reducing budget deficits and external debt by decreasing government involvement in providing services in agriculture such as extension, research, credit, inputs, price-based support mechanisms, and agricultural subsidies (Ajwang, Atela, and Arora 2019). This is reflected in, for example, the Agricultural Sector Development Strategy (ASDS) of 1997, as implemented through the Agricultural Sector Investment Program (Odhiambo, Nyangito, and Nzuma 2004). While reforms in the liberalization era benefited export-oriented farming, smallholders were affected negatively by the dismantling of state services in marketing and production (Ajwang, Atela, and Arora 2019) in the context of a poorly developed private sector.

Policies in the 2000s have focused on transforming agriculture from subsistence to commercial and market-oriented production. These have included the Strategy for Revitalizing Agriculture (SRA) 2004–2014, the ASDS 2010–2020, and the Agriculture Sector Transformation and Growth Strategy (ASTGS) 2019–2029. They underscore different roles of the private sector and government. Private sector input suppliers have largely replaced parastatals and are increasingly providing diverse agricultural inputs. For instance, the ASTGS recognizes that agricultural transformation requires access to more affordable and higher-quality inputs but embodies a shift in subsidy provision. Under this, provision of inputs is the responsibility of the private sector, and subsidies are targeted only at high-need farming households, which access various inputs from multiple private sector providers through an e-voucher system.

This chapter highlights key agricultural input demand–supply systems, how policy has affected their development, lessons learned, policy implications, and research gaps to fill to ensure the smooth functioning of these systems.

The fertilizer system

Policy and institutional interventions

The fertilizer system in Kenya can be understood in the context of the policy and institutional interventions that have shaped trends and patterns in fertilizer supply and use. These interventions can be organized into three periods, as outlined in Table 8.1 (Ariga and Jayne 2011; Nduati et al. 2015). The pre-1990 period was marked by the government's heavy involvement in agricultural production and marketing activities. Fertilizer trade, including importation and distribution, was restricted to a few state agencies, and fertilizer prices were controlled. The government imported and distributed fertilizers through the Kenya Grain Growers Cooperative Union, Kenya Farmers Association, and Kenya National Trading Corporation. Poor coordination of fertilizer importation led to surpluses and deficits in fertilizer supply in the local market (Ariga and Jayne 2011).

As part of the structural adjustment program during the early to mid-1990s, the fertilizer sector was gradually liberalized. Fertilizer imports, distribution

TABLE 8.1 Evolution of Kenya's fertilizer sector policies, pre-1990–2022

Period	Policy intervention
Pre–1990	Emphasis on government control • Imposition of import licensing quotas • Fertilizer price controls • Fertilizer donations by external donor agencies • Allocation of foreign exchange
1991–2006	Fertilizer market gradually liberalized • Elimination of import licensing quotas • Removal of government price controls • Phasing-out of fertilizer donations by external donor agencies • Liberalization of foreign exchange regime • Private trade in fertilizer (including importation and distribution) permitted • Removal of custom duty and value added tax on fertilizer
2007–2022	Resurgence of government involvement in fertilizer importation and distribution • Targeted fertilizer subsidy through the National Accelerated Agricultural Input Access Programme • Subsidized fertilizer importation and distribution through the National Cereals and Produce Board • Fertilizer subsidies by county governments

Source: Authors using Ariga and Jayne (2011); Nduati et al. (2015).

restrictions, and price controls were eliminated. Private trade in fertilizer was permitted and the import duty and value-added tax were eliminated. The immediate result of relaxing these restrictions was an increase in the number of private traders in fertilizer, estimated at 12 major importers, 500 wholesalers, and 5,000 retailers (Allgood and Kilungo 1996). By 2000, there were 7,000–8,000 fertilizer retailers in Kenya (IFDC 2001).

A newly elected government in 2002 prioritized increasing fertilizer use to improve agricultural productivity and to foster broader economic growth, given the central role of the agriculture sector in Kenya's economy and its dismal performance at that time. Hence, the government developed a three-tiered fertilizer cost reduction strategy, comprising bulk procurement of fertilizer, domestic fertilizer blending and packaging, and establishment of a fertilizer manufacturing plant to meet national/regional fertilizer needs (Kenya, Ministry of State Planning 2007). It also initiated the National Accelerated Agricultural Input Access Program (NAAIAP) in response to the 2006 Abuja Declaration on Fertilizer for the African Green Revolution, in which the African Union Member States resolved to increase fertilizer usage from an average of 8 kg/ha to at least 50 kg/ha by 2015 (African Union 2006).

The NAAIAP had several components, including free provision of fertilizers for maize cultivation to resource-poor farmers in a phased manner. The targeted fertilizer subsidy was distributed through private agrodealers and thus supported private sector trade in fertilizer. The program distributed 50,800 metric tons[1] of fertilizer, valued at KSh 2.73 billion, to about 533,000 smallholder farmers in a span of seven years up to 2015. This subsidy component was terminated in 2017 and re-engineered as a more effective and efficient e-voucher program under the National Value Chain Support Programme (NVSP), discussed at the end of this subsection.

In 2008, high world prices of fertilizer combined with the post-election violence that followed Kenya's disputed presidential elections in December 2007 contributed to a doubling of fertilizer prices in the domestic market. The violence disrupted activities, including transportation and trading, in the local supply chains of goods and services. As a result of these factors, for example, the average retail price of 50 kg of diammonium phosphate (DAP), the most widely applied fertilizer, increased from KSh 2,250 in 2007 to KSh 4,500 in 2008 (Ministry of Agriculture, Livestock, Fisheries, and Cooperatives 2010a).

1 Tons refers to metric tons throughout this volume.

In response, the government introduced a "blanket" fertilizer subsidy program in 2008, referred to as the National Fertilizer Price Stabilization Plan (Kenya, Ministry of State Planning 2008; 2013). This subsidy was different from the NAAIAP in that it aimed at reducing and stabilizing local fertilizer prices to make fertilizers affordable for farmers, and it was not targeted to a specific group of farmers. Importation of the fertilizer was through competitive tendering managed by the National Cereals and Produce Board (NCPB), while distribution was through the latter's depots. Implemented through 2019, this subsidy program procured and distributed 1.27 million tons of fertilizers at a cost of KSh 28 billion over a decade. The distribution of the subsidized fertilizer through the NCPB depots made the subsidized fertilizers inaccessible for farmers who were far away, and at the same time "crowded out" some private trade in fertilizer. It is not clear whether the program was permanently discontinued or whether it was just suspended because of the fiscal constraints imposed by the COVID-19 pandemic. However, the sharp increase in global fertilizer prices in 2022 led the government to allocate money to a fertilizer subsidy for the long rainy season; this, as before, is administered through the NCPB.

In 2013, as part of the devolution of government functions, most agriculture functions were transferred to the county level . The national government has retained the role of policy development for the agriculture sector, while county governments are responsible for implementation of policies and programs. As a result, some county governments have been providing fertilizer subsidies to farmers. However, there is no coordination among the counties concerning these, and they follow no common criteria, since each county prepares its own development plan to guide its investments.

To support greater access of smallholder farmers to agricultural inputs, the national government initiated the NVSP in 2019. This aims to make a wide range of inputs accessible to targeted smallholder farmers through a nationwide e-voucher input subsidy model over the next six years. It seeks to improve the efficiency and effectiveness of delivery of the subsidy and targets 1.4 million *high-need* farming households with a subsidized package of inputs, including fertilizers, agricultural lime, certified seeds, agrochemicals, soil testing services, agricultural insurance services, and livestock and fish feed. The pilot phase began in the 2019/20 cropping year and targeted 309,076 smallholder farmers in 12 pilot counties, focused on coffee, Irish potatoes, maize, and rice. The NVSP is different from the subsidy programs described above, which emphasized fertilizer, but did not have the capacity-building component and did not involve private traders, especially agrodealers, as much.

Fertilizer demand

Kenya's aggregate fertilizer consumption has risen steadily over the past three decades (Figure 8.1). Between 1990 and 2006, the period when fertilizer trade was largely in the hands of the private sector, annual growth in aggregate fertilizer consumption averaged 4.3 percent, while between 2007 and 2021, the period when the government intervened in the fertilizer market with subsidies, annual growth averaged 4.1 percent. A dip in fertilizer consumption in 2018/19 reflects a reduction in the national government's fertilizer supply under the "blanket" subsidy program (see Table 8.3).

Available evidence indicates tremendous strides in fertilizer use on farms after the market reforms. Panel household survey data collected by the Tegemeo Institute show a remarkable increase in the percentage of households that applied fertilizer on crops between 2000 and 2010 among both smallholders (cultivating less than 3 ha) and medium- to large-scale farmers (cultivating at least 3 ha) (Table 8.2). The same data also indicate that the average distance a household traveled to a fertilizer retail shop declined by 58 percent, from 8.1 to 3.4 km. However, the fertilizer application rate has remained stable, at 73–78 kg/ha, with smallholders applying at higher rates than medium- and large-scale farmers (Table 8.2).

As such, the increase in fertilizer consumption over the past years is arising more from the increased number of farmers using fertilizers and less from an

FIGURE 8.1 Trends in aggregate fertilizer consumption, 1990/91–2020/21

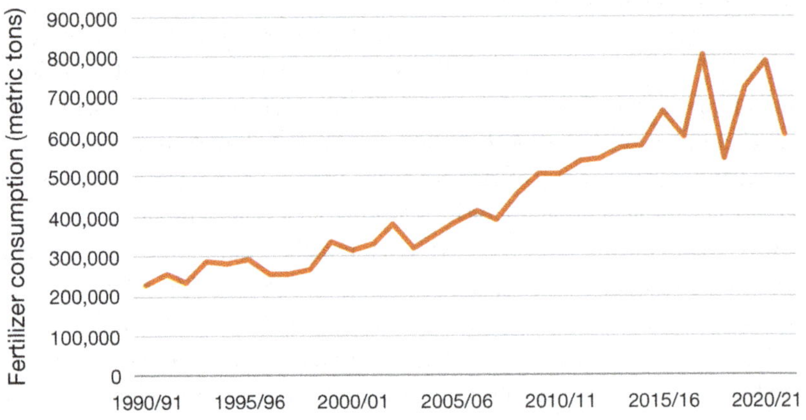

Source: Authors using data from Ministry of Agriculture and Food Security (2010a), Ministry of Agriculture, Livestock, and Fisheries (2015); AfricaFertilizer (https://vifaakenya.org/#/kenya/use).

TABLE 8.2 Share of farm households applying fertilizers and average application rate, 2000–2014

Year	Cultivated less than 3 ha			Cultivated 3 ha or more			Whole sample		
	Sample size	% of households applying	Mean application rate (kg/ha)	Sample size	% of households applying	Mean application rate (kg/ha)	Sample size	% of households applying	Mean application rate (kg/ha)
2000	1,315	67.7	78	163	66.9	59	1,478	67.59	75
2004	1,254	71.2	75	143	72.0	59	1,397	71.30	73
2007	1,219	75.4	79	116	85.3	77	1,335	76.25	78
2010	1,224	74.3	75	83	86.7	66	1,307	75.06	74
2014	6,179	66.6	75	114	57.9	63	6,293	66.45	75

Source: Authors using Tegemeo Institute household survey data.

Note: Data for 2000, 2004, 2007, and 2010 are panel data, and the sample was mainly in the high rainfall areas of Kenya. Data for 2014 are cross-sectional and had wider geographical coverage, including in semiarid areas of Kenya.

FIGURE 8.2 Share of main fertilizer types in aggregate fertilizer consumption, average for 2010–2021

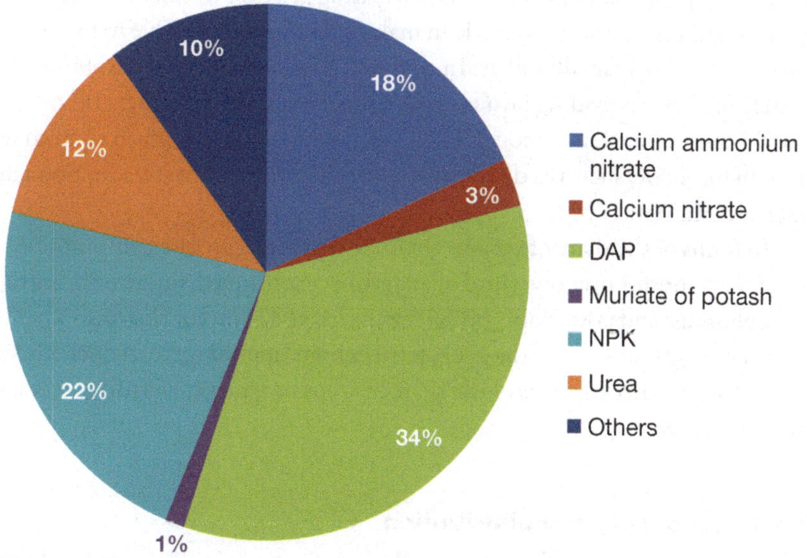

Source: Authors using data from AfricaFertilizer (https://vifaakenya.org/#/kenya/use).

FIGURE 8.3 Share of crops in aggregate fertilizer consumption, average for 2011–2016

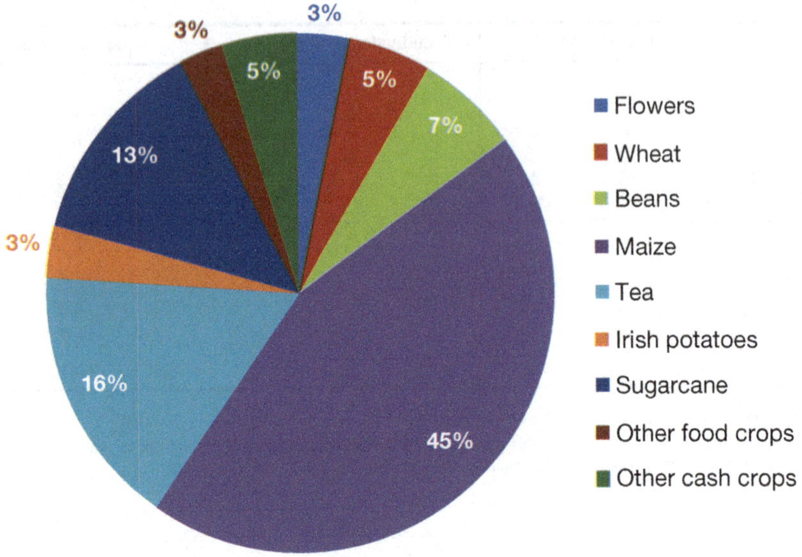

Source: Authors using data from AfricaFertilizer (https://vifaakenya.org/#/kenya/use).

increased application rate. However, increasing prices of fertilizers in recent years may dampen the gains made in making fertilizers accessible to more farmers, especially smallholders. In addition, Duflo, Kremer, and Robinson (2011) find that providing farmers with information about the benefits of fertilizer use alone is not enough to encourage adoption; but, when coupled with a small incentive, such as a discount on the price of fertilizer, the adoption rate increases substantially.

In terms of the share of various fertilizer types in aggregate consumption, DAP accounts for over one-third of all fertilizer consumed, followed by nitrogen, phosphorous, and potassium (NPK) fertilizers, at 22 percent (Figure 8.2). It is worth noting that most of the NPK fertilizers are applied on tea, which ranks second after maize in terms of the share of crops in aggregate fertilizer consumption (Figure 8.3).

Fertilizer supply and distribution

Kenya relies on imports for virtually all its fertilizer needs. The only fertilizer manufactured in the country is the single super phosphate, while fertilizer blending is currently done by only four companies. In 2012, the government commissioned a feasibility study to determine the potential for local fertilizer manufacturing. This concluded that, for fertilizer manufacturing to be a viable investment in the country, the required raw materials (natural gas and phosphate rock) had to be available in commercial quantities and the internal rate of return to the investment needed to be above 15 percent. It found that the government would need to invest at least 30 percent in capital expenditure to incentivize a strategic investor in fertilizer manufacturing. Also, it would be cheaper for Kenya to continue importing fertilizers rather than importing raw materials to manufacture the commodity locally. These findings made the government reconsider its position on fertilizer manufacturing in the near term.

Fertilizer imports show a general increasing trend over the three decades and reflect increasing demand (Figure 8.4). Kenya has approximately 18 active fertilizer importers, with the largest 3 accounting for 55.7 percent of the imports (Oseko and Dienya 2015). Fertilizer distribution is through three main channels: the commodity-based interlinked input–credit–output marketing system implemented mainly by the Kenya Tea Development Agency, for tea; a network of private importers, wholesalers, and retailers; and, government distribution through the subsidy programs. The network of private players comprises the 18 active importers, 150 hubs (wholesale agrodealers), and 8,000 retail agrodealers. These private traders are driven by profit rather than an interest in farm development, and thus mainly stock a few fertilizer types that are familiar to farmers and that sell quickly.

FIGURE 8.4 **Trends in fertilizer imports, 1990/91–2020/21**

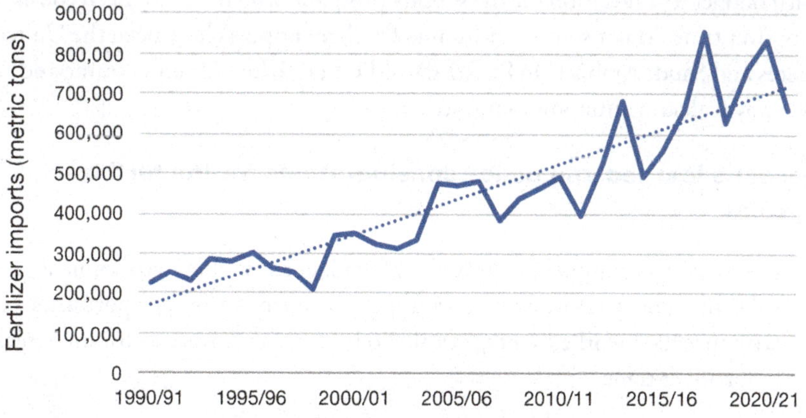

Source: Authors using data from Ministry of Agriculture, Livestock, and Fisheries (2010a, 2015); AfricaFertilizer (https://vifaakenya.org/#/kenya/use).

TABLE 8.3 Quantity of fertilizer procured by government vs. total imports, 2008/09–2018/19

Year	Total fertilizer imported (metric tons)	Quantity of fertilizer procured by government (metric tons)	Share of government procurement in total imports (%)
2008/09	440,689	129,746	29.4
2009/10	465,674	16,624	3.6
2010/11	493,567	96,000	19.5
2011/12	395,774	94,155	23.8
2012/13	522,595	66,276	12.7
2013/14	684,448	171,750	25.1
2014/15	494,718	205,955	41.6
2015/16	556,433	147,926	26.6
2016/17	659,834	177,600	26.9
2017/18	855,044	160,900	18.8
2018/19	626,419	44,250	7.1
Total	**6,195,195**	**1,266,931**	**20.5**

Source: Authors using data from Ministry of Agriculture, Livestock, and Fisheries (2010, 2015); AfricaFertilizer (https://vifaakenya.org/#/kenya/use).

Fertilizer procured by the government between 2008 and 2019 through the National Fertilizer Price Stabilization Plan accounted for 21 percent of Kenya's total fertilizer imports (Table 8.3). This does not include fertilizers procured by county governments for their subsidy programs, which we estimate to be 44,403 tons between 2013 and 2017 (a period of five years) based on the achievement reports contained in the 2018–2022 County Integrated Development Plans for individual counties.[2] The government's share of fertilizer imports has varied with budgetary allocation for the subsidy program, which has not been consistent over time. Major sources of Kenya's fertilizer imports have been the United States and Saudi Arabia (DAP), Russia and Ukraine (urea), and Ukraine and Norway (calcium ammonium nitrate).

Lessons learned and policy considerations for the fertilizer system

- Kenya achieved impressive growth in fertilizer use without government subsidies. Policy reforms and government investments in transport infrastructure have worked synergistically to incentivize private sector investment in fertilizer trade.

2 These are available at Council of Governors (nd).

- While a range of fertilizer types exist in the Kenyan market, farmers have largely stuck to DAP despite soil acidity concerns, which have motivated increased investment in fertilizer blending. There is a need to prioritize farmer learning through more effective extension services. The NVSP, which includes capacity-building for extension staff, farmers, and agrodealers, should be supported.
- Among the challenges encountered in the National Fertilizer Price Stabilization Plan were delayed procurement and delivery of inputs to farmers, poor targeting of beneficiaries, inability of many farmers to reach NCPB depots because of long distances, and inadequate and inconsistent budgetary allocations. These challenges can be addressed to make the NVSP more efficient and effective.
- While some county governments provide fertilizer subsidies, information about targeting, quantity, pricing, fertilizer types, and mode of delivery is not publicly available. Lack of access to such information may hinder proper planning by fertilizer market players.
- Because Kenya relies on imports for its fertilizer needs, and there is considerable instability in fertilizer world prices, fertilizer subsidies will likely continue to be on the government's development policy agenda for the foreseeable future. Therefore, there is a need to ensure that Kenya's gains in expanding private sector trade in fertilizer are sustained.

Seed systems

Seed delivery systems in Kenya are categorized as formal, informal, seed aid, and mixed (CTA 2014). The formal sector refers to the production, processing, packaging, labeling, and marketing of certified seed by registered producers, whereas the informal system denotes production, processing, marketing, and distribution of seed by unregistered seed producers. The formal system started with the establishment of the Kenya Seed Company in 1956 to produce pasture seed for the colonial settlers. Liberalization of the industry in 1996 paved the way for several companies to enter the formal sector, and by 2005 there were 50 registered seed companies (Ministry of Agriculture 2010b). The big companies focused on seeds for cereals (maize, wheat, barley, oats, triticale, and sorghum), oil crops (rapeseed and sunflower), pulses, pastures, horticultural crops, and Irish potatoes. The government provided resources to support the sector to equip technology incubation centers and train technical services teams on testing, inspection, and certification procedures.

The informal seed sector is the major source of planting material for farmers, though the exact source may not be known and the quality of the seed may be questionable (Ministry of Agriculture 2010b). Informal seed sources include "roadside" nurseries for forest and fruit trees, farm-saved seed, farmer-to-farmer exchange, local markets, nongovernmental organizations (NGOs), and community-based organizations (CBOs). Flower companies are also engaged in the informal seed sector through importation and/or local multiplication of the planting material for their own use and sale to other local growers, although they are not registered as seed dealers.

Additionally, Seed Aid (emergency seed) started as a collaborative program between the government, NGOs, CBOs, farmers, and development agencies to supply seeds to communities facing acute seed shortages following drought-related stress in 1992. Although this program was intended to be a limited, one-time intervention, it has become a regular seed source for affected communities. However, its operations are poorly linked with the formal research system.

The mixed seed system combines elements from the formal and informal systems. It is operated by small seed companies and/or commercially oriented individual seed producers that may or may not be registered, with part of the seed produced locally under a certification scheme or imported and packaged locally. Such producers may provide seed for emergency aid, often by cleaning, dressing, and packaging commodities not intended for seed use. An integrated seed delivery system is proposed to respond to the limitations associated with the current seed delivery systems.

The ASTGS acknowledges that providing quality and affordable seeds is an important entry point in promoting agricultural productivity and resilience among smallholder farmers and pastoral communities in Kenya. However, the seed sector faces challenges including limited adoption of new varieties, erosion of plant genetic resources, inadequate seed security stocks, national seed policies that are based on institutions rather than farmers, lack of a clear seed strategy, inefficient extension services, deficient marketing policies, and limited collaboration within the seed sector (FAO 2001). Seed counterfeit issues have been solved through the implementation of regulations by the Kenya Plant Health Inspectorate Service (KEPHIS).

The trends and patterns of seed and supply vary by the type of seed system in use and the agents involved in the seed supply chain. The major roles in the formal seed sector include research and breeding, variety release and regulation, breeder and foundation seed production, seed production, processing and packaging, capacity building, and distribution (Kuhlmann and Zhou 2015).

Trends and patterns in seed usage

The value of seed (mainly certified) demanded in Kenya has been stable over the years (except for in 2016 and 2018), with the highest value recorded in 2015 (Figure 8.5). Official statistics represent seed that is handled by the formal sector and hence is certified seed. There is limited certified seed of pigeon pea, cowpea, sorghum, millet, or green gram produced by private companies, with the demand for such crops often filled by open pollinated varieties. This is because their demand is considered unreliable and insufficient to make a viable business for many seed companies; it is often argued that farmers will buy seeds once and use farm-saved seeds in subsequent seasons.

Overall, a substantial percentage of seeds in Kenya comes from the informal sector, with the main sources being "own seed," followed by local market, neighbors, farmer groups, and seed companies (Figure 8.6). Field days and agricultural shows are the most common sources of seed information within the sector, representing 68 percent and 50 percent, respectively.

FIGURE 8.5 Value of seed planted, 2015–2020

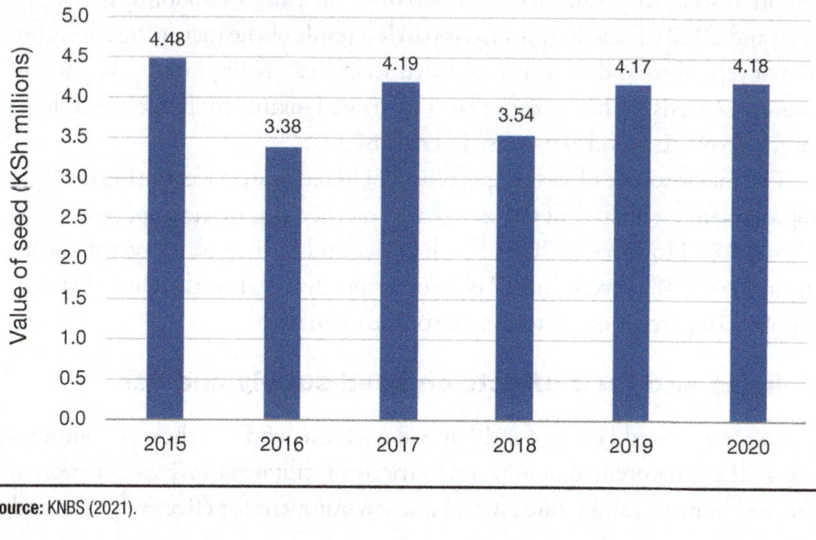

Source: KNBS (2021).

FIGURE 8.6 Main sources of seed as per the share of farmers served

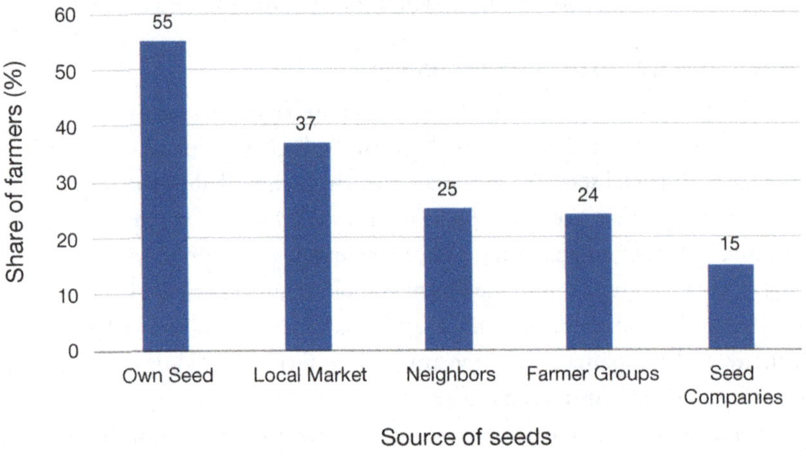

Source: Recha and Recha (2018).

Trends and patterns in seed supply

The value of imported seeds has remained stable, except in 2020, when it almost doubled from the previous year (Figure 8.7). This increase is due to the costly and time-consuming exemption of plants, seeds, and seedlings from the Pre-Shipment Verification of Conformity requirement. Vegetable seeds are the most imported seed; their value has increased over time, almost doubling between 2019 and 2020 (Trade Map). This is partly a result of the increasing consumption of vegetables and the use of certified seeds that are imported. The value of imports of seeds for herbaceous plants cultivated mainly for flowers and those for forage plants has also increased (Trade Map).

The major source of seed imports is the United States, accounting for approximately a quarter of the seed (based on the value of the imports) (Figure 8.8). However, in 2020, South Africa and India took a larger share of the market at 18 percent and 17 percent, respectively, while the United States supplied 20 percent of the total value of the imports.

Policies and their effects on seed supply and use

The National Seed Policy of 2010 provides a framework for the seed industry to realize the full potential of improved varieties, facilitate effective regulation of the seed industry, and create an enabling environment for effective public and

FIGURE 8.7 Value of seed imported to Kenya, 2015–2020

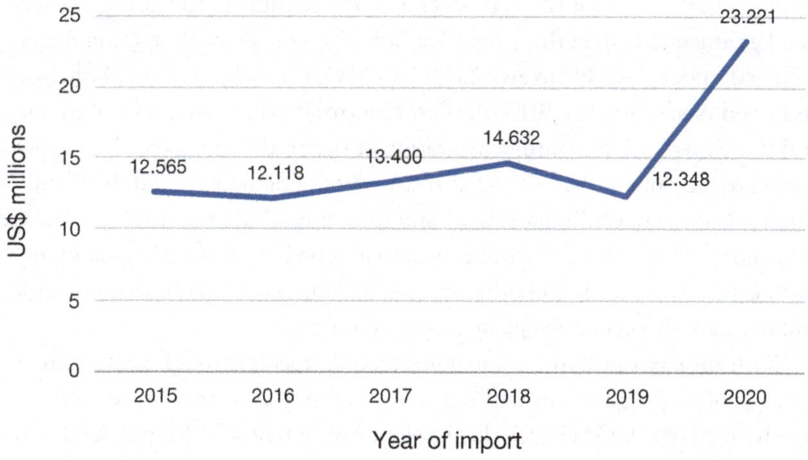

Source: Trade Map (https://www.trademap.org/).

FIGURE 8.8 Proportion of imported seed from major source countries, 2016–2020

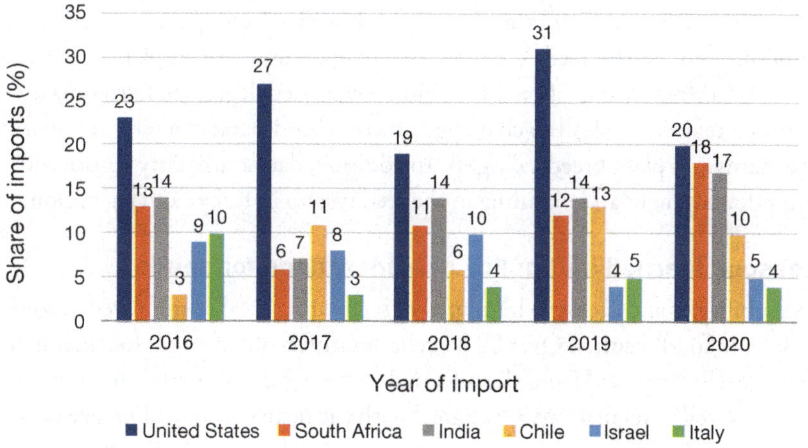

Source: Trade Map (https://www.trademap.org/).

private sector participation in the production and use of quality seeds. The seed industry is regulated by KEPHIS, while the Seed Traders Association of Kenya brings seed companies together with members of the private and public sectors

(Access to Seeds Foundation nd). In addition to the National Seed Policy, Kenya regulates the seed sector through several legal instruments, including the Seed Act, last amended 2012; the Crops Act 2013; and the Plant Protection Act, enforced under the KPHIS Act 2012. In addition, the Agriculture, Fisheries and Food Authority Act 2013; the Pest Control Products Act, revised in 2012; and those involved in the implementation of the Seed Policy Directives apply related regulations such as the Seed and Varieties Regulations and the Plant Variety Protection Ordinance. Laws and regulations have recently been changed to authorize certain seed certification activities on behalf of the regulator, to include forest tree seeds and other species, including wild plant domestication, and to establish a center for plant genetic resources.

With the entry into force of regional seed protocols through the harmonization of seed regulations within the Common Market for Eastern and Southern Africa (COMESA), the development potential of Kenya's seed system will increase, as will the possibility of challenges in implementing laws and regulations (Kuhlmann and Zhou 2015). Kenya is also a signatory to other international treaties, including the World Trade Organization Agreement on Trade-Related Aspects of Intellectual Property Rights, and has been a member of the International Union for the Protection of New Varieties of Plants since May 1999 and thus has adopted the Seed and Variety Regulations, Subordination to the Seed Act (Cap 326) to grant and protect plant breeders' rights (Kuhlmann and Zhou 2015). However, varied degrees of harmonization between regions can lead to challenges in cross-border trade in terms of quality standards and plant breeders' rights. In addition, data availability constraints limit development and planning in the seed system in Kenya and the region.

Lessons learned and policy considerations for seeds

Availability of and access to high-quality seeds are critical to improved productivity and food security. Over 80 percent of farmers obtain seeds for their main crops (excluding corn) from the informal sector, while infrastructure supporting R&D as well as regulations are aligned with the formal system. The size of the informal system suggests a need to implement an integrated seed system that combines the improved seed technologies in the formal sector and the responsiveness of the informal sector in supplying seeds for rare crops, legumes, native vegetables, and tubers (Munyi and de Jonge 2015). This requires (1) integration of the informal seed system into government investment plans in R&D and (2) development of a policy and regulatory framework responsive to the seed subsector.

Plant protection inputs

The agriculture sector has faced increasingly frequent pest and disease outbreaks, as a result of climate change, which has made the environment friendlier for some of the vectors, migratory pests, and also increased transboundary trade that makes countries with lax regulations a risk to others (Skendžić et al. 2021). For instance, the fall armyworm was first reported in Western Kenya in 2016 but is now a major pest in the country, leading to a loss of about 33 percent in annual maize production, estimated at about 1 million tons (De Groote et al. 2020). To address these challenges, the use of plant protection products is necessary. This section explains trends in pesticide use and the policy environment governing the use and disposal of pesticides in Kenya.

Trends in pesticide supply and use

Kenya imports about 90 percent of its chemical pesticides (GIZ 2019). The main classes of pesticides imported are herbicides, pesticides, fungicides, acaricides, fumigants, and nematicides. Figure 8.9 shows trends in imported pesticides between 2010 and 2019. Fungicides were the largest class of pesticides imported during this period, accounting for about one-third of the volumes. The other major classes were insecticides and herbicides. These three classes accounted for about 80 percent of the imported volumes (AAK 2021). China

FIGURE 8.9 Imported pesticide volumes, 2010–2019

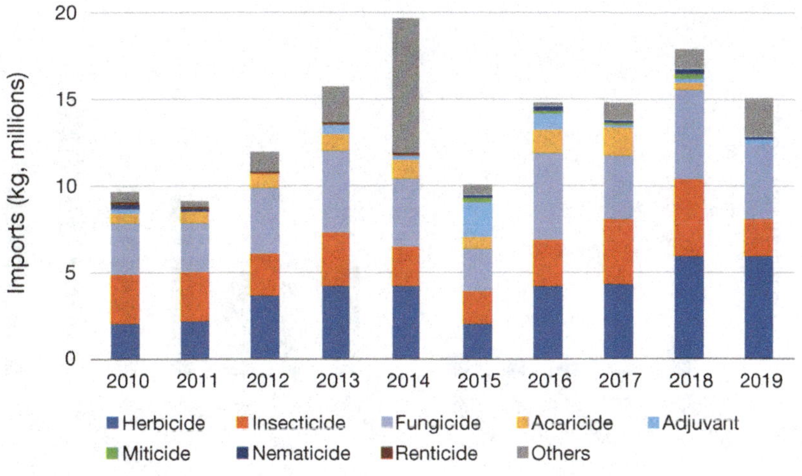

Source: Data from AAK (2020).

and the European Union are the leading sources of pesticides, at 42 percent and 30 percent, respectively.

As expected, use of agrochemicals has increased significantly in the past five years in direct response to the increased incidence of pests and diseases, particularly maize lethal necrosis disease, fall armyworm, and tomato leaf miner. For key commodities such as coffee, horticultural products, and flowers, pest and disease management continue to be a critical production challenge, hence the need to balance increased productivity and safe use and disposal of pesticides. About 18 percent of pesticides used in Kenya are regarded as counterfeit, mainly imported through Uganda and Tanzania (Sarkar et al. 2021). Also, the current disposal of pesticide products after use, or those that are unsold or expired, raises safety concerns.

Pesticide use patterns

Data availability on pesticide use per hectare is limited, and this remains a key gap in addressing food safety issues. It has been estimated that pesticide use in Kenya is less than 1 kg/ha (Figure 8.10). This is consistent with other sub-Saharan African countries that have very low use, such as Rwanda, Sudan, Zimbabwe, and Malawi (Sharma 2019). Conversely, pesticide use in Europe is significantly higher, ranging from 3.35 kg/ha in Spain to 9.86 kg/ha in the Netherlands. Low

FIGURE 8.10 Per hectare pesticide use in several African and European countries, 2014

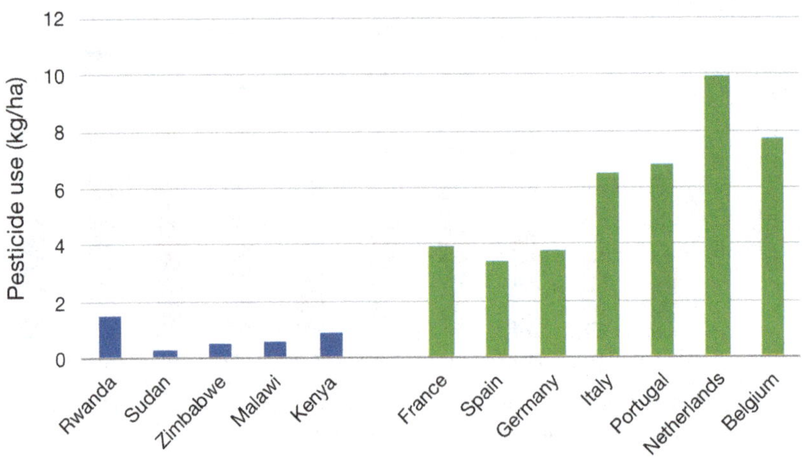

Source: Authors using secondary data (WHO 2019).

utilization in Kenya is consistent with the majority of producers being smallholder subsistence farmers. As the country moves toward the commercialization of agriculture, consumption of plant protection products is expected to increase.

Pesticide policy and regulatory framework

The regulatory framework governing pesticide use in Kenya is widely regarded as one of the most rigorous in Africa, and one of the closest to global benchmarks. Several bodies are involved in the regulation of the use of pesticides and agrochemicals. These include the Pest Control Products Board (PCPB), KEPHIS, and the National Environmental Management Agency (NEMA). The industry also has a self-regulatory mechanism under the Agrochemical Association of Kenya (AAK), which by the end of 2020 had 68 members that were small and medium-sized enterprises involved in the packaging and distribution of pesticides. The Ministry of Agriculture and Livestock Development also plays a critical role in shaping policy on agrochemicals.

PCPB is the primary regulator for pesticide products in the country, mainly responsible for the approval and registration of plant protection products. KEPHIS supports the monitoring of pesticide residues in agricultural products and of threats to agricultural production from pests and diseases. NEMA's mandate is to ensure environmental health for the various land uses through monitoring and enforcement of environmental protection laws. The national government, through the Ministry of Agriculture, Livestock, Fisheries, and Cooperatives, leads in policy development for the sector, including policies on the use and disposal of pesticide products. The Ministry of Health plays a vital role in food safety policies, while county governments ensure the safe use and disposal of pesticide products through the extension function.

AAK has made significant investments to promote the safe use and disposal of pesticides. Industry players developed a brand/mark of quality for pesticide products imported by members of the industry association. Association members invest in field days and demonstrations to train farmers on how to safely use and dispose of pesticide products. This includes training of service providers, who are also equipped with protective gear. Other investments by the industry are in compliance with packaging and labeling requirements and in traceability systems, especially for horticulture products.

Altogether, these institutions are expected to ensure the safe use of pesticide products from production and supply chains until they reach consumers. This includes strict compliance with registration requirements, use and disposal of products, and monitoring of residues in agricultural products. The Pest

Control Products Act, revised in 2012, and its regulations govern the use of pesticide products.

Policy challenges affecting the use of pesticides

Implementation of the regulations on pesticides in the country has faced several challenges. Surveillance and monitoring systems for pesticide use and disposal are weak. This stems from weak data collection systems, lack of investment in modern laboratories to undertake pesticide toxicology testing, and limited funding for R&D to support the regulations. Lack of financial resources has hampered staffing capacities for the regulatory bodies and for key functions such as routine monitoring and surveillance. Additionally, knowhow on pesticide use and disposal among users, especially smallholders, is constrained by the collapsed public extension system on which the majority of farmers relied for information on farming practices. Although the country has established a standard, the Kenya Good Agricultural Practice (KenyaGAP)[3] (Carey 2008), its adoption remains low among farmers, and domestic food markets do not incentivize its adoption through the pricing of products.

Significant gaps remain on the regional front. Key among them is that most of the COMESA Member States have underdeveloped capacity to address trade constraints related to pesticide maximum residue limits (MRLs). This poses difficulties in the production of safe food for both domestic and international markets. Often, the absence of MRL assessment results from a lack of residue data for the particular crop/pesticide combination. Also, most COMESA countries cannot generate high-quality data to establish international trade standards. Regional collaboration on the registration of pesticides and the sharing of registration procedures and data are widely reported not to be robust within the COMESA region or greater sub-Saharan Africa (USDA 2016).

Lessons learned and policy considerations for pesticides

Per hectare use of pesticides in Kenya is much lower compared with use in European countries but similar to that for countries within the region. However, per hectare use is likely to rise as the country moves toward the commercialization of agriculture, and as incidence of pests and diseases increases. This will imply the need to balance increased productivity, and safe use and disposal of pesticides. Strategies to attain this balance include adoption of improved management practices such as varieties resistant to specific pests, diseases, and weeds;

3 KenyaGAP is benchmarked to the internationally recognized GLOBALGAP standard for fruit, vegetables, and flowers.

integrated pest and disease management practices; GAPs; climate-smart crop production practices and technologies; crop rotation to break vector reproduction cycles; and biological control of vectors.

Knowledge and information systems

Low productivity in sub-Saharan African agriculture is attributed to many factors (World Bank 2018; Bonilla-Cedrez, Chamberlin, and Hijmans 2021), key among them being low input use. For instance, low yields in potato farming in Kenya have been blamed on a failure to use clean seeds, fertilizers, fungicides, and irrigation (Wang'ombe and van Dijk 2013)—farmers adopting such technology have more than doubled their yields. Low adoption of technologies has been associated with insufficient knowledge on best practices among intended users and insufficient information to assess benefits associated with using a product, as well as on liquidity constraints, access (Simtowe et al. 2021), and the risks associated with the product. This section discusses the importance of information and extension services and their effect on the supply and usage of agricultural inputs, as well as policies that have shaped information systems and extension services over the years.

Agricultural extension and information systems and input supply and usage

Information helps farmers understand the benefits associated with specific product use (Mastenbroek, Sirutyte, and Sparrow 2021). Thus, when lack of knowledge about the product or lack of information is the barrier to adoption (Shiferaw et al. 2015; Shikuku 2019), relevant information can increase the adoption rate. When such knowledge and information remain inaccessible to different stakeholders, or are not packaged in a form suitable for use by farmers, this affects both supply and demand for agricultural inputs. In Kenya, agricultural information has predominantly comprised technical information on production knowhow and agronomic practices; crop varieties or animal breeds; types of fertilizer, feeds, and pesticides; and their application and where to source them. It has also comprised market information, weather forecasts, and weather-informed advisories on pest incidence, control, and management. More recently, sources of information have expanded to include providers of early warning alerts on risks emanating from pests and weather and climate shocks, and agro-advisories based on weather realities. Considering that farmers are generally unwilling to pay for this information, it is predominantly supplied by the public sector, mainly research institutes and government agencies, which is justified by market failures in the provision of this public good (Anderson and Feder 2004; Maffioli et al. 2011). Regional and international research institutes, the private sector, civil society, and farmer organizations are also important sources of information.

Extension services enable research-to-farm technology diffusion and help overcome the information barrier to adoption by transferring information from research to farmers (Kondylis, Mueller, and Zhu 2017). They also support farmers' decisions on the adoption of technologies by providing information on the advantages, value, or risks associated with a technology (Anderson and Feder 2004; Wanyoike 2019). Extension service providers are also well placed to build capacities of farmers in good agronomic practices (AGRA nd). In some extension models, farmers are linked to other actors in the economy, including input dealers, agro-processors, marketers, and financial institutions (AGRA nd; Farmingtech Solutions nd; Kuza Biashara nd; One Acre Fund nd). Hence, extension service is a critical change agent required to transform smallholder agriculture (Kenya, Agricultural Sector Coordination Unit 2012a). Although proximity to extension service providers is important in the uptake of productivity-enhancing inputs (Muyanga and Jayne 2006), there is now a shift toward digitally enabled extension such as Digital Green and Kuza Biashara. Kenya is a leading agrotechnology hub, with about 60 scalable disruptive agricultural technologies operational in the country (Jeehye et al. 2020).

Policies affecting the agricultural information system and extension services

The government recognizes that transformation of the agriculture sector is knowledge-intensive. However, the nonexistence of appropriate policy and infrastructure for ICT and knowledge management has been a major cause for low technology uptake and the broadening gap between knowledge and application (Kenya, Agricultural Sector Coordination Unit 2012b, 2021). Hence, the expectation is that various public sector-driven information systems will enhance the use of information and the application of technology by various actors.

The National Agricultural Research System Policy (Kenya, Ministry of Agriculture, Livestock, Fisheries and Cooperatives 2021) aims at creating an efficient and effective agricultural knowledge management system by developing an Integrated Agricultural Management Information System as a strategy for improving the processing, storing, and management of knowledge. This is envisioned as a one-stop-shop for all actors, holding all information and knowledge related to agricultural research. To accomplish this, capacities in brokerage and the deployment of knowledge and technology to end-users will be strengthened, and the development and testing of innovative extension approaches will be encouraged (Kenya, Ministry of Agriculture, Livestock, Fisheries and Cooperatives 2021).

The critical role of extension services in the development of Kenya's agriculture sector was first espoused in Sessional Paper 10 of 1965. Determined to revolutionize agriculture, the government identified extension services and training as key in smallholder production (Kenya, Republic of Kenya 1965). Farmer and pastoralist training centers were established across the country to provide extension services to smallholders. Since then, government policies and strategies have influenced the level and stability of funding for agricultural extension services; methods and approaches to extension service provision; and the role of extension providers.

For instance, the World Bank-supported "training and visit extension approach" was introduced in 1982 to supplement the existing extension (Evanson and Mwabu 1998). This was characterized by frequent and continuous training of extension workers, regular and scheduled visits to farmers' fields, and linkages with research. It was later abandoned for being too demanding on resources and for not allowing farmers to articulate their needs, and hence as unsustainable (Kenya, Agricultural Sector Coordination Unit 2012b, 2021). A "commodity-specialized approach" was also in place, led by agricultural commodity boards and private companies. In subsequent years, the government's emphasis has been on intensification through the use of purchased inputs and the commercialization of agriculture as articulated in the SRA, ASDS and, currently, ASTGS.

For years, the government was the sole supplier of extension services to smallholder farmers. This public extension service was, however, criticized for being unable to serve the demands of a modernizing agriculture sector (Muyanga and Jayne 2006; Anderson and Feder 2004). A subsequent decision to freeze public employment and reduce funding in the early 1990s resulted in a massive reduction in public sector extension staff and facilitation (Evanson and Mwabu 1998; Muyanga and Jayne 2006). This greatly affected coverage by extension services, resulting in ineffective service delivery (Kenya, Agricultural Sector Coordination Unit 2012a; Simtowe et al. 2021). Gaps in public extension services led to the emergence of other extension service providers, including faith-based organizations, CBOs and NGOs, the private sector, and, more recently, start-ups that are revolutionizing extension services through digitalization.

Other notable policies have included the National Agricultural Extension Policy, implemented through the National Agriculture and Livestock Extension Programme, which intended to address the aforementioned challenges by promoting pluralistic, efficient, effective, and demand-driven extension services (Muyanga and Jayne 2006). Implementation of an information service—the National Farmers Information Service—was affected by poor funding.

The National Agricultural Sector Extension Policy was later introduced to provide a plan for information and communication, and directions for improved technology delivery to end-users. The policy aimed to see the agriculture sector served by commercialized extension services, while the government would continue providing subsidized services for non-market enterprises and in disadvantaged communities, or partially charge for the services offered. For quality assurance, registration and licensing for extension service providers was to be instituted, and standards enforced. The government would then promote decentralized extension service in line with the devolved structures, empower farmers to organize themselves, and link them to critical resources. It would also encourage the use of ICT and mass media to enhance information sharing. The ASDS supported this policy by focusing on "strengthening and reforming provision of extension services using coordinated, decentralized, multi-sectoral and multidisciplinary approaches that respond to user demand" (nd). Gradual privatization of extension services and capacity building for other extension service providers were key.

Extension services were devolved in 2013 with the aim at taking services closer to the farmers. Realization of the vision for extension was, however, negatively affected by poor coordination between the two levels of government, and inadequate resourcing for agriculture departments in counties. The National Agricultural and Rural Inclusive Growth Project and the Kenya Climate Smart Agriculture Project are designed to overcome these challenges, through better coordination of extension services and partnering with private service providers within counties to deliver the services.

The ASTGS seeks to reduce the ratio of extension officers to farmers by increasing the number of extension personnel (to 3,000 digitally enabled youth extension officers and 200 transformation leaders) (Kenya, Ministry of Agriculture, Livestock, Fisheries and Irrigation 2019). The NVSP is already deploying digitally enabled extension agents in its e-voucher program.

Lessons learned and policy considerations for knowledge and information systems

- Kenya's policies on knowledge and information systems have been progressive, highly adaptive, and responsive to challenges encountered in the management and dissemination of information. Their full effects on technology adoption in smallholder farms have, however, been curtailed by the partial implementation of these policies as a result of poor coordination and under-resourcing.
- Entry of multiple actors and digitalization have pushed the frontier in extension service provision, with tangible effects on input supply and use on smallholder farms.
- Continual improvement of the information system and extension service necessitates the documentation of impacts of policies, and lessons learned being used to inform subsequent policies and programs.

Conclusion

Kenya's policies on inputs have largely been progressive, and have played a critical role in influencing the dynamics of demand and supply of agriculture inputs. However, the policy environment has faced challenges of an incoherent policy mix and negative effects of some policies. For instance, although agricultural input markets are largely liberalized, the country has continued to use subsidies to reduce the cost of inputs, and these have crowded out private sector investment.

The supply of inputs is hampered by unreliable distribution networks and high costs. Hence, there is a need to improve transport infrastructure and eliminate nontariff barriers (for example, delays at roadblocks and weighbridges) and multiple and burdensome regulations, charges, and taxes. The presence of counterfeit products has affected the quality of inputs. This can be addressed through digitally enabled mechanisms to verify the quality of fertilizers and seeds. To improve crop response to fertilizer, and the profitability of its use, a holistic approach is required that focuses on interventions to enhance soil health rather than just access to inorganic fertilizer. The NVSP e-voucher program will support this, since the subsidy package has lime and soil testing services, among other components that will enhance productivity.

Pesticide use is low but is expected to increase as the country moves toward the commercialization and intensification of agriculture. This needs to be accompanied by the safe use and disposal of pesticides through awareness creation and training through extension systems.

Kenya is a leading agrotechnology hub, which presents an opportunity to use digital solutions to link farmers to input and service providers (extension, agro-weather, and so on) and to scale up these solutions for greater impact.

References

Access to Seeds. 2019. "Access to Seeds Index: Kenya." Last updated March 2019.

African Union. 2006. *Report on the Special Summit on Fertilizer for the African Green Revolution.* Addis Ababa.

AAK (Agrochemicals Association of Kenya/CropLife Kenya). 2021. *Annual Report, 2021.* Nairobi.

AGRA (Alliance for a Green Revolution in Africa). (undated). Extension Strategy: AGRA's Private Sector-Led Approach to Extension. Nairobi. https://agra.org/extension-capacity-building/

Ajwang, F., J. Atela, and S. Arora. 2019. *Agricultural Policy Making in Kenya: Why Must Smallholders' Agency be Made Central?* Famler, UK: STEPS Centre.

Allgood, J.H., and J. Kilungo. 1996. *An Appraisal of the Fertilizer Market in Kenya and Recommendations for Improving Fertilizer Use Practices by Smallholder Farmers: A Field Report.* Muscle Shoals, AL: International Fertilizer Development Center (IFDC).

Argwings-Kodhek. 1996. "The Evolution of Fertiliser Marketing in Kenya." Paper presented at the conference Fine-Tuning Market Reforms for Improved Agricultural Performance, Policy Analysis Project. Njoro, Kenya: Egerton University.

ASDP (Agricultural Sector Development Support Programme). Ministry of Agriculture, Livestock, Fisheries and Cooperatives. Accessed 2022. https://asdsp.kilimo.go.ke/

Ariga, J., and T. Jayne. 2011. "Fertilizer in Kenya: Factors Driving the Increase in Usage by Smallholder Farmers." In *Yes Africa Can: Success Stories from a Dynamic Continent,* eds. P. Chuhan-Pole and M. Angwafo, 269–288. Washington, DC: World Bank.

Carey, C. 2008. *Kenya and the KenyaGAP Standard for Good Agricultural Practice.* www.isealalliance.org

Council of Governors. (undated). County Integrated Development Plans 2018–2022.

CTA (Technical Centre for Agricultural and Rural Cooperation). 2014. *Seed Systems, Science and Policy in East and Central Africa.* Wageningen, Netherlands.

De Groote, H., S.C. Kimenju, B. Munyua, S. Palmas, M. Kassie, and A. Bruce. 2020. "Spread and Impact of Fall Armyworm (*Spodoptera frugiperda* J.E. Smith) in Maize Production Areas of Kenya." *Agriculture, Ecosystems & Environment* 292: 106804.

Duflo, E., M. Kremer, and J. Robinson. 2011. "Nudging Farmers to Use Fertilizer: Theory and Experimental Evidence from Kenya." *American Economic Review* 101 (6): 2350–2390.

Evanson, R.E., and G. Mwabu. 1998. "The Effects of Agricultural Extension on Farm Yields in Kenya." Centre Discussion Paper No. 798. Economic Growth Center, New Haven, CT.

FAO (Food and Agriculture Organization of the United Nations). 2001. *Seed Policy and Programmes for the Central and Eastern European countries, Commonwealth of Independent States and other Countries in Transition*. Rome.

Farmingtech Solutions. (undated). "Digicow Dairy App." https://digicow.co.ke/our-products/

GIZ (German Society for International Cooperation). 2019. *Chemical Products-Pesticides Kenya*. Bonn.

Gulati, A., Y. Zhou, A. Tal, and R. Juneja. 2021. *From Food Scarcity to Surplus: Innovations in Indian, Chinese and Israeli Agriculture*. Singapore: Springer Nature.

HLPE (High Level Panel of Experts). 2017. *Nutrition and Food Systems*. Rome.

IFDC (International Fertilizer Development Center). 2001. "An Assessment of Fertiliser Prices in Kenya and Uganda: Domestic Prices vis-a-vis International Market Prices." IFDC Paper Series PCD-27. IFDC, Muscle Shoals, AL.

Jeehye, K., P. Shah, J.C. Gaskell, A. Prasann, and A. Luthra. 2020. *International Development in Focus: Scaling Up Disruptive Agricultural Technologies in Africa*. Washington, DC: World Bank.

Jock, A., and G. Feder. 2004. "Agricultural Extension: Good Intentions and Hard Realities." *The World Bank Research Observer* 19 (1): 41–60.

Kamande, D.M. 2009. "Determinants of Agriculture Sector Performance in Kenya." Master's thesis, School of Economics, University of Nairobi, Kenya.

Kelly V., A.A. Adesina, and A. Gordon. 2003. "Expanding Access to Agricultural Inputs in Africa: A Review of Recent Market Development Experience." *Food Policy* 28 (4): 379–404.

Kenya, Ministry of Agriculture, Livestock, Fisheries, and Cooperatives. 2010a. *Economic Review of Agriculture*. Nairobi.

Kenya, Ministry of Agriculture, Livestock and Fisheries. 2010b. *National Seed Policy*. Nairobi.

Kenya, Ministry of Agriculture. 2011. *National Accelerated Agricultural Inputs Access Program (NAAIAP) Final Evaluation Report*. Unpublished Report.

Kenya, Republic of Kenya. 1965. *African Socialism and Its Application to Planning in Kenya*. Sessional Paper No. 10. Nairobi.

Kenya, Ministry of State Planning. 2007. *Kenya Vision 2030: The Popular Version*. Nairobi.

Kenya, Ministry of State Planning. 2008. *Kenya Vision 2030: First Medium Term Plan, 2008–2012*. Nairobi.

Kenya, Agricultural Sector Coordination Unit. 2012a. *National Agriculture Sector Extension Policy (NASEP)*. Nairobi.

Kenya, Agricultural Sector Coordination Unit. 2012b. *National Agricultural Research System Policy, 2012*. Nairobi.

Kenya, Ministry of State Planning. 2013. Kenya Vision 2030: Second Medium Term Plan, 2013–2017. Nairobi.

Kenya, Ministry of Agriculture, Livestock, Fisheries and Irrigation. 2019. *Agricultural Growth and Transformation Strategy (ASGTS): Towards Sustainable Agricultural Transformation and Food Security in Kenya, 2019–2029.* Nairobi.

Kenya, Ministry of Agriculture, Livestock, Fisheries and Cooperatives. 2021. *National Agricultural Research System Policy 2021, Third Draft.* Nairobi.

KNBS (Kenya National Bureau of Statistics). 2021. *Statistical Abstract 2021.* Nairobi.

Kondylis, F., V. Mueller, and J. Zhu. 2017. "Seeing is Believing? Evidence from an Extension Network Experiment." *Journal of Development Economics* 125: 1–20.

Kuhlmann, K., and Y. Zhou. 2015. "Seed Policy Harmonization in the EAC and COMESA: The Case of Kenya." Working paper. Syngenta Foundation, Basel, Switzerland.

Kuza Biashara. 2021-2022. Kuza. Revolutionalizing Rural Businesses: Creating Livelihood Opportunities and Jobs. Accessed May 5, 2022. https://www.kuza.one/about/

Maffioli, A., D. Ubfal, G. Vázquez Baré, and P. Cerdán-Infantes. 2011. "Extension Services, Product Quality and Yields: The Case of Grapes in Argentina." *Agricultural Economics* 42 (6): 727–734.

Mastenbroek, A., I. Sirutyte, and R. Sparrow. 2021. "Information Barriers to Adoption of Agricultural Technologies: Willingness to Pay for Certified Seed of an Open Pollinated Maize Variety in Northern Uganda." *Journal of Agricultural Economics* 72 (1): 180–201.

Munyi, P., and B. De Jonge. 2015. "Seed Systems Support in Kenya: Consideration for an Integrated Seed Sector Development Approach." *Journal of Sustainable Development* 8 (2): 161–173.

Muyanga, M., and T.S. Jayne. 2006. "Agricultural Extension in Kenya: Practice and Policy Lessons." Tegemeo Working Paper No. 26/2006. Tegemeo Institute of Agricultural Policy and Development, Nairobi.

Nduati, M., P. Ncube, S. Roberts, and T. Vilakazi. 2015. *Non-Confidential Final Report: Market Inquiry on Fertilizer in Kenya.* Nairobi: Competition Authority of Kenya; Johannesburg: Africa: Centre for Competition, Regulation, and Economic Development.

Nin-Pratt, A. 2016. "Inputs, Productivity and Agricultural Growth in Sub-Saharan Africa." In *Productivity and Efficiency Analysis,* eds. W. Greene, L. Khalaf, R. Sickles, M. Veall, and M.C. Voja, 175–201. Cham, Switzerland: Springer.

Odhiambo, W., H. Nyangito, and J. Nzuma. 2004. "Source and Determinant of Agricultural Growth and Productivity in Kenya." KIPPRA Discussion Paper No. 34. University of Nairobi, Kenya.

One Acre Fund. 2022. Our Model. We supply Africa's Farmers with everything they need to grow more food and earn more money. Accessed March 2, 2022. https://oneacrefund.org/what-we-do/our-model/

Oseko, E., and T. Dienya. 2015. *Fertilizer Consumption and Fertilizer Use by Crop (FUBC) in Kenya.* Study conducted for Africafertilizer.org.

Recha, T., and J. Recha. 2018. "Seed Networks for Climate Change Adaptation in Kenya, Uganda and Tanzania." *CCAFS News,* June 25.

Sarkar, S., J. Dias Bernardes Gil, J. Keeley, N. Möhring, and K. Jansen. 2021. *The Use of Pesticides in Developing Countries and Their Impact on Health and the Right to Food*. Strasbourg, France: European Parliament.

Sharma, A., V. Kumar, B. Shahzad, M. Tanveer, G.P. Singh Sidhu, N. Handa, S. Kaur Kohli, et al. 2019. "Worldwide Pesticide Usage and Its Impacts on the Ecosystem." *SN Applied Sciences* 1: 1446.

Shiferaw, B., T. Kebede, M. Kassie, and M. Fisher. 2015. "Market Imperfections, Access to Information and Technology Adoption in Uganda: Challenges of Overcoming Multiple Constraints." *Agricultural Economics* 46 (4): 475–488.

Shikuku, K.M. 2019. "Information Exchange Links, Knowledge Exposure, and Adoption of Agricultural Technologies in Northern Uganda." *World Development* 115: 94–106.

Simtowe, F., D. Makumbi, M. Worku, H. Mawia and D. Bahadur Rahut. 2021. "Scalability of Adaptation Strategies to Drought Stress: The Case of Drought Tolerant Maize Varieties in Kenya." *International Journal of Agricultural Sustainability* 19 (1): 91–105.

Skendžić, S., M. Zovko, I.P. Živković, V. Lešić, and D. Lemić. 2021. "The Impact of Climate Change on Agricultural Insect Pests" *Insects* 12 (5): 440.

USDA (US Department of Agriculture) Foreign Agricultural Service. 2016. *Food and Agricultural Import Regulations and Standards-Narrative Kenya*. FAIRS County Report. Washington, DC.

Wang'ombe, J.G., and M.P. van Dijk. 2013. "Low Potato Yields in Kenya: Do Conventional Input Innovations Account for the Yields Disparity?" *Agriculture & Food Security* 2: 14.

Wanyoike, S. 2019. "Seeing is Believing: Helping the Kenyan Farmer Adopt Good Agricultural Practice." *LDRI Blog,* October 14.

World Bank. 2018. *Kenya Poverty and Gender Assessment 2015/16: Reflecting on a Decade of Progress and the Road Ahead*. Washington, DC.

WHO (World Health Organization) and FAO (Food and Agriculture Organization of the United Nations). 2019. *Global Situation of Pesticide Management in Agriculture and Public Health: Report of a 2018 WHO-FAO Survey*. Geneva: WHO; Rome: FAO.

Zeleza, T. 1991. "Economic Policy and Performance in Kenya since Independence." *Transafrican Journal of History* 20: 35–76.

Chapter 9

MECHANIZATION OF AGRICULTURAL PRODUCTION IN KENYA: CURRENT STATE AND FUTURE OUTLOOK

Pascal Kaumbutho and Hiroyuki Takeshima

Agricultural mechanization is the use of machinery, equipment, and implements—rather than human or animal power—to carry out agricultural practices. When the use of mechanization is sufficiently high, it can help improve the overall efficiency of food systems, reduce the costs of producing outputs and providing services, enhance economies of scale, and raise labor productivity and incomes (FAO and AUC 2018; Diao, Takeshima, and Zhang 2020). While mechanized practices are traditionally thought of in terms of tilling, seed drilling, and spraying, in recent years mechanization has been considered to include broader applications along the food system, such as irrigation, postharvest cleaning of harvests, cold storage, value addition, and processing.

Agricultural mechanization is considered one of the critical tools to achieve Kenya's broader agricultural development strategies (Kenya, Ministry of Agriculture, Livestock, Fisheries, and Irrigation 2019). In particular, agricultural mechanization is expected to contribute to the scaling-up of larger commercial production; increasing the productivity and incomes of small farmers, pastoralists, and fisherfolks; supporting aging farmers and attracting youth into modern farming; and facilitating the growth of industrial agroprocessing. It may also facilitate the adoption of other modern inputs and practices like irrigation, improvements in food system resilience, stability of food supplies, sustainable use of land and natural resources, and climate change adaptation (ibid.). Also, suppliers of mechanized equipment, service providers, and processors are regarded as critical change agents to improve market access to inputs and offtake (including transportation).

As part of its agricultural development strategy, Kenya's latest National Agricultural Mechanization Policy (Kenya, Ministry of Agriculture, Livestock, Fisheries, and Cooperatives 2021) prioritizes a range of approaches to modernize mechanization. These include the development of machine and equipment value chains, fostering the machine development and fabrication industries, stimulating machine investments through e-voucher-based subsidies and supplementary financing mechanisms, and quality assurance for machines through enhanced regulatory capacity. The policy also emphasizes increased R&D on mechanization; technology development; modern mechanization; sustainable and climate-smart mechanization; and improving knowledge and skills on machines, machine operations, and mechanization practices. In addition, it promotes the expansion of high-quality mechanization service provision, in combination with e-voucher-based subsidies and registration of service providers eligible for voucher redemption, as key instruments to improve access to mechanization technologies for smallholders. Lastly, the policy highlights the goal of addressing gender equality in mechanization use.

However, optimal promotion of agricultural mechanization in Kenya requires careful assessments of demand levels, and the identification of an appropriate public sector role in mechanization sector growth if such demand is sufficient. Overall demand for agricultural mechanization depends on the level of farming system intensification, which affects returns to more frequent land preparation and tillage, as well as relative costs of mechanization compared with substitutes like human labor and draft animals (Diao, Takeshima, and Zhang 2020). Desirable forms of mechanization can vary considerably across different food production environments. And increased mechanization may be beneficial under only certain conditions. Mechanization may not make economic sense in a more subsistence-based system (temporal). As later sections discuss, shifting cultivation and fallow remain more productive methods than tillage in parts of Kenya where the population density remains low. Excessive mechanization that displaces labor can reduce labor productivity and income. Mechanization adoption also often follows a particular sequence depending on the farming operation. For example, it typically starts with the mechanization of power-intensive, static operations, and then involves the mechanization of more control-intensive, mobile operations (Pingali, Bigot, and Binswanger 1987).[1]

1 Agricultural mechanization processes generally move from manual operations to animal-powered and onward to motorized (engine-powered) applications. This progression taps into the existing technological and industrial advancements and market demand prevailing at a given time. For example, FAO and AUC (2018) illustrate the mechanization process in six stages. In the first stage, draft animals or relatively simple machines assist in hand-tool (manual) operations. Mechanization in stage 2 substantially replaces manual operations, except for in control-intensive operations like weeding. Stage 3 mechanization advances into precision planting and spraying operations, calling for machine calibration. This approach can accommodate or influence cropping patterns, from multiple crops to monocropping systems, for operational efficiency. Stage 4 mechanization involves adapting the farming system and production environment in response to mechanization. Land management that exploits economies of scale becomes important, sometimes promoting the consolidation of small farms. This stage also often involves mechanization for poultry and livestock production. Approaches like conservation agriculture systems of reduced soil manipulation and retaining cover on the soil surface may become more relevant. Stage 5 mechanization involves adapting crops to improve machine performance efficiency. Crops are bred for better height and less lodging to be suitable for mechanical threshing and to resist bruise damage during mechanical harvesting. Stage 6 involves automation leading to workable higher levels of mechanization, with many production operations tapping into machine intelligence. Mechanized operations at this level are used for feeding systems in livestock production, automated application of fertilizers and other chemicals, and GPS-positioned machinery in planting, spraying, and the like.

While this chapter focuses primarily on the potential roles of mechanization in Kenya and suitable strategies for the Kenyan government's mechanization promotion policies, these are relevant only if demand for each stage of mechanization has risen sufficiently in the relevant food production environment. Where demand is indeed sufficient, appropriate supply-side strategies are required to identify and address key market failures in the agricultural mechanization sector while minimizing government failures so the industry remains efficient and competitive.

With this background, this chapter describes current challenges and strategies for an optimal mechanization approach in Kenya. The first section reviews recent trends in mechanization in Kenya and general mechanization practices in comparison with other African countries. The chapter then discusses key drivers of mechanization and potential roles in the future. Finally, it describes key policy issues with regard to mechanization growth in Kenya, including functions of the public sector and key principles of support approaches.

Mechanization status in Kenya

Farmer typologies and mechanization

Agricultural mechanization in Kenya dates back to the 1954 Swynnerton Plan, famous for the land demarcation policy implemented to commercialize small-scale agriculture in the country. In this plan, native farmers were settled on newly cleared 10–20 acre farms in the Highlands, considered to be of sufficient size to be viably mechanized (Swynnerton 1954; IBRD 1960). Farmers were given deeds to their newly acquired land. This plan facilitated the establishment of vibrant agroenterprises and industries anchored on small and medium-scale farms after independence. Since then, the Kenyan agriculture sector has consisted mainly of four broad types of farmers—namely, peasant subsistence farmers, small-scale commercial farmers, medium-scale farmers, and large-scale farmers.

Peasant subsistence farmers typically cultivate less than 2 ha. They rely on family labor and hand-tool technologies for land preparation and crop husbandry tasks. Few of these farmers hire tractors or draft animal power for land preparation. Small-scale commercial farmers, who have a bit more land (cultivating 2–10 ha), typically use draft animal power where available, or sometimes tractors and machines for rotating paddy fields, planting, or shelling maize. Some may own a four-wheel tractor (4WT) and offer a tractor-hiring service to other small-scale commercial farmers.

FIGURE 9.1 Area under different farm sizes in Ghana, Kenya, Tanzania, and Zambia, 2015

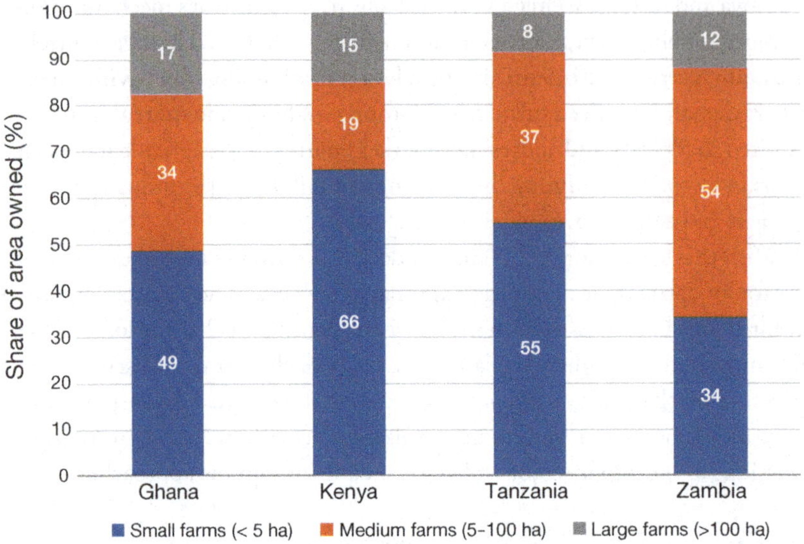

Source: Adapted from Jayne and Ameyaw (2016).

Medium-scale farmers typically cultivate 10–100 ha. They are more likely to own a 4WT and an assortment of implements, although some rely on hired services if available. If they own a 4WT, they often offer a tractor-hiring service, including off-farm activities such as transportation, since they are unlikely to attain economical use rates on their farms alone.

Large-scale farmers (typically cultivating 50–2,000 ha) often own a range of 4WTs with their assorted implements, and operate a significant proportion of the tractor fleet in the country (De Groote, Marangu, and Gitonga 2020). They may also hire specialized machinery like combine harvesters. Also, they may offer a tractor-hiring service on a contract farming basis. Historically, large-scale farmers have produced cash and/or industrial crops such as coffee, sisal, tobacco, pyrethrum, flowers, and horticultural products like tea, maize, rice, wheat, dairy, beef, and sugarcane, among others (Mayne 1955; Eicher and Baker 1982). At independence in the 1960s, large farmers were predominately settler farmers and transnational corporations. During the 1970s and 1980s, they consisted of many government-owned state and private farms. After the economic structural adjustment programs of the 1990s, most state farms were privatized.

At independence in 1963 and immediately thereafter, most cultivated land was owned by peasant subsistence farmers, small commercial farmers, and large colonial settlers. However, this situation changed over time, especially from the beginning of the 21st century, as in other African countries. Medium- and large-scale farmers have recently grown to account for one-third of farmland in Kenya, following patterns occurring in countries like Ghana, Tanzania, and Zambia (Figure 9.1). The growth in the medium-scale farmer group has been driven by effective demand for agricultural products generated by urbanization, income growth, off-farm employment opportunities, rising wage rates, and enhanced demand for mechanization (Clarke and Bishop 2002). All of these types of farmers—peasant, small commercial, medium- and large-scale—coexist in Kenya, with neither small nor large farms dominating (Muyanga and Jayne 2019).

Stocktaking of mechanization in Kenya

As measured by horsepower per hectare of farmland, the mechanization level in Kenya has remained relatively stable during the past six decades (Figure 9.2). It

FIGURE 9.2 Growth of major agricultural machinery inventories in Kenya and other selected countries/regions

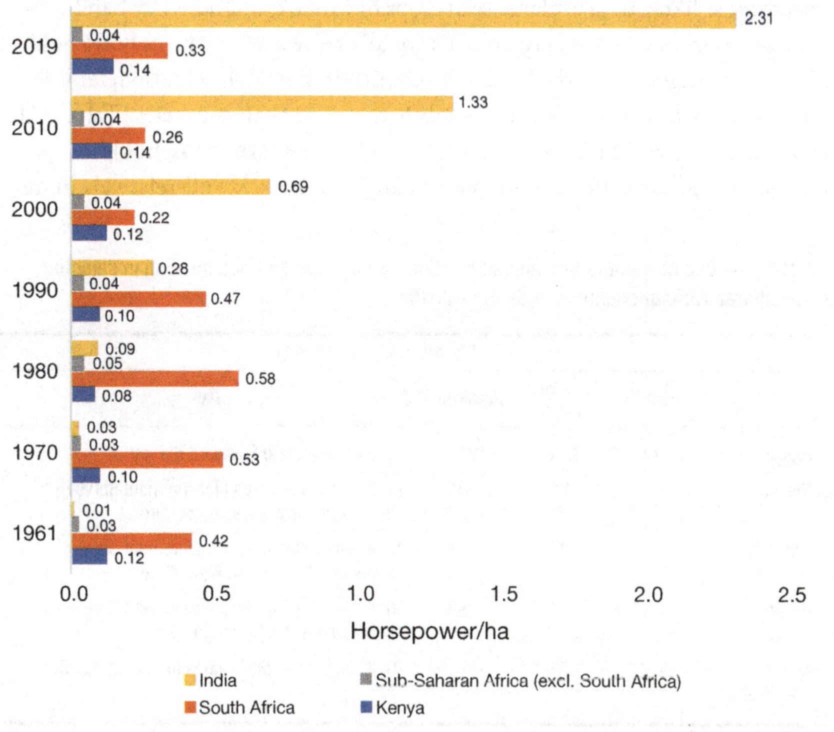

Source: Authors' calculations using USDA (2022).

Note: Cultivated land includes cropland and area under permanent crops. Major machineries include tractors, combine threshers, and milking machines.

increased slightly from 0.08 hp/ha in 1980 to 0.14 hp/ha in 2019, mechanizing slightly faster than the whole of sub-Saharan Africa (excluding South Africa). However, the rate of change has been much lower compared with that in Asian countries like India.

MECHANIZATION OF LAND PREPARATION

Most Kenyan smallholder farmers still rely on hoe-based operations or oxen-hauled mouldboard plows (De Groote, Marangu, and Gitonga 2020). A survey in 2012 based on representative samples of farm households in a significant maize production region in Kenya indicated that only 2 percent of maize farmers owned tractors, 33 percent owned oxen, and 28 percent owned plows. However, the last two shares exhibit a significant increase since 1992, from 12 and 17 percent, respectively (De Groote et al. 2020). The use of oxen and plow has been more common in dry or moist mid-altitude zones, while the use of tractors has historically been concentrated in highland maize areas (De Groote et al. 2020, Figure 12.4). Except in places like the North Rift region, smallholders are most likely to plant large seed crops like maize and beans by hand.

Figures from selected countries' nationally representative farm household survey data suggest that the level of mechanization in Kenya is also relatively consistent with that in other major sub-Saharan African countries (Table 9.1). In Ghana, Nigeria, Tanzania, and Uganda, shares of farm households using tractors in the late 2010s are not more than 13 percent. While relatively more

TABLE 9.1 Use of tractors and animal traction for all crops and rice farming in selected sub-Saharan African countries in the mid-2010s

	Share of farming households (%)			
	Tractors	Animal traction	Hand hoe	Data
Ghana	13	N/A	N/A	2017 Ghana Living Standard Survey
Nigeria	10	25	65	2018 Living Standards Measurement Survey: Integrated Surveys on Agriculture
Tanzania	9	37	54	2014 Living Standards Measurement Survey: Integrated Surveys on Agriculture
Uganda	< 1	7	93	2018 Living Standards Measurement Survey: Integrated Surveys on Agriculture
Kenya (ownership)	2	33	< 65	2012 (De Groote, Marangu, and Gitonga 2020)

Source: Authors using De Groote, Marangu, and Gitonga (2020); Takeshima and Mano (2022).

Note: Figures for Kenya are the share of farm households owning tractors and animal traction tools, rather than the share of farm households using them, and representative within maize production areas only (De Groote, Marangu, and Gitonga 2020).

common, the share of farm households using animal traction remains about one-third or less in these countries. Mechanized land preparation, used in Kenya, often combines disc plow and harrow. Several runs of the plow are often applied, followed by several passes of the harrow to break the clods. Tractors are used mainly by cereal growers, although some medium-scale farmers may have mechanized maize or wheat planters to follow the harrowing.

MECHANIZATION OF LAND MANAGEMENT

Agricultural mechanization of operations other than land preparation has lagged in Kenya. However, small hand-held motorized tools are gradually replacing manual labor, allowing more focus on land management activities that are more attractive to youth (Kaumbutho 2016). These tools include brush cutters, motorized knapsack sprayers, fence cutters, weeders, and power tillers in irrigated rice schemes (Diao, Takeshima, and Zhang 2020; Mano, Njagi, and Otsuka 2022).

Typically, control-intensive farming operations like weeding are conducted primarily by hand, mainly using family or hired women's labor in developing countries in Africa (Takeshima and Diao 2021), and this is expected to be the case in Kenya as well. Mechanized crop maintenance, like hiller weeding in potato farming and spraying herbicides and pesticides, is rare for smallholder farmers in Kenya. Mechanized spraying through knapsack sprayers is occasionally done for pest and disease control. This has attracted some youth to form "spray gangs" to serve farmers for a fee. Some have progressed to using motorized knapsack sprayers and brush cutters for weeding and tidying farms with overgrown bushes (Kaumbutho 2016).

MECHANIZATION OF POSTHARVEST HANDLING PRACTICES

Harvest and postharvest mechanization varies widely, depending on the size of the enterprise, the value chain involved, and the market. Combine harvesters are used chiefly for cereal crops, mainly maize, wheat, and barley, and recently rice. Some combine harvester service providers travel 200–300 km to service customers. Similar custom hiring services for mechanized harvesting provided by migratory service providers to smallholders have been emerging in Asia and other African countries like Ethiopia, albeit to a limited extent in the latter (Diao, Takeshima, and Zhang 2020). Some youths have developed businesses ferrying around small shellers and threshers made locally by informal sector artisans to provide services. A few companies have introduced harvesting machines for tuber crops like Irish potatoes (Kaumbutho 2016).

The *tuk-tuk* (three-wheeler) and *bodaboda* (motorbike) transport service sectors have also grown in the postharvest transport arena (ibid.).

Modern rice milling machines have also spread, albeit slowly, in rice production regions, including in the Mwea Irrigation Scheme in Kenya (Mano, Njagi, and Otsuka 2022; Takeshima and Mano 2023). These modern milling facilities, equipped with destoners and grading machines, among other features, have often been imported by private sector entrepreneurs. The machines significantly reduce per unit milling costs and improve milling quality, increasing the supply of better-quality milled rice at lower prices (Mano, Njagi, and Otsuka 2022). Importantly, however, a steady and sufficient supply of raw materials (paddy) enabled through well-functioning irrigation infrastructure has been critical to the viability of these modern mills. Outside a few successful irrigation schemes within sub-Saharan Africa, an insufficient supply of paddy has often crippled the economic viability of modern milling, for example, in Nigeria (Gyimah-Brempong, Johnson, and Takeshima 2016).

MECHANIZATION IN THE FORAGE SECTOR

The dairy industry in Kenya has progressed well, and by large strides compared with neighboring countries. Mechanization application for forage production and harvesting is also an important part of the initial mechanization of the livestock sector (for example, Gürsoy 2017; Benoit and Mottet 2023). A farmer with fewer than three animals may manage with only a machete and a chuff cutter for the hay or Napier grass, brachiaria, sorghum, or lucerne harvests. A more advanced farmer may have the services of a hayseed broadcaster, mower, rake, baler, or chopper for animal feed. Recently, it has become possible to chop fodder and pack it directly without the need for ensiling. More modern farmers may even access a feed mixer or a mini or full-size milking machine.

MODALITY OF MECHANIZATION SERVICE PROVISION

All these mechanization activities in Kenya are enabled through private sector custom hiring mechanization services. *Nomadic service providers* use tractors and disc plows to serve farmers in several localities. They traverse the land following rainfall isohyets in search of land needing plowing. These providers can travel some 700 km across the country and be away from home for around eight months a year. Elsewhere in sub-Saharan Africa, mechanization service providers of similar characteristics typically serve on the order of 100 ha in a year, such as in Nigeria (Takeshima et al. 2015). In Kenya, too, they are likely to remain critical providers of mechanization services to smallholders who have demands but cannot own machines themselves. Nomadic service providers are likely to continue to dominate in the short term, where high-tech service provision is still underdeveloped.

Some *government service providers* also offer mechanization services within their respective county boundaries, using county government-owned tractors and implements. These government-operated providers may be less efficient than the nomadic providers, as has been experienced in other African countries (Diao, Takeshima, and Zhang 2020).

In recent years, digital technologies have enabled a new type of service provider: *the innovative and modern tech-backed provider*. These service providers work in a given locality and particular value chains. They often establish a working base near farmers and use technology to manage and coordinate services, serve farmers more effectively, and provide more presence and accountability to their customers. These private sector service providers, particularly modern tech-backed providers, have significant potential to improve the availability, accessibility, affordability, accountability, and climate-smartness of mechanization services (Balyamujura, Kaumbutho, and Karu 2018). Innovative service providers in Kenya include Hello Tractor Ltd and Agrimech Ltd. Hello Tractor uses GPS-based monitoring devices to remotely monitor tractors and digital booking platforms that match farmers with the nearest tractors (Daum et al. 2021). Agrimech Ltd. manages the fleet of these device-installed tractors. In Nigeria, where Hello Tractor operates similar services, these modern methods have significantly reduced transaction costs in mechanization service provisions (Daum et al. 2021). While such modern tech-based service provision is still nascent in Kenya, it has potential to improve the affordability and accessibility of mechanization services for smallholders.

Considerations for agricultural mechanization into the future

Future drivers of demand for mechanization

FARMING SYSTEM INTENSIFICATION

Demand growth for mechanization in Kenya in the near future depends on the intensification levels of farming systems. In parts of Africa where the intensification level is low, such that most farmland is cultivated only once in a few years and left fallow most of the time, shifting cultivation remains cheaper than mechanized tillage to clear land (Diao, Takeshima, and Zhang 2020). Based on an index called R-value (which proxies the level of farming system intensification), farming systems in Kenya are less intensive than in many other African

FIGURE 9.3 R-values in African countries and developing regions, 1960–2017

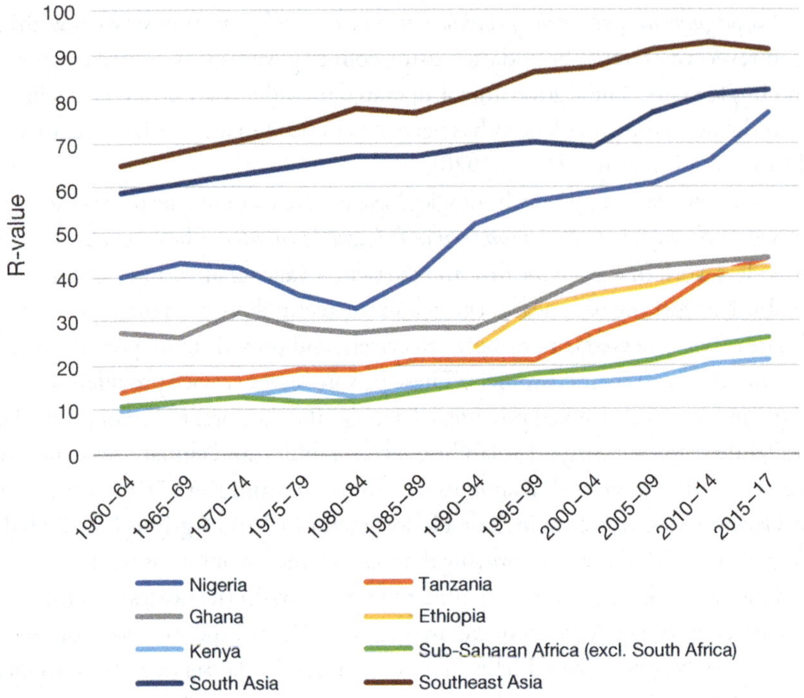

Source: Authors' calculations using FAOSTAT (2022).
Note: Rvalue = (Harvested area of all crops summed) / (Arable land + Permanent pasture and meadows) * 100.

countries. Still, the intensification level is gradually increasing (Figure 9.3). This may explain why, for a majority of land in Kenya, animal traction remains more viable than tractors for the foreseeable future (De Groote, Marangu, and Gitonga 2020).

However, the national figures mask considerable heterogeneity in farming systems. There are pockets within Kenya where farming systems are sufficiently intensive to generate significant returns from investment into mechanization. For example, the maize agroecological zones and the low tropics zone (the coastal area in southeast Kenya) have seen relative growth in tractor ownership in recent years, while ownership has stagnated in other zones (De Groote, Marangu, and Gitonga 2020). Better understanding the heterogeneity of farming systems and varying mechanization potential within Kenya is a crucial part of designing mechanization support efforts. Recent studies have also identified pockets of areas like Machakos and Kisii counties in Kenya, where the

FIGURE 9.4 Population growth (rural and urban) and urban share in Kenya

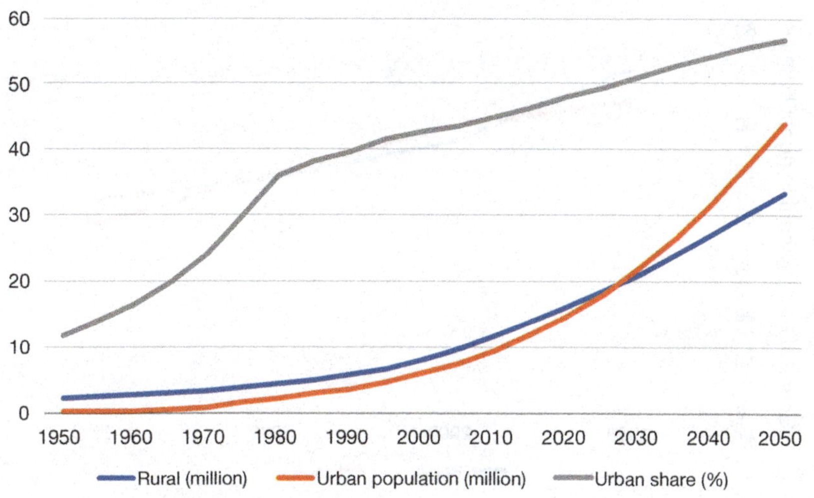

Source: UN DESA Population Division (2018).

need for more sustainable land management becomes vital as a result of rising population density (for example, past 600 persons/km^2) (Willy, Muyanga, and Jayne 2019).

DEMOGRAPHIC OUTLAY

Urbanization and labor movement out of the agriculture sector will intensify and drive demand for mechanization. The urban population in Kenya, which currently accounts for around 48 percent of the total population, is expected to reach 57 percent by 2050, putting pressure on rural labor to be ever more productive (Figure 9.4).

Figure 9.5 shows that the employment share of the agriculture sector in Kenya has been declining over time. This pattern has been consistent with other sub-Saharan African countries where the share of full-time employment in farming has dropped to less than 50 percent, with an increasing number of jobs shifting to off-farm employment such as agroprocessing and trade (Jayne et al. 2016). Figure 9.5 also shows that the share has been comparable with that in India, where mechanization has grown considerably (with 90 percent of land preparation currently done by tractors) (Diao, Takeshima, and Zhang 2020). These trends in Kenyan agricultural transformation suggest that, demographically speaking, the potential of mechanization is rising.

FIGURE 9.5 Employment share of the agriculture sector in Kenya and India

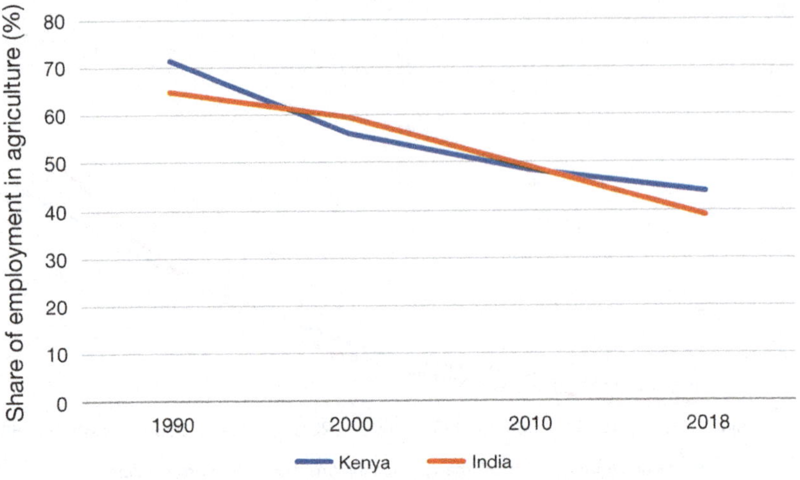

Source: de Vries et al. (2021).

Growing urbanization in Kenya can also affect investments in land and machines by local entrepreneurs and foreign investors, including emerging medium-scale "absentee" or "telephone" farmers. These "farmers" notice that the original farmer is aging and unable to perform many manual tasks, as evidenced by the statistics of falling food volumes. They also have an in-built passion for agribusiness but carry the label of a corporate worker, "stuck" at their office job accumulating resources that can be invested in farming, alone or with partners (Kaumbutho 2016). Such absentee farmers are also consumers of organized and accessible mechanization services and inputs on hire. Mechanization, combined with irrigation and other rural infrastructure such as road improvement and storage facilities, can help well-resourced midlife farmers to fill the labor shortage gap. These farmers will likely create a more enabling environment for investments in agricultural mechanization in Kenya in the near future (ibid.).

Evolving roles of mechanization to address emerging challenges

PRODUCTIVITY, EFFICIENCY, ECONOMIES OF SCALE, AND SCOPE

Agricultural mechanization can provide various benefits in appropriate contexts of sufficient demand and affordable costs of machines and hiring services. These benefits include reduced food production costs (Diao, Takeshima, and Zhang 2020), increased productivity through better land preparation timing

(Kosura-Oluoch 1983), increased returns to intensive tillage, and reduced harvesting losses and/or higher processing and milling efficiency (Takeshima and Liu 2020; Zheng, Ma, and Zhou 2021; Mano, Njagi, and Otsuka 2022). Mechanization can enhance economies of scale and returns to specialization (see examples from Ghana in Takeshima, Houssou, and Diao 2018). Recent studies also show that mechanization can improve economies of scope in a particular farm setting, facilitating crop diversification and, thus, resilience (see examples from Nigeria in Takeshima, Diao, and Aboagye 2020).

SUSTAINABILITY AND CLIMATE CHANGE RESILIENCE

Exploring the roles of mechanization in achieving environmental sustainability is an important consideration. Climate change, carbon dioxide emissions, and how these are related to overall farm production, specifically mechanization technologies, chemical use, and precision farming, will all need to be tracked to ensure the sustainability and resilience of food system goals are met (Mrema and Rolle 2003). Climate changes, particularly rising temperatures, are found to increase weeds on maize plots in Kenya (Jagnani et al. 2021), where suppressing weeds through mechanized land preparation could play a greater role. In parts of Kenya, community-based electric micro-grids have enabled cold storage (Kirubi et al. 2009), which, if combined with renewable energy like solar power, could be another example of mechanization enhancing climate resilience during storage through sustainable energy sources.

Soil degradation has been contributing to the loss of agricultural productivity by 1 percent a year or more in extensive areas in Kenya (World Bank 2008). Conservation agriculture (CA) is considered a resource-preserving method that can benefit from improved mechanization technologies (Friedrich 2013). CA is found to improve on conventional practices in drier climates with seasonal rainfall at less than 600 mm and seasonal temperatures above 20°C, and in other humid subtropics, given good management (Laborde et al. 2020), which can apply to much of Kenya. Recently, the Conservation Agriculture Sustainable Agriculture Rural Development Project, in collaboration with African Conservation Tillage Network, piloted the use of CA machinery in Kenya (Kenya, Ministry of Agriculture, Livestock, Fisheries, and Cooperatives 2021). The Kenya Draught Animal Technology Project promoted draft animal use and/or small-motorized CA machinery (ibid.). The Common Market for Eastern and Southern Africa (COMESA) has supported using land preparation CA equipment. Further knowledge bases should be built through these initiatives, including assessing where CA technologies are particularly effective and viable in Kenya.

Once such niches are identified, business models for CA mechanization hiring services can be promoted. In the Indo-Gangetic Plain in Asia, for example, zero tillage has spread considerably among smallholders in recent decades: around 5 million ha out of 13 million ha under the rice-wheat system have adopted such practices (Chauhan et al. 2012; Keil, D'Souza, and McDonald 2016). There, zero tillage technologies have been provided to smallholders through custom hiring services using zero till drills attached to 4WTs (Keil, D'Souza, and McDonald 2016). The experience in India can serve as a roadmap for Kenya. It is, however, important to note that CA mechanization technologies may not be suitable in areas where shifting cultivation is still common and tillage practices are rare (Pingali 2007).

GENDER AND YOUTH INCLUSIVITY IN RURAL DEVELOPMENT

If successfully promoted, mechanization can contribute to gender- and youth-inclusive rural development in Kenya. Optimally empowering rural youth through mechanization is worth exploring to boost agricultural production (Mrema et al. 2017; Daum and Birner 2020). Historically, tractor operations worldwide have been considered masculine and are often targeted at men (for example, Brandth 1995). Consequently, mechanization strategies should pay more attention to providing mechanical technologies suitable for women. Gender-sensitive mechanization programs can increase women's labor productivity and reduce the struggle associated with on-farm and postharvest operations typically performed by women. For example, some women in Kenya grow crops on smaller, fragmented plots for risk mitigation purposes (wa Gĩthĩnji et al. 2014), and mechanization technologies that are more suitable for such production practices may be needed for these women. Some focus group-based studies indicate that women in Kenya may value better land preparation and weed suppression, increased time for nonfarm activities, reductions in drudgery, and reduced production costs as key benefits of mechanization, while more men expect farmland expansion and higher crop diversity as main benefits (Daum et al. 2020). Gender-inclusive access to mechanization is also important because, in Kenya, households headed by single, divorced, or widowed women have less access to mechanization services compared with women with male household members, who have a better social network to access such services (Wanjiku et al. 2007). Importantly, promoting mechanization in a gender-inclusive way in Kenya should consider potential benefits in broader rural development and smallholder livelihoods that include the nonfarm sector, in addition to on-farm activities. Recent evidence from sub-Saharan Africa and Asia suggests mechanization (such as increased use of tractors) may benefit farm households by releasing women from arduous farm labor activities to engage more in nonfarm activities (see Takeshima and Diao 2021).

The food sector is expected to continue providing a critical source of employment for youth in Africa in the near future (Christiaensen, Rutledge, and Taylor 2021). In some African countries, like Ghana, access to mechanization has allowed youth to enhance their autonomy in farm decision-making by reducing the need to secure support from village chiefs and/or elders (Amanor and Iddrisu 2022). In Kenya, anecdotal evidence shows that youths become more interested in working in agriculture because of mechanization opportunities (Marechera and Muinga 2017; Makini et al. 2020). Training programs designed to build capacity to access and effectively and profitably operate and maintain mechanization equipment can be targeted at youth. Supporting and strengthening training programs, including technical vocational education and training (TVET), can help youth master knowledge of intensive farming and postharvest handling operations, machine operations, and mechanization service provision (Makini et al. 2020).

Policy guidelines for mechanization programs and projects

The role of government

The Kenyan government's mechanization policy goals as described earlier can facilitate mechanization to fulfill various potentials, also described above. In addition to providing specific support, the government's general roles are to address critical market failures while minimizing the risks of government failures.

Several key policy lessons for mechanization can be gleaned from experiences in Asia and other more advanced countries in sub-Saharan Africa over the past five decades (Mrema, Baker, and Kahan 2008; Singh 2013; FAO 2014). Mechanization should be viewed strategically within a longer-term timeframe as part of broader economic development and agro-industrialization strategies (Diao, Takeshima, and Zhang 2020). Mechanization planning in Kenya, therefore, should ideally be through cross-ministry partnerships among ministries like agriculture, environment, trade, industry, ICT, and others (FAO and AUC 2018).

At the same time, the government may need to recognize that its capacity may be limited with regard to customizing support in different ways for heterogeneous agroecological and socioeconomic conditions. Elsewhere, successful mechanization development has not depended on the government's direct involvement in machinery supply, financing, or mechanization hire service provision. Instead, governments have succeeded by supplying information, knowledge, and institutional infrastructures and encouraging market-based competition (Diao, Takeshima, and Zhang 2020). These general principles should guide the Kenyan government as well. Below, we discuss specific applications of these principles to the Kenyan government's mechanization policy goals.

Supporting hire services

Promoting the growth of mechanization hiring services is one of the policy goals of the Kenyan government. One of the most influential roles for the government lies in strengthening support to training on machine operations, field inspections for problems like tree stumps that can cause machine breakage, and maintenance/repair, which can enhance the efficiency and reduce the costs of service provision, as experienced in Ghana (Diao, Takeshima, and Zhang 2020). Growing evidence points to significant variations in efficiency among service providers in African countries, implying that there is scope for enhancing overall efficiency through appropriate training on operation skills in Kenya (see Takeshima et al. 2015 on mechanization service providers in Nigeria and Diao, Takeshima, and Zhang 2020 on Ghana). As stipulated in the National Agricultural Mechanization Policy (Kenya, Ministry of Agriculture, Livestock, Fisheries, and Cooperatives 2021), this training can be provided in TVET institutes, universities, and research institutes, in particular the Agricultural Mechanization Training Institute.

In doing so, it is important to incorporate the knowledge of existing private sector hiring service providers, whose experiences and expertise are specific to their local business environments. This is because the viability of hiring service businesses depends on there being a sufficient annual use rate (Diao, Takeshima, and Zhang 2020), and existing private sector providers often have the knowhow to achieve high rates (for example, Takeshima et al. 2015). This is critical in Kenya, where available tractors are typically large, with an average of 100 hp (Diao et al. 2016), requiring sufficient use rates to achieve viability.

For financial support to machine investments, value chain financing (credit provided by agricultural equipment dealers to buyers) has been one of the most common models used elsewhere, and should be promoted in Kenya (Animaw et al. 2016). Subsidies for machines should generally be avoided. If needed, they should be made available to a broad type of equipment (while keeping subsidy rates low to keep subsidy budgets minimal) so that the equipment market remains competitive. Similar subsidy models have been used in India, leading to growth in mechanization (Diao, Takeshima, and Zhang 2020).

The government can also invest in gathering and sharing information about promising models of innovative financing mechanisms emerging in and outside Kenya. For example, asset and finance leasing and pay-as-you-go schemes that allow lessors to monitor the repair and maintenance of the machinery closely have been found to be promising (Balyamujura, Kaumbutho, and Karu 2018).

Supporting local industry

The Kenyan government is also prioritizing the development of machine and equipment value chains as well as fostering machine development and fabrication industries as part of its mechanization policies (Kenya, Ministry of Agriculture, Livestock, Fisheries, and Cooperatives 2021). In this effort, it is critical to understand the expected sequence of the manufacturing sector's growth process. Based on historical experience, manufacturers of spare parts and simple machine attachments/implements extend gradually to cover more sophisticated equipment and machines because greater skills and knowledge are required for the latter stage (Ito 1986; Adubifa 1993; Diao, Takeshima, and Zhang 2020). In other words, viable development of the machine manufacturing sector is unlikely before the development of spare parts and simple attachment manufacturing.

Currently, most tractor dealers sell fewer than 30 tractors yearly but many more spare parts are sold in Kenya (Kaumbutho 2016). The Kenyan government should therefore focus on developing the local manufacturing of spare parts first. For example, developing local spare parts manufacturing for commonly used types of tractors can contribute significantly to increasing the availability of spare parts, which would otherwise have to be imported with added transaction costs. Doing so is also critical for some old but popular tractors marketed in domestic secondhand markets, for which spare parts may no longer be manufactured in foreign countries (Diao, Takeshima, and Zhang 2020). Developing local spare parts manufacturing is also important for other technologies, including mechanical irrigation like drip irrigation technologies in Kenya (Kulecho and Weatherhead 2005; Malabo Montpellier Panel 2018).

Promoting social capital

Effectively mechanizing smallholder farmers requires their exploitation of the power of aggregating their demand, as observed in other sub-Saharan African countries like Ghana (for example, Diao et al. 2018). Higher levels of mechanization can disadvantage smallholder farmers and other small actors within value chains in favor of large, commercialized farmers. Such players must come together to lower the costs of access to and service of machinery.

Once organized in groups, smallholders can access larger and more organized markets. For example, in Ghana and parts of Asia, recent studies show, smallholders increasingly use their internal and external social capital to access mechanization services collectively (Kansanga 2017; Diao et al. 2018; Müller 2020). Promoting and mobilizing social capital appears to be more successful in coordinating such mechanization timing than other modes like joint ownership of machines, which often suffer from overuse and insufficient maintenance (Diao, Takeshima, and Zhang 2020). Improving communication using ICTs is also found to facilitate collective action (Binswanger and Deininger 1997). For example, in China, local governments worked with mobile companies to set up group message platforms for mechanization service providers who provide migratory services together. This reduced communication costs among service providers significantly (Zhang, Yang, and Reardon 2017). Similar support to facilitate collective action among various stakeholders along agricultural mechanization value chains by local governments in Kenya could be effective.

Social capital also plays an essential role in knowledge diffusion. Promoting such social capital among potential adopters of mechanization could be important given that recent studies in low-income countries indicate significant exposure gaps in potential mechanization benefits among farmers who could otherwise consider adopting the technology (for example, Brown, Paudel, and Krupnik 2021).

Supporting applied research and development

The public sector plays a significant role in R&D related to agricultural engineering, economics, and a broader set of related technologies and institutions. Holistic mechanization research efforts should be designed involving departments of agriculture (mechanization research, soils, postharvest, irrigation, and so on), trade and industry (industrial research, manufacturing, patenting, standards, trade licensing, and so on), energy (energy generation and distribution, alternative fuels, and so on), and higher education (research and education on all aspects of mechanization in schools of agriculture and engineering). Linkages between the public and the private sectors in R&D activities must be strengthened.

Legislative changes should be guided by impact assessments and the collection, compilation, and analysis of gender-disaggregated data on labor, income, decision-making, control of resources, and other indicators. Enhancing institutional capacity on mechanization R&D remains key, including through a tractor census or the collection of additional data to monitor the status of various aspects of mechanization sector development, as well as the capacity development of professional bodies of agricultural engineers and agricultural economists who play essential roles in policy advocacy for agricultural mechanization (Diao, Takeshima, and Zhang 2020).

Outside Kenya, sub-Saharan African governments have assigned local institutions general R&D tasks, including engineering or the development of new equipment designs (Takeshima, Hatzenbuehler, and Edeh 2020). For example, mechanization units of state-level agricultural and extension organizations in Nigeria and the National Centre for Agricultural Mechanization have been mandated to coordinate R&D. International organizations such as the United Nations Industrial Development Organization and the United Nations Economic Commission for Africa contribute to similar R&D efforts (Takeshima et al. 2020), which can potentially provide knowledge and information that individual countries like Kenya can use for their further adaptive R&D on agricultural mechanization.

In Kenya, establishment of the National Agricultural Mechanization Data Management Information System, as prescribed in the National Agricultural Mechanization Policy (Kenya, Ministry of Agriculture, Livestock, Fisheries, and Cooperatives 2021), should be pursued. In addition, the capacity to conduct cost-benefit analyses of mechanization adoption for different types of farms based on size and cropping systems, among others, should be strengthened at multiple research agencies, including the Agricultural Mechanization Research Institute, Agricultural Technology Development Centers, Kenya Marine and Fisheries Research Institute, and Kenya Forest Research Institute, all of which the National Agricultural Mechanization Policy mentions as key research agencies.

Lastly, Kenya needs to support research on how public–private partnerships can foster the development of sustainable agricultural mechanization. An understanding of the supply of and effective demand for machinery and mechanization services and how to link mechanization to output marketing services is needed. National and county research should determine what works best under prevailing conditions. This should be guided by the three "Ws": *Who* has been successful, *Where* was the success, and *What* made it successful. Government bodies like the Agricultural Finance Corporation could collaborate with private companies and banks such as Quipbank, Rentworks, and Rentco, to facilitate impact studies in the agricultural mechanization finance sector. Such studies could reduce uncertainty among these entities with regard to extending mechanization-related loans.

References

Adubifa, A.O. 1993. "Technology Policy in National Development: A Comparative Study of the Automobile Industry in Nigeria and Brazil." *Journal of Asian and African Studies* 28 (1–2): 42–53.

Amanor, K.S., and A. Iddrisu. 2022. "Old Tractors, New Policies and Induced Technological Transformation: Agricultural Mechanisation, Class Formation, and Market Liberalisation in Ghana." *Journal of Peasant Studies* 49 (1): 158–178.

Animaw, A.T., J.A.M. Nkanya, J.M. Nyakiba, and T.H. Woldemariam. 2016. *Agricultural Mechanization and South-South Knowledge Exchange: What Can Ethiopian and Kenyan Policymakers Learn from Bangladesh's Experience?* ESSP Policy Note 47. Washington, DC: International Food Policy Research Institute (IFPRI).

Balyamujura, H., P. Kaumbutho, and T. Karu. 2018. *Working Business Models for Sustainable Agricultural Mechanization in Africa*. Unpublished report submitted to AGRA.

Benoit, M., and A. Mottet2023. "Energy Scarcity and Rising Cost: Towards a Paradigm Shift for Livestock." *Agricultural Systems* 205: 103585.

Binswanger, H.P., and K. Deininger. 1997. "Explaining Agricultural and Agrarian Policies in Developing Countries." *Journal of Economic Literature* 35 (4): 1958–2005.

Brandth, B. 1995. "Rural Masculinity in Transition: Gender Images in Tractor Advertisements." *Journal of Rural Studies* 11 (2): 123–133.

Brown, B., G.P. Paudel, and T.J. Krupnik. 2021. "Visualising Adoption Processes through a Stepwise Framework: A Case Study of Mechanisation on the Nepal Terai." *Agricultural Systems* 192: 103200.

Chauhan, B.S., G. Mahajan, V. Sardana, J. Timsina, and M.L. Jat. 2012. "Productivity and Sustainability of the Rice-Wheat Cropping System in the Indo-Gangetic Plains of the Indian Subcontinent: Problems, Opportunities, and Strategies." *Advances in Agronomy* 117 (1): 315–369.

Christiaensen, L., Z. Rutledge, and J.E. Taylor. 2021. "The Future of Work in Food." *Food Policy* 99: 101963.

Clarke, L., and C. Bishop. 2002. "Farm Power—Present and Future Availability in Developing Countries." Agricultural Engineering International: the *CIGR Journal of Scientific Research and Development* vol. 4. Invited Overview Paper. Presented at the Special Session on Agricultural Engineering and International Development in the Third Millennium. ASAE Annual International Meeting/CIGR World Congress, July 30, 2002, Chicago.

Daum, T., and R. Birner. 2020. "Agricultural Mechanization in Africa: Myths, Realities and an Emerging Research Agenda." *Global Food Security* 26: 100393.

Daum T., Y.P. Adegbola, G. Kamau, C. Daudu, R.C. Zossou, G.F. Crinot, P. Houssou, et al. 2020. "Perceived Effects of Farm Tractors in Four African Countries, Highlighted by Participatory Impact Diagrams." *Agronomy for Sustainable Development* 40 (6): 1–19.

Daum, T., R. Villalba, O. Anidi, S.M. Mayienga, S. Gupta, and R. Birner. 2021. "Uber for Tractors? Opportunities and Challenges of Digital Tools for Tractor Hire in India and Nigeria." *World Development* 144: 105480.

De Groote, H., C. Marangu, and Z.M. Gitonga. 2020. "Evolution of Agricultural Mechanization in Kenya." In *An Evolving Paradigm of Agricultural Mechanization Development: How Much Can Africa Learn from Asia?* eds. X. Diao, H. Takeshima, and X. Zhang, 401–422. Washington, DC: IFPRI.

Diao, X., J. Agandin, P. Fang, S.E. Justice, D. Kufoalor, and H. Takeshima. 2018. "Agricultural Mechanization in Ghana: Insights from a Recent Field Study." IFPRI Discussion Paper 01729. IFPRI, Washngton, DC.

Diao, X., J. Silver, and H. Takeshima. 2016. "Agricultural Mechanization and Agricultural Transformation." Background Paper for African Economic Transformation Report, submitted to the African Center of Economic Transformation.

Diao, X., H. Takeshima, and X. Zhang. 2020. *An Evolving Paradigm of Agricultural Mechanization Development: How Much Can Africa Learn from Asia?* Washington DC: IFPRI.

de Vries, G., L. Arfelt, D. Drees, M. Godemann, C. Hamilton, B. Jessen-Thiesen, A.I. Kaya, H. Kruse, E. Mensah, and P. Woltjer. 2021. *The Economic Transformation Database (ETD): Content, Sources, and Methods.* WIDER Technical Note 2/2021. Helsinki, Finland: UNU-WIDER.

Eicher, C.K., and D. Baker. 1982. "Research on Agricultural Development in Sub-Saharan Africa: A Critical Survey." MSU International Development Paper Number 1. Michigan State University, East Lansing, MI.

FAO (United Nations Food and Agriculture Organization). 2022. FAOSTAT database. http://faostat.fao.org.

FAO-RAP (FAO Regional Office for Asia and the Pacific). 2014. *Report of the High Level Multi-Stakeholder Consultation on Sustainable Agricultural Mechanization Strategy (SAMS) for the Asia and the Pacific Region convened by FAO and UNESCAP/CSAM.* Bangkok: FAO-RAP.

FAO and AUC (African Union Commission). 2018. "Sustainable Agricultural Mechanization: A Framework for Africa. Addis Ababa." Rome: FAO; Addis Ababa: AUC.

Friedrich, T. 2013. "Agricultural Mechanization and the Environment." In *Mechanization for Rural Development: A Review of Patterns and Progress from Around the World*, eds. J. Kienzle, J.E. Ashburner, and B. Sims, Integrated Crop Management vol. 20-2013, 181–204. Rome: FAO.

Gürsoy, S. 2017. "Evaluating the Mechanization of Livestock Production in Turkey." Scientific Papers Series-*Management, Economic Engineering in Agriculture and Rural Development* 17 (4): 143–148.

Gyimah-Brempong K., M. Johnson, and H. Takeshima. 2016. *The Nigerian Rice Economy: Policy Options for Transforming Production, Marketing, and Trade.* Philadelphia: University of Pennsylvania Press for IFPRI.

IBRD (International Bank for Reconstruction and Development). 1960. *The Economic Development of Tanganyika.* Baltimore, MD: Johns Hopkins University Press for IBRD.

Ito, S. 1986. "Modifying Imported Technology by Local Engineers: Hypotheses and Case Study of India." *Developing Economies* 24 (4): 334–348.

Jagnani, M., C.B. Barrett, Y. Liu, and L. You. 2021. "Within-Season Producer Response to Warmer Temperatures: Defensive Investments by Kenyan Farmers." *Economic Journal* 131 (633): 392–419.

Jayne, T., and D.S. Ameyaw. 2016. "Africa's Emerging Agricultural Transformation: Evidence, Opportunities and Challenges." In *African Agriculture Status Report 2016: Progress towards Agriculture Transformation in Sub Saharan Africa,* Chpt. 1, 1–20. Nairobi: AGRA.

Jayne, T.S., J. Chamberlin, L. Traub, N. Sitko, M. Muyanga, F.K. Yeboah, W. Anseeuw, et al. 2016. "Africa's Changing Farm Size Distribution Patterns: The Rise of Medium-Scale Farms." *Agricultural Economics* 47 (S1): 197–214.

Kansanga, M.M. 2017. "Who You Know and When You Plough? Social Capital and Agricultural Mechanization under the New Green Revolution in Ghana." *International Journal of Agricultural Sustainability* 15 (6): 708–723.

Kaumbutho, P.G. 2016. *Elements Contributing to the Framework for Sustainable Agricultural Mechanization Strategies for the East Africa Sub-region.* Addis Ababa: FAO Sub-regional Office for Eastern Africa.

Keil, A., A. D'Souza, and A. McDonald. 2016. "Growing the Service Economy for Sustainable Wheat Intensification in the Eastern Indo-Gangetic Plains: Lessons from Custom Hiring Services for Zero-Tillage." *Food Security* 8 (5): 1011–1028.

Kenya, Ministry of Agriculture, Livestock, Fisheries and Cooperatives. 2018. *National Agricultural Mechanization Policy.* Nairobi: Republic of Kenya.

Kenya, Ministry of Agriculture, Livestock, Fisheries, and Irrigation. 2019. *Agricultural Sector Transformation and Growth Strategy.* Nairobi.

Kirubi, C., A. Jacobson, D.M. Kammen, and A. Mills. 2009. "Community-Based Electric Micro-grids Can Contribute to Rural Development: Evidence from Kenya." *World Development* 37 (7): 1208–1221.

Kosura-Oluoch, W. 1983. "An Economic Analysis of Small Farm Mechanization in Western Province, Kenya." PhD thesis, Cornell University.

Kulecho, I.K., and E.K. Weatherhead. 2005. "Reasons for Smallholder Farmers Discontinuing with Low-Cost Micro-irrigation: A Case Study from Kenya." *Irrigation Drainage System* 19 (2): 179–188.

Laborde, J.P., C.S. Wortmann, H. Blanco-Canqui, G.A. Baigorria, and J.L. Lindquist. 2020. "Identifying the Drivers and Predicting the Outcome of Conservation Agriculture Globally." *Agricultural Systems* 177: 102692.

Makini, F., L. Mose, G. Kamau, N. Wawire, B. Salasya, W. Mulinge, M. Makelo, S. Makokha, and E. Thuranira. 2020. *Mechanization and Skills Development for Productivity Growth, Employment and Value Addition: Insights from KENYA*. Accra, Ghana: Forum for Agricultural Research in Africa (FARA).

Malabo Montpellier Panel. 2018. *WATER-WISE: Smart Irrigation Strategies for Africa*. Dakar, Senegal: IFPRI and Malabo Montpellier Panel Report.

Mano, Y., T. Njagi, and K. Otsuka. 2022. "An Inquiry into the Process of Upgrading Rice Milling Service: The Case of Mwea Irrigation Scheme in Kenya." *Food Policy* 106: 102195.

Marechera, G., and G. Muinga. 2017. "Value Chain Approaches to Mechanization in Cassava Cultivation and Harvesting in Africa Foundation (AATF), Kenya." In *Achieving Sustainable Cultivation of Cassava Volume 1*, ed. C.H. Hershey, Chapter 17. Cambridge, UK: Burleigh Dodds Science Publishing.

Mayne, J.E. 1955. "Progress in the Mechanization of Farming in the Colonial Territories." *Tropical Agriculture* 32 (3): 95–99.

Mrema G.C., and R.S. Rolle. 2003. "Status of the Post-Harvest Sector and its Contribution to Agricultural Development and Economic Growth in Value-Addition to Agricultural Products." Paper present at the 9th JIRCAS International Symposium in Tokyo, March 2003.

Mrema, G.C., D. Baker, and D. Kahan. 2008. "Agricultural Mechanization in Sub Saharan Africa: Time for a New Look." AGSF Occasional Paper No. 22. FAO, Rome.

Mrema G.C., P.G. Kaumbutho, M. Fonteh, M. Tapela, E.A. Ajav, and A. Jerome. 2017. *A Framework for Sustainable Agricultural Mechanization Strategies in Africa: Sending the Hand-Hoe to the Museum*. Addis Ababa: FAO Sub-regional Office for Eastern Africa.

Müller, M. 2020. "Leadership in Agricultural Machinery Circles: Experimental Evidence from Tajikistan." *Australian Journal of Agricultural and Resource Economics* 64 (2): 533–554.

Muyanga, M., and T.S. Jayne. 2019. "Revisiting the Farm Size-Productivity Relationship Based on a Relatively Wide Range of Farm Sizes: Evidence from Kenya." *American Journal of Agricultural Economics* 101 (4): 1140–1163.

Pingali, P.L. 2007. "Agricultural Mechanization: Adoption Patterns and Economic Impact." In *Handbook of Agricultural Economics,* eds. R. Evenson and P.L. Pingali, 2779–2805. Amsterdam: Elsevier Science.

Pingali, P.L., Y. Bigot, and H.P. Binswanger. 1987. *Agricultural Mechanization and the Evolution of Farming in Sub-Saharan Africa.* Baltimore, MD: Johns Hopkins University Press.

Singh, G. 2013. "Agricultural Mechanization in India." In *Mechanization for Rural Development: A Review of Patterns and Progress from Around the World,* eds. J. Kienzle, J.E. Ashburner, and B.G. Sims, *Integrated Crop Management* Vol. 20-2013. Rome: FAO.

Swynnerton, R.J.M. 1954. *A Plan for Intensification of African Agriculture in Kenya.* Nairobi: Government Printer.

Takeshima, H., and X. Diao. 2021. "Agricultural Mechanization and Gendered Labor Activities across Sectors: Micro-evidence from Multi-country Farm Household Data." IFPRI Discussion Paper 2066. IFPRI, Washington, DC.

Takeshima, H., X. Diao, and P. Aboagye. 2020. "Policies for Competitive and Sustainable Agricultural Production Systems: A Case Study of Ghana's Recent Mechanization Interventions." In Annual *Trends* and *Outlook Report*: Sustaining Africa's *Agrifood System Transformation*: The *Role* of *Public Policies*, eds. D. Resnick, X. Diao, and G. Tadesse, 45–56. Washington, DC: IFPRI.

Takeshima, H., E. Edeh, A. Lawal, and M. Ishiaka. 2015. "Characteristics of Private-Sector Tractor Service Provisions: Insights from Nigeria." *Developing Economies* 53 (3): 188–217.

Takeshima, H., P. Hatzenbuehler, and H. Edeh. 2020. "Effects of Agricultural Mechanization on Economies of Scope in Crop Production in Nigeria." *Agricultural Systems* 177: 102691.

Takeshima, H., N. Houssou, and X. Diao. 2018. "Effects of Tractor Ownership on Agricultural Returns-to-Scale in Household Maize Production: Evidence from Ghana." *Food Policy* 77: 33–49.

Takeshima, H., and Y. Liu. 2020. "Smallholder Mechanization Induced by Yield-Enhancing Biological Technologies: Evidence from Nepal and Ghana." *Agricultural Systems* 184: 102914.

Takeshima, H., and Y. Mano. 2023. "Intensification of Rice Farming: The Role of Mechanization and Irrigation." In *Rice Green Revolution in Sub-Saharan Africa*, eds. K. Otsuka, Y. Mano, and K. Takahashi, Chpt. 7, 143–159. Springer.

UN DESA (United Nations Department of Economic and Social Affairs), Populations Division. 2018. *World Urbanization Prospects: The 2018 Revision*. Geneva.

USDA (United States Department of Agriculture). 2022. Economic Research Service Agricultural Productivity Project. https://www.ers.usda.gov/data-products/international-agricultural-productivity/. Accessed April 29, 2022.

wa Gĩthĩnji, M., C. Konstantinidis, and A. Barenberg. 2014. "Small and Productive: Kenyan Women and Crop Choice." *Feminist Economics* 20 (1): 101–129.

Wanjiku, J., J.U. Manyengo, W. Oluoch-Kosura, and J.T. Karugia. 2007. "Gender Differentiation in the Analysis of Alternative Farm Mechanization Choices on Small Farms in Kenya." In *Food Insecurity, Vulnerability and Human Rights Failure,* eds. B. Guha-Khasnobis, S.S. Acharya, B. Davis, and A.F. Shorrocks, 194–218. London: Palgrave Macmillan.

Willy, D.K., M. Muyanga, and T.S. Jayne. 2019. "Can Economic and Environmental Benefits Associated with Agricultural Intensification Be Sustained at High Population Densities? A Farm Level Empirical Analysis." *Land Use Policy* 81: 100–110.

World Bank. 2008. *World Development Report 2008: Agriculture for Development.* Washington, DC: World Bank.

Zhang, X., J. Yang, and T. Reardon. 2017. "Mechanization Outsourcing Clusters and Division of Labor in Chinese Agriculture." *China Economic Review* 43: 184–195.

Zheng, H., W. Ma, and X. Zhou. 2021. "Renting-in Cropland, Machinery Use Intensity, and Land Productivity in Rural China." *Applied Economics* 53 (47): 5503–5517.

PART 4

Toward More Resilient Food Systems

Part 4

TOWARD MORE RESILIENT FOOD SYSTEMS

Climate change, locusts, global price volatility, and COVID-19 have placed the Kenyan food system under strain in recent years. These external drivers disrupt the capacity of the food system to operate productively and supply healthy foods at all times. A resilient food system can withstand and recover from these shocks and adapt to changing conditions. As shocks become more prevalent with the changing climate and a rapidly evolving global environment, building food system resilience has become increasingly important. Part 4 addresses food system resilience in general and presents several financial solutions to boosting the resilience of smallholder farmers in particular.

The Kenyan food system has gradually become more resilient over the past 20 years. Chapter 10 presents the results of input–output modeling to measure food system resilience. Food system resilience is linked to its sophistication and diversification. As such, production sectors with more downstream linkages (that is, to processors, manufacturers, or retailers) and more diversified systems help build food system resilience. From the policy standpoint, developing market linkages (that is, value chain development) and promoting diversification can be promising avenues to build food system resilience. However, this brings to the forefront the challenge of simultaneously diversifying and intensifying agricultural production. Small producers, who are often the most vulnerable to shocks, may need other measures, such as insurance, to de-risk their production.

Insurance can help build resilience for small producers and protect them against risks that on-farm practices, technologies, and diversification cannot (for example, severe drought). However, as Chapter 11 indicates, traditional indemnity-based insurance can be expensive to scale because in-person visits are needed to assess damage. Digital solutions can enable alternatives, such as index- or picture-based insurance. However, insurance products cannot be one-size-fits-all and must address the varying needs of farmers, herders, women, the poor, and the non-poor. Insurance should also be bundled with other interventions, as it is not a silver bullet.

One example of bundling insurance, discussed in Chapter 12, is risk-contingent credit (RCC). RCC is insurance bundled with credit, whereby loan repayments are essentially insured for farmers facing climate risk. Such a solution overcomes the problem of risk rationing in credit markets and solves liquidity constraints to adopting insurance (as discussed in Chapter 11). Chapter 12 shows that RCC can increase credit uptake and has potential to scale up but is hindered by a lack of reliable yield data. The public sector can play a supporting role by promoting large-scale data collection so that researchers can better estimate yields nationwide.

In sum, Part 4 discusses several issues related to food system resilience in Kenya. Creating more diverse production systems and promoting more value chain linkages can help build climate resilience for the food system. At the farm level, innovative financial products can help de-risk production while also supporting more intensified production. However, these solutions cannot be pursued in isolation, and must also work in unison with enhancing productivity through on-farm technology adoption (Part 3) and ensuring production meets the demand for healthy diets (Part 2).

Chapter 10

ASSESSING THE RESILIENCE OF KENYA'S FOOD SYSTEM: A PRODUCTION APPROACH

John M. Ulimwengu, Juneweenex Mbuthia, and Lensa Omune

A food system includes all elements (environment, people, inputs, processes, infrastructures, institutions, etc.) and activities that relate to the production, processing, distribution, preparation, and consumption of food, and the outputs of these activities, including socioeconomic and environmental outcomes (HLPE 2017). Thus, a food system links society and nature (Blesh and Wittman 2015). Resilience is "the ability of people, households, communities, countries and systems to mitigate, adapt to, and recover from shocks and stresses in a manner that reduces chronic vulnerability and facilitates inclusive growth" (USAID 2018). Applied to food systems, resilience is defined by the Alliance for a Green Revolution in Africa (AGRA) as the ability to withstand major shocks and stressors emanating from climate/weather, conflict, disease, external economic shocks, and other sources, which, if not prevented or mitigated, would delay, or limit economic progress, transformation, prosperity, and self-reliance (AGRA 2021). In this sense, resilience of a food system may be considered a system property that plays a critical role in its sustainability (Jacobi et al. 2018), thus ensuring sustained food security. This chapter adopts this definition with the objective of assessing the resilience of Kenya's food system and its components using systemwide metrics. Specifically, we use a production approach based on input–output linkages.

A consensus emerging across the globe is that building resilient and sustainable food systems is crucial to ensuring sustainable economies and achieving the Sustainable Development Goals (SDGs) and the Africa Agenda 2063 goals (AGRA 2021). However, ongoing and recent shocks to food systems emanating from climate change crises such as droughts, famines, floods, and locust invasions, as well as civil conflicts and the COVID-19 pandemic, are delaying the progress made in achieving these targets. For this reason, building the resilience of food systems to endure such constant shocks becomes even more important. It also requires coordination and partnership at every level of the system, to ensure the system's efficient functioning.

Over the years, Kenya has taken big strides in building the foundations needed to transform its food system and boost household food resilience. Article 43 of the Constitution of Kenya recognizes the basic human right to freedom from hunger. To achieve its food and nutrition security aspirations, Kenya seeks to transform the agriculture sector in line with its commitments to the Comprehensive Africa Agriculture Development Program (CAADP), the SDGs, the Big Four Agenda, and, by extension, Medium-Term Plan III of Vision 2030. Vision 2030 identifies agriculture as a key sector to transform the economy, given its significant contribution to GDP (Figure 10.1). This important contribution further confirms the need to strengthen synergies between agriculture and Kenya's economy. When agriculture grows, its extensive linkages with the off-farm stages of the food system and nonfarm sectors expand employment and livelihoods in the rest of the economy (AGRA 2021).

The policy framework for the implementation of agricultural transformation includes the Agricultural Sector Transformation and Growth Strategy (ASTGS) 2019–2029 as well as a short-term National Agriculture Investment

FIGURE 10.1 **Kenya's annual GDP and agricultural value-added growth trend, 2000–2019**

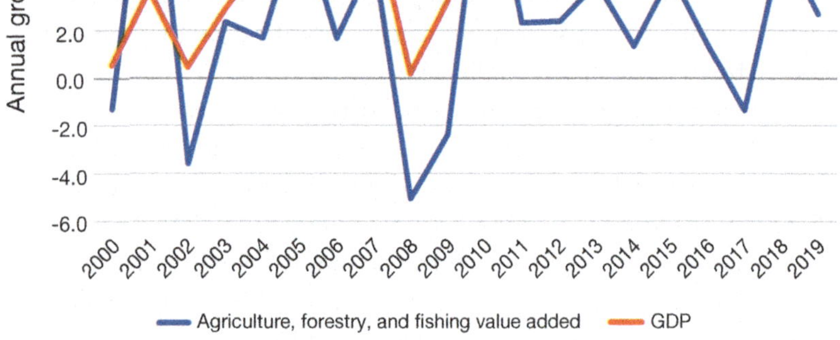

Source: Authors using World Bank's World Development Indicators.

Plan (NAIP) for the period 2019–2024. The ASTGS and the NAIP are premised on a transformed, vibrant, commercial, modern, and equitable agriculture sector to achieve complete food and nutrition security by 2030, and to sustainably support economic growth. The framework has identified three overarching targets, closely aligned with global (SDGs), regional (CAADP, Agenda 2063), and national (Big Four Agenda) aspirations. The first target is to increase small-scale farmer, pastoralist, and fisherfolk incomes in line with SDG target 2.3. The second is to increase agricultural output and value added, directly derived from CAADP's target of annual agricultural GDP growth of 6 percent. The third, increased household food resilience, aligns with the Big Four Agenda commitments of 100 percent food and nutrition security.

CAADP is a policy framework for stimulating production and bringing about food security among the populations of Africa. It was launched in 2003 under the Maputo Declaration on Agriculture and Food Security. In 2014, African heads of state and government adopted the Malabo Declaration on Accelerated African Agricultural Growth and Transformation for Shared Prosperity and Improved Livelihoods, in which they recommitted to the principles and values of CAADP and set ambitious targets in five broad areas (enhancing agricultural investment, ending hunger, reducing poverty, boosting intra-African agricultural trade, and enhancing resilience of livelihoods and production systems). The member states track and monitor their progress on commitments through Biennial Review reports. Kenya signed to the CAADP Compact in July 2010, thereby forming the basis for its NAIPs and sector transformation strategies.

The current 10-year CAADP tracks seven key commitment areas: recommitment to the CAADP process; investment finance in agriculture; ending hunger; eradicating poverty through agriculture by 2025; boosting intra-African trade in agricultural commodities and services; enhancing resilience to climate variability; and enhancing mutual accountability for actions and results. Figure 10.2 shows Kenya's progress relative to its peers on five of these commitment areas, drawing on the 2021 Biennial Review report (AU 2022).

Notably, Kenya lags on most of the commitment areas. However, the country shows five key areas of strong performance: 75 percent of youth engaged in new job opportunities in agricultural value chains; a 63.2 percent reduction in postharvest losses for national agricultural commodities; a 100 percent budget allocation of the total (CAADP 2015–2025) requirement for social protection for vulnerable social groups from the government budget; a 109.8 percent increase in agricultural value added per agricultural worker; and a 126.7 percent increase in agricultural value added per hectare of arable land.

FIGURE 10.2 The current 10-year tracking of CAADP commitment areas in selected member countries

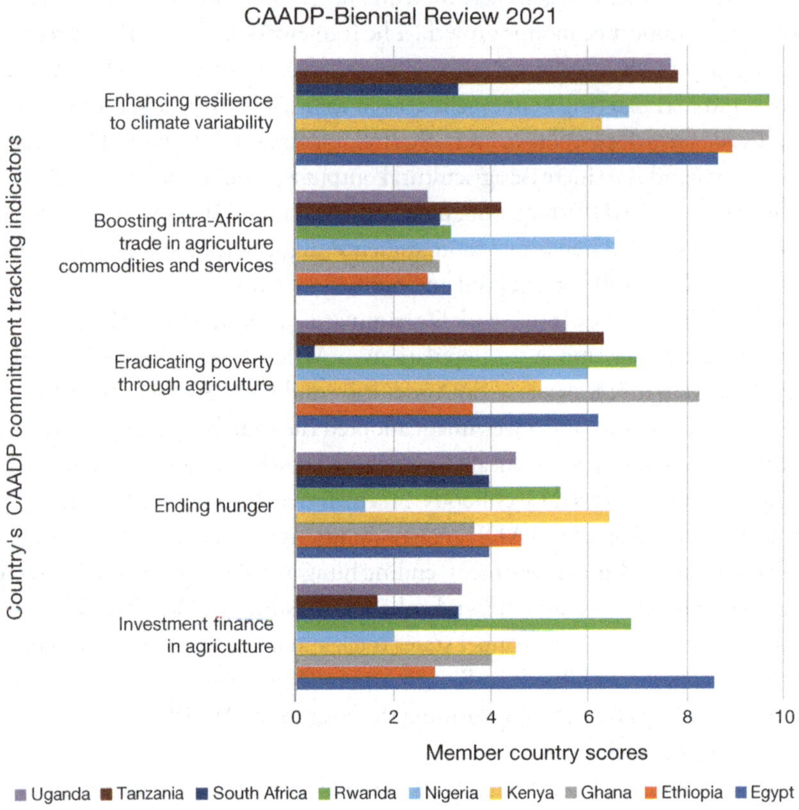

Source: AU (2022).

The 2021 CAADP Biennial Review report maps its indicators to the five Action Tracks of the 2021 United Nations Food Systems Summit. Figure 10.3 presents the change in the performance indicators between the Biennial Reviews of 2019 and 2021 (AU 2020, 2022).

Figure 10.3 shows that, with respect to the indicators under Action Track 1 of the UN Food Systems Summit, 10 countries were on track in 2019 compared with only 5 in 2021. On Action Track 2, the numbers had fallen from 15 to only 6. On Action Track 3 only one country was on track in 2021, up from zero in 2019. Numbers remained at zero for both years on Action Track 4. With respect to indicators under Action Track 5, five countries were on track in 2019 compared with only three in 2021.

FIGURE 10.3 CAADP implementation by UN Food Systems Summit action tracks

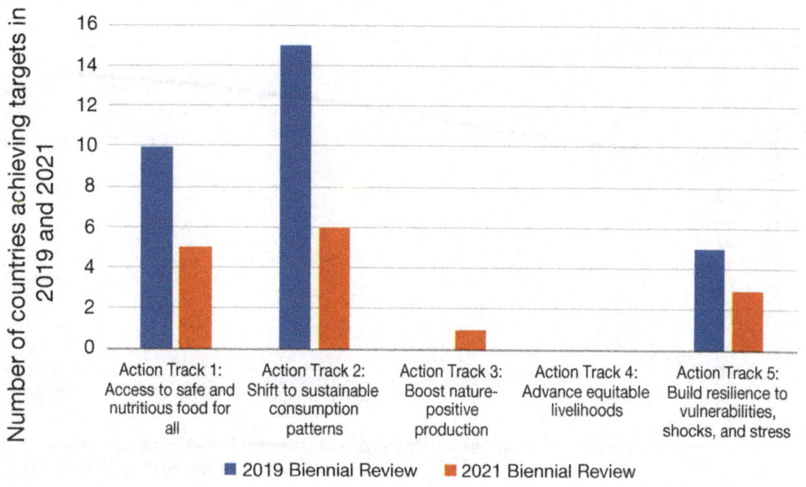

Source: AU (2022).

The results suggest that Kenya, like most African countries, has not made significant progress in transforming its food system to ensure wealth creation, food and nutrition security, poverty alleviation and prosperity, and resilience and sustainability. The momentum created by the UN Food Systems Summit is therefore an opportunity to substantially improve on implementation of the CAADP agenda.

Despite progress made in transforming the food system in Kenya, it remains fragile and vulnerable to climate shocks such as drought and changes in rain patterns, as most of its production is rainfed. In recent years, agricultural production has faced a desert locust invasion in the northern, eastern, and central zones; drought; delayed short and long rains; and the adverse impacts of the COVID-19 pandemic, which has caused disruptions in every component of the food system. Indicators suggest that, even before the pandemic, undernourishment and food insecurity were on the rise (Figure 10.4).

The UN Food Systems Summit 2021, through its Action Tracks, underscored the need to have food systems that can maintain functionality, recover from the adverse effects of shocks and stresses, and build back better, and thus that are more resilient to future shocks (UNFSS 2021). Resilience is thus critical at all stages (both upstream and downstream) of the food system.

FIGURE 10.4 Undernourishment and food insecurity trends in Kenya, 2015–2019

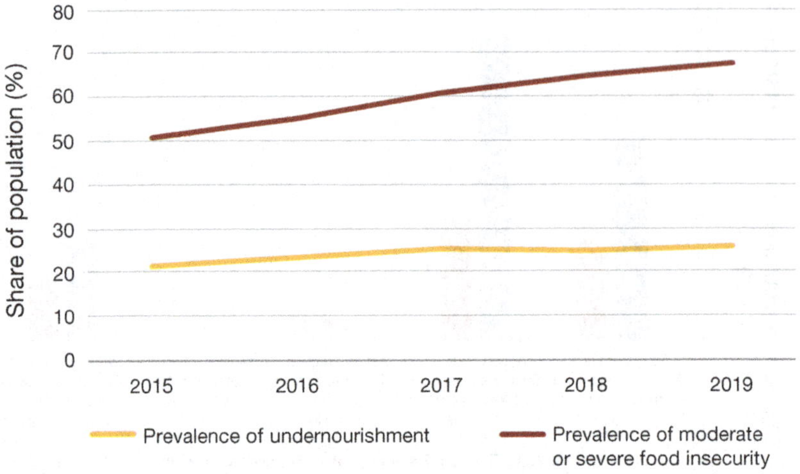

Source: Authors using World Bank's World Development Indicators.

Modeling food system resilience

Resilience is notoriously difficult to measure yet the concept is gaining steam among development partners and policymakers because it captures not just the ability to "bounce back" so much as to "bounce forward" after a shock alters a system. The need to build such resilience drives many development interventions, drawing on a breadth of disciplines, including ecology, medicine, and psychology (Holling 1973; Walker et al. 2004, 2009; Fleming and Ledogar 2008; Béné 2012).

A multitude of practical approaches to measuring resilience, using different types of data, have thus been developed in the past decade. One of the most prominent is the Resilience Index Measurement and Analysis II (RIMA-II) methodology developed by the Food and Agriculture Organization of the United Nations (FAO), which, along with its predecessor RIMA, has been implemented in some 15 African countries to estimate households' ability to maintain well-being in the face of shocks (FAO 2016). The RIMA/RIMA-II approach has gained considerable momentum and has been incorporated into the monitoring and evaluation of resilience-building efforts under the CAADP agenda.

RIMA-II estimates a resilience capacity index at the household and location/community level that can be used to rank households and identify the least resilient. First, factor analysis is used to construct four pillars (interventions) that are expected to contribute to overall resilience, based on a larger set of underlying variables. Second, structural equation modeling is used to estimate resilience capacity as a latent variable, based on the pillar values and on outcome variables, usually reflecting food security. The resilience capacity score generated for each household is standardized so that values fall between 0 and 1, with higher scores indicating greater resilience capacity.

Not all current resilience metrics measure system resilience, which requires accounting for the interdependence of food system components. In other words, a metric for food system resilience cannot be implemented unless the food system is defined as a network of connected components working together to achieve the desired outcome.

This chapter focuses on assessing the resilience of the food system in Kenya. We use the production approach of Leontief, whereby an economy is described through a set of specialized production units. Each of these units relies on the flow of inputs from its suppliers to produce its own output, which in turn is routed toward other downstream units forming a production network (Carvalho and Voigtländer 2014). Hence, a shock affecting only a particular node along the chain can propagate throughout the economy and shape the network outcomes. Such shock propagation will ultimately affect the main outcome of the food system, which is food and nutrition security.

Following Barzel and Barabási (2013) and Barzel, Liu, and Barabási (2015), we model the food system as a network of N sectors whose activities are interconnected. Given the interdependence of sectors, the network effective state can be characterized using the average nearest-neighbor activity, as follows (Barzel, Liu, and Barabási, 2015):

$$x_{\text{eff}} = \frac{\langle s^{out} x \rangle}{\langle s \rangle} \qquad 1$$

where $s^{out} = (s_1^{out}, \dots, s_N^{out})^T$ is the vector of outgoing weighted degrees, $s^{in} = (s_1^{in}, \dots, s_N^{in})^T$ is the vector of incoming weighted degrees, and $\langle s^{out} x \rangle = (1/N)\sum_{i=1}^{N} s_i^{out} x_i$, $\langle s \rangle$ is the average weighted degree.

The slope of the network effective state, which determines the system resilience, is given by:

$$\beta_{\text{eff}} = \frac{\langle s^{out} s^{in} \rangle}{\langle s \rangle} \qquad 2$$

Equation 2 can be rewritten as:

$$\beta_{\text{eff}} = \langle s \rangle + S\mathcal{H} \qquad 3$$

where $\langle s \rangle$, S, and $\mathcal{H}$ represent three characteristics (density, symmetry, and heterogeneity) of the system adjacency matrix A_{ij}. It follows that the resilience index (β_{eff}) dependence on the network density $\langle s \rangle$ indicates that a denser network has a large β_{eff}. The system heterogeneity in the s^{in} and s^{out} is captured by $\mathcal{H} = \sigma_{in}\,\sigma_{out}/\langle s \rangle$, where σ_{in}^2 and σ_{out}^2 are the variance of the marginal probability density functions $P(s^{in})$ and $P(s^{out})$ of in and out weighted degrees, respectively. Finally, the symmetry between s^{in} and s^{out} is given by $S = (\langle s^{in}\, s^{out} \rangle)/(\sigma_{in}\,\sigma_{out})$, which is the in–out weighted degree correlation coefficient.

To help understand how food systems can be more resilient, Piters and colleagues (2021) subdivide resilience capacities according to five phases of a shock/stressor scenario that we group in three: (1) anticipation and prevention, which relate to the phase prior to the occurrence of any shocks, (2) absorption, which plays the largest role during the occurrence of a shock, and (3) adaptation and transformation, which are most relevant in the aftermath of the shock and influence the recovery toward post-shock food and nutrition security. They then define four properties of system resilience building:

- Agency—the means and capacities of people to mitigate risks and to respond to shocks;
- Buffering—resources to fall back on in the face of shocks and stressors;
- Connectivity—the interconnection of and communication between actors and market segments; and
- Diversity at different scales and in different places, from production to consumption and from farm level to regional diversity.

The metrics derived from Equation 3 fall under "connectivity" and "diversity". Connectivity refers to the nature and strength of the interactions between the various components of a given food system. It follows that maintaining and building connectivity helps build resilience and guard against negative outcomes (Love et al. 2020). As Piters and colleagues point out, improved connectivity in agricultural value chains improves a food system's capacity to respond to shocks and stressors and is an essential contributor to adaptation and transformation capacities. Regarding diversity, evidence suggests that resilient systems are diverse systems, as the loss of one resource may be compensated by another (Levia et al. 2020; Benton et al. 2021). For example, Piters and colleagues report that more diverse farming systems have greater capacity to absorb the effects of shocks and stressors, and this capacity stabilizes food supplies throughout value chains to consumer markets.

Data

We use Kenya's input–output tables (IOTs) compiled by IFPRI for 2003, 2013, and 2019 (IFPRI 2021). IOTs describe transaction flows within an economy for a given period; they involve sales and purchases between producers and consumers and reconcile the supply and use of goods and services. Each IOT illustrates flows between final and intermediate sales and purchases of industry outputs or those of product outputs. The actual I–O analysis, also referred to as "inter-industry analysis," measures the relationships between various sectors in the economy. We extracted food systems by adding every other sector to food sectors with a non-zero interaction with food sectors. This allows us to avoid the bias of truncated food systems where the food system is analyzed outside of the overall economic system. For example, we know that fertilizer, which is in the chemicals sector, is a key ingredient in the primary production component of the food system. In another example, the transport sector is important to the food system when it facilitates the movement of raw materials and intermediate and final outputs to markets. Further, the processing sector is vital to the food system since it demands raw materials and intermediate goods and supplies final products to the food system. In the latter case, looking specifically at maize farming, maize is supplied to flour processing firms and in turn these firms sell animal feed as a byproduct from the processing activity to the food system. Hence, analyzing food system resilience without these sectors will likely lead to biased results. Table 10.1 presents key characteristics of

TABLE 10.1 Key characteristics of Kenya's food system, 2003, 2013, and 2019

Period	Number of sectors	Number of links	Density	Average clustering coefficient
2003	45	193	0.097	0.113
2013	52	293	0.099	0.187
2019	42	192	0.111	0.266

Source: Authors' calculations using OITs compiled by IFPRI (2021).

Note: Density describes the portion of the potential connections in a network that are actual connections. A "potential connection" is a connection that could potentially exist between two "sectors"—regardless of whether it actually exists. It is estimated as the ratio of actual connections over potential connections, which is equal to nx(n–1)/2 for a network of n sectors. A sector's clustering coefficient measures how close its neighborhood is to a complete network in terms of the relative density of links in its neighborhood.

FIGURE 10.5 Visual representations of Kenyan food system, 2003–2019

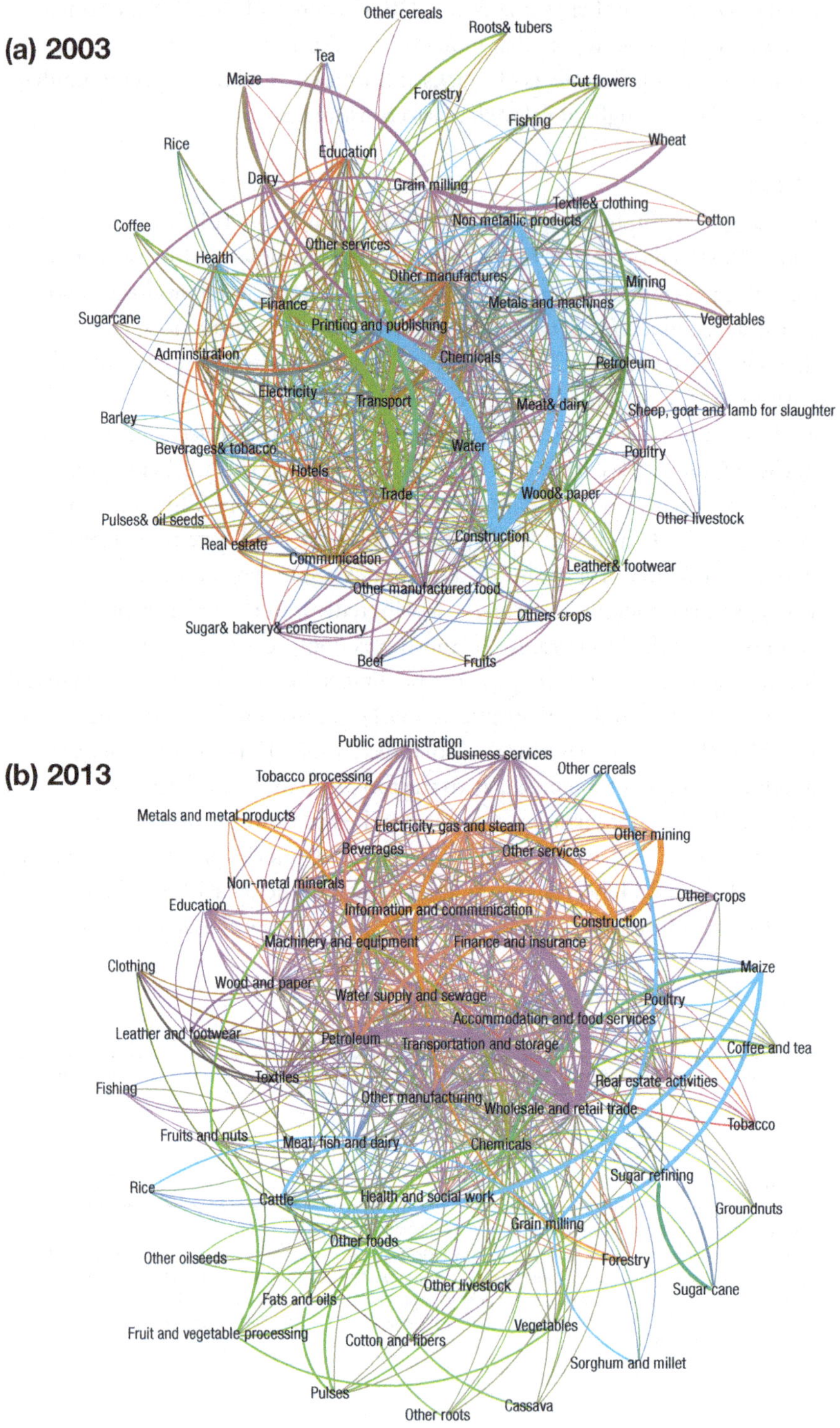

FIGURE 10.5 Visual representations of Kenyan food system, 2003–2019 *continued*

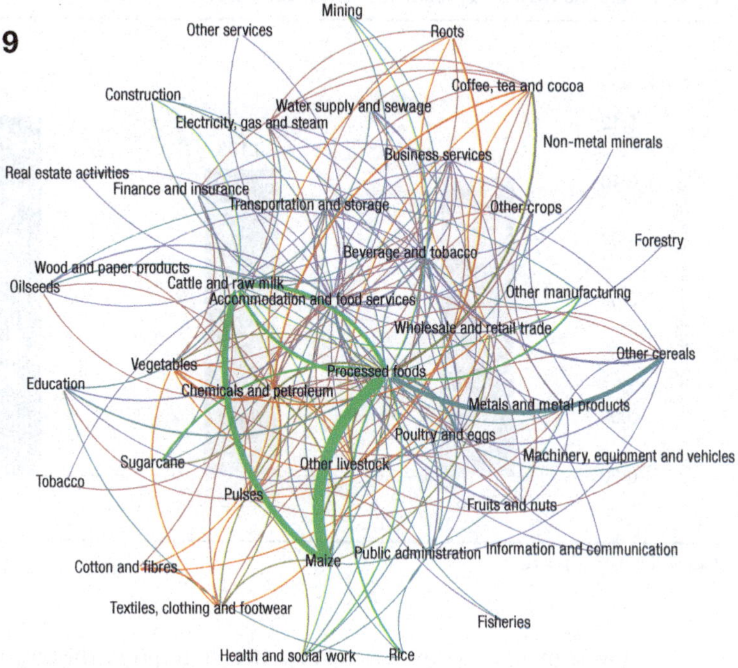

Source: Authors using https://gephi.org/.

Note: Colors represent different network modularity classes; nodes/edges with the same color belong to the same class. Modularity measures the strength of division of a network into clusters or communities (see Newman 2006 for a brief presentation). The size of the arrow represents the value of the interactions between sectors; for example, in 2003, the highest transaction occurred between "Maize" and "Grain milling."

Kenya's food system for periods for which we have IOTs. Figure 10.5 provides visual representations of the system in different years.

The increased number of sectors could be explained partly by the new government that came into place in 2002 and the subsequent bold structural and economic reforms as elaborated in the Economic Recovery Strategy (ERS) covering the 2003–2007 period that paved way toward the adoption of Kenya's Vision 2030 in 2008. The ERS was anchored in three key pillars—namely, restoration of economic growth, rehabilitation, and expansion of infrastructure; equity and poverty reduction; and improving governance. Examples of the reforms include—but are not limited to—the introduction of free primary education; reforming the public transport sector, especially by tightening rules and regulations for minibus operators and by reconstructing roads; and improving efficiency and productivity in the coffee, pyrethrum, sugar, and cooperative subsectors.

FIGURE 10.6 Kenya's food system resilience, 2003–2019

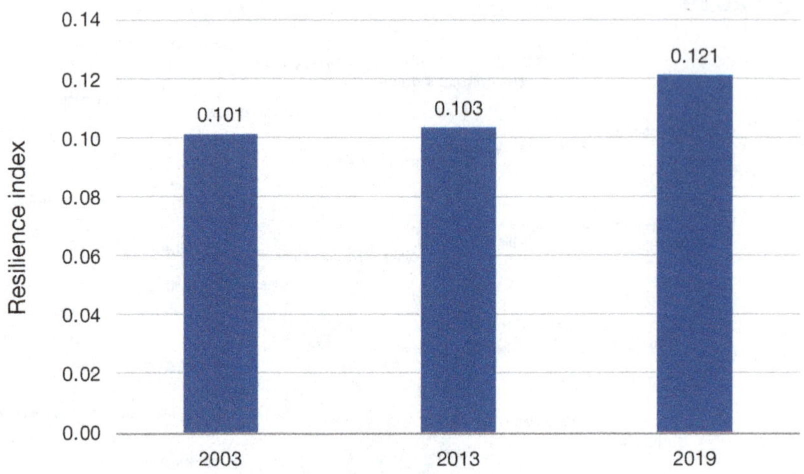

Source: Authors' calculations.

The government also took several measures to improve the business environment and stimulate production, including enhancing the tax incentives first introduced in 2003/04, which included duty waivers on capital goods, industrial plant, and equipment. The incentives largely attracted firms to export promoting zone firms (OECD 2007).

Overall, the ERS entailed the adoption of a growth strategy based in the sectors that would generate employment most rapidly and that would provide more income-generating opportunities for the poor. The sectors identified included agriculture, tourism, trade and industry, ICT, forestry, and mining (Kenya, Republic of Kenya 2007).

Results and discussions

Our findings suggest a slow but gradual increase of Kenya's food system resilience from 2003 to 2019 (Figure 10.6). Despite this, the system remains fragile and vulnerable to shocks, which are tending to occur more often, reversing gains from the increase in resilience. Examples of shocks include climate shocks, crop diseases, the COVID-19 pandemic, the global oil crisis, and the recent Russia-Ukraine conflict. Of the three drivers of system resilience derived above—namely, density, heterogeneity, and symmetry—it appears

FIGURE 10.7 Kenya's food system density, 2003–2019

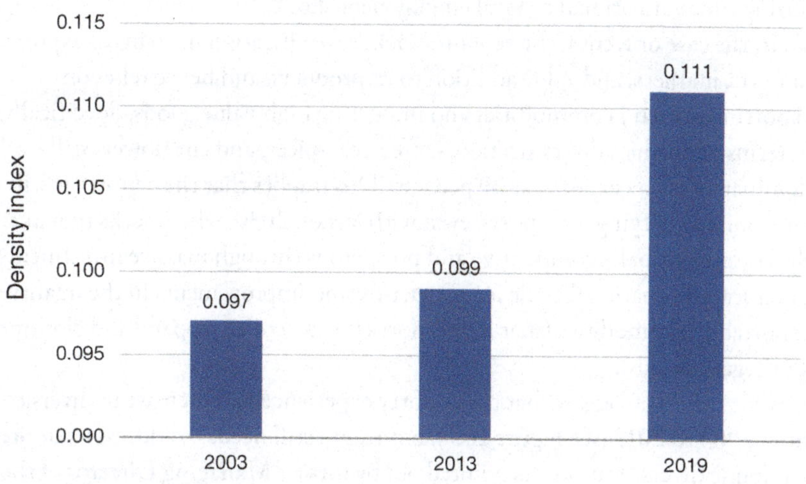

Source: Authors' calculations.

that heterogeneity may explain the substantial increase in resilience in 2019 compared with 2003 and 2013.

A dense food system can be understood as a web with integrated strings, which makes it more resilient compared with a sparse food system (see Figure 10.7). In the economic literature, network density also reflects economic diversification. As Usman and Landry (2021) point out, there are several ways of measuring economic diversification, including variety-, quality-, and output-based approaches. This chapter uses the variety-based approach (number of connected nodes), which measures the diversity of economic activities regardless of their quality.

There is evidence of a significant relationship between density in particular industries and the ability to withstand shock (Brown and Greenbaum 2017). Exploring the extent to which a less concentrated meat processing sector in the United States would be less vulnerable to the risks of temporary plant shutdowns, Ma and Lusk (2021) find that, when each plant in the industry faces a chance of shutdown equal to 10–30 percent, a more concentrated packing sector performs better than a diffuse or sparse packing sector in ensuring a relatively high level of output. Using fixed effects models, Brown and Greenbaum (2017) examined the influence of industrial diversity and concentration on unemployment rate stability in Ohio counties between 1977 and 2011. Their results indicate that counties with more diverse industry structures fare better during times of national or local employment shock.

In the case of Kenya, the economy lacks diversification in terms of exports, imports, markets, and value addition to its products, and hence relies on exporting primary commodities and importing high-value goods. Specifically in terms of product diversification, coffee, tea, spices, and cut flowers still dominate Kenya's exports, at 29 percent. This implies that the country is still far from diversifying its exports, even with Vision 2030, which seeks to make the country globally competitive and prosperous through massive investments in critical international trade infrastructure and improvements in the quality of exportable commodities (that is, tea, horticulture, coffee, apparel and clothing, and vegetable oils).

Our findings suggest that the country experienced an increase in diversity from 2003 to 2019 (see Figure 10.7), but Kenya still needs to address its limited economic diversification. As pointed out by former Managing Director of the International Monetary Fund Christine Lagarde, "We know that economic diversification is good for growth. Diversification is also tremendously important for resilience" (Usman and Landry 2021, 1).

In addition to containing interconnected nodes (sectors), heterogeneous systems are characterized by links of different types. Heterogeneity of a system also reflects the sophistication of its sectors; for example, a food system that manages to produce oil, alcohol, animal feeds, and biofuel is more sophisticated and resilient than a system that produces only maize. Economic sophistication is defined as the ability to produce complex products that require specific skills and tacit knowledge embedded in labor (Arif 2021). A key component of a country's growth process and ultimately its resilience is an increase in this "sophistication" of the country's production, which may evolve either through an increase in the quality of previously produced goods or through a move into new, more complex products (Anand, Mishra, and Spatafora 2012). Rodrik (2007) notes that countries that can produce and export more sophisticated goods grow faster. According to Hausmann and colleagues (2021), economic growth is driven by diversification into new products that are incrementally more complex and less ubiquitous.

Figure 10.8 presents food system heterogeneity in Kenya from 2003 to 2019. Kenya's economic sophistication has been low, given that the largest share of the economy's output is from primary agriculture, which does not require special skills or advanced technology. The heterogeneity index is the highest in 2019, after a significant decrease between 2003 and 2013. If the 2019 trend is sustained, this will show that Kenya's food system has become more complex.

FIGURE 10.8 Kenya's food system heterogeneity, 2003–2019

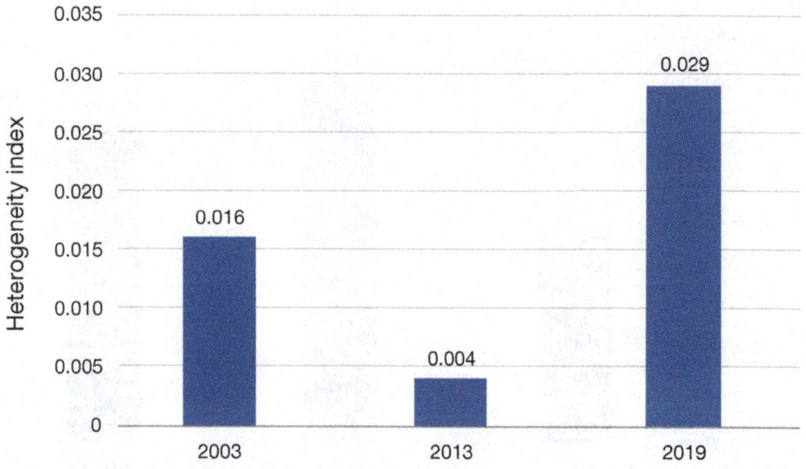

Source: Authors' calculations.

However, the increase in the heterogeneity index in 2019 may be explained partly by the adoption of technology in the country, especially in the implementation of Vision 2030, whose main objective is to transform Kenya into a newly industrializing, middle-income country with a key role for science, technology, and innovation (Kenya, Republic of Kenya 2007).

While sectors and their interactions are the building blocks of a food system, the types of relationships that the interactions represent play a significant role in the system dynamics. Indeed, understanding the nature of the interaction between two sectors is important to capture system resilience. In particular, symmetry is a crucial attribute that determines the resilience of a food system. This has practical implications that have not yet been fully explored, nor systematically exploited by network practitioners (Sánchez-García 2020). Acemoglu and colleagues (2012) have shown that aggregate volatility is observed from sectoral idiosyncratic shocks only if significant asymmetry exists in the interactions between sectors. Crowcroft (2015) points out that a symmetric network, with sectors offering as well as demanding resources, maximizes diversity. Similarly, symmetric interactions amplify network heterogeneity. One example would be the interaction between maize and energy sectors, where the first supplies maize to the second to produce biofuel, which is then used by the first sector to power equipment used in its production process. In the case of Kenya's

FIGURE 10.9 Kenya's food system symmetry, 2003–2019

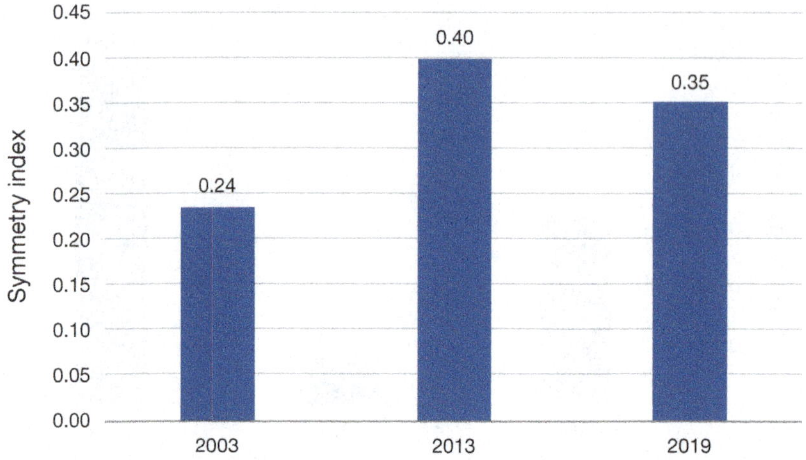

Source: Authors' calculations.

food system, the symmetry index in 2003 was 0.236 (see Figure 10.9); it reached 0.399 in 2013, but decreased by 12 percent in 2019.

The food system resilience index and its components discussed above are all systemwide metrics that highlight how input–output linkages play a crucial role in transmitting shocks between economic actors (Carvalho et al. 2020). However, the "central" sectors are the ones responsible for the amplification of idiosyncratic shocks (Contreras and Fagiolo 2014). There are various measures of system centrality, depending on how connected a given sector is (degree centrality) or how far on average it is from any other sectors in the system (closeness centrality), or how crucial a given sector is in connecting other sectors (betweenness centrality). This chapter uses the eigenvector centrality developed by Bonacich (1987) to identify central sectors. This is also called the "influence measure" of centrality, whereby sectors are relatively more central in the system if their neighbors are themselves well-connected sectors (Carvalho 2014).

Table 10.2 presents the top 10 central sectors with their respective centrality measure. Because of constant changes in the IOT nomenclature, it is impossible to follow the dynamics of each top central sector from 2003 to 2019. However, using 2003 as a reference period, some relevant trends can be observed. For example, meat and dairy, along with poultry, have consistently been among the top 5 while beverages and tobacco have gone from the top spot in 2003 to

TABLE 10.2 Top 10 central sectors in Kenya, 2003, 2013, and 2019

2003		2013		2019	
Sector	Eigen centrality	Sector	Eigen centrality	Sector	Eigen centrality
Beverages and tobacco	1	Accommodation and food services	1	Accommodation and food services	1
Meat and dairy	0.685685	Meat, fish, and dairy	0.77386	Processed foods	0.667106
Poultry	0.504577	Beverages	0.72147	Beverage and tobacco	0.418471
Other manufactured food	0.491296	Other foods	0.58553	Poultry and eggs	0.390752
Sugar, bakery, and confectionary	0.441409	Poultry	0.36538	Cattle and raw milk	0.366997
Dairy	0.422082	Fruit and vegetable processing	0.33508	Other livestock	0.328605
Grain milling	0.403196	Cattle	0.33025	Vegetables	0.295296
Beef	0.276732	Fats and oils	0.32399	Coffee, tea, and cocoa	0.22337
Sheep, goats, and lambs for slaughter	0.276732	Other livestock	0.28725	Pulses	0.204533
Others crops	0.25363	Grain milling	0.26838	Fruits and nuts	0.187565

Source: Authors' calculations.

number three in 2019. It is worth noting the appearance of accommodation and food services in first place in both 2013 and 2019. This sector includes hotels, restaurants, and fast food. Given their role, the top 5 or 10 central sectors should be the targets of policies aimed at building a resilient food system in Kenya.

As part of measures to contain the spread of COVID-19, the Kenyan government, in addition to setting a daily curfew from 8 pm to 4 am, prohibited all public gatherings and in-person meetings, and bars and restaurants remained closed (only allowing takeaway) (GAIN 2021). Therefore, we implemented an extreme scenario assuming that these COVID-19-related restrictions took out the entire accommodation and food services sector. More specifically, the simulated shock corresponds to the removal of the sector from the 2019 food system. As Table 10.3 shows, the impact is felt in all types of interactions across the food system—those incoming (in degree) as well as those outgoing (out degree). The values of these interactions are also affected. It goes without saying that, if no mitigation measure is implemented, such an impact will ultimately reduce system production and eventually food and nutrition outcomes. With respect to incoming transactions, the most affected are construction, finance and insurance, information and communication, wholesale and retail

TABLE 10.3 **Simulated effect of the removal of accommodation and food services as a result of the COVID–19 shock**

	In degree	Out degree	Weighted in degree	Weighted out degree
Beverage and tobacco	0.0	–14.3	0.0	–83.3
Business services	–50.0	–12.5	–83.3	–61.2
Cattle and raw milk	0.0	–16.7	0.0	–12.2
Chemicals and petroleum	0.0	–5.6	0.0	–1.8
Coffee, tea, and cocoa	0.0	–50.0	0.0	–7.5
Construction	–100.0	–33.3	–100.0	–50.0
Cotton and fibers	0.0	0.0	0.0	0.0
Education	–16.7	0.0	–28.6	0.0
Electricity, gas, and steam	0.0	–12.5	0.0	–43.3
Finance and insurance	–100.0	–12.5	–100.0	–26.9
Fisheries	0.0	–50.0	0.0	–70.0
Forestry	0.0	0.0	0.0	0.0
Fruits and nuts	0.0	–33.3	0.0	–68.6
Health and social work	–16.7	0.0	–15.2	0.0
Information and communication	–100.0	–50.0	–100.0	–92.9
Machine, equipment, and vehicles	0.0	–25.0	0.0	–28.6
Maize	0.0	–20.0	0.0	–0.2
Metals and metal products	0.0	0.0	0.0	0.0
Mining	0.0	–33.3	0.0	–13.3
Non–metal minerals	0.0	–50.0	0.0	–33.3
Oilseeds	0.0	–25.0	0.0	–38.2
Other cereals	0.0	0.0	0.0	0.0
Other crops	0.0	–33.3	0.0	–9.8
Other livestock	0.0	–16.7	0.0	–4.2
Other manufacturing	0.0	0.0	0.0	0.0
Other services	0.0	0.0	0.0	0.0
Poultry and eggs	0.0	–20.0	0.0	–32.0
Processed foods	0.0	–11.1	0.0	–12.1
Public administration	–14.3	0.0	–41.8	0.0
Pulses	0.0	–20.0	0.0	–40.0
Real estate activities	0.0	–100.0	0.0	–100.0
Rice	0.0	–25.0	0.0	–22.5
Roots	0.0	–50.0	0.0	–6.9
Sugarcane	0.0	0.0	0.0	0.0
Textiles, clothing, and footwear	0.0	–16.7	0.0	–8.3
Tobacco	0.0	0.0	0.0	0.0
Transportation and storage	–50.0	–7.1	0.0	–11.9
Vegetables	0.0	–33.3	0.0	–41.7
Water supply and sewage	0.0	–14.3	0.0	–23.7
Wholesale and retail trade	–100.0	0.0	–100.0	0.0
Wood and paper products	0.0	0.0	0.0	0.0

Source: Authors' calculations.

trade, and transportation and storage. With respect to outgoing transactions, real estate; coffee, tea, and cocoa; fisheries; information and communication; non-metal minerals; and roots experience the most negative effect.

Concluding remarks and policy implications

Defined as the ability to maintain an acceptable level of the desired outcome despite stressors or shocks, resilience is inherently a dynamic concept—which makes it difficult to measure. Still, the concept is gaining interest among development partners and policymakers because it also captures the ability to bounce forward after a shock alters the system. To make matters more complicated, the resilience of a system is not a mere sum of the resilience of its components. As Cerqueti, Ferraro, and Iovanella (2019) point out, system disruption depends on the magnitude of the shock and its propagation across the system because of the interconnectivity of its components.

This chapter, drawing on the ecological and engineering literature, has used systemwide metrics to measure food system resilience and its components. We used a production approach based on input–output linkages. Our results suggest that the resilience of a food system is driven by its density, heterogeneity, and symmetry. In economic terms, this means that economic diversification (more food sectors) and sophistication (high-value food sectors) are the main drivers of a resilient food system.

There was a sharp decline in Kenya's food system resilience between 2001 and 2003, followed by a slow but gradual increase from 2003 to 2019. These resilience dynamics were driven mostly by loss of density and symmetry. Overall, though, the country's food system density increased from 2003 to 2019. The symmetry index reached 0.351 in 2019, up from 0.236 in 2003. As Carvalho (2014) points out, central production sectors—those with more direct or indirect downstream interactions—are relatively more important in determining aggregate system volatility. We identified and ranked these central sectors for each period. Frequent changes in the IOT nomenclature mean we could not properly analyze the dynamics of each top central sector from 2003 to 2019. Nonetheless, given their role, the top 5 or 10 central sectors should be the targets of policies aimed at building a resilient food system in Kenya.

Waha and colleagues (2018) have demonstrated that diversification plays an essential role in ensuring food security and stabilizing food production in Africa. Other research (such as Jones, Shrinivas, and Bezner-Kerr 2014) shows clear relations between farming diversity and food security, and a linkage to nutritional diversity, but conclusions on how market orientation influences the relationship vary (Sibhatu, Krishna, and Qaim 2015). This suggests the need for incentives to promote diversification while intensifying production systems. Households may still be limited in their ability to diversify because of soil, labor, input, or land constraints or because of their remote location without access to extension services that provide support for new crops or crop management techniques (Waha et al. 2018).

Given the critical role of livestock in supplying food of animal origin, Vision 2030, which is Kenya's development blueprint, identifies this as one of the eight priority sectors within its economic pillar, with various programs for the period 2018–2022. These include the Livestock Production Program, the Smallholder Productivity and Agroprocessing Program, and the Pastoral Resilience Building Program (Kenya, Republic of Kenya 2018).

Overall, to reinforce, maintain, and improve sectors' interconnectivity for increased system resilience, a systemic policy approach is needed to prevent the build-up of vulnerabilities and reduce exposure to shocks. Such a policy should cover relevant institutions, infrastructures, regulations, and markets.

As many have suggested before, what is required to build a resilient food system in a country such as Kenya is a fundamentally different model of agriculture based on diversifying farms and farming landscapes, optimizing biodiversity, and stimulating interactions between different sectors for a sustainable healthy diet for all. Together, a varied and balanced diet, a wide range of crops and foodstuffs, and a diverse system of production and distribution make for a more resilient, stable, and healthy food system (EC 2020).

Finally, the concept of a food system is very appealing because it emphasizes connectivity and interdependence of activities, actors, and institutions to achieve a sustainable healthy diet for all. This calls for coordination and partnership at every level to ensure the whole system functions efficiently and yields the expected outcomes. In this, the development of food system modeling frameworks is critical in integrating the complex interactions between food, ecology, economy, and society, to provide evidence on trade-offs when diversifying food systems to improve their resilience (Hertel et al. 2021). This will require substantial investments in building statistical systems that capture the main participants in a food system along with related operations and connections.

References

Acemoglu, D., V.M. Carvalho, A. Ozdaglar, and A. Tahbaz-Salehi. 2012. "The Network Origins of Aggregate Fluctuations." *Econometrica* 80 (5): 1977–2016.

AGRA (Alliance for a Green Revolution in Africa). 2021. *A Decade of Action: Building Sustainable and Resilient Food Systems in Africa (Issue 9).* Africa Agriculture Status Report. Nairobi.

Anand, R., S. Mishra, and N. Spatafora. 2012. "Structural Transformation and the Sophistication of Production." IMF Working Paper WP/12/59. International Monetary Fund (IMF), Washington, DC.

Arif, I. 2021. "Productive Knowledge, Economic Sophistication, and Labor Share." *World Development* 139: 105303.

AU (African Union). 2020 *Second Biennial Review Report of the African Union Commission on the Implementation of the Malabo Declaration on Accelerated African Agricultural Growth and Transformation for Shared Prosperity and Improved Livelihoods.* Addis Ababa.

AU. 2022. *3rd CAADP Biennial Review Report 2015–2021.* Addis Ababa.

Barzel, B., and A-L. Barabási. 2013. "Universality in Network Dynamics." *Nature Physics* 9: 673–681.

Barzel, B., Y-Y. Liu, and A-L. Barabási. 2015. "Constructing Minimal Models for Complex System Dynamics." *Nature Communications* 6: 7186.

Béné, C., R. Godfrey-Wood, A. Newsham, and M. Davies. 2012. "Resilience: New Utopia or New Tyranny?: Reflection about the Potentials and Limits of the Concept of Resilience in Relation to Vulnerability Reduction Programmes." IDS Working Paper 405. Institute of Development Studies, Brighton, UK.

Benton, T.G., C. Bieg, H. Harwatt, R. Pudasaini, and L. Wellesley. 2021. *Food System Impacts on Biodiversity Loss: Three Levers for Food System Transformation in Support of Nature.* London: Chatham House.

Blesh, J., and H. Wittman. 2015. "'Brasilience: Assessing Resilience in Land Reform Settlements in the Brazilian Cerrado." *Human Ecology* 43 (4): 531–546.

Bonacich, P. 1987. "Power and Centrality: A Family of Measures." *American Journal of Sociology* 92 (5): 1170–1182.

Brown, L., and R.T. Greenbaum. 2017. "The Role of Industrial Diversity in Economic Resilience: An Empirical Examination across 35 Years." *Urban Studies* 54 (6): 1347–1366.

Carvalho, V.M. 2014. "From Micro to Macro via Production Networks." *Journal of Economic Perspectives* 28 (4): 3–48.

Carvalho, V.M., and N. Voigtländer. 2014. "Input Diffusion and the Evolution of Production Networks." NBER Working Paper 20025. National Bureau of Economic Research (NBER), Cambridge, MA.

Carvalho V.M., M. Nirei, S. Yu, and A. Tahbaz-Salehi. 2020. "Supply Chain Disruptions: Evidence from the Great East Japan Earthquake." *Quarterly Journal of Economics* 136 (2): 1255–1321.

Cerqueti, R., G. Ferraro, and A. Iovanella. 2019. "Measuring Network Resilience through Connection Patterns." *Reliability Engineering & System Safety* 188: 320–329.

Contreras, M.G.A., and G. Fagiolo. 2014. "Propagation of Economic Shocks in Input–Output Networks: A Cross-Country Analysis." *Physical Review E* 90 (6): 062812.

Crowcroft, J. 2015. "On the Duality of Resilience and Privacy." *Proceedings of the Royal Society A Mathematical, Physical, and Engineering Sciences* 471 (2175): 20140862.

EC (European Commission). 2020. *Resilience and Transformation: Report of the 5th SCAR Foresight Exercise Expert Group.* Luxembourg: Publications Office of the European Union.

FAO (Food and Agriculture Organization of the United Nations). 2016. *RIMA-II: Resilience Index Measurement and Analysis II.* Rome.

Fleming, J., and R.J. Ledogar. 2008. "Resilience, an Evolving Concept: A Review of Literature Relevant to Aboriginal Research." *Pimatisiwin* 6 (2): 7–23.

GAIN (Global Alliance for Improved Nutrition). 2021. *Impact of COVID-19 on Kenya's Food Systems: A Situation Report.* Geneva.

Hausmann, R., M.A. Santos, C. Macchiarelli, and R. Giacon. 2021. *What Economic Complexity Theory Can Tell Us about the EU's Pandemic Recovery and Resilience Plans.* Cambridge, MA: Harvard University Growth Lab.

Hertel, T., I. Elouafi, M. Tanticharoen, and F. Ewert. 2021. "Diversification for Enhanced Food Systems Resilience." *Nature Food* 2: 832–834.

High Level Panel of Experts (HLPE) on Food Security and Nutrition. 2017. *Nutrition and Food Systems. A Report by the High-Level Panel of Experts on Food Security and Nutrition of the Committee on World Food Security.* Rome: FAO.

Holling, C.S. 1973. "Resilience and Stability of Ecological Systems." *Annual Review of Ecology and Systematics* 4: 1–23.

IFPRI (International Food Policy Research Institute). 2021. 2019 Social Accounting Matrix for Kenya database. Harvard Dataverse, Version 1. https://doi.org/10.7910/DVN/ALUXSI

Jacobi, J., S. Mukhovi, A. Llanque, H. Augstburger, F. Käser, C. Pozo, M. Ngutu Peter, J.M.F. Delgado, B.P. Kiteme, S. Rist, C. Ifejika Speranza. 2018. "Operationalizing Food System Resilience: An Indicator-Based Assessment in Agroindustrial, Smallholder Farming, and Agroecological Contexts in Bolivia and Kenya." *Land Use Policy* 79: 433–446.

Jones, A.D., A. Shrinivas, and R. Bezner-Kerr. 2014. "Farm Production Diversity is Associated with Greater Household Dietary Diversity in Malawi: Findings from Nationally Representative Data." *Journal of Food Policy* 46: 1–12.

Kenya, Republic of Kenya. 2007. *Kenya Vision 2030: Economic, Social and Political Pillars.* Nairobi.

Kenya, Republic of Kenya. 2018. *Third Medium Term Plan 2018-2022, Transforming Lives: Advancing Socio-economic Development through the "Big Four."* Nairobi.

Levia, D.F., I.F. Creed, D.M. Hannah, K. Nanko, E.B. Boyer, D.E. Carlyle-Moses, N. van de Giesen, et al. 2020. "Homogenization of the Terrestrial Water Cycle." *Nature Geoscience* 13: 656–658.

Love, D., E.H. Allison, F. Asche, B. Belton, R.S. Cottrell, H.E. Froehlich, J.A. Gephart, et al. 2020. "Emerging COVID-19 Impacts, Responses, and Lessons for Building Resilience in the Seafood System." *Global Food Security* 28: 100494.

Ma, M., and J.L. Lusk. 2021. "Concentration and Resilience in the U.S. Meat Supply Chains." NBER Working Paper No. 29103. NBER, Cambridge, MA.

Newman, M.E.J. 2006. "Modularity and Community Structure in Networks." *Proceedings of the National Academy of Sciences* 103 (23): 8577–8582.

OECD/African Development Bank. 2007. *African Economic Outlook 2007.* Paris: OECD Publishing.

Piters, B., E. Termeer, D. Bakker, H. Fonteijn, and H. Brouwer. 2021. *Food System Resilience Towards a Joint Understanding and Implications for Policy.* Wageningen Economic Research Policy paper. Wageningen, Netherlands: Wageningen University & Research.

Rodrik, D. 2007. "Industrial Development: Stylized Facts and Policies. Industrial Development for the 21st Century." In *Sustainable Development Perspectives*, 7–28. New York: United Nations.

Sánchez-García, R.J. 2020. "Exploiting Symmetry in Network Analysis." *Communications Physics* 3 : 87.

Sibhatu, K.T., V.V. Krishna, and M. Qaim. 2015. Production Diversity and Dietary Diversity in Smallholder Farm Households." *PNAS* 112 (34): 10657–10662.

UNFSS. 2021. UN Secretary-General's Chair Summary and Statement of Action on the UN Food Systems Summit. https://www.un.org/en/food-systems-summit/news/making-food-systems-work-people-planet-and-prosperity

USAID (United States Agency for International Development). 2018. *Resilience Evidence Forum Report.* Washington, DC: Center for Resilience.

Usman, Z., and D. Landry. 2021. *Economic Diversification in Africa: How and Why It Matters.* Washington, DC: Carnegie Endowment for International Peace.

Waha, K., M.T. Van Wijk, S. Fritz, L. See, P.K. Thornton, J. Wichern, and M. Herrero. 2018. "Agricultural Diversification as an Important Strategy for Achieving Food Security in Africa." *Global Change Biology* 24 (8): 3390–3400.

Walker, B.H., N. Abel, J.M. Anderies, and P. Ryan. 2009. "Resilience, Adaptability and Transformability in the Goulburn-Broken Catchment, Australia." *Ecology and Society* 14 (1): 12.

Walker, B.H., C.S. Holling, S.R. Carpenter, and A.P. Kinzig. 2004. "Resilience, Adaptability and Transformability." *Ecology and Society* 9 (2): 5.

World Bank. World Development Indicators (daraabase). https://databank.worldbank.org/source/world-development-indicators

Chapter 11

CLIMATE INSURANCE: OPPORTUNITIES FOR IMPROVING AGRICULTURAL RISK MANAGEMENT IN KENYA

Berber Kramer

Climate change represents a major challenge to food systems. It is associated not only with rising average temperatures but also with less predictable weather and changes in humidity, with severe consequences for agricultural production, input markets, aggregation, processing, distribution, and consumption. Negative impacts on food production can raise consumer prices, potentially leading to social unrest and conflict; increased temperatures and changes in humidity require stronger cold chains and improved storage facilities to avoid postharvest damage (de Brauw and Pacillo 2022).

This chapter highlights several innovations in climate insurance that were developed and tested in Kenya with the aim of improving smallholder farmers' ability to manage the production risks associated with climate change.

In Kenya, farmers and herders are facing increasingly unpredictable and unreliable rainfall patterns, resulting in agricultural losses from drought and excess rain, as well as pests and disease. Climate change will continue to negatively affect crop and livestock production and food security (Kogo, Kumar, and Koech 2021). Increased incidence of droughts and other natural hazards reduces agricultural productivity in two ways. The most visible channel is that, when these shocks occur, they limit crop growth and the amount of food and water available for livestock, with a negative impact on production and food security. But even in the absence of such a shock, the mere possibility of a natural hazard occurring will discourage risk-averse farmers and herders from investing in agriculture (Vargas Hill et al. 2019), and lenders from financing these investments (Carter, Cheng, and Sarris 2016). Thus, an increase in agricultural production risks as a result of climate change lowers agricultural productivity both ex post, when a shock occurs, and ex ante, when the farmer merely anticipates the risk of a shock occurring; and these ex ante impacts are estimated to be twice as large as the impacts of shocks themselves (Elbers, Gunning, and Kinsey 2007).

In this context, adapting to climate change is important to meet growing food demand and to improve vulnerable smallholder farmers' livelihoods. Farmers and herders can adopt new technologies or practices to reduce their exposure to climate-induced risks, and adapt to climate change, but this requires investments in rural and agricultural development—particularly in resilience technologies such as drought-tolerant cultivars, irrigation and soil and water conservation, or improved methods to store and preserve livestock feed (Bryan et al. 2013). Unfortunately, adoption of these solutions remains slow, in part because they require upfront investments while not providing full protection from all types of climate risk. For instance, a farmer purchasing drought-tolerant cultivars will spend more on seeds than will farmers growing more common varieties, and although the drought-tolerant cultivars offer protection from moderate droughts, they do not offer as much protection from more extreme droughts and other natural hazards. As a result, farmers may still lose their crops or livestock. The risk of being unable to recuperate investments in risk-reducing technologies will discourage farmers from investing in these technologies.

Climate insurance can help reduce these risks that farmers cannot manage through better practices and technologies. Providing farmers with insurance for more severe droughts and other natural hazards will help protect their investments, and potentially increase investments in adaptive measures (Kramer and Ceballos 2018). There are, however, several challenges in the provision of climate insurance (Carter et al. 2017; Kramer et al. 2019).

On the one hand, there are indemnity-based insurance products that aim to settle claims based on actual losses experienced by a farmer. Such insurance is typically unavailable to smallholder farmers; if it is available, it will be unaffordable. Indemnity-based products carry high transaction costs because the value that a smallholder farmer seeks to insure is small relative to the cost of in-person visits required to verify that damage has indeed occurred in case of a claim. Such products are also known to be expensive because of asymmetric information between insurers and the insured. Providing compensation for a farmer's actual loss can induce moral hazard: it lowers the farmer's incentives to minimize damage, as payouts are made only when crops are damaged regardless of someone's effort, and efforts to minimize risk reduce the chances of a farmer receiving an insurance payout. It can also induce adverse selection: at a given premium, only farmers with relatively higher risk exposure will enroll in insurance, which will drive up expected insurance payouts, and insurers will reflect such selection by raising their premiums (Just, Calvin, and Quiggin 1999; Gunnsteinsson 2020; Ceballos and Kramer 2021).

On the other hand, index-based insurance settles claims based on objective measurements of an index, which has been designed to proxy for agricultural losses, for instance rainfall. The advantage of index insurance products is that, in principle, they can be implemented at a relatively low cost, since no in-person visits are required to verify damage. In addition, the perfect observability of the index helps overcome any information asymmetries, eliminating concerns around moral hazard or adverse selection driving up insurance premiums. Index-based insurance therefore helps address some of the key challenges associated with the provision of indemnity-based insurance (Barnett and Mahul 2007). However, these products often suffer basis risk, meaning that the index and thus insurance payouts do not correlate adequately with the actual losses that a farmer or herder experiences. Too often, insurance beneficiaries experience damage for which the index does not trigger a payout, given that the index is a proxy only for agricultural losses. Such basis risk, combined with limited understanding of how index insurance works, also results in poor trust and low take-up of insurance products (Clarke 2016).

Another major challenge in the provision of agricultural insurance, regardless of whether a scheme provides indemnity- or index-based insurance, is that limited market intelligence goes into the design of these insurance schemes. Programs often fail to differentiate across different types of farmers, who require different solutions (Hansen et al. 2019; Kramer et al. 2022). For instance, more commercially oriented farmers have more risk-absorbing capacity because of their greater wealth levels and will hence be less risk-averse (Clarke 2016; Vargas Hill, Hoddinott, and Kumar 2013). Because of the risk-absorbing capacity and the lower vulnerability levels of this type of farmer, partial insurance—covering, for instance, only their inputs—will likely be sufficient, and they will not have to insure the full value of production or income over a season. Subsistence-oriented farmers are more vulnerable, with lower risk-absorbing capacity. This is why they will benefit from more complete insurance, covering not only investments in inputs but also the forgone profits from selling crops in the case of a bad harvest. Nonetheless, insurance policies offered to smallholder farmers are typically designed to cover only a portion of their investments in inputs, for instance only seeds. Existing solutions rarely protect households from the full loss in consumption or income that they experience in a bad season.

Finally, insurance providers are challenged by farmers' ability to pay and liquidity constraints, the sharing of risks across value chain actors, and the broad nature of risks experienced. More subsistence-oriented farmers may be able to purchase insurance for a small portion of their investments but they will not be able to afford an unsubsidized commercial insurance premium for their full value of production, or the full loss that they may experience in a bad year, even though this would provide them with the type of social protection that would enable increased investments in their farms (Kramer et al. 2022). More commercially oriented farmers may be able to afford insurance premiums, but only at the time of harvest, when they have cash on hand. For them, overcoming liquidity constraints to premium payment will be critical to create demand. Collecting the full insurance premium from vulnerable smallholder farmers can also be difficult when the risk is shared across value chains. A weather shock that lowers productivity will affect not only farmers but also their lenders, input providers, aggregators, and laborers. Farmers' decision to take up insurance can hence benefit these other actors, and innovative mechanisms are required to share the costs of insurance premiums across them. It is, however, important to recognize that insurance cannot cover all risks, including reduced market prices for farmers' produce and other postharvest risks, which can, in some cases, affect farmers' incomes more than production risks (Ceballos, Kannan, and Kramer 2021).

The remainder of this chapter describes examples of innovations that have been tested in Kenya, along with the shortcomings and implications for Kenya's national agricultural insurance scheme. Given that insurance design often neglects the role of gender and social inclusion, the chapter includes a separate section discussing how to advance gender and social equity through climate insurance. The chapter focuses on Kenya; for reviews of innovations in climate insurance from a more global perspective, see Carter and colleagues (2017), Kramer and colleagues (2019), and Kramer and colleagues (2022).

Innovations in climate insurance with a focus on solutions tested in Kenya

This section discusses potential solutions to the challenges introduced above. The discussion focuses on technological innovations as well as those in the institutional space, including, for instance, bundling insurance with other vital services for smallholder farmers, providing macro-level insurance coverage, and integration in social safety nets. This is not a comprehensive overview of solutions in the agricultural insurance sector, since the section focuses on innovations developed and tested in the context of Kenyan agriculture. However, many solutions have been either developed or tested in Kenya, and the innovations provided below are therefore a valuable starting point.

On the technological side, remote sensing has been used to make low-cost index-based livestock insurance (IBLI) available for Kenyan pastoralists. IBLI was designed by linking semi-annual seasons of longitudinal household data on livestock mortality with remote sensing data to construct a vegetation index-based proxy for livestock mortality. The Kenya Livestock Insurance Program (KLIP) provides fully subsidized IBLI coverage for up to five tropical livestock units for eligible pastoralists, and more commercially oriented herders can purchase additional unsubsidized coverage.[1] IBLI was predicted to have positive welfare effects (Chantarat et al. 2013). Empirical impact evaluations indeed show that, for many households, it substantially reduces exposure to covariate risk, or risks that affect households in a wider region at the same time (Jensen, Barrett, and Mude 2016). Also, among households experiencing drought, it reduces reliance on costly coping strategies such as selling assets to smooth consumption or reducing consumption to protect assets (Janzen and Carter 2019). Yet, basis risk remains a challenge: IBLI policyholders are left to manage on average 69 percent of their original risk because of idiosyncratic livestock mortality (Jensen, Barrett, and Mude 2016).

In the area of crop insurance, remote sensing has been used to improve farmers' options to manage risk as well. ACRE Africa is a private company headquartered in Kenya that offers solutions for smallholder farmers to manage climate risks. It designed an index insurance product that uses remote sensing-based data on soil moisture to proxy for crop health. ACRE Africa and its partners state that the product has been successfully piloted and scaled out to thousands of farmers but there are no peer-reviewed publications to validate product quality or the welfare impacts of this insurance scheme. While remote sensing-based crop insurance products may achieve substantial reductions in spatial basis risk compared with weather index-based insurance products that use weather stations to proxy for crop losses, they may still face considerable design basis risk, where indexes derived from satellite spectral bands are a limited proxy for individual crop damage. In addition, remote sensing products suffer from limited visibility of ground conditions owing to cloud cover, and decreasing signal-to-noise ratios as spatial resolution increases. They also require considerable processing power and storage capacity at very high resolutions (IFAD 2017), with providers such as PlanetLabs now offering a global 3–5 meter resolution of daily red-green-blue images. To address such basis risk, ACRE Africa has started introducing picture-based insurance into its index insurance solutions (Ceballos, Kramer, and Robles 2019, see Box 11.1).

1 Tropical livestock units are livestock numbers converted to a common unit, whereby larger animals are converted into a larger number and smaller animals into a smaller number. As a result, two cows, for instance, will result in a greater measure of tropical livestock units than will one cow and one sheep.

BOX 11.1 Picture-based insurance

Picture-based insurance (PBI) is a new way of delivering affordable multi-peril crop insurance. By relying on smartphone pictures taken from a farmer's field, PBI settles claims based on plot-level damage, resembling indemnity insurance without having insurance agents visit fields to verify losses. Sending in pictures can also make the insurance process more engaging, comprehensible, and accessible to small farmers. Here is how it works:

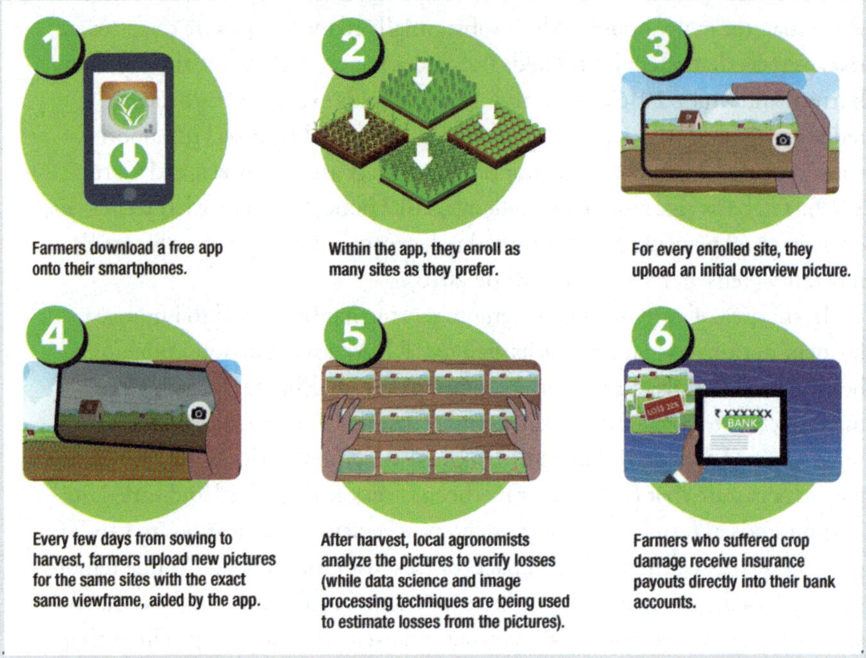

Source: https://www.ifpri.org/project/PBInsurance

IFPRI tested this approach in Kenya with ACRE Africa, Kenya Agricultural and Livestock Research Organization (KALRO), and Wageningen University, first offering it as a free trial and later at commercial insurance premiums. An initial proof of concept (Ceballos, Kramer, and Robles 2019) shows PBI to be feasible and sustainable: insurance experts can quickly and accurately process claims for sites where farmers report crop damage remotely; the research team has been able to use machine learning to partially automate image processing and damage classification for claims settlement (Hufkens et al. 2019); and limited smartphone ownership has been overcome by relying on village-based agents to send in pictures on behalf of insured farmers in their villages.

Given that PBI revolutionizes the insurance–client relationship, enabling insurers to directly observe the farm, PBI could be implemented as indemnity-based insurance, whereby costly in-person visits by insurance agents are replaced by inspections of images of the damaged crop. In addition, the wealth of information in field pictures and the direct communication channel with farmers enabled by a smartphone app could be further capitalized to design more comprehensive risk management solutions for smallholder farming.

An institutional innovation that has been piloted in the context of insurance programs in Kenya is bundling insurance with other vital services for smallholder farmers. First, insurance has been offered in combination with seeds. In fact, this is how ACRE Africa—the organization mentioned above—started. Its originator, the Kilimo Salama project, launched by the Syngenta Foundation for Sustainable Agriculture and the Global Index Insurance Facility, provided weather index-based insurance to farmers purchasing certified maize seeds, which they could activate at the time of planting using a scratch card found in the seed bag, using the same methods as in topping up airtime. In case of drought or excess rainfall, Kilimo Salama would send an insurance payout to the mobile money account linked to the phone number that was used to activate insurance coverage. In its inception, the program subsidized the premiums of these so-called replanting guarantees, with the idea that, over time, seed companies would start paying the premiums directly as they would see the value of providing farmers with insurance (as it would increase farmers' loyalty and thus demand for their products).

The program shows that uptake of Kilimo Salama was positively and significantly associated with the use of chemical fertilizer and improved seeds, and also with crop yield (Sibiko and Qaim 2020). A key challenge, however, lay in understandings of product benefits. Many farmers were not aware that the seeds they were purchasing came with insurance coverage, or they did not know how to activate the coverage. Farmers were also unaware of how rainfall measurements and thresholds were linked to insurance payouts, suggesting a need for more transparency (Sibiko, Veetill, and Qaim 2018). Moreover, removing subsidies for seed companies to bundle their products with insurance appears to have crowded out their interest in insurance. Companies do not have the marketing budgets to include insurance free of charge and they fear that adding insurance premiums to the costs of seeds will reduce their competitiveness.

Bundling seeds with insurance could, however, have positive impacts. Bulte and colleagues (2020) use a randomized trial in Kenya to show that farmers increase effort—increasing total investments and taking more land in production—when receiving a free hybrid crop insurance product conditional on purchasing certified seeds. In addition to adopting more certified seeds, they invest more in complementary inputs such as fertilizer and hired-in farm machinery and nonfarm labor. They also find that being provided with insurance free of charge increases ex post willingness to pay for the insurance product. This suggests that learning about the benefits of (subsidized) insurance outweighs any anchoring effects on the zero price during the pilot study. A study in Mozambique and Tanzania finds favorable resilience impacts of bundling insurance with drought-tolerant maize: the drought-tolerant maize protects farmers from moderate droughts while insurance provides payouts in case of a severe drought (Boucher et al. 2021). Bundling with seeds offers a natural entry point for increasing insurance coverage. A policy recommendation, then, is to explore instruments that encourage seed companies to bundle their seeds with insurance.

Second, there have been successful initiatives bundling insurance with contract farming. Casaburi and Willis (2018) tested this strategy through a randomized controlled trial in Kenya. They find that such bundling can increase the demand for insurance because the buyer of crops can deduct the insurance premium from farmer revenues at harvest time (just like the cost of any other inputs). This offers a mechanism for farmers to pay insurance premiums in the future, which can help overcome challenges around limited trust, present bias, and liquidity constraints. The One Acre Fund, a large-scale program that provides agricultural service packages including inputs such as seeds and fertilizers, often on credit, and typically along with extension, offers crop insurance as part of its standard agricultural service package to farmers in five African countries, including Kenya. This allows the One Acre Fund to reimburse farmers for a portion of their upfront investments in seeds and fertilizer when weather harms their crops, easing the immediate financial pressure of a poor harvest. Insurance has also been bundled with credit alone; see more on this in Chapter 12 on risk-contingent credit.

The third set of successful insurance programs has been operating at the macro level, with governments funding insurance coverage out of their social protection, disaster relief, or agriculture budgets. One example of macro insurance provided in the African continent is African Risk Capacity (ARC), which was launched in 2012 to provide African Union member states with parametric insurance to finance disaster risk management operations, facilitating faster delivery of assistance to beneficiary households (Kramer, Rusconi, and Glauber 2020). Kenya was part of ARC's first two risk pools (2014–2015 and 2015–2016) but has opted out since then, after having developed a national disaster risk management plan—which Kenya perceived as the main benefit in joining ARC. However, the KLIP has been integrated with Kenya's major social safety net, the Hunger Safety Net Program (HSNP). Jensen, Ikegami, and Mude (2017) find that this integration of insurance into a cash transfer program improves cost-effectiveness, as it protects the vulnerable non-poor from falling into poverty and requiring permanent cash transfers. The Alliance to Feed the Earth in Disasters (ALLFED) is developing bonds or insurance products that trigger financing when there is a risk of a locust outbreak in the Horn of

Africa; the funding, when a payout is triggered, is used to finance locust control operations for the Food and Agriculture Organization of the United Nations (FAO), so that FAO together with in-country partners can intervene through a coordinated response and prevent locust outbreaks from causing major damage (ALLFED 2021).

Gender and social equity: How to make climate insurance more equitable?

Given public investments in climate insurance, an important question relates to the extent to which insurance programs achieve gender and social equity. Most studies on gender and agricultural insurance focus on gender inclusivity by analyzing gender gaps in insurance *reach*, for instance by comparing the number of men and women enrolled, and by studying how to increase uptake among women. Altamirano and Beers (2018), for instance, discuss the inclusiveness of the Kilimo Salama program that was bundling insurance with seeds. In the short term, the use of mobile technology and remote sensing was likely not inclusive, given limited access to and understanding of this phone-based system, and particularly control over mobile money accounts, among the most vulnerable. The authors argue that the inclusiveness of this innovation may increase in the longer term. However, gender inclusiveness is not enough to create equity; to *benefit* or *empower* socially excluded groups, more attention needs to be paid to the distribution of insurance outcomes—that is, the extent to which insurance benefits and empowers women as much as men (Timu and Kramer 2021).

In theory, insurance programs can promote gender equity in *benefits* from insurance by addressing both demand- and supply-side constraints to uptake. This also involves providing quality insurance products that are beneficial to both men and women; targeting not only primarily male-controlled livestock and crops but also crops that are more commonly produced and commercialized by women; focusing on those risks that are difficult to manage for women, such as covariate risks that reduce the capacity of women's informal risk-sharing networks, including rotating savings and credit associations, to provide a safety net; and working through gender-inclusive distribution channels. Unfortunately, to date, there is limited research on how to create more gender-equitable insurance schemes. IFPRI's insurance research in Kenya is working to address these evidence gaps. This includes, for instance, using rigorous randomized controlled trials to analyze whether agricultural insurance has equally large impacts on technology adoption, agricultural productivity, and dietary diversity for women as for men. It also includes looking at which agricultural risk management bundles—insurance alone, insurance offered in combination with seeds of stress-tolerant varieties, or insurance combined with remote advisory services—have the greatest impacts for women vis-à-vis men.

Other examples of IFPRI research, focused primarily on supporting women's *empowerment* through climate insurance, include efforts to ensure that contracts purchased by women are registered under their names and that payouts are subsequently paid to their accounts; initiatives that bundle insurance with empowerment tools such as behavioral change communication and "edutainment" around gender roles; and working through local champion farmers, who preserve and connect with informal mutual assistance group activities, to promote insurance. To track the extent to which insurance benefits empower men and women alike, it is important to collect gender-disaggregated data. Given that many women are members of male-headed households, this type of research needs to move beyond a comparison of impacts for the primary decision-maker in male- and female-headed households. Rather, impact evaluations should track outcomes measured for different types of male and female household members, to shed light on the distribution of insurance outcomes within a household.

An important area in which careful consideration of gender and social equity is necessary is the digitization of insurance. The rapid expansion of mobile phones and cellular coverage in rural Africa has led to increased interest in digital financial instruments for the agriculture sector. This includes a large push to improve agricultural finance and insurance for smallholder farmers through digital innovations. At the same time, mobile technologies are not yet revolutionizing agriculture in Kenya, and there is still a long way to go to reap the full benefits from digitization. Parlasca, Johnen, and Qaim (2021), for instance, find that more than 80 percent of farmers use mobile money but only 15 percent use this innovation for agriculture-related payments; less than 1 percent of farmers use mobile loans for agricultural investments. Indicative of a digital divide, ownership and use of digital agricultural services are lowest among the most vulnerable. Koo and colleagues (2022) note that rural areas, especially in sub-Saharan Africa, are underserved when it comes to digital infrastructure, with limited access to digital services especially among women. Bridging this digital divide will require policy incentives and public–private partnerships to accelerate investments in digital infrastructure, and investments in agricultural service providers' capabilities to strengthen their digital services in a socially equitable way.

Shortcomings and implications for Kenya's national agricultural insurance scheme

The innovations discussed above, while promising, also come with a few key challenges. This section discusses these, and how the public sector can address them by creating a more enabling environment for insurance schemes.

A first challenge is the lack of historical data to design and underwrite insurance products. Designing high-quality index insurance products that are correlated with agricultural losses and do not suffer major basis risk requires access to sound historical data on both the proposed index and agricultural losses. Such data are hard to come by. This is a key reason why agricultural insurance often faces basis risk, even when high-resolution satellite imagery is limiting basis risk in the spatial domain. Further, absence of such data makes it difficult for insurers and reinsurers to determine expected payouts from a product, which inflates premiums.

To address this challenge, it is important for the public sector to increase data availability. The government could, for instance, improve access to geo-referenced yield data, which are collected already through multiple initiatives. These include the World Bank-funded One Million Farmer Platform, led by KALRO, other KALRO initiatives, and Kenya's national agricultural insurance scheme. Moreover, initiatives can create incentives for the private sector to share data. For instance, the Lacuna Fund awarded ACRE Africa a grant to publish its labeled smartphone images of targeted crops, collected as part of its picture-based insurance operations (see Box 11.1), and make the data publicly available. This could, in fact, be made a requirement for start-up companies accepting grant money. Overall, the problem is not so much that the data are not there, but that the infrastructure, capacity, and incentives for sharing these data are not in place. Infrastructure is needed to access data and associated satellite imagery in an ethically sound way, whereby exact GPS coordinates—because they are personal identifiers—are kept confidential.

A second challenge requiring government intervention in the context of agricultural insurance is in the regulatory space. Regulators should play an important role in quality control and consumer protection. Capacity development is required to enable regulators to assess insurance products and to ensure they are of high quality, with limited basis risk and reasonable insurance premiums. The United States Agency for International Development's Innovation Lab for Markets, Risk, and Resilience has developed a tool that allows regulators to calculate the expected welfare impacts of an insurance product and to assess the extent to which the product meets basic quality standards, prior to approving its commercial sale.

Moreover, if bundled with other services, insurance premiums should not be hidden in the total price of the product bundle. For instance, when insurance is bundled with loans, farmers should be made aware of the fact that the loan comes with insurance, how to file a claim under that policy, and by how much the bundling with insurance is increasing their interest rate. Too often, bundling insurance with other services and inputs helps programs increase uptake but customers are not aware of their coverage, limiting the impacts on resilience. Regulators could monitor whether adequate consumer education is in place.

At the same time, regulation in an early stage of piloting new products could discourage innovation. Innovation in product design and delivery will therefore require regulatory sandboxes, making it possible to pilot new products without having to go through the longer regulatory approval processes that would be desirable for more mature products offered at a larger scale (Jenik and Lauer 2017).

A third challenge facing insurance markets is the existence, and increasing incidence, of systemic catastrophic risks that are too expensive to insure through micro or meso insurance. For instance, an individual farmer will most likely not be able to access affordable micro insurance against the risk of a locust outbreak or a large-scale drought that causes food insecurity throughout the country. These types of scenarios, which can affect millions of farmers at once, are well suited to be insured through macro insurance schemes, for instance the above-mentioned ARC (Kramer, Rusconi, and Glauber 2020) and the ALLFED initiative to insure response operations for locust outbreaks (ALLFED 2021). An important question for a next generation of insurance research to address is what impacts the public sector can have on micro and meso insurance markets by insuring these catastrophic risks. It is likely that a transfer of catastrophic risks out of the country, to international insurance markets, will create a more favorable environment to offer insurance for the more moderate risks. These complementarities between different levels of insurance coverage have, however, not been studied sufficiently; additional research in this area is crucial to better assess the costs and benefits of innovations in catastrophic disaster risk management.

A fourth challenge is low demand for new products that farmers and other value chains actors will not have seen at work yet. On the one hand, for more commercially oriented farmers, who should in principle be able to afford unsubsidized insurance, the government can help increase demand by providing smart subsidies, mainly to stimulate adoption in the short run. These smart subsidies enable insurers to provide products at a discount (or even free of charge) on a promotional basis, giving potential clients the product experience that is a prerequisite for creating demand (Vargas Hill et al. 2014; Hazell and Varangis 2020). For commercially oriented farmers, a program would want to clearly communicate that the policies are being sold at a discount, to encourage early adoption but prevent farmers from using the discounted premium as an anchor for what future premiums should be. These subsidies would need to be phased out over time. On the other hand, for more subsistence-oriented farmers, subsidies may be required in the long run, given that these farmers likely cannot afford unsubsidized insurance premiums. Moreover, subsistence-oriented farmers will require higher levels of insurance coverage, given their lower absorptive capacity. For these types of farmers, one would want to look at protecting their full income loss in a bad year, not only investments. In addition, insurance should not be tied to the purchase of inputs, given that farmers may not be able to plant or invest in more expensive inputs when the rains fail.

The HSNP, the KLIP, and IBLI schemes provide an interesting example of how Kenya has integrated social safety nets, targeting different types of subsidies and solutions toward different types of herders. Some (the relatively poorest, including laborers who do not have their own land or livestock) are eligible to receive regular cash transfers through the HSNP. Others (the vulnerable non-poor, but subsistence-oriented) receive subsidized livestock insurance through the KLIP for up to a given number of tropical livestock units. More commercially oriented herders have the option to buy additional insurance for their larger herds. In the context of crop insurance, a similar system could be applied. The poorest would be eligible for cash transfers; the vulnerable non-poor engaged in subsistence farming would be able to obtain subsidized crop insurance with a sufficiently large amount covered to protect not only investments but also their incomes from extreme weather shocks; and, finally, more commercially oriented farmers seeking higher coverage levels could purchase additional unsubsidized insurance, facilitated through value chain actors such as input providers, aggregators, and contract farming schemes. Such a layered approach is also essential to create social equity; adequate targeting of insurance subsidies will help ensure the benefits accrue not only to commercial farmers who can afford unsubsidized premiums but also to the most vulnerable, who will not have the financial means to pay insurance premiums.

The solutions provided above illustrate how a government can invest in an enabling environment for scale-up of agricultural insurance. This will, however, require public investments. It is therefore important to expand the evidence base on what impacts insurance programs have on the resilience and welfare of beneficiary communities, what costs and benefits public investments in these programs will have, and how these costs and benefits compare with the cost-benefit ratios of other types of social protection, for instance cash transfer programs. Demand for an insurance product, or levels of insurance payouts, should not be seen as an indicator of impact; to study impacts of an insurance program, and get at true costs and benefits, factors such as agricultural incomes, investments, and gender and social equity need to be considered and quantified. It is important to build in these impact evaluations and cost-benefit analyses, including metrics of gender equity and social inclusion, from the onset of an insurance program. Monitoring and evaluation plans would ideally move beyond a comparison of impacts for male- and female-headed households to investigate how the program interacts with gender norms, and how an insurance program affects different members *within* a household, considering that the vast majority of women reside in male-headed households and are overlooked in a comparison based on the gender of a household head (Timu and Kramer 2021).

Finally, a challenge with designing and implementing agricultural insurance programs is that smallholder farmers face a multitude of risks, including not only droughts but also other weather-related hazards, biotic stressors such as pests and disease, and price fluctuations. By design, an index-based insurance program cannot insure farmers from all these risks at once. Moreover, insurance is not the appropriate instrument for every type of risk; it is the right instrument to manage more severe and extreme risks but not to manage more moderate risks, for instance a moderate drought that affects only a small portion of the harvest or herd. Farmers can manage those moderate risks at a lower cost using other financial instruments such as savings and credit, as well as risk-reducing technologies and practices such as crop diversification, stress-tolerant cultivars, or better food and water storage facilities. Insurance should complement these other risk management instruments, rather than crowd out their use. This also involves designing insurance subsidies such that they do not discourage farmers and herders from adopting unsustainable practices. Heavy premium subsidies can result in environmentally risky production behaviors such as growing high-risk crops on unsuitable land (for example, growing maize in an area where rainfall has become generally inadequate to grow maize), increasing pesticide and fertilizer use beyond socially optimal levels (Weber, Key, and O'Donoghue 2016; Möhring et al. 2020; Dougherty, Gallenstein, and Mishra 2021), rapid livestock accumulation disrupting natural pasture recovery dynamics (John et al. 2019), or disincentivizing climate change adaptation in agriculture (Wang, Rejesus, and Aglasan 2021). Insurance programs need to be carefully designed to avoid such distortionary effects.

Conclusion

In conclusion, climate insurance is an important financial instrument that can help improve agricultural risk management for smallholder farmers, herders, and other value chain actors. In the face of the present climate crisis, it will be crucial for Kenya to improve agricultural risk management, and in particular smallholder farmers' capacity to adapt to increasing incidence of weather extremes. At the same time, prioritizing investments by the public sector to create a more enabling insurance environment can come at the expense of investing in other agricultural development programs or social safety nets. Before prioritizing investments that aim to roll out agricultural insurance programs at scale, policymakers will need evidence in the form of rigorous impact assessments and cost-benefit analyses, and a comparison of the cost-effectiveness of insurance programs versus other social protection instruments.

It is therefore important to differentiate between different types of farmers and herders, distinguishing at the very least between the poorest farmers, typically landless laborers; the vulnerable non-poor, typically subsistence-oriented farmers; and more commercially oriented smallholder farmers. Each of these groups will require different insurance or social protection solutions (Kramer et al. 2021). This will also involve putting social and gender equity at the forefront of impact assessments, to ascertain that, among the most vulnerable segments of the population, insurance programs reach, benefit, and empower women and men alike.

Agricultural insurance is also not a stand-alone solution but, if treated that way, a program will have limited impacts on technology adoption, market participation, and other indicators of structural transformation. If not carefully designed, insurance could even induce the adoption of environmentally unsustainable practices and technologies. A more holistic approach to insurance program design acknowledges that farmers and herders have their existing informal insurance solutions, such as saving for a rainy day or sharing risks within their social networks; and that this, along with financial instruments as well as adaptive technologies and practices, can help them manage relatively moderate risks. Insurance is then designed to provide financial protection from more severe and extreme risks, and could even be offered to existing risk-sharing groups. As shown by an experiment in Ethiopia, this could help crowd in demand for formal insurance (Dercon et al. 2014). In this light, insurance programs, including premium subsidies, need to be "climate-smart," and encourage farmers and herders to use complementary risk management instruments along with insurance.

Finally, digitalization offers great potential to improve the cost-effectiveness of agricultural insurance. Remote sensing and the use of smartphone photos and other sensors increase the amount of data available on smallholder farming systems, which will allow insurers to better monitor smallholder farms, reduce basis risk, and lower the cost of insurance products. Groundbreaking work is being done in this area in Kenya, with several pilots underway that can shed light on how to better embed agricultural insurance into adaptation and disaster risk management plans. At the same time, most solutions are still in a pilot phase, and the hard digital infrastructure and softer capabilities to use digital technologies are not yet sufficiently developed to transform the insurance sector at scale. Efforts to digitize agricultural insurance will not be socially inclusive and will potentially widen gender gaps if nothing is done to bridge the digital divide. Investments will be required in digital infrastructure to enhance phone ownership and use, mobile coverage, and internet access among the most vulnerable smallholder farmers and herders in Kenya.

Summarizing, there are exciting areas of innovation in climate insurance, with great promise to improve smallholder farmers' ability to manage climate change-related production risks. However, investments to scale these innovations will need to undergo rigorous cost-benefit analysis, positioning in a broader risk management strategy, and, when involving digitalization, complementary efforts to address digital divides. This will take time, and thus climate insurance will not provide a short-term solution that can be widely and successfully adopted by farmers at scale.

References

ALLFED (Alliance to Feed the Earth in Disasters). 2021. *The Potential for a Novel Bond to Safeguard Communities Against Future Locust Outbreaks: A Scoping, Financing, and Implementation Overview.* High-Level Guidance Document. London: ALLFED.

Altamirano, M.A., and C.P.V. Beers. 2018. "Frugal Innovations in Technological and Institutional Infrastructure: Impact of Mobile Phone Technology on Productivity, Public Service Provision and Inclusiveness." *European Journal of Development Research* 30 (1): 84–107.

Barnett, B.J., and O. Mahul. 2007. "Weather Index Insurance for Agriculture and Rural Areas in Lower-Income Countries." *American Journal of Agricultural Economics* 89 (5): 1241–1247.

Boucher, S.R., M.R. Carter, J.E. Flatnes, T.J. Lybbert, J.G. Malacarne, P. Marenya, and L.A. Paul. 2021. "Bundling Stress Tolerant Seeds and Insurance for More Resilient and Productive Small-Scale Agriculture." NBER Working Paper 29234. National Bureau of Economic Research (NBER), Cambridge, MA.

Bryan, E., C. Ringler, B. Okoba, J. Koo, M. Herrero, and S. Silvestri. 2013. "Can Agriculture Support Climate Change Adaptation, Greenhouse Gas Mitigation and Rural Livelihoods? Insights from Kenya." *Climatic Change* 118 (2): 151–165.

Bulte, E., F. Cecchi, R. Lensink, A. Marr, and M. Van Asseldonk. 2020. "Does Bundling Crop Insurance with Certified Seeds Crowd-In Investments? Experimental Evidence from Kenya." *Journal of Economic Behavior & Organization* 180: 744–757.

Carter, M.R., L. Cheng, and A. Sarris. 2016. "Where and How Index Insurance Can Boost the Adoption of Improved Agricultural Technologies." *Journal of Development Economics* 118: 59–71.

Carter, M., A. de Janvry, E. Sadoulet, and A. Sarris. 2017. "Index Insurance for Developing Country Agriculture: A Reassessment." *Annual Review of Resource Economics* 9: 421–438.

Casaburi, L., and J. Willis. 2018. "Time Versus State in Insurance: Experimental Evidence from Contract Farming in Kenya." *American Economic Review* 108 (12): 3778–3813.

Ceballos, F., and B. Kramer. 2021. "The Role of Asymmetric Information in Multi-peril Picture-Based Crop Insurance: Field Experiments in India." IFPRI Discussion Paper 2088. International Food Policy Research Institute (IFPRI), Washington, DC.

Ceballos, F., B. Kramer, and M. Robles. 2019. "The Feasibility of Picture-Based Insurance (PBI): Smartphone Pictures for Affordable Crop Insurance." *Development Engineering* 4: 100042.

Ceballos, F., S. Kannan, and B. Kramer. 2021. "Crop Prices, Farm Incomes, and Food Security during the COVID-19 Pandemic in India: Phone-Based Producer Survey Evidence from Haryana State." *Agricultural Economics* 52 (3): 525–542.

Chantarat, S., A.G. Mude, C.B. Barrett, and M.R. Carter. 2013. "Designing Index-Based Livestock Insurance for Managing Asset Risk in Northern Kenya." *Journal of Risk and Insurance* 80 (1): 205–237.

Clarke, D.J. 2016. "A Theory of Rational Demand for Index Insurance." *American Economic Journal: Microeconomics* 8 (1): 283–306.

de Brauw, A. and G. Pacillo. 2022. "Food Value Chains: Increasing Productivity, Sustainability, and Resilience to Climate Change." In *2022 Global Food Policy Report: Climate Change and Food Systems*, 100–105. Washington, DC: IFPRI.

Dercon, S., R.V. Hill, D. Clarke, I. Outes-Leon, and A.S. Taffesse. 2014. "Offering Rainfall Insurance to Informal Insurance Groups: Evidence from a Field Experiment in Ethiopia." *Journal of Development Economics* 106: 132–143.

Dougherty J.P., R.A. Gallenstein, and K. Mishra. 2021. "Impact of Index Insurance on Moral Hazard in the Agricultural Credit Market: Theory and Evidence from Ghana." *Journal of African Economies* 30 (5):418–446.

Elbers, C., J.W. Gunning, and B. Kinsey. 2007. "Growth and Risk: Methodology and Micro-Evidence." *World Bank Economic Review* 21 (1): 1–20.

Gunnsteinsson, S. 2020. "Experimental Identification of Asymmetric Information: Evidence on Crop Insurance in the Philippines." *Journal of Development Economics* 144: 102414.

Hansen, J., J. Hellin, T. Rosenstock, E. Fisher, J. Cairns, C. Stirling, C. Lamanna, J. van Etten, A. Rose, and B. Campbell. 2019. "Climate Risk Management and Rural Poverty Reduction." *Agricultural Systems* 172: 28–46.

Hazell, P., and P. Varangis. 2020. "Best Practices for Subsidizing Agricultural Insurance." *Global Food Security* 25: 100326.

Hufkens, K., E.K. Melaas, M.L. Mann, T. Foster, F. Ceballos, M. Robles, and B. Kramer. 2019. "Monitoring Crop Phenology Using a Smartphone Based Near-Surface Remote Sensing Approach." *Agricultural and Forest Meteorology* 265: 327–337.

IFAD (International Fund for Agricultural Development). 2017. *Remote Sensing for Index Insurance: Findings and Lessons Learned for Smallholder Agriculture*. Rome.

Janzen, S.A., and M.R. Carter. 2019. "After the Drought: The Impact of Microinsurance on Consumption Smoothing and Asset Protection." *American Journal of Agricultural Economics* 101 (3): 651–671.

Jenik, I., and K. Lauer. 2017. *Regulatory Sandboxes and Financial Inclusion*. Washington, DC: Consultative Group to Assist the Poor (CGAP).

Jensen, N.D., C.B. Barrett, and A.G. Mude. 2016. "Index Insurance Quality and Basis Risk: Evidence from Northern Kenya." *American Journal of Agricultural Economics* 98 (5): 1450–1469.

Jensen, N., M. Ikegami, and A. Mude. 2017. "Integrating Social Protection Strategies for Improved Impact: A Comparative Evaluation of Cash Transfers and Index Insurance in Kenya." *Geneva Papers on Risk and Insurance-Issues and Practice* 42 (4): 675–707.

John F, R. Toth, K. Frank, J. Groeneveld, and B. Müller. 2019. "Ecological Vulnerability through Insurance? Potential Unintended Consequences of Livestock Drought Insurance." *Ecological Economics* 157: 357–368.

Just, R.E., L. Calvin, and J. Quiggin. 1999. "Adverse Selection in Crop Insurance: Actuarial and Asymmetric Information Incentives." *American Journal of Agricultural Economics* 81 (4): 834–849.

Kogo, B.K., L. Kumar, and R. Koech. 2021. "Climate Change and Variability in Kenya: A Review of Impacts on Agriculture and Food Security." *Environment, Development and Sustainability* 23 (1): 23–43.

Koo, J., B. Kramer, S. Langan, A. Ghosh, A. Gardeazabal Monsalue, and T. Lunt. 2022. "Digital Innovations: Using Data and Technology for Sustainable Food Systems." In *2022 Global Food Policy Report: Climate Change and Food Systems,* 106–113. Washington, DC: IFPRI.

Kramer, B., and F. Ceballos. 2018. "Enhancing Adaptive Capacity through Climate-Smart Insurance: Theory and Evidence from India." Paper presented at the 2018 International Association of Agricultural Economics (IAAE) Conference, July 28–August 2.

Kramer, B., J. Hellin, J. Hansen, A. Rose, and M. Braun. 2019. "Building Resilience through Climate Risk Insurance: Insights from Agricultural Research for Development." CCAFS Working Paper No. 287. Wageningen, Netherlands: CGIAR Research Program on Climate Change, Agriculture and Food Security.

Kramer, B., R. Rusconi, and J.W. Glauber. 2020. "Five Years of Regional Risk Pooling: An Updated Cost-Benefit Analysis of the African Risk Capacity." IFPRI Discussion Paper No. 1965. IFPRI, Washington, DC.

Kramer, B., P. Hazell, H. Alderman, F. Ceballos, N. Kumar, and A.G. Timu. 2022. "Is Agricultural Insurance Fulfilling Its Promise for the Developing World?" *Annual Reviews of Resource Economics* 14: 291–311.

Möhring, N., T. Dalhaus, G. Enjolras, and R. Finger. 2020. "Crop Insurance and Pesticide Use in European Agriculture." *Agricultural Systems* 184: 102902.

Parlasca, M., C. Johnen, and M. Qaim. 2022. "Use of Mobile Financial Services among Farmers in Africa: Insights from Kenya." *Global Food Security* 32: 100590.

Sibiko, K.W., and M. Qaim. 2020. "Weather Index Insurance, Agricultural Input Use, and Crop Productivity in Kenya." *Food Security* 12: 151–167.

Sibiko, K.W., P.C. Veettil, and M. Qaim. 2018. "Small Farmers' Preferences for Weather Index Insurance: Insights from Kenya." *Agriculture & Food Security* 7 (1): 1–14.

Timu, A.G., and B. Kramer. 2021. "Gender-Inclusive, -Responsive and -Transformative Insurance: A Literature Review." CCAFS Working Paper 417. Wageningen University, Wageningen, Netherlands.

Vargas Hill, R., J. Hoddinott, and N. Kumar. 2013. "Adoption of Weather-Index Insurance: Learning from Willingness to Pay among a Panel of Households in Rural Ethiopia." *Agricultural Economics* 44 (4-5): 385–398.

Vargas Hill, R., G. Gajate-Garrido, C. Phily, and A. Dalal. 2014. "Using Subsidies for Inclusive Insurance: Lessons from Agriculture and Health." Microinsurance Paper No. 29. Microinsurance Innovation Facility, Geneva.

Vargas Hill, R., N. Kumar, N. Magnan, S. Makhija, F. de Nicola, D.J. Spielman, and P.S. Ward. 2019. "Ex Ante and Ex Post Effects of Hybrid Index Insurance in Bangladesh." *Journal of Development Economics* 136: 1–17.

Wang, R., R.M. Rejesus, and S. Aglasan. 2021. "Warming Temperatures, Yield Risk and Crop Insurance Participation." *European Review of Agricultural Economics* 48 (5): 1109–1131.

Weber, J., N. Key, and E. O'Donoghue. 2016. "Does Federal Crop Insurance Make Environmental Externalities from Agriculture Worse?" *Journal of the Association of Environmental and Resource Economics* 3 (3): 707–742.

Chapter 12

TRANSFORMING FOOD SYSTEMS THROUGH RISK-CONTINGENT CREDIT IN RURAL AFRICA: DEVELOPMENT, EXPERIMENTATION, AND EVALUATION[1]

Apurba Shee, Michael Ndegwa, Calum G. Turvey, and Liangzhi You

Throughout Africa, climate change is posing severe challenges to agricultural production and food security. Agricultural risks—particularly those associated with drought—are a major cause of low agricultural productivity in most African countries, including Kenya. According to the Government of Kenya, four consecutive years (2008–2011) of drought caused US$12.1 billion in losses, accounting for about 8 percent of GDP, including losses in assets and disruptions to the economy across sectors (Kenya, Ministry of Agriculture, Livestock and Fisheries 2014). Currently, Kenya is in the middle of an acute drought following three consecutive poor rainy seasons. This has led to a drop in crop production nationally of about 70 percent, which has disproportionately exposed the communities of arid and semi-arid lands to hunger and malnutrition.

New technologies, such as improved crop varieties, fertilizers, and disease and pest control approaches, provide one avenue by which to increase productivity and improve farm incomes, hence reducing the vulnerability of farming households to drought and other shocks. Yet many farmers cannot access sufficient credit to adopt these technologies. Lenders in Kenya's credit markets limit the supply of credit to borrowers because of seemingly uninsurable weather

1 This research was funded through the German Federal Ministry for Economic Cooperation and Development, commissioned and administered through the German Agency for International Cooperation Fund for International Agricultural Research Grant 81260864. The research was also supported partly by the Agricultural Insurance Thematic Window of the International Initiative for Impact Evaluation and the Global Resilience Partnership through the Round 1 Global Resilience Challenge, supported by the United States Agency for International Development in the context of a project titled Satellite Technologies, Innovative, and Smart Financing for Food Security. The work was undertaken as part of the CGIAR Research Program on Policies, Institutions, and Markets led by IFPRI. We appreciate the generous hospitality provided by our project partner Equity Bank Kenya Ltd. The opinions expressed in this chapter do not necessarily reflect the views of our donors or partners. Any errors that remain are the authors' responsibility.

risks tied to frequent failures in seasonal rainfall and related heat stress. In cases where farmers can access credit, loans are made under relatively high collateral restrictions, which farmers are reluctant to accept. Consequently, borrowers voluntarily withdraw from the credit market (risk rationing), which, in turn, suppresses incentives for lenders to expand financial services (quantity rationing). Ultimately, incomplete financial markets result in suboptimal, welfare-reducing credit access that forces most farmers—especially women farmers—to adopt low-risk, low-return activities. This, in turn, is considered a key driver of persistent poverty and jeopardizes food security (Barrett, Carter, and Chavas 2019; Santos and Barrett 2019).

There have been efforts by private sector financial institutions to expand credit access to farmers by setting up branches in rural areas, but frequent droughts across and within years put agricultural productivity at risk for many farms and regions simultaneously. Such covariate risks can stress lenders' portfolios through increased loan defaults and have been identified as a significant contributing factor in lenders' reluctance to offer credit to farmers. Additionally, the absence of insurance markets results in more inelastic credit demands and supplies than might otherwise occur with a targeted (for example, weather-based) insurance product in place. A fundamental hinderance to increasing the depth and breadth of more inclusive financial policies for farmers in drought-prone states is the absence of an effective risk-transferring mechanism that simultaneously addresses the problem of weather-related business risks facing farmers and the financial/credit risks facing lenders.

Over the past two decades, index insurance has been promoted to help farmers manage weather-related risks but low demand and uptake have hampered its utility (Turvey 2001; Binswanger 2012; Miranda and Farrin 2012; Marr et al. 2016; Cole and Xiong 2017; Jensen and Barrett 2017). Economists and practitioners are now turning their attention to new financially engineered approaches to the design and implementation of insurance programs that can be bundled with credit in a way that reduces the high default risk facing lenders when exposed to widespread droughts and other natural (or market) catastrophes (Shee and Turvey 2012; Gallenstein et al. 2019; Mishra et al. 2021a, 2021b).

To address these challenges, we developed an innovative, market-based credit solution referred to as risk-contingent credit (RCC). RCC is a financial product that embeds within its structure an insurance contract that, when triggered, offsets loan payments, providing a risk-efficient balance between business and financial risks. Because the insurance coverage substitutes for collateral, it is more financially inclusive than conventional credit and has the potential to bring quantity- and risk-rationed farmers into the credit market, with expected benefits from higher productivity and improved livelihoods.

This chapter presents the development and evaluation of RCC for smallholder farmers with a special focus on the innovative pre-experimental and experimental methods to develop, test, adapt, and evaluate the intervention. We describe the scientific bundling of insurance with credit, how we approached communicating the RCC concept with farmers and stakeholders, and ways we incorporated their feedback to incrementally co-develop the full RCC product and its subsequent implementation in Kenya. We also provide an overview of the impact evaluation of RCC through a variety of in-the-field activities, including formative evaluation through choice experiments, survey design, and a randomized controlled trial (RCT). The chapter concludes with a discussion of the broader implications of RCC and its potential for scale-up to contribute to transforming African food systems.

Development of risk-contingent credit for smallholder farmers

Drought-related climate risk is the largest source of risk to agricultural productivity and consumption in most of sub-Saharan Africa, including Kenya (Giné and Yang 2009; Dercon and Christiaensen 2011; Cole et al. 2012). Drought imposes a considerable and costly constraint on capital and wealth accumulation for those engaged in agricultural activities or with livelihoods tied to the well-being of the farming sector (Barrett et al. 2007). Credit access compounds this problem, especially among smallholder households that lack adequate collateral. This exacerbates both supply-side quantity rationing and demand-side risk rationing (Boucher, Carter, and Guirkinger 2008; Verteramo-Chiu, Khantachavana, and Turvey2014). Faced with high loan default risk, banks use restrictive lending policies such as high interest rates, harsh collateral requirements, and credit denial to those deemed risky borrowers, among others, which effectively crowd out smallholders from the credit fold (Stiglitz and Weiss 1981; Boucher, Carter, and Guirkinger 2008; Boucher, Guirkinger, and Trivelli 2009).

There have been efforts to mitigate the impacts of drought on agricultural productivity in this region, mostly by governments and donors. Most of them have been developed under the climate-smart agriculture umbrella, with adaptive capacity, sustainable intensification, and greenhouse gas mitigation as the main goals (Branca 2012). They range from conservation agriculture to stress-tolerant crop varieties, pest and disease management methods, promotion of adaptable alternative crops and animals, capacity building, and gender mainstreaming, as well as innovative ways of sharing the right information with farmers in a timely manner, mostly through mobile phones. Such interventions mostly aim to develop and promote new and efficient farming technologies and practices. However, the large economic cost of drought-related climate shocks (and, by extension, flooding in some regions) cannot be financed by government and the donor community alone. There is a need for market-based interventions to enhance farmers' capacity to capitalize on the benefits of agricultural technologies and/or practices and at the same time to provide them with protection against the major agricultural downside risks (Shee, Turvey, and You 2019; Ndegwa et al. 2020).

The idea behind the RCC mechanism is to avoid low uptake of agricultural insurance. The function of any insurance is to transfer (smooth) income across good and bad states. In dealing with poor agricultural farmers, the upfront payment of a premium in standard insurance not only imposes liquidity constraints but also transfers income across time (Casaburi and Willis 2018). Literature shows that liquidity issues and time preferences are important constraints to insurance uptake. By removing both the liquidity constraint and the effects of climate risk, the RCC mechanism can achieve better targeting of poorer farmers. Since insurance can substitute for collateral, the RCC mechanism has the potential to encourage otherwise risk-rationed farmers to take up RCC loans. And since the indemnity from the embedded insurance is applied to the underlying debt obligation, the RCC can reduce the probability of default and build trust that can boost uptake.

We designed RCC that bundles insurance with credit to achieve the dual advantages of coverage against covariate risks and enhanced adoption of technologies that would lead to improved livelihoods. Unlike standard indemnity- and index-based insurance products,[2] RCC does not require farmers to pay premiums upfront and out of pocket, and hence removes liquidity constraints and targets poorer farmers more effectively (Shee, Turvey, and You 2019). The insurance component with RCC substitutes for collateral and hence makes it more financially inclusive than conventional credit products. We developed and implemented RCC in Machakos county of Kenya in

2 With indemnity-based crop insurance, payouts are based on the actual losses experienced by the farmer, regardless of the cause. They are all-peril in nature and have high administrative costs because of the farm follow-up required to ascertain losses. On the other hand, the objective of weather index insurance is to establish a trigger below (or above) which the weather peril is highly correlated with yield loss. The most common index is based on cumulative rainfall, although other notable indices have been proposed, including average area yield, soil moisture, heat index, vegetation indexes such as the normalized difference vegetation index and enhanced vegetation index, and commodity prices (Ndegwa et al. 2022).

FIGURE 12.1 The RCC business model and institutional setup

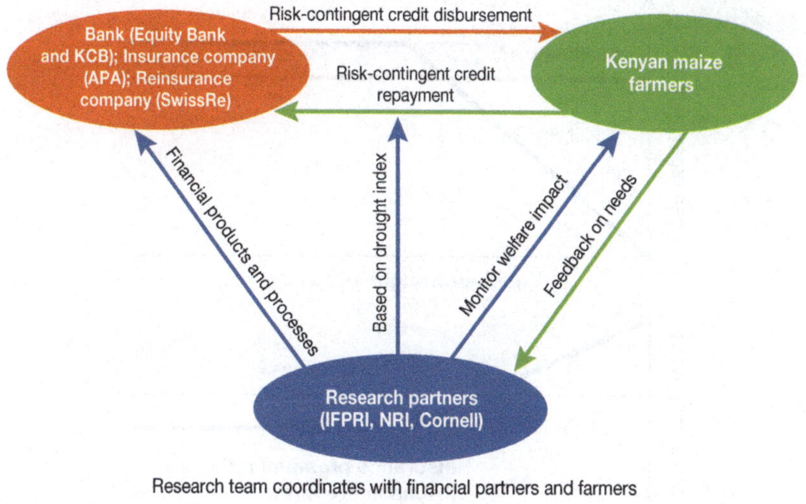

Source: Authors' illustration.

collaboration with our private sector partners Equity Bank Kenya Ltd., Kenya Commercial Bank (KCB), and APA Insurance, along with reinsurance offered by SwissRe. Equity Bank is responsible for farmers' training, loan disbursement, and monitoring in Kenya. KCB does a similar job outside our experimental area. APA Insurance underwrites the weather index insurance embedded in RCC. Figure 12.1 depicts the RCC business model including institutional setup.

Below we provide a brief description of RCC and how it can protect farmers from drought-related production risks. In Figure 12.2, the upper graph shows loan repayment and the lower graph illustrates the underlying insurance payout, which depends on weather conditions (to the left). If the underlying risk (weather-related) worsens and crosses a certain threshold or trigger, the total repayment obligation of the farmer falls linearly, with the difference deposited directly into the farmer's loan account at the bank by the insurer. On the other hand, if the underlying risk is not triggered, the loan has to be repaid, along with the cost of insurance. RCC therefore has the unique characteristic that, even though farmers have to pay a risk premium during normal circumstances, they are insured against adverse circumstances. RCC is designed with an actuarially fair interest rate that is interlinked with the underlying weather risk. Over the years we have developed, fine-tuned, and improved the scientific design of RCC (for

FIGURE 12.2 A schematic illustration of risk-contingent credit

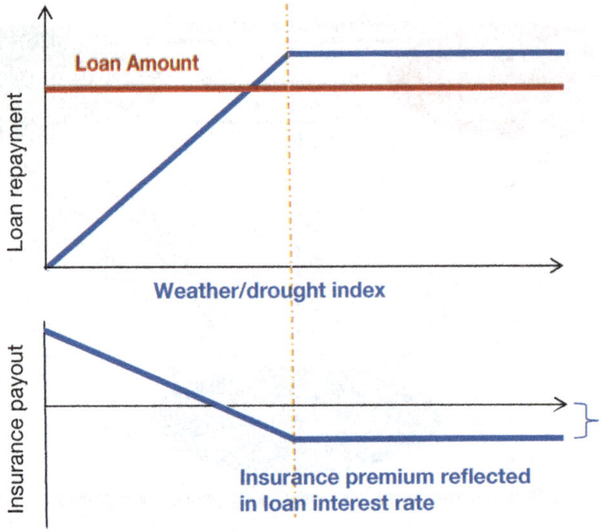

Source: Authors' illustration.

details on the evolution of the scientific RCC design, see Shee and Turvey 2012; Shee, Turvey, and You 2019; Turvey, Shee, and Marr 2019).

Our underlying insurance product is weather (rainfall) index insurance that provides protection against rainfall shortage in the long-rain cropping season. The index was constructed based on Climate Hazards Group InfraRed Precipitation with Station data (CHIRPS) rainfall measures for the traditional long rain season in Machakos county from October 15 to January 15. The study involved two rounds of loan disbursement among randomly selected smallholder households from Machakos county, with some significant improvements in the insurance component between rounds. During the first round of implementation, from September 2017 to January 2018, historical dekadal (10-daily) rainfall data from 1981 to the present were collected for each of the 11 sub-counties in Machakos county. Seasonal cumulative rainfall measures were fit to a PERT distribution, with a cumulative rainfall "trigger" set at the 20th percentile for each sub-county (Shee, Turvey, and You 2019). A spatially correlated Monte Carlo simulation was used to compute actuarial rates assuming a KSh 300 tick value for every 1 mm of rainfall below the trigger. Although the trigger value and probability distributions differed by sub-county, the actual premiums averaged about 12 percent across districts. With a 25 percent loading factor imposed by the insurer,[3] the yield as a percentage of the loan amount was set at 14 percent for each sub-county, although each of them had a distinct trigger against which indemnities were to be calculated.

At the end of the season, the product suffered significant basis risk in terms of false negatives and intra-seasonal basis risk, warranting some improvements to better capture actual losses experienced by farmers. We modified the insurance component of RCC to better represent the within-season patterning of rainfall. To accomplish this, we re-engineered RCC from a seasonal cumulative rainfall model to a multiple-event model based on 21-day cumulative rainfall, setting the trigger at 50 percent of the historical average rainfall measure over the years and the tick value at KSh 50. We define specific events as 21 fixed days that are overlapping in measurement but non-overlapping in indemnity, implying a multi-event dynamic trigger that can indemnify only once within a period of 21 days (Turvey, Shee, and Marr 2019).[4] Up to four nonoverlapping events, each paying an indemnity, make up the total indemnity.

This structure resolves several problems that have plagued the implementation of rainfall-based index insurance in the past. It directly addresses the problem of intraseasonal drought conditions while providing indemnities to rainfall across phenological growth stages throughout the growing season. Furthermore, it reduces the incidence of type I (receiving an indemnity with no crop loss) and type II (receiving no indemnity, when crop loss occurs) errors. In both instances, we captured loss measure by determining the difference between the actual rainfall and the trigger value by $Max\ (0, Z - R(T))$ where the trigger level is Z and the level of rainfall for the specific period T is $R(T)$. To capture the economic loss, we multiply the above loss measure by a "tick" value, determined in consultation with the participating insurance company. The tick value was determined by the amount required to pay off the loan principal in a worst-case scenario.

Focus group discussions and communication through participatory games

After developing a prototype RCC, we needed local information to decide on different product parameters and to assess feasibility. We first conducted

3 The loading factor is the percentage added above the actuarially fair premium by the insurer to cover their administrative costs and markup.

4 We have also developed a multi-event product that considers crop water requirements/evapotranspiration as the triggering event for rainfall index insurance, a promising product but yet to be empirically evaluated in the field (Ndegwa et al. 2022).

scoping missions in June 2016, holding six focus group discussions (FGDs) in three counties in Kenya—namely, Machakos, Makueni, and Uasin Gishu—to assess farmers' responses and to select our project pilot area. We included diverse counties to assess the suitability of RCC in a variety of agroecologies. The scoping mission aimed to understand the local situation, needs, and the likely reception of RCC. Through qualitative discussions, we collected information on agricultural practices and on historical weather shocks, drought in particular, and the impacts of this on farming households. The FGDs helped us gather information on the timeline and crop growth cycle of maize—the main crop grown in the area—which included sowing time, vegetative growth period (germination to panicle initiation), reproductive growth period (panicle initiation to flowering), ripening period (flowering to mature grain), and harvest time. This detailed information on the maize crop growth cycle helped us incorporate crop phenology to develop and improve RCC. The information on historical drought events reported by the FGD participants also helped the team ground-truth historical remote sensing rainfall indexes. The FGDs also helped us understand farmers' demand for agricultural loans, in which months they need loans, for what purpose, and the best time to repay.

Perhaps the greater challenge lay in communicating the functioning of RCC to farmers who had only basic education. We developed a participatory and pictorial game as an innovative approach to community engagement to communicate RCC to farmers (Shee, Turvey, and Woodard 2015). We assessed the uptake and impact potential of RCC using the game played in a series of FGDs.

The purpose of the participatory role playing game was to convey the design features of RCC and to demonstrate how RCC works and how it can provide

FIGURE 12.3 **Participatory RCC games played by farmers in focus groups**

Source: Apurba Shee (photographs taken during SATISFy project field visit).

downside risk protection. Farmers selected one option from a set of three pictorially depicted choices with their potential outcomes in good and bad seasons: traditional practice with no need for credit, high-potential practice with traditional credit, and high-potential practice with RCC (a detailed description of the game is presented in Shee, Turvey, and Woodard 2015). Traditional practice referred to agricultural practices where no loan was required; high-potential practice, on the other hand, required a loan (a traditional loan or RCC) for investment in a high-quality improved crop production practice. Participants also were given a realized weather (risk) condition through a random draw of a marker from a bag with one red (represents bad weather) and four green (represents normal weather) markers. In the end, participants calculated the end-of-season earnings according to the realized good and bad weather conditions. Looking at the end-of-the-season earnings for the three choices, farmers realized that they were not subject to negative seasonal earnings with RCC but that, with traditional credit, they had negative earnings in the bad season. Hence, the game showed that, even though farmers earned less with RCC than with traditional credit during a normal season, they were able to be protected in a bad season. Using the participatory role play, farmers were able to see how RCC effectively reduced their downside risk. In the game, about 80 percent of participants opted for high-potential practice with RCC, which revealed strong farmer interest in RCC.

After the scoping mission, the team selected Machakos county for pilot interventions, for two main reasons. First, Machakos is a semiarid and hilly terrain area that receives very low annual rainfall of around 700 mm per year, with average rainfall in the long and short rain seasons being 315 and 266 mm, respectively (Kenya, Ministry of Agriculture, Livestock and Fisheries 2014). Smallholder farmers practice agriculture, with maize the primary staple crop. We therefore expected RCC to be attractive and capable of providing large benefits to the smallholders in the county. Second, our key implementing partner, Equity Bank, has several branches in Machakos county, which could directly facilitate RCC disbursement. Makueni county was also a good fit for the study, but the Bank's presence and coverage were better in Machakos. Uasin Gishu county was not ideal for RCC since drought is not a major production risk there, and hence drought insurance is not a viable intervention.

Behavioral experiment and formative evaluation

Although the qualitative investigation described above generated strong interest in and support for RCC among farmers, the question remained as to whether

FIGURE 12.4 Choice experiment attributes and corresponding levels

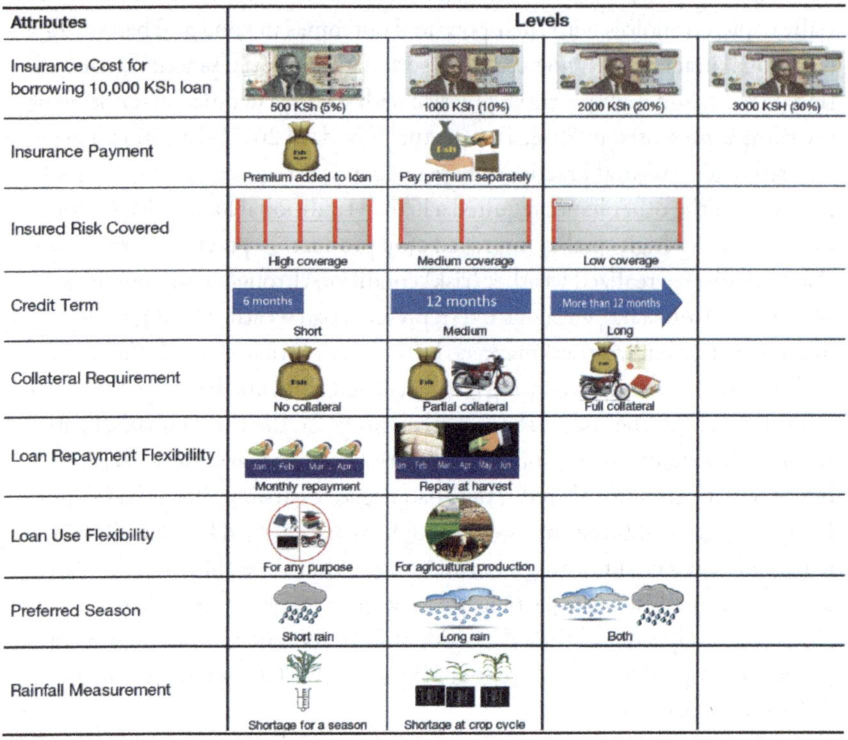

Source: Authors.

RCC could meet their demand and whether financial institutions could supply them. We conducted a formative evaluation through a choice experiment[5] not only to assess the demand and supply feasibility but also to test and adapt the RCC product by incorporating feedback received from supply- and demand-side stakeholders. The detailed choice experiment method is presented in Shee, Turvey, and Marr (2020).

Since RCC is a combination of insurance and credit products, it became a very complex product with many attributes of both. Identifying the preferred attributes of both farmers and financial institutions was central to developing an improved RCC product representing an optimal mix of insurance and credit attributes. Through consultations with stakeholders, the team identified nine attributes for choice experiments: (1) insurance premium; (2) insurance payout;

5 Choice experiments help in eliciting consumer preferences, the theoretical explanation of which is rooted in utility maximization, where individuals derive their utilities from the attributes of a good or service (Lancaster 1966).

(3) insured risk coverage; (4) credit terms; (5) collateral requirement; (6) loan repayment flexibility; (7) loan use flexibility; (8) preferred season for a loan; and (9) rainfall measurement. The first three attributes relate to the insurance component and the remaining ones are associated with the loan component. Figure 12.4 presents the choice experiment attributes and their levels of coverage.

We collected choice data from 330 smallholder farmers and 39 supply-side managers from key insurance companies and banks in Kenya. We analyzed these using the maximum simulated likelihood estimation of a mixed logit model (Train 2009). Demand-side results suggested that farmers preferred credit for both seasons, longer-term credit, no or partial collateral loans, lower risk premiums, high risk coverage, and loans to be used for multiple purposes. Supply-side results suggested that bank and insurance company managers preferred the risk premium to be added with the loan, loan repayment to be done after harvest, credit to be provided for both seasons, loans to be utilized only for agricultural purposes, and loans to be partially or fully collateralized. Hence, we found some conflicts between demand- and supply-side preferences regarding credit term, loan use flexibility, and collateral requirement. While farmers preferred medium- to long-term loans, this was not the case for financial institutions. While farmers preferred loans to be used for multiple purposes, bank managers preferred loans to be used only for agricultural production. While farmers disliked the collateral requirement, bank managers preferred this.

Overall, the choice experiment method of the formative evaluation helped us adapt actuarial RCC design to optimal bundling of attributes by marketing a partial collateral RCC contract, adding insurance premium with loans, and allowing loans to be offered in both seasons with a postharvest repayment schedule. We later piloted this optimal design of the RCC product to smallholder farmers in Machakos county through a full-fledged RCT (presented in the next section).

Apart from the above-mentioned choice experimental evidence of stakeholders' preference on RCC attributes, we conducted a formative evaluation of farmers' demand for RCC, and the socioeconomic factors influencing this. Using baseline household survey data from 1,170 Machakos households during the first phase of RCC implementation (April 2017 to June 2018), we empirically tested factors that could potentially influence the uptake of RCC. In our baseline survey, we adapted and modified the direct elicitation method proposed by Boucher, Guirkinger, and Trivelli (2009) to capture households' credit rationing status; the process is illustrated by Shee, Pervez, and Turvey (2018).

In the baseline survey, a module for eliciting credit rationing status asked the respondents about a set of credit-related experiences and used the responses to classify the credit rationing status of households into the following three groups: (1) unconstrained status, consisting of farmers who were being approached by banks to take credit, who would receive the full credit they requested, or who did not require credit; (2) quantity-rationed, implying financial institutions rationing their credit supply, with farmers either getting less credit than they requested or being denied credit altogether; and (3) risk-rationed, with farmers voluntarily withdrawing themselves from the credit market for fear of losing collateral. Out of the total sample of 1,170, we found 48 percent to be unconstrained, 11 percent to be quantity-rationed, and 41 percent to be risk-rationed. Hence, we found credit rationing to be pervasive in our baseline sample households in Kenya, with about half of the sample credit-rationed in some fashion.

From the first phase of RCC implementation, the average credit uptake rate was 33 percent, with the uptake of RCC significantly higher than that of traditional credit (34 percent and 31 percent, respectively). We then estimated a probit model to identify the factors of credit uptake (uptake = 1, otherwise = 0) (see Ndegwa et al. 2020). We found that quantity rationing had a small positive effect on uptake. This implied that farmers who were quantity rationed were potentially more likely than unconstrained households to take the credit if offered.

However, we also found that risk rationing harmed credit uptake, which implied that risk-rationed farmers were less likely to take the offered credit compared with their unconstrained counterparts. This finding supported the existing theory that, even when provided with agricultural credit, a risk-rationed farmer may still choose to withdraw from the credit market. Risk-rationed farmers do not participate in the credit market because they are afraid of losing collateral or undergoing other defaulting implications.

Among the socioeconomic variables, training attendance, food expenditure, maize labor requirement, hired labor, livestock revenue, and access to credit were found to influence credit uptake positively, whereas nonfood expenditure was negatively related to credit uptake.

Experimental design and impact evaluation

To evaluate the uptake, investment, productivity, and welfare benefits of RCC, we implemented a multi-arm RCT where we compared RCC to traditional credit with no drought insurance attached to it. The RCT design was constructed to provide a statistically valid, representative, and unbiased assessment of the uptake of RCC and its impact on agricultural investment, productivity, and farmer welfare, and to compare it with the effect of traditional credit. In the

RCT, we randomly assigned 1,150 households to one of three research groups: traditional credit (TC) (treatment 1; 350 households), RCC (treatment 2; 350 households), or control (no credit; 350 households). Treatment assignment was carried out in location-level public lotteries: representatives from all the selected households in a location (90 in total) were invited for an initial training on RCC and basic maize agronomic practices, after which they were invited to blindly draw a printed chip from an urn to determine which experimental group they fell in—either RCC, TC, or control.[6] As is common with such RCTs, those who randomly fell into the RCC and TC groups were offered the respective credit types. Credit uptake was not enforced but left to the household's volition.

The evaluation study involved a baseline survey, two phases of project implementation (marketing of RCC and TC loans), and two follow-up surveys. The baseline survey was conducted in May–June 2017 and was followed by the first phase of implementation, when input loans were offered, in October 2017, for the long rain season running from October 2017 to January 2018. The first follow-up (midline) survey was conducted in May–June 2018, roughly eight months from the time the first phase loans had been disbursed. Input loans for the second phase of implementation were offered in October 2019 to be used in the long rain season from October 2019 to January 2020. The end-line survey was planned for May–June 2020 (to maintain consistent timing) but this was disrupted by the COVID-19 pandemic; it was postponed until August–September 2020. Further, whereas baseline and midline surveys were through face-to-face interviews using a long-form structured questionnaire, the end-line survey was conducted via telephone, and the questionnaire itself was revised and significantly reduced so that the interview could be completed within roughly 30 minutes. However, we made extensive efforts to ensure we tracked all key variables needed to complete the study.

Our primary objective in this study was to provide evidence of the efficacy of RCC to engender a transformation and modernization of agriculture in sub-Saharan Africa. To do so, we tested whether RCC and TC yielded ex ante and ex post impacts on agricultural households in our sample area and whether the effects were differential. For ex ante impacts, we considered treatment

---

6 To confirm that our random assignment to experimental groups was effective, we conducted checks for baseline differences, individually regressing the outcome and control variables against treatment assignments. We started by comparing both treatments to the control group, then compared RCC with TC households. The results indicate that, apart from in the dependency ratio, the control group was not systematically different from the treatment groups. Further, the two treatment arms (RCC and TC) were similar on all outcome and control variables. This suggests that the treatment assignment was balanced and hence selection bias is not a major concern.

effects for on-farm investment in improved maize varieties and chemical fertilizers, as well as the intensity with which these were used. For ex post impacts, we considered household productivity and welfare outcomes.

With considerably high imperfect compliance among the treatment households (credit uptake was 32 percent and 16 percent at midline and end-line, respectively), local average treatment effect (LATE) is the ideal empirical strategy to estimate the impact of RCC and TC on households' outcomes. We estimated LATE by implementing the instrumental variables technique to account for imperfect compliance. We employed the two-stage least squares approach where uptake of either RCC or TC was instrumented by random assignment to either RCC, TC, or control group, respectively.

Ex ante investment impact evaluation results

For indicators of households' agricultural investment, we considered five outcomes: use of fertilizer on maize as a binary choice variable equal to one if chemical fertilizer was used on maize and zero otherwise; chemical fertilizer use intensity proxied by the amount of money (in Kenyan shillings) spent on chemical fertilizers per acre for maize; adoption of improved maize

TABLE 12.1 Results of ex ante impact evaluation

Outcomes	LATE1	LATE2
Panel A: Binary response fertilize use		
RCC	0.135*	0.137
TC	0.132	0.216
RCC=TC: P-value	0.971	0.529
Panel B: Fertilizer spending		
RCC	1,922.156***	1,937.909***
TC	741.569*	969.931**
RCC=TC: P-value	0.004	0.049
Panel C: Improved maize seed use—binary		
RCC	0.069	0.043
TC	−0.044	0.026
RCC=TC: P-value	0.118	0.994
Panel D: Share of land under improved maize seed		
RCC	4.321	1.599
TC	−8.458	−2.045
RCC=TC: P-value	0.089	0.925
Panel E: Area under maize		
RCC	0.237	0.342
TC	0.462	0.291
RCC=TC: P-value	0.572	0.919

Source: Authors.

Note: LATE1 and LATE2 show the LATE estimated with analysis of covariance (ANCOVA) and fixed effects (FE), respectively. ***P<0.01, **P<0.05, *P<0.1.

varieties as a binary choice variable equal to one if improved maize varieties were planted at all and zero otherwise; improved maize adoption intensity measured as the share of maize fields under improved varieties; and area under maize overall. Table 12.1 presents the treatment effect estimates of ex ante investment.

Panel A shows a substantial positive effect of uptake of RCC on the decision to use chemical fertilizer. RCC uptake effects ranged from 13.5 percent to 13.7 percent and were significant in the ANCOVA column. This implies that, compared with others, households that were offered and took RCC were roughly 14 percentage points more likely to use chemical fertilizers on their maize fields. On the other hand, the TC uptake effect ranged between 13.2 percent and 21.6 percent, which, although substantial, was not statistically significant. We could not, however, reject the null hypothesis that the difference between RCC and TC effects (RCC=TC: P-value) was equal to zero. Panel B shows an even higher impact of RCC on fertilizer use intensity where statistical significance was much higher ($p<0.01$) across all models. Uptake of RCC had a substantial and significant effect on maize fertilizer usage whereby those who were offered and took RCC spent roughly KSh 1,900 more on fertilizer per acre than the rest ($p<0.001$). The effects of TC uptake, on the other hand, ranged between KSh 740 and KSh 970, with lower statistical significance. Further, we found strong evidence that RCC effects on maize fertilizer spending were significantly larger than the TC effects (RCC=TC: P-value <0.05).

The LATE effects of TC and RCC on the decision to use improved maize seeds and the intensity of improved maize seed adoption are presented in panels C and D, respectively. Across all the specifications, those who were offered and took RCC were more likely to use improved maize varieties and had a higher share of their maize fields under these. These improvements were, however, marginal and statistically insignificant. On the other hand, those who were offered and took TC appear to have been less likely to use improved maize varieties and had less of a share of their maize fields under these. This negative effect was, however, statistically insignificant. Further, we did not find strong grounds to reject the null hypothesis that RCC and TC effects were statistically not different.

Panel E presents LATE effects on the overall area cultivated with maize. The aim here was to investigate whether RCC and TC had led to the expansion of land cultivated with maize. The estimates were positive but small in magnitude and statistically insignificant. This suggests that both RCC and TC did not lead to any significant change in the area cultivated with maize. We then can conclude that uptake of RCC promoted agricultural intensification but not extensification. We also did not find a significant difference between RCC and TC effects on maize area expansion.

Ex post productivity and welfare impact evaluation results

As for indicators of households' productivity, we considered maize yields and acreage revenue from maize and its main intercrops—common beans and cowpeas. LATE estimates of RCC and TC effects on maize yield and acreage revenue are presented in panels A and B, respectively. We divided the farmer-reported yields by the inverse of the number of reported intercrops to adjust for intercropping in the yield variable. We computed farm revenue by summing the value of acreage production of the three main crops (maize, normal beans, and cowpeas) grown by almost all households in the study area.[7]

For indicators of household welfare, we evaluated the food security situation using the coping strategy index (CSI) and dietary diversity using the household dietary diversity index (HDDI). CSI, as defined by Maxwell and Caldwell (2008), measures the unavailability and insufficiency of food, while HDDI measures the diversity and hence balance of diets. Table 12.2 presents treatment

7 The value of the crops used was the actual market prices of the commodities at Machakos market at the end of each season. We obtained the market prices from the Kenyan Ministry of Agriculture, Livestock, and FIsheries.

TABLE 12.2 Results of ex post impact evaluation

Outcomes	LATE1	LATE2
Panel A: Maize yield		
RCC	–268.911	–116.984
TC	–553.808*	–564.813
RCC=TC: P-value	0.295	0.560
Panel B: Acreage revenue		
RCC	–5,628.192	–5,466.198
TC	–3,644.003	–10,250.567
RCC=TC: P-value	0.678	0.924
Panel C: Food insecurity		
RCC	1.392	2.027
TC	1.350	2.591
RCC=TC: P-value	0.9693	0.7108
Panel D: HDDI		
RCC	–0.165	–0.283
TC	–0.226	–0.288
RCC=TC: P-value	0.855	0.619

Note: LATE1 and LATE2 show the LATE estimated with ANCOVA and FE, respectively. ***P<0.01, **P<0.05, *P<0.1.

effect estimates of ex post investment. To calculate the CSI, we selected the eight most common food insecurity coping strategies and asked respondents the number of days in the past seven they may have employed each. Our scale ranges from 0 to 56, or simply the product of the maximum possible number of days (7) and coping strategies (8). The higher the score, the more times a household had to employ one of the coping strategies, and hence the worse off they were regarding food insecurity. To construct the HDDI, we asked respondents if in the past 7 days they had consumed a whole host of food items, categorized into 15 discrete food groups. We then created binary response variables corresponding with the 15 food groups, whose values were 1 if at least 1 food item in the group was consumed at least once in the past 7 days, and 0 otherwise. We then summed up the food groups for each household, which gave us their score on a 15-point scale. The higher the score, the more diverse the diet; hence, positive effects were desired. LATE effects on CSI and HDDI are presented in panels C and D, respectively.

The results in Table 12.2 indicate that uptake of both RCC and TC did not lead directly to any significant improvement in the ex post outcomes (both productivity and welfare) of interest. However, we use the structural equation modelling (SEM)[8] approach to assess if such benefits of RCC and TC could be reached but via intermediary outcomes. Figure 12.5 presents the hypothesized impact pathways between the treatments, the intermediate, and the main outcomes. Here, we consider maize yield and overall acreage revenue as our productivity indicators, and CSI as our welfare indicator. As such, we have three main SEM models, one for each outcome. We consider credit rationing and investment ability as the intermediate factors between credit uptake and households' productivity. Both variables (credit rationing and investment ability) are further considered as mediating for welfare, where one of the productivity indicators (revenue) is included as an additional mediator between credit uptake and households' welfare. Table 12.3 presents the results of the three SEM models. Each model bears the title (as indicated in the column heading) of the main productivity or welfare outcome whose impact pathways are being assessed. A model comprises simultaneous regression equations; in the table, the outcome for each individual equation is bolded while the predictors are indented.

First, all three models exhibit average fitted uptake of RCC and TC (27.6 percent and 23.3 percent, respectively) across the two intervention implementation phases. Further, credit-rationed status did not significantly or

---

8 See Heckert, Olney, and Ruel (2019) for a recent and closely comparable application of SEM in impact and mediation analysis.

FIGURE 12.5 Structural equation model showing the causal pathways from treatments (RCC and TC) to the intermediate and main (productivity and welfare) outcomes

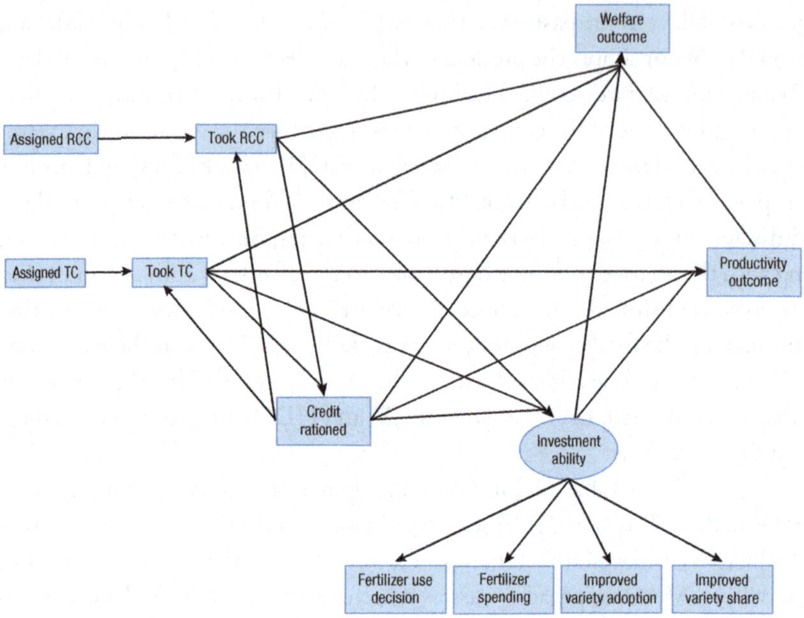

Source: Authors.

substantially affect the uptake of the credit offered, whether RCC or TC. This eliminates reverse causality and simultaneity concerns when credit rationing is used as an intermediary variable for other impact outcomes. Across all models, uptake of either RCC or TC had a desirable and statistically significant effect on the main intermediate variables, namely credit rationing and investment ability. Both RCC and TC reduced credit rationing while increasing investment ability. The effect of credit rationing on investment ability, although negative across all the models, was small in size and statistically insignificant, implying that credit rationing did not significantly moderate the impact of RCC or TC on households' investment ability.

Looking at the productivity rows under the yield model column, the direct paths from credit uptake (either RCC or TC) were positive but statistically insignificant. Also, credit rationing had a negative and significant effect on maize yield while investment ability had a positive and significant effect on maize yield. These results suggest that the impact of RCC and TC on maize yield is fully mediated by credit rationing and investment ability. Results

TABLE 12.3 Results of structural equation models for yield, revenue, and food insecurity

	Yield model		Revenue model		Food insecurity (CSI) model	
	Coefficient	SE	Coefficient	SE	Coefficient	SE
Uptake of RCC						
Assigned RCC	0.276***	0.009	0.276***	0.009	0.276***	0.009
Credit-rationed	−0.003	0.016	−0.003	0.016	−0.003	0.016
Uptake of TC						
Assigned TC	0.233***	0.009	0.233***	0.009	0.233***	0.009
Credit-rationed	0.01	0.017	0.01	0.017	0.01	0.017
Credit rationing						
Took RCC	−0.17**	0.081	−0.17**	0.081	−0.17**	0.081
Took TC	−0.278***	0.097	−0.278***	0.097	−0.278***	0.097
Investment ability						
Took RCC	0.13***	0.025	0.13***	0.025	0.13***	0.025
Took TC	0.121***	0.027	0.12***	0.027	0.121***	0.027
Credit-rationed	−0.009	0.012	−0.009	0.012	−0.009	0.012
Productivity outcome						
Took RCC	209.1	183.212	297.17	3245.433	287.387	3245.499
Took TC	243.558	217.09	7523.47**	3867.181	7512.314**	3867.151
Credit-rationed	−149.485***	42.668	−3227.554***	757.191	−3227.587***	757.097
Investment ability	535.75***	66.885	7243.822***	1188.38	7236.618***	1187.758
Welfare outcome - CSI						
Took RCC					−0.386	0.441
Took TC					−0.635	0.479
Credit-rationed					1.491***	0.209
Investment ability					−0.788**	0.348
Revenue					−0.001***	0.000

Note: ***P<0.01, **P<0.05. All coefficients are unstandardized. SE = standard error. In bold are lefthand variables and indented are righthand variables for each equation.

indicate that the negative effect that credit rationing has on maize yields can be abated with expanded access to credit. Similarly, the positive effects that investment ability has on maize yields can be augmented further through enhanced access to credit. The case is similar with revenue, where acreage revenue declines with credit rationing and increases with investment ability, and the two intermediates can be enhanced with credit access, which will eventually lead to better farm revenues. The only difference here is that, although the impact of RCC uptake on-farm revenue is fully mediated by credit rationing and investment ability (a positive but insignificant coefficient), that of TC uptake is partially mediated by the same variables. It is possible that there could be other mediation routes between credit uptake and farm revenue that are not included in this analysis. Finally, looking at the welfare outcome rows under the food insecurity column, we see that the uptake effect of both RCC and TC on the food insecurity index (CSI) is fully mediated by credit rationing, investment ability, and household productivity, which in this model is proxied by acreage revenue. Food insecurity reduces with an increase in investment ability and farm revenue and increases with credit rationing.

Scaling potential

Outside of our experimental design, we conducted commercial implementation of RCC in Kenya, where any farmers interested in RCC can apply for RCC loans from respective local banks (KCB). The outcome of our scientific approach to RCC implementation through RCT, plus the commercial application, has provided strong grounds for extending RCC to many countries in Africa. There is considerable potential for scaling up within and beyond semiarid areas of Kenya as many countries in sub-Saharan Africa often face severe agricultural risks (resulting from extreme weather conditions such as droughts and floods) and insufficient access to credit. Our study area, Machakos county, is characterized by a semiarid climate that is shared by many sub-Saharan African countries, especially those in the Horn of Africa, but also including many Sahelian countries and several southern African countries like Botswana, Mozambique, Namibia, South Africa, and Zimbabwe. In addition to similarities in climate and agricultural risks faced, farmers in other contexts also find themselves quantity- or risk-constrained in their access to credit, thereby limiting their ability to invest in modern agricultural inputs that could help them escape pernicious poverty traps.

We foresee the following scale-up opportunities for RCC:

Innovation on enhancing uptake: Because credit is interlinked with agricultural production risks, loan default risk will also be reduced significantly. RCC also eliminates the drawbacks of stand-alone index insurance products by not requiring farmers to pay a premium upfront. This mechanism of RCC eliminates farmers' liquidity constraint (which is considered the main barrier to insurance uptake) and could enhance RCC uptake significantly.

Reducing basis risk: We developed multi-event index insurance and RCC products to incorporate intraseasonal variations and integrated crop water needs to further improve the triggering mechanism. We also plan to use satellite-derived soil moisture and vegetation data along with rainfall measurements to develop a composite weather index to efficiently capture the risk associated with extreme weather situations. This will reduce the design-related basis risk significantly and increase the value proposition of the product.

Scalable index and RCC design: At the heart of designing RCC, the objective of an index is mapping a signal (for example, weather or remotely sensed data) to an index that can predict crop yield accurately. We have developed a spatial econometric framework that allows maximum information extraction in the presence of missing data for investigating the construction and design of scalable RCC indexes.

Developing methodology for less data-intensive RCC development: Currently, the data requirements for the design of RCC are quite high. For the RCC to be commercially scalable, we plan to develop design tools directly from biophysical variables that can be modeled for wider applications. Although in this model it could be difficult to conceptualize basis risk, this will be advantageous for scaling-up, particularly for the areas where household survey data on crop production are not available.

Efforts to reduce the transaction cost of RCC delivery: Developing a mechanism that is relatively cheap but reliable is important for reducing the transaction costs of RCC delivery. In this regard, we plan to use a mobile-based application to create cost-effective information dissemination and trust in RCC products. Also, since our local implementers already have local branches, village banking agents, and distribution channels, we can create a robust and efficient delivery mechanism for RCC.

Effective financial education and extension: Along with local partners, we put special effort into financial education and extension with farmers. We plan to enhance our strategic partnerships to expand financial education and awareness.

Synergies with government and complementary interventions: Our project objectives are in line with the Kenya Rural Development Program's goal of managing drought and food security. RCC can also build synergy with sustainable intensification interventions by responding to farmers' investment needs.

Commercial viability and welfare impact: We believe RCC structures could provide the proper incentives not only to entice banks (for example, Equity Bank) to increase the supply of credit but also to attract farmers to the credit market. We understand that long-run commercial sustainability will depend on the effective assessment of the social and economic benefits of RCC. Our impact evaluation results justify the long-term effect of RCC.

Conclusion

Over the past two decades, index insurance has gained some global success in managing weather risks faced by farmers but has been hampered by low demand. New insurance approaches are needed that bundle index insurance with term credit to reduce the downside risk facing farmers and the default risk facing lenders. The upfront premium required for standard insurance imposes significant liquidity constraints on poor farmers and also transfers income across time. RCC can overcome this by removing the liquidity constraint, reducing the effects of climate risk, and increasing the uptake of agricultural insurance via better targeting of poor farmers. Since insurance can substitute (at least partly) for collateral, the RCC mechanism has the potential to encourage otherwise risk-rationed farmers to take up RCC loans. Since the indemnity from the embedded insurance is applied to the underlying debt obligation, RCC can reduce the probability of default and build trust that can boost uptake. However, literature on the benefits of insurance bundled with credit remains very thin, especially evidence based on field experiments.

This chapter has discussed the development, testing, and adaptation of RCC through innovative pre-experimental methods, along with impact evaluation through implementing a full-fledged randomized experiment. After developing a prototype of scientific bundling of insurance with credit, we incorporated feedback from the community and stakeholders and incrementally co-developed the full RCC product, and subsequently implemented it in Kenya.

Through formative evaluation, we found that linking insurance to credit improved the uptake of agricultural credit. This confirms that downside risk is a hindrance to credit uptake, limiting the already liquidity-constrained farmers' options with regard to raising the capital needed to enhance their productivity and welfare. Our impact evaluation results indicate that access to credit, regardless of the type, can improve farmers' agricultural investment, which helps transform their livelihoods and quality of life. Further, RCC effects were consistently higher than TC effects, and the difference was statistically significant in the adoption of chemical fertilizers, a key investment indicator. These additional benefits, coupled with enhanced uptake, suggest that RCC could be an effective way to promote both credit and insurance uptake, which in return can lead to enhanced household productivity and welfare.

Our findings suggest that developing policies that hedge smallholders against systemic shocks, such as drought, is one way of enhancing access to credit. The use of formal insurance markets is a viable policy since it transfers the risk outside the household and hence protects its collateral. Bundling insurance with credit also minimizes the risk of default by smallholder borrowers, which lessens financial risks to lenders that threaten their business stability—a common phenomenon when rural agricultural production systems experience systemic shocks such as drought. Financial institutions keen on growing their agricultural lending portfolio may then consider insurance-linked products to offer cheaper and safer loans to farmers. However, insurance and credit providers should ensure greater transparency for farmers by providing continuous index updates using a simplified pictorial tool. In this regard, digital technology such as mobile phones could be effective. In a situation where climate change is expected to put unprecedented strains on smallholder agricultural systems, RCC can not only help increase the resilience of the food system but also improve smallholders' productivity and growth outcomes.

Although we scientifically designed RCC as a product that can reduce intertemporal basis risk by incorporating crop phenology in the insurance pricing framework, lack of reliable historical crop yield data has been a hindrance to the creation of an improved insurance mechanism that can reduce basis risk significantly. Ideally, estimating and predicting yield using weather indexes would be preferrable. This is common in low-income economies. We therefore recommend governments promote large-scale data collection and make such data available for researchers. In this regard, big data and machine learning techniques could improve index construction by inducing improved model specifications and predictions of expected crop yield. Further research in this area is warranted, particularly on trying alternative indexes, building composite indexes, or developing multiple crop and/or whole-farm index insurance and insurance-linked products.

Productivity, welfare, and resilience outcomes are higher-level benefits that may be harder to achieve and to estimate by means of the limited periods in an experiment. Also, this study may have limited predictive power because of the low uptake of credit offered. As such, although we made every effort to control for any form of endogeneity, we cautiously interpret and claim causality and recommend that the impact results and interpretations be treated as indicative. We hope this motivates further investigation, especially with longer panels, a variety of credit and insurance products, and higher uptake.

Further, financial literacy remains very low among smallholder farmers in low- and middle-income country settings. Although financial education through Equity Bank was part of the RCC protocol, we recommend investigation of the ideal education and extension methods to train farmers on such products, ensuring that key details are covered while maintaining a level of simplicity congruent with low literacy and numeracy among rural smallholders in the region. In this regard, providing financial education using simplified games through extension services at the county level could be an effective government policy.

Finally, as the recent literature suggests, women are important agricultural producers, especially in sub-Saharan Africa, but face greater credit constraints than men. It is, therefore, worth exploring how to include more women farmers in the RCC program, and how this would improve women's access to credit and empowerment in society.

References

Barrett, C., B. Barnett, M. Carter, S. Chantarat, J. Hansen, A. Mude, D. Osgood, J. Skees, C. Turvey, and M.N. Ward. 2007. *Poverty Traps and Climate Risk: Limitations and Opportunities of Index-Based Risk Financing.* IRI Technical Report No 07-02. Palisades, NY: International Research Institute for Climate and Society (IRI).

Barrett, C.B, M.R. Carter, and J.P. Chavas. 2019. *The Economics of Poverty Traps.* Chicago: University of Chicago Press.

Binswanger, H.P. 2012. "Is There Too Much Hype About Index-Based Agricultural Insurance?" *Journal of Development Studies* 48 (2): 187–200.

Boucher, S.R., M.R. Carter, and C. Guirkinger. 2008. "Risk Rationing and Wealth Effects in Credit Markets: Theory and Implications for Agricultural Development." *American Journal of Agricultural Economics* 90 (2): 409–423.

Boucher, S.R., C. Guirkinger, and C. Trivelli. 2009. "Direct Elicitation of Credit Constraints: Conceptual and Practical Issues with an Application to Peruvian Agriculture." *Economic Development and Cultural Change* 57 (4): 609–640.

Branca, G., T. Tennigkeit, W. Mann, and L. Lipper. 2011. *Identifying Opportunities for Climate-Smart Agriculture Investments in Africa.* Rome: FAO.

Casaburi, L., and J. Willis. 2018. "Time versus State in Insurance: Experimental Evidence from Contract Farming in Kenya." *American Economic Review* 108 (12): 3778–3813.

Cole, S.A., and W. Xiong. 2017. "Agricultural Insurance and Economic Development." *Annual Review of Economics* 9 (1): 235–262.

Cole, S., G.G. Bastian, S. Vyas, C. Wendel, and D. Stein. 2012. *The Effectiveness of Index-Based Micro-insurance in Helping Smallholders Manage Weather-Related Risks.* London: EEPI-Centre.

Dercon, S., and L. Christiaensen. 2011. "Consumption Risk, Technology Adoption and Poverty Traps: Evidence from Ethiopia." *Journal of Development Economics* 96 (2): 159–173.

Gallenstein, R.A., K. Mishra, A.G. Sam, and M.J. Miranda. 2019. "Willingness to Pay for Insured Loans in Northern Ghana." *Journal of Agricultural Economics* 70 (3): 640–662.

Giné, X., and D. Yang. 2009. "Insurance, Credit, and Technology Adoption: Field Experimental Evidence from Malawi." *Journal of Development Economics* 89 (1): 1–11.

Heckert, J., D.K. Olney, and M.T. Ruel. 2019. "Is Women's Empowerment a Pathway to Improving Child Nutrition Outcomes in a Nutrition-Sensitive Agriculture Program? Evidence from a Randomized Controlled Trial in Burkina Faso." *Social Science and Medicine* 233: 93–102.

Jensen, N., and C. Barrett. 2017. "Agricultural Index Insurance for Development." *Applied Economic Perspectives and Policy* 39 (92): 199–219.

Kenya, Ministry of Agriculture, Livestock and Fisheries. 2014. *KENYA: Situation Analysis for a National Agricultural Insurance Policy (NAIP).* Nairobi: Ministry of Agriculture, Livestock and Fisheries.

Lancaster, K. 1966. "A New Approach to Consumer Theory." *Journal of Political Economy* 74: 132–157.

Marr, A., A. Winkel, M. van Asseldonk, R. Lensink, and E. Bulte. 2016. "Adoption and Impact of Index-Insurance and Credit for Smallholder Farmers in Developing Countries: A Systematic Review." *Agricultural Finance Review* 76 (1): 94–118.

Maxwell, D., and R. Caldwell. 2008. *The Coping Strategies Index: Field Methods Manual.* 2nd ed. Atlanta, GA: CARE USA.

Mishra, K., R.A. Gallenstein, M.J. Miranda, A.G. Sam, P. Toledo, and F. Mulangu. 2021a. "Insured Loans and Credit Access: Evidence from a Randomized Field Experiment in Northern Ghana." *American Journal of Agricultural Economics* 103 (3): 923–943.

Mishra, K., R.A. Gallenstein, M.J. Miranda, A.G. Sam, P. Toledo, and F. Mulangu. 2021b. "Do Index-Insured Loans Spur Technology Adoption? Evidence from a Randomized Control Trial in Northern Ghana." Selected poster presented at the 2022 Agricultural & Applied Economics Annual Meeting, Anaheim, CA, July 31–August 2.

Miranda, M.J., and K. Farrin. 2012. "Index Insurance for Developing Countries." *Applied Economic Perspectives and Policy* 34 (3): 391–427.

Ndegwa, M., A. Shee, C. Turvey, and L. You. 2020. "Uptake of Insurance-Embedded Credit in Presence of Credit Rationing: Evidence from a Randomized Controlled Trial in Kenya." *Agricultural Finance Review* 80 (5): 745–766.

Ndegwa, M., A. Shee, C. Turvey, and L. You. 2022. "Sequenced Crop Evapotranspiration and Water Requirement in Developing a Multi-trigger Rainfall Index Insurance and Risk-Contingent Credit." *Weather, Climate, and Society* 14 (1): 19–38.

Santos, P., and C.B. Barrett. 2019. "Heterogeneous Wealth Dynamics: On the Roles of Risk and Ability." In *The Economics of Poverty Traps*, eds. C.B. Barrett, M.R. Carter, and J-P. Chavas, 265–290. Chicago: University of Chicago Press.

Shee, A., S. Pervez, and C.G. Turvey. 2018. "Heterogeneous Impacts of Credit Rationing on Agricultural Productivity: Evidence from Kenya." Paper presented at the Agricultural & Applied Economics Association Annual Meeting, Washington, DC, August 5–7.

Shee, A., and C.G. Turvey. 2012. "Collateral-Free Lending with Risk-Contingent Credit for Agricultural Development: Indemnifying Loans against Pulse Crop Price Risk in India." *Agricultural Economics* 43 (5): 561–574.

Shee, A., C.G. Turvey, and A. Marr. 2020. "Heterogeneous Demand and Supply for an Insurance-linked Credit Product in Kenya: A Stated Choice Experiment Approach." *Journal of Agricultural Economics* 72 (1): 244–267.

Shee, A., C.G. Turvey, and J.D. Woodard. 2015. "A Field Study for Assessing Risk-Contingent Credit for Kenyan Pastoralists and Dairy Farmers." *Agricultural Finance Review* 75 (3): 330–348.

Shee, A., C.G. Turvey, and L. You. 2019. "Design and Rating of Risk-Contingent Credit for Balancing Business and Financial Risks for Kenyan Farmers." *Applied Economics* 51 (50): 5447–5465.

Stiglitz, J.E., and A. Weiss. 1981. "Credit Rationing in Markets with Imperfect Information." *American Economic Review* 71 (3): 393–410.

Train, K.E. 2009. *Discrete Choice Models with Simulation.* Cambridge, UK: Cambridge University Press.

Turvey, C.G. 2001. "Weather Derivatives for Specific Event Risks in Agriculture." *Applied Economic Perspectives and Policy* 23 (2): 333–351.

Turvey, C.G., A. Shee, and A. Marr. 2019. "Addressing Fractional Dimensionality in the Application of Weather Index Insurance and Climate Risk Financing in Agricultural Development: A Dynamic Triggering Approach." *Weather, Climate, and Society* 11 (4): 901–915.

Verteramo-Chiu, L.J., S.V. Khantachavana, and C.G. Turvey. 2014. "Risk Rationing and the Demand for Agricultural Credit: A Comparative Investigation of Mexico and China." *Agricultural Finance Review* 74 (2): 278–270.

PART 5

Toward More Inclusive Food Systems

Part 5

TOWARD MORE INCLUSIVE FOOD SYSTEMS

Vulnerable and marginalized groups are at risk of being left behind in food systems transformation. Cultural norms, demographic shifts, and economic structures can systematically leave out certain groups from having healthier, more productive, and more resilient futures. Such groups include (but are not limited to) women, youth, small farmers, people living in conflict zones, and the urban poor. Designing policy that recognizes their situations, constraints, and opportunities not only is an end in itself but also can strengthen the broader economy. Part 5 discusses inclusion in food systems with a focus on women, youth, and small farmers.

Even though women's labor contributes substantially to food production, women still face barriers to accessing resources, credit, markets, and inputs. The indigenous chicken value chain is a clear case of the challenges facing women in food systems. Chapter 13 shows that women are responsible for rearing indigenous chicken but men often dominate decision-making regarding vaccinations, feed purchases, and selling, because men typically have control over household resources and are the household heads. Decision-making skewed toward men can lead to suboptimal decisions regarding the use of inputs, such as vaccinations. Chapter 13 suggests that vaccination of indigenous chickens can be improved through training and education targeted to both men and women but also through providing women with control over resources via microcredit interventions targeted to women or through women-led collectives. Further, promoting women's involvement along the entire value chain can address systematic barriers to their participation in indigenous chicken value chains.

Promoting off-farm jobs for women can help address another issue facing Kenya—the growing youth population, or youth bulge. Indeed, the food system has the potential to generate both off- and on-farm jobs for youth in the coming years. However, policymakers face challenges in ensuring there are enough high-quality jobs for the growing youth population in the country. Chapter 14 describes the challenges and opportunities for youth in food systems. It shows that youth are engaged in the food system in a variety of ways—developing off-farm businesses, innovating on the farm, and leading climate action. However, the land tenure system, limited knowledge, household dynamics, and lack of job opportunities can prevent youth participation in food systems, especially for young women. For Kenya to properly support and leverage such participation, Chapter 14 suggests, policy should help youth build capacity and access capital for agribusiness, revive existing school-based programs to improve agricultural education (for example, 4K Clubs), and generate more evidence to better understand youth's heterogenous needs.

With the right approach, strengthening value chains can deliver not only opportunities for women and youth but also value for smallholders. While Kenya is one of the world's largest exporters of avocados, including smallholder farmers in this export value chain remains a challenge. Chapter 15 presents a study on the avocado export value chain and shows that smallholder participation is limited by lack of technical knowledge, trust, access to market information, and contract enforcement. Policy can support smallholder inclusion by overcoming these barriers through improved extension and training services, disseminating market information, and enforcing contracts through third parties. Further, contract farming should be mainstreamed; this requires a stronger and clearer legal framework that provides adequate protection for smallholders.

Part 5 addresses various issues related to inclusion in food systems, specifically for women, youth, and smallholder farmers. While strengthening value chains has been a recurring theme throughout this book, Part 5 points to the need for a conscious effort to develop value chains in a way that benefits marginalized groups. Improved inclusion not only is a desirable outcome in itself but also improves the food system's ability to be healthier, more productive, and more resilient.

Chapter 13

GENDER AND FOOD SYSTEMS: A CASE STUDY OF THE POULTRY VALUE CHAIN IN EASTERN KENYA

Salome A. Bukachi, Mariah Ngutu, Dalmas O. Omia, Mercy M. Musyoka, Judith K. Chemuliti, and Isaac K. Nyamongo

Women are key stakeholders in sustainable and resilient food systems, given their roles as primary food producers and household caretakers (Visser and Wangu 2021). Understanding how gendered roles affect food security and women's well-being is essential for pursuing sustainable development (Angel-Urdinola and Wodon 2010; Doss, Meinzen-Dick, and Quisumbing 2018; Meinzen-Dick et al. 2019). Their participation in agriculture is documented widely, but there is a need for more gendered data on the roles of men and women in different contexts and agricultural value chains, including livestock value chains (Micere Njuki et al. 2016; Richardson 2018; Doss and Rubin 2021; Njuki et al. 2021).

Men and women play different roles within food systems. In terms of production, evidence shows women contribute approximately 40 percent of agricultural labor and in some countries more, for example Tanzania, where women contribute 53 percent (Palacios-Lopez, Christiaensen, and Kilic 2017). On top of providing a substantial share of agricultural labor, women typically have more domestic responsibilities than men, which can impose severe constraints on their time and limit their engagement in off-farm components of value chains, such as marketing. Further, when it comes to ownership of the factors of production (such as land), there is a notable gender gap in much of sub-Saharan Africa (SSA), which can make women less likely to access productive inputs and less connected to upstream segments of value chains. As a result, there is a substantial gender gap in productivity between plots managed by women and those managed by men in SSA (UNDP 2015). Gender difference are also evident in the types of value chains in which men and women participate. Men are more likely to be involved in cash crops and food production intended for sale, while women are more likely to work on food crops intended for home consumption

(UNDP 2016). Women are thus less connected to downstream segments of the value chain. This leads to differential control over income, which has serious implications for food security, as women are more likely to spend income on household goods, while men are more likely to spend on private goods.

Social norms and gender inequalities exacerbate the situation, with women facing more normative challenges than men with regard to access to and control over productive resources, credit, and markets. Farhall and Rickards (2021) note that these norms go beyond the community and household level and are deeply entrenched in international food organizations, including those supporting agriculture for development. Consequently, these norms affect policy and legal framework formulation and implementation, decision-making processes, and resource distribution at both the global and the local scale, as well as relevant institutional management and governance (Bell 2021). To address these biases, the literature points to the need to focus on all aspects of food systems, from input supply to consumption, to incorporate the commonly hidden trade-offs and sacrifices that different food actors make. As Bell (2021) notes, food system programs need to shift from encouraging equal participation of women and men toward transformative change and behavior if they are to tackle systematic gender inequalities.

Following the 2030 Agenda for Sustainable Development principle of leaving no one behind, Kenya developed a Country Programming Framework 2018–2022 with the overall objective of achieving 100 percent food and nutrition security by 2022. This had four priority areas: (1) developing an enabling policy and investment environment, (2) strengthening inclusive value chains, (3) increasing resilient food and livelihood systems, and (4) improving governance of natural resources.[1] Through the second and third priority areas, the Government of Kenya, working with the Food and Agriculture Organization of the United Nations (FAO), attempts to develop and improve women's positions in several value chains while also strengthening the connections between various value chain actors.

However, the goal of achieving complete food security by 2022 has not been met (IPC 2023). A 2021 food systems summit in Kenya revealed shortcomings in the policy design and implementation, resulting from inadequate consideration of the needs of smallholders, limited funding, and knowledge gaps among food producers along the value chain (United Nations 2021). Moreover, recent studies show that policies and frameworks overemphasize the role of women's

1 FAO, "FAO in Kenya." Accessed June 2023. https://www.fao.org/kenya/programmes-and-projects/en/

access to resources such as inputs and technologies and overlook the importance of cultural rules that deny or limit their use of such resources (Mkandawire et al. 2021). To inform the agenda for sustainable food systems and gender equality, including addressing the persistent gender gaps in food systems, a deeper understanding is needed of how gender interacts with food systems as well as on ways to strengthen women's roles and voice to make food systems more inclusive, efficient, and resilient (CARE 2021; Njuki et al. 2021). There is an urgent need for nuanced and comprehensive data that illuminate the different needs of men and women smallholder farmers and how they intersect with their families, communities, and institutions to drive gender relations. These data will make it possible to measure change over time, identify the drivers of change, examine heterogeneity of impact, and establish best practices in increasing food security and gender equality (Doss, Meinzen-Dick, and Quisumbing 2018; Njuki et al. 2021).

This chapter provides a gendered study of how farmers interact with various nodes along the indigenous chicken value chain to better understand the needs of men and women smallholder chicken farmers in eastern Kenya. Indigenous chicken plays a key role in the food system in Kenya as a source of income and nutritious food (Chapter 2). As of the 2019 census, 27 percent of all households in Kenya were raising indigenous chickens (KNBS 2019), making this an important case study for understanding the role of gender in the country's food systems. We look into gender dynamics along the value chain to understand the current state of participation at each node of the chain and related challenges and opportunities, as well as potential solutions that may contribute to a more viable, inclusive, and equitable poultry value chain that supports the needs and livelihoods of all smallholder farmers.

The first part of the chapter provides an overview of the literature on chicken value chains in SSA, and particularly in Kenya, with an eye to the gender differences in ownership, labor, and decision-making power. The second section reviews the results of a recent gender analysis conducted in Makueni County, Kenya, in preparation for implementing a gender-inclusive project to promote vaccination of indigenous chickens. This analysis shows substantial participation by women in both ownership of indigenous chickens and labor related to their production, with more limited participation by men, but also shows that men take on a larger role in decision-making over vaccines, chick purchases, and chicken sales. Our results indicate that identifying and addressing the different needs and priorities of men and women smallholder farmers in relation to food production challenges, such as livestock diseases and their management and access to and control of financial resources, could increase productivity, improve food availability and security, and reduce related inequalities—helping to close the gender gap. The final section provides policy recommendations.

Literature Review

Overview of indigenous chicken value chains

Poultry value chains consist of all the linkages between poultry activities (from production to marketing and consumption) and the actors (input suppliers, producers, traders, processors, end-users) who are involved in value addition to poultry and poultry products until delivery to the final consumer (Mensah-Bonsu, Lartey, and Kuwornu 2019). The components of the poultry value chain include specific inputs (feeds, vaccines, drugs), breeding (stock, hatching, brooding), production (feeding, housing, chick care, disease management), collection and processing (eggs, chickens), transportation and marketing (handling of eggs/birds, transactions through brokers, traders, market outlets), and consumption (processing, packaging of poultry products) (KIT, Faida MaLi, and IIRR 2006).

Chicken value chains encompass both direct and indirect actors. The direct actors include farmers (producers), processors, and consumers. The indirect actors are the supporting systems, such as financial entities, research institutions, extension officers, and credit facilities (KIT, Faida MaLi, and IIRR 2006; Bulama et al. 2019). Value chain analysis can be used to understand the motives, incentives, and challenges that exist at various nodes of an indigenous chicken value chain and how different actors network and compete for optimum benefits.

Poultry production can be divided into three main systems: traditional extensive production systems that comprise mainly indigenous chickens, and semi-intensive and intensive systems that can comprise either indigenous or non-indigenous chicken. The prevalence of these different systems in East Africa is shown in Figure 13.1 (Mujyambere et al. 2022). Eighty percent of Kenya's chicken production comes from extensive systems, and the remaining 20 percent from semi-intensive systems. The majority of chicken-producing households in low- and middle-income countries keep only 5–15 adult chickens (de Bruyn et al. 2015), but there has been significant growth in production through a gradual transition from backyard to more commercialized systems in East Africa (Vernooij, Masaki, and Meijer-Willems 2018).

In Kenya, the chicken value chain plays a significant role in the food system, with potential to contribute substantially to poverty reduction,

FIGURE 13.1 Chicken production systems in East African countries

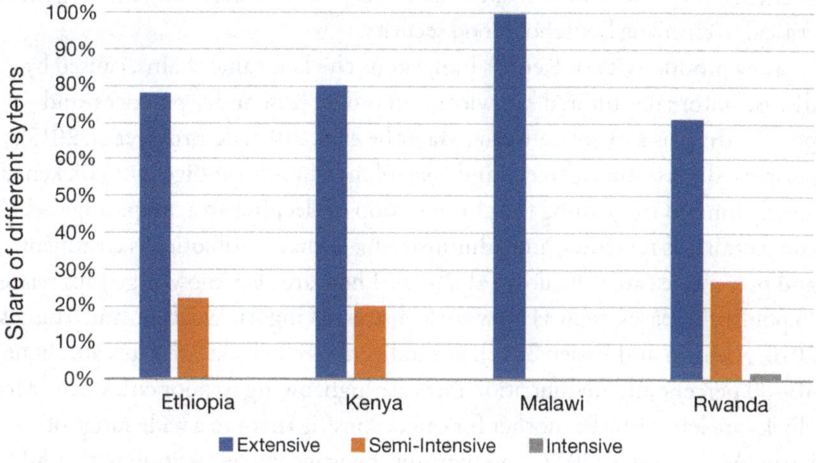

Source: Adapted from Mujyambere et al. (2022).

improved diets, and economic growth (Chapter 2). According to Kenya's 2019 census, 30 percent of Kenyan households keep chickens in their household (KNBS 2019). Indigenous chicken, in particular, plays a key role in the food system—92 percent of all households with chickens keep indigenous chicken (a total of 27 percent of all households in Kenya) (KNBS 2019). The high percentage of households keeping indigenous chickens has been found in other settings in SSA as well. For example, in some areas of Zambia, 99 percent of chicken farmers keep indigenous chickens (Bwalya 2014).

The prevalence of indigenous chicken in Kenya, and elsewhere, reflects both demand and supply factors. On the demand side, indigenous chicken is often preferred for its unique taste qualities and its sociocultural import in local communities (Okello et al. 2010). Indigenous chicken characteristics such as plumage color, weight, size, and overall body constitution help determine chicken prices (Bett et al. 2012). These dynamics are not unique to Kenya—in China, Padhi (2016) notes increasing consumer demand for slow-growing chicken. In Ethiopia, birds with red and white plumage or pea-shaped combs are reported to fetch prices approximately 15–35 percent above prices for those without (Mengesha and Tsega 2011). On the supply side, indigenous chickens are often preferred for their natural immunity to diseases; their ability to survive in harsh climatic conditions; and because they are ideal mothers to their chicks and require relatively little management capital and care compared with other species (Padhi 2016). Indigenous chickens even serve as an important source of emergency income in Kenyan households (Okello et al. 2010), making them critical to ensuring household food security.

Low productivity of Kenya's indigenous chicken value chains, caused by disease outbreaks, limited knowledge on proper husbandry practices, and poor feeding, is a major concern (Magothe et al. 2012; de Bruyn et al. 2015). Farmers' disease management and control measures for indigenous chickens are largely limited to cleaning the chicken coop or sleeping area, preparing herbal concoctions as remedies, and administering human antibiotics as treatment and preventive care (Okello et. al. 2010). There are also knowledge gaps related to poultry diseases, such as Newcastle disease (Kingori, Wachira, and Tuitoek 2010; Akinola and Essien 2011). For indigenous chicks, death rates are about 40–80 percent and malnutrition rates are high, owing to poor chick care. Most chicks are left with the mother for care, exposing them to a wide range of harm (Mahoro et al. 2017) and most must engage in competition with adult chickens for scarce feed (Kingori, Wachira, and Tuitoek 2010; Akinola and Essien 2011). Okello and colleagues (2010) note that indigenous chickens are generally left to scavenge for feed, with kitchen leftovers and grains and cereals as supplementary feed.

Strengthening upstream and downstream value chain linkages can help overcome productivity challenges for indigenous chicken producers. Upstream, poultry farmers need access to vaccines for disease control and proper feed for supplementation as well as knowledge on proper housing (Kingori, Wachira, and Tuitoek 2010; Akinola and Essien 2011; Mujyambere et al. 2021). Downstream, better market linkages can help farmers receive more revenue from their chicken production, which can help finance productive inputs, such as vaccines. For successful value addition downstream, producers need access to both good husbandry knowledge and market information to be able to meet end-user quality demands as well as to benefit from the market structure in terms of fair pricing for their chicken (Mensah-Bonsu, Lartey, and Kuwornu 2019). The main actors in the marketing value chain are brokers (middlemen/women), traders/wholesalers, and chicken producers (Relucio 2021), with the brokers benefiting more than the producers and retailers. Even when local producers have informal engagements with urban buyers, through either long-term transactions or regular delivery agreements, brokers have more bargaining power in negotiating the selling and buying prices than either producers or retailers (Okello et al. 2010). As such, chicken producers are unable to capture the rents from the sale and production of chicken, and these dynamics are often gendered, as we discuss below.

The role of gender in poultry value chains

Gender plays an important role in terms of labor, decision-making, marketing, and access to productive inputs and services within poultry value chains, including in indigenous chicken value chains. For smallholders and rural residents, the local sale of chicken from traditional production systems is valued as a source of animal protein and revenue (Kingori, Wachira, and Tuitoek 2010; Kristjanson et al. 2014; Mathiu, Ndirangu, and Mwangi 2021; Mujyambere et al. 2021). How gender plays a role in production and marketing is often context specific.

From a labor perspective, women tend to be more involved at the production stage, particularly in the extensive production systems that are prevalent in low- and middle-income countries. For example, in Burkina Faso, where chicken-raising has traditionally been considered the preserve of women, women are generally in charge of day-to-day activities of poultry management (Eissler et al. 2020). A study from Ethiopia shows that, while 77.7 percent of women participated in poultry management in men-headed households and 80.7 percent in women-headed households, only 2.2 percent and 5.1 percent of the men in men- and women-headed households, respectively, were involved (Gebremedhin et al. 2016). Likewise, on smallholder farms in Ethiopia, women (64 percent) are primarily in charge of caring for and maintaining chickens (Hailemichael et al. 2016). In western Kenya, women are largely responsible for feeding (75 percent), cleaning (75 percent), and tending to sick hens (60 percent). Although children also contribute (10 percent), men are usually in charge of building chicken coops (85 percent) (Okitoi et al. 2007). The high share of labor from women in extensive systems can also lead to time constraints. Bulama and colleagues (2019) show that in Maiduguri, Nigeria, women are the majority (60 percent) of producers involved in stocking, feeding, hygiene, and day-to-day care. Women are also involved in other gendered roles in different spheres (notably domestic duties) and therefore face time constraints when dealing with chickens.

In semi-intensive and intensive production systems, the gender dynamics of labor are different. As chicken production evolves toward a more intensive production system, offering an alternative source of livelihood to crop farming, men are increasingly involved in production (Mathiu, Ndirangu, and Mwangi 2021; Mujyambere et al. 2021). In intensive poultry production that involves larger-scale commercial chicken and egg production, men and women provide labor at different nodes of the production chain. For example, in Kericho, Kenya, men provide a large share of labor for logistics and cash-related activities, including selling culls (46.8 percent), buying and transporting feed (64.1 percent), and paying hatching collectives and individuals to reserve chicks and transport them (73.4 percent) (Ngeno et. al. 2011). Studies in Ghana and Nigeria indicate that men dominate the producer (75.9 percent) and processing (83.3 percent) nodes of the commercial poultry value chain, while women dominate the marketing node (70.1 percent) (Mensah-Bonsu, Lartey, and Kuwornu 2019; Adeyonu et al. 2022).

Gendered decision-making and the control of benefits from the production and sale of poultry follow a pattern similar to the labor dynamics, with women contributing more in extensive systems and men more in semi-intensive and intensive systems. Studies from across Kenya suggest that women dominate indigenous chicken production in the traditional extensive system (95 percent) and have considerable access to and control of benefits from ownership and sale of the indigenous chicken produced on a small scale (Okitoi et al. 2007; Ngeno et al. 2011). In a study conducted in central and western Kenya, the majority of women (54.6 percent) and girls (57 percent) were found to make decisions about the type of poultry kept and the type of feed used (Waithanji et al. 2020).

In semi-intensive and intensive systems, however, men tend to dominate decision-making and control of benefits. Gammage (2009) indicates that chicken value chain operations reflect existing gender norms and inequalities in bargaining power (and decision-making), hence women are engaged at the lower nodes of the chain for relatively lower wages. For instance, in Burkina Faso, women are the main caretakers of poultry, but contestations arise between men and women on decision-making on the commercial disposal of birds. Women are thus found to have limited decision-making power in selling chicken in large quantities (Eissler et al. 2020). Intensified production, aimed at maximizing profits over the social value of chicken production, leads to inequalities especially for women. Intensification has often resulted in labor exploitation, diminishing decision-making power for upstream and downstream value chain segments, and little control over income generated from chicken and egg sales. Adeyonu and colleagues (2021) further argue that, despite variations in participation in the chicken value chain, men add more value than women, and benefit more, at the different nodes of the value chain because they operate on a larger scale.

Men tend to have better access to resources and knowledge than women in poultry value chains. The literature suggests that men chicken producers are more likely than women chicken producers to obtain credit (Mensah-Bonsu, Lartey, and Kuwornu 2019; Ndirangu, Mbogoh, and Mbatia 2018). This has been attributed in part to women's lack of collateral for securing credit, since they have limited or no say over productive resources such as land. Women's limited mobility owing to normative constraints and unpaid care work limits their access to information through trainings and group networks, markets and marketing information, and chicken vaccination information, affecting their value addition as traders, producers, and processors (Eissler et al. 2020). These limitations can have important consequences. Mathiu, Ndirangu, and Mwangi (2021) show that in men-headed households, production is 19.69 percent higher than in women-headed households in Meru County, Kenya.

Gender analysis of indigenous chicken value chain in eastern Kenya

Given the different roles that men and women play at different nodes of the poultry value chain, gender analysis of value chains can provide insights needed for creating more productive and inclusive poultry systems. A gender analysis of the poultry value chain reveals the various nodes of the value chain that women and men occupy (Mutua, Njuki, and Waithanji 2014) and provides understanding of the value addition process (Adeoye, Adeolu, and Ibrahim 2013).

This study conducts a gender analysis of the indigenous chicken value chain in eastern Kenya in the context of the Gender Inclusive Vaccine Ecosystem (GIVE) project. To empower women smallholder farmers and improve their access to agricultural technologies, GIVE was funded under the Livestock Vaccines Innovative Fund by the International Development Research Centre and the Bill & Melinda Gates Foundation from 2019 to 2023. The project set out to enhance vaccine distribution and delivery systems in response to two endemic livestock diseases, Newcastle disease (ND) in chicken and contagious caprine pleuropneumonia (CCPP) in sheep and goats in eastern Kenya. A gendered value chain analysis was conducted to uncover and understand the factors that drive structural gender inequalities, particularly related to vaccine uptake, within the livestock value chain in rural Kenya.

For the analysis, a cross-sectional study design was used along with qualitative methods. Eighteen sex-segregated focus group discussions (FGDs) were conducted with a total of 153 participants in Makueni County (Table 13.1). These activities produced data on the management of chickens and specific roles, duties, and bargaining power positions, including decision-making, among women and men in smallholder farming households.

Gender-disaggregated data were collected on various aspects of rural indigenous chicken management, such as housing, feeding, and disease control, including vaccination, ownership, and decision-making. Using FGD data on gender roles, an analysis was carried out of the gender differentials and the intensity of involvement of different household members in indigenous chicken management in the study setting, which is dominated by extensive production systems.

FIGURE 13.2 Map of Makueni County

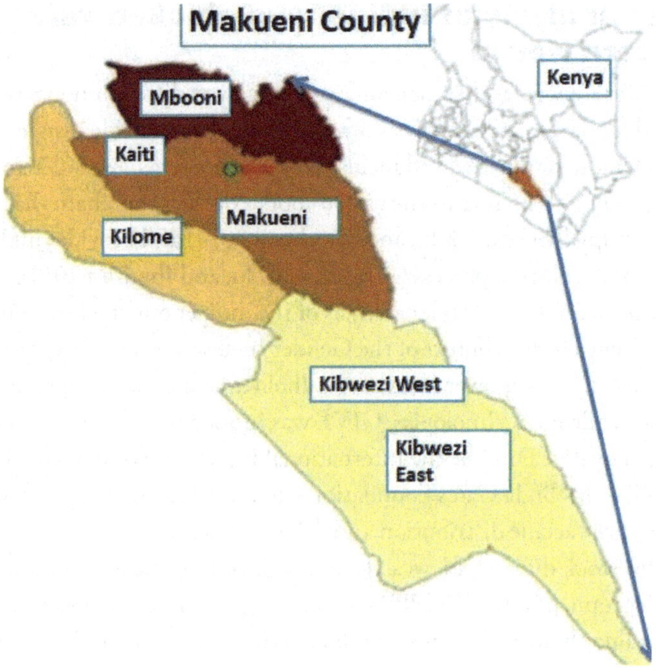

Source: Musyoka and Mutia (2016).

TABLE 13.1 Gender focus group discussions conducted in Makueni County, 2023

Sub-county	Focus Group Discussions		
	Total sessions	No. of men	No. of women
Makueni	6	22	27
Kibwezi East	6	31	25
Kibwezi West	6	24	24
Total	**18**	**77**	**76**

Source: Authors.

Household indigenous chicken ownership and decision-making

The participatory exercise results are shown in Table 13.2. The FDG responses indicated that 51.2 percent of the chickens are owned by women, and 8.8 percent are owned by men. The remaining 40 percent are owned by children (11.6 percent), youth (9.5 percent), and jointly by couples (18.9 percent). Decision-making in the value chain is gendered: men and women participate in decision-making for different activities. In half of the focus groups, women were reported to be able to make decisions about selling chickens and their products (eggs and meat) as well as about their consumption. Men were reported to be involved to some extent in making decisions on selling chickens but less so on the sale of eggs. Men were reported to be most involved in decisions about the purchase of drugs and vaccination of the chickens. In all, women were reported to be the primary decision-makers for all activities, as shown in Table 13.2.

Ownership of chickens is linked directly with participation in decision-making regarding selling, consuming, distributing gifts, confining, and vaccinating the birds. Table 13.2 shows that women are the main actors in the different decisions related to chicken production and management, although men also participate significantly.

Figure 13.3 shows differences in men's and women's perceptions about ownership of indigenous chickens. Men regard their position as owners in rural indigenous chicken production and management to be almost equal to that of women. Women, however, consider themselves the majority stakeholders in terms of ownership of rural indigenous chickens, and their production

TABLE13.2 Responses from FGDs on household chicken ownership and decision-making for indigenous chicken production (N=153 participants)

Household members	Decision-making (%)								
	Owning chickens	Selling eggs	Selling chickens	Consuming eggs	Consuming chickens	Gifting to visitors	Confining chickens	Vaccinating	Purchasing drugs
Men	8.8	5.1	23.3	7.6	16.9	27.2	27.9	32.9	37.4
Women	51.2	79.6	53.1	89.9	70.1	64.9	60.8	38.8	42.2
Youth	9.5	3.3	11.5	2.0	2.1	1.0	3.6	1.2	0.0
Children	11.6	0.0	1.4	0.5	2.3	0.0	2.2	1.3	0.6
Joint family	18.9	12.0	10.3	0.0	8.6	6.9	5.6	4.6	3.2

Source: Authors.

and management. Men also reported more joint ownership than women, which may imply that men perceive chicken-raising in their household as a joint venture, while women may have a different perspective. These are important differences in perspective to take into consideration when developing or implementing programs that touch on chicken value chains in the

FIGURE 13.3 Gendered perceptions of ownership of indigenous chickens

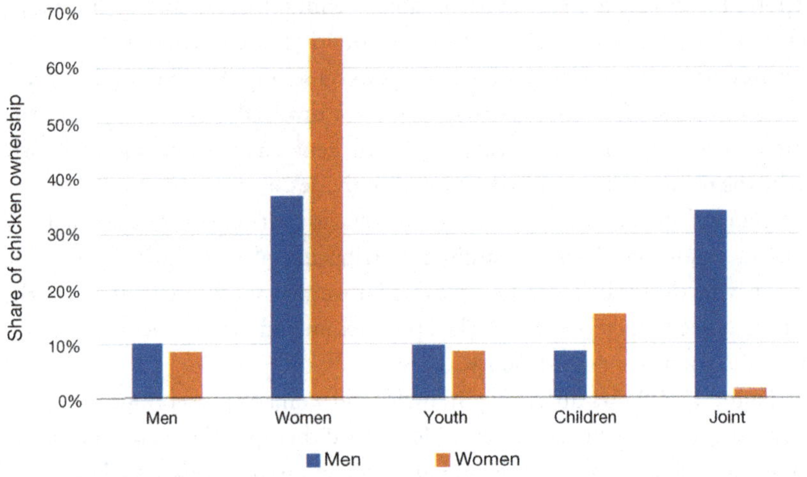

Source: Authors.

TABLE13.3 Responses from FGDs on division of labor among household members in indigenous chicken management activities (N= 18 FGDs)

Household members	Construction of coop Mean (%)	Cleaning of poultry house Mean (%)	Feeding Mean (%)	Watering Mean (%)	Treatment Mean (%)	Buying of poultry Mean (%)	Selling of poultry Mean (%)	Protecting of chicks Mean (%)	Vaccination Mean (%)	Purchase of feed Mean (%)	Purchase of drugs Mean (%)
Men	44.7	1.9	13.5	17.4	35.2	27.5	22.5	16.7	25.8	34.6	43.8
Women	28.3	55.9	51.7	53.5	51.3	64.4	64.6	33.6	38.3	35.2	40.7
Youth	15.7	20.3	11.3	9.2	9.8	6.9	10.1	16.3	6.9	10.8	9.4
Children	3.9	16.3	9.4	7.0	2.7	0.0	0.0	10.6	5.7	2.0	0.0
Joint family	7.9	5.6	8.4	7.3	1.0	1.8	2.9	22.8	1.7	0.7	6.1

Source: Authors.

community. While ownership positions show the increasing role of men, there remains an imbalance in the division of labor, as presented in Table 13.3 and discussed below.

Division of labor

Men, women, youth, and children all provide labor in indigenous chicken management, but women were reported to provide the bulk of the labor in almost all the chicken management activities. Men's role was larger only in the construction of chicken houses and purchase of drugs. Women and children were said to do most of the time-consuming daily management work, while men engage in activities that may be demanding but one-off, such as constructing chicken coops.

Construction of chicken houses. The task of constructing chicken houses was reported to fall largely on men, with 44 percent participation reported. Depending on the type of chicken house, women also participate in the construction, with 28 percent participation reported. Joint family participation was reported at 8 percent. Youth and children were also mentioned as helping with construction, at 15 percent and 4 percent, respectively. Young men especially help find tree branches that are used to construct the houses and structures; in some households, they also build chicken houses on their own.

Cleaning of chicken houses. Cleaning chicken houses and structures was reported to be the job of women, youth, and children. This is because the women are mostly left at home, and youth and children play that role during weekends and school holidays. Women's participation was reported as 56 percent, youth participation at 20 percent, and children's at 16 percent. Joint family participation was reported at 6 percent, while men's participation was just 2 percent because they were away from the house most of the time to find employment.

Feeding. Feeding chickens was reported as a task performed by all household members. Women played the major role, with 52 percent participation reported. Men's participation was reported at 14 percent. Youth and children were mentioned as participating in feeding on the days they did not attend school. Their participation was reported as 11 percent and 9 percent, respectively. Joint family participation was reported at 8 percent.

Watering. Participation in providing water is reported to be similar to feeding, with women's participation reported at 54 percent and men's participation at 17 percent. Youth participation was reported at 9 percent, that of children at 7 percent, and that of the joint family at 7 percent.

Treatment of chicken. The FGD participants mentioned that the administration of medication to treat chicken diseases is mostly done when they are being given feed and water. Women were reported to participate at 51 percent. Men's participation (35 percent) was slightly higher for treatment than their participation in feeding and watering. Youth participation was reported at 10 percent, children at 3 percent, and the joint family at 1 percent.

From these results, we see that women and children were responsible for most of the daily management routines for indigenous chickens, including feeding and watering the chickens and cleaning the chicken houses. Focus group participants noted that men provide support in these duties when their wives are unwell or away from home, and the children are at school.

Decision-making for vaccination of chickens

The proportion of women who make the decisions about vaccination of chickens is slightly larger than the proportion of men who do so. According to the 18 participatory exercises we carried out, 32 percent of men and 38 percent of women made decisions regarding vaccinations, respectively. Focus group participants suggested that women are slightly more likely to make vaccination decisions than men because they are the primary owners of chickens; however, men also frequently make these decisions, because they usually provide the money to buy vaccines, which gives them the decision-making power. Discussants in four of the FGDs reported that they do not vaccinate their chickens at all due to their perception that vaccination is only needed for improved or hybrid chickens. These farmers use herbal remedies to manage the health of indigenous chicken but use commercial vaccines for their hybrid chickens. As one woman explained, *"I vaccinate the commercial chickens which I keep here at the market but for the ordinary chickens at home I only use herbal remedies"* (FGD#5). This reflects variations in management for different breeds of chicken.

Conclusions and recommendations

The poultry value chain is a driver of food security and sustainable development and represents a viable pathway out of poverty for the poor, especially women. It plays a significant role in the Kenyan agriculture sector, with 30 percent of all households keeping chickens (KNBS 2019). Traditional extensive production, which comprises mainly indigenous chickens and occurs mostly in rural settings among smallholder farmers, could contribute toward transforming food systems to empower women and enhance gender equality. Women already play a key role in the indigenous chicken value chain, although various sociocultural factors often limit their contribution. The value chain provides unique opportunities for women especially, with significant implications for household food security and socioeconomic well-being (Grace et al. 2014). However, an analysis of the value chain using the case study presented here highlights some gender considerations that are crucial to achieving a sustainable food system.

The indigenous chicken value chain is gendered along aspects of access to and control over resources. It is mostly women who are the owners of chickens, and thus they have more control over sales and income. However, joint ownership exists in some coupled households, which could influence intrahousehold bargaining dynamics when it comes to decision-making and have negative impacts on women's agency if not handled in an equitable manner. Nevertheless, women are the main decision-makers in indigenous chicken production. However, it is interesting to note that men's involvement is greater in decision-making concerning sales of chicken but not sales of eggs. This could imply that men compete with women where they know there is greater benefit; indeed, it has been noted that, with commercialization or intensification of other crops (such as groundnuts), men tend to take more control of the value chain than when the practice is small in scale (Forsythe et al. 2015).

Women reported that they have some power to make decisions about vaccination, while men indicated that they are the decision-makers because they provide the money to purchase vaccines and because they are the household heads. Although women are the main owners of indigenous chickens, their limited access to finance and other resources can be a constraining factor in their capacity to engage and scale up their production into a higher-value agri-enterprise, which is seen in cassava value chains (Forsythe et al. 2016).

In terms of division of labor, women are more involved in roles that seem light (cleaning chicken coops, feeding, watering) but are repetitive, whereas men construct coops in a one-off engagement. Women's roles can thus be time-consuming, adding to their multiple household responsibilities, and may hinder their engagement in other productive activities. Value chain interventions should take into consideration the time burden that may be placed on men and women by new technologies, hence the need for doing cost and time benefit analysis.

Given that women engage in the marketing of chicken and related products, improved marketing strategies that enable women smallholder farmers to obtain premium prices for their products would go a long way toward improving their economic and livelihood outcomes as well as contributing indirectly to improved dietary outcomes. Indigenous poultry farming can make a positive contribution to household economies and subsistence, and can contribute toward a resilient food system in the face of climate change. Tackling gender inequalities in access to resources for indigenous chicken production and management and in marketing dynamics will advance women's empowerment and household income.

Finally, noting that various factors may interact to create inequalities, it is important to include an intersectionality lens in any gender review of food system value chains. Areas for future research include the collection of sex-, gender- and intersectionality-disaggregated data on participation and roles in the poultry value chain to provide empirical evidence to inform inclusive policy and programming, as well as to set up indicators for monitoring gender aspects in the poultry value chain.

References

Adeoye, P.A., R.A. Adeolu, and H.M. Ibrahim. 2013. "Appraisal of Rural Water Supply: Case Study of Kwara State, North Central Nigeria." *International Journal of Basic and Applied Science* 1 (4): 816–826.

Adeyonu, A., A. Okunola, M. Alao, E. Oyawoye, and C. Okonkwo. 2021. "An Assessment of Broiler Value Chain in Nigeria." *Open Agriculture* 6 (1): 296–307.

Ahmed, S., M.Beum, A. Khatun, R. Gofur, Md.TA. Azad, A. Kabir, and T.S. Haque. 2021. "Family (FP) as a Tool for Improving Gender Equity and Women's Empowerment in Developing Countries: Evidence from Bangladesh." *European Journal of Agriculture and Food Sciences* 3 (2): 37–44.

Angel-Urdinola, D., and Q. Wodon. 2010. "Income Generation and Intra-household Decision Making: A Gender Analysis for Nigeria." In *Gender Disparities in Africa's Labor Market,* eds. J.S. Arbache, A. Kolev, and E. Filipiak, 381–406. Washington, DC: World Bank.

Akinola, L.A.F., and A. Essien. 2011. "Relevance of Rural Poultry Production in Developing Countries with Special Reference to Africa." *World's Poultry Science Journal* 67 (4): 697–705.

Bell, J. 2021. *Evidence Pathways to Gender Equality and Food Systems Transformation.* Rome: World Food Programme (WFP).

Bett, H.K., R.C. Bett, K.J. Peters, and W/ Bokelmann. 2012. "Linking Utilisation and Conservation of Indigenous Chicken Genetic Resources to Value Chains." *Journal of Animal Production Advances* 2 (1): 33–51.

Bulama, Y.M., U. Bukar, B.G. Shettima, and S. Maina. 2019. "Analysis of Chicken Value Chain in Maiduguri Metropolitan Council, Borno State, Nigeria." *Nigerian Journal of Animal Science and Technology* 2 (3): 95–102.

Bwalya, R., and T. Kalinda. 2014. "An Analysis of the Value Chain for Indigenous Chickens in Zambia's Lusaka and Central Provinces." *Journal of Agricultural Studies* 2 (2): 32–51.

CARE. 2021. *Enabling Inclusive Food Systems: Food and Water Systems Annual Report.* Atlanta.

De Bruyn, J.D., J. Wong, B. Bagnol, B. Pengelly, and R. Alders. 2015. "Family Poultry and Food and Nutrition Security." *CABI Reviews* 10 (13): 1–9.

Doss, C., and D. Rubin. 2021. "Exploring Women's Empowerment Using a Mixed Methods Approach." IFPRI Discussion Paper 2035. International Food Policy Research Institute (IFPRI), Washington, DC.

Doss, C., R. Meinzen-Dick, A. Quisumbing, and S. Theis. 2018. "Women in Agriculture: Four Myths." *Global Food Security* 16: 69–74.

Eissler, S., A. Sanou, J. Heckert, E. Myers, S Nignan, E. Thio, L.A. Pitropia, R. Ganaba, A. Pedehombga, and A. Gelli. 2020. "Gendered Participation in Poultry Value Chains: Qualitative Findings from an Impact Evaluation of Nutrition-Sensitive Poultry Value Chain Intervention in Burkina Faso." IFPRI Discussion Paper 1928. IFPRI, Washington, DC.

Fan S., and J. Swinnen. 2020. "Reshaping Food Systems: The Importance of Inclusion." In *2020 Global Food Policy Report: Building Inclusive Food Systems*. Washington, DC: IFPRI.

Farhall, K., and L. Rickards. 2021. "The 'Gender Agenda' in Agriculture for Development and Its (Lack of) Alignment with Feminist Scholarship." *Frontiers in Sustainable Food Systems* 5: 573424.

Farrell, D. 2013. "The Role of Poultry in Human Nutrition." *Poultry Development Review*. Rome: FAO.

Forsythe, L., M. Nyamanda, A. Mbachi Mwangwela, and B. Bennett. 2015. "Beliefs, Taboos and Minor Crop Value Chains: The Case of Bambara Groundnut in Malawi." *Food, Culture & Society* 18 (3): 501–517.

Forsythe, L., H. Posthumus, and A. Martin. 2016. "A Crop of One's Own? Women's Experiences of Cassava Commercialization in Nigeria and Malawi." *Journal of Gender, Agriculture and Food Security* 1 (2): 110–128.

Gammage, S., C. Manfre, and K. Cook. 2009. *Gender and Pro-Poor Value Chain Analysis: Insights from the GATE Project Methodology and Case Studies*. Washington, DC: USAID.

Gebremedhin, B., E. Tesema, A. Tegegne, D. Hoekstra, and S. Nicola. 2016. "Value Chain Opportunities for Women and Young People in Livestock Production in Ethiopia: Lessons Learned." LIVES Working Paper 24. International Livestock Research Institute (ILRI), Addis Ababa.

Gondwe, T.N., and C.B.A. Wollny. 2007. "Local Chicken Production System in Malawi: Household Flock Structure, Dynamics, Management and Health." *Tropical Animal Health and Production* 39 (2): 103–113.

Grace, D., K. Roesel, E. Kang'ethe, B. Bonfoh, and S. Theis. 2015. "Gender Roles and Food Safety in 20 Informal Livestock and Fish Value Chains." IFPRI Discussion Paper 1489. IFPRI, Washington, DC.

Hailemichael, A., B. Gabremedhin, S. Gizaw, and A. Tegegne. 2016. "Analysis of Village Poultry Value Chain in Ethiopia: Implications for Action Research and Development." LIVES Working Paper 10. International Livestock Research Institute, Nairobi.

Hirwa, A., D.R. Kugonza, A. Kayitesi, T. Murekezi, F. Semahoro, G. Uwimana, and R. Habimana. 2019. "Phenotypes, Production Systems and Reproductive Performance of Indigenous Chickens in Contemporary Rwanda." *International Journal of Livestock Production* 10 (10): 213–231.

IPC (Integrated Food Security Phase Classification). 2023. *IPC Kenya Acute Food Insecurity and Malnutrition March–June 2023*. Nairobi: National Drought Management Authority.

Kingori, A.M., A.M. Wachira, and J.K. Tuitoek. 2010. "Indigenous Chicken Production in Kenya: A Review." *International Journal of Poultry Science* 9 (4): 309–316.

KIT (Royal Tropical Institute), Faida MaLi (Faida Market Link), and IIRR (International Institute of Rural Reconstruction). 2006. *Chain Empowerment: Supporting African Farmers to Develop Markets*. Amsterdam: KIT; Arusha, Tanzania: Faida MaLi; and Nairobi: IIRR.

KNBS (Kenya, National Bureau of Statistics). 2019. *Kenya Population and Housing Census 2019*. Nairobi.

Kristjanson, P., A. Waters-Bayer, N. Johnson, A. Tipilda, J. Njuki, I. Baltenweck, D. Grace, and S. MacMillan. 2014. "Livestock and Women's Livelihoods." In *Gender in Agriculture*, eds. A.R. Quisumbing, R. Meinzen-Dick, T.L. Raney, A. Croppenstedt, J.A. Berhman, and A. Peterman, 209–233. Dordrecht, the Netherlands: Springer.

Machina, H., and M. Lubungu. 2019. "Understanding Intra-household Gender Disparities of Smallholder Livestock Production in Zambia." *Journal of Gender, Agriculture and Food Security* 4 (2): 11–24.

Magothe, T.M., T.O. Okeno, W.B. Muhuyi, and A.K. Kahi. 2012. "Indigenous Chicken Production in Kenya: I. Current Status." *World's Poultry Science Journal* 68 (1): 119–132.

Mahoro, J., T.K. Muasya, F. Mbuza, R. Habimana, and A.K. Kahi. 2017. "Characterization of Indigenous Chicken Production Systems in Rwanda." *Poultry Science* 96 (12): 4245–4252.

Mapiye, C., M. Mwale, J.F. Mupangwa1, M. Chimonyo, R. Foti1, and M.J. Mutenje. 2008. "A Research Review of Village Chicken Production Constraints and Opportunities in Zimbabwe." *Asian-Australian Journal of Animal Sciences* 21 (11): 1680–1688.

Mathiu, E.M., S.N. Ndirangu, and S.C. Mwangi. 2021. "Production of Indigenous Poultry among Smallholder Farmers in Tigania West Meru County, Kenya." *African Journal of Agricultural Research* 17 (5): 705–713.

Meinzen-Dick, R.S., D. Rubin, M. Elias, A.A. Mulema, and E. Myers. 2019. "Women's Empowerment in Agriculture: Lessons from Qualitative Research." IFPRI Discussion Paper 1797. IFPRI, Washington, DC.

Mengesha, M., and W. Tsega. 2011. "Phenotypic and Genotypic Characteristics of Indigenous Chickens in Ethiopia: A Review." *African Journal of Agricultural Research* 6 (24): 5398–5404.

Mensah-Bonsu, A., N.N. Lartey, and J.K. Kuwornu. 2019. "Gender-Segregated Analysis of the Poultry Value Chain in Ghana." *Gender, Technology and Development* 23 (2): 130–164.

Micere Njuki, J., A. Wyatt, I. Baltenweck, K. Yount, C. Null, U. Ramakrishnan, A.W. Girard, and S. Sreenath. 2016. "An Exploratory Study of Dairying Intensification, Women's Decision Making, and Time Use and Implications for Child Nutrition in Kenya." *European Journal of Development Research* 28 (4): 722–740.

Mkandawire, E., M. Mentz-Coetzee, M.N. Mangheni, and E. Barusi. 2021. "Enhancing the Glopan Food Systems Framework by Integrating Gender: Relevance for Women in African Agriculture." *Sustainability* 13 (15): 8564.

Mottet, A., and G. Tempio. 2017. "Global Poultry Production: Current State and Future Outlook and Challenges." *World's Poultry Science Journal* 73 (2): 245–256.

Mujyambere, V., K. Adomako, S.O. Olympio, M. Ntawubizi, L. Nyinawamwiza, J. Mahoro, and A. Conroy. 2022. "Local Chickens in East African Region: Their Production and Potential." *Poultry Science* 101 (1): 101547.

Musyoka, S.N., and G.M. Mutia. 2016. "The Status of Fish Farming Development in Arid and Semi-arid Counties of Kenya: Case Study of Makueni County." *European Journal of Physical Sciences* 4 (3): 2056–5879.

Mutua, E., J. Njuki, and E. Waithanji. 2014. *Review of Gender and Value Chain Analysis, Development and Evaluation Toolkits.* Nairobi: International Livestock Research Institute.

Ndirangu, S.N., S.G. Mbogoh, and O.L.E. Mbatia. 2018. "Evaluation of the Elasticity of Farm Output among Smallholder Farmers in Selected Agro-Ecological Zones of Embu County, Kenya." *Asian Journal of Agricultural Extension, Economics & Sociology* 26 (3): 1–10.

Ngeno, V., B.K. Langat, R. Wendi, and M.J. Kipsat. 2011."Gender Aspect in Adoption of Commercial Poultry Production among Peri-Urban Farmers in Kericho Municipality, Kenya." *Journal of Development and Agricultural Economics* 3 (7): 286–301.

Njuki, J., S. Eissler, H.J. Malapit, R.S. Meinzen-Dick, E. Bryan, and A.R. Quisumbing. 2021. *A Review of Evidence on Gender Equality, Women's Empowerment, and Food Systems.* United Nations Food System Summit Brief. Bonn: United Food Systems Summit; Washington, DC: IFPRI.

Njuki, J., and P.C. Sanginga. 2013. "Gender and Livestock: Key Issues, Challenges and Opportunities." In *Women, Livestock Ownership and Markets: Bridging the Gender Gap in Eastern and Southern Africa,* eds. J. Njuki and P.C. Sanginga, 1–8. New York: Routledge.

Ochieng, J., G. Owuor, and B.O. Bebe. 2013. "Management Practices and Challenges in Smallholder Indigenous Chicken Production in Western Kenya." *Journal of Agriculture and Rural Development in the Tropics and Subtropics* 114 (1): 51–58.

Okello, J.J., Z. Gitonga, J. Mutune, R.M. Okello, M. Afande, and K.M. Rich. 2010. "Value Chain Analysis of the Kenyan Poultry Industry: The Case of Kiambu, Kilifi, Vihiga, and Nakuru Districts." Africa/Indonesia Team Working Paper 24. IFPRI, International Livestock Research Institute, and Royal Veterinary College, Washington, DC.

Okitoi, L.O., H.M.J. Udo, E.A. Mukisira, R. De Jong, and R.P. Kwakkel. 2007. "Evaluation of Low Input Invention for Improved Productivity of Indigenous Chickens in Western Kenya." *Agricultura et Subtropica* 39 (3):179–182.

Padhi, M.K. 2016. "Importance of Indigenous Breeds of Chicken for Rural Economy and Their Improvements for Higher Production Performance." *Scientifica* 2016: 2604685.

Palacios-Lopez, A., L. Christiaensen, and T. Kilic. 2017. "How Much of the Labor in African Agriculture is Provided by Women?" *Food Policy* 67: 52–63.

Pius, L.O., P. Strausz, and S. Kusza. 2021. "Overview of Poultry Management as a Key Factor for Solving Food and Nutritional Security with a Special Focus on Chicken Breeding in East African Countries." *Biology* 10 (8): 810.

Relucio, J. 2021. "Value Chain Analysis of Native Chicken in Selected Barangay of Partido." *Technium: Romanian Journal of Applied Sciences and Technology* 3 (6): 50–66.

Richardson, R.A. 2018. "Measuring Women's Empowerment: A Critical Review of Current Practices and Recommendations for Researchers." *Social Indicators Research* 137 (2): 539–557.

Shahbandeh, M. 2022. *"Global Number of Chickens 1990-2020."* Statista. Accessed May 21, 2022. https://www.statista.com/statistics/263962/number-of-chickens-worldwide-since-1990/

United Nations. 2021. *Kenya's Pathway to Sustainable Food Systems.* Kenya National Food Systems Dialogue Series. https://www.unfoodsystemshub.org/docs/unfoodsystemslibraries/national-pathways/kenya/2021-09-27-en-kenya-fss-dialogue-series-final-report-.pdf?sfvrsn=109d50c_1

UNDP (United Nations Development Programme). 2015. *The Cost of the Gender Gap in Agricultural Productivity.* New York.

UNDP. 2016. *Gender and Climate Change: Gender, Climate Change and Food Security.* New York.

Vernooij, A., M.N. Masaki, and D. Meijer-Willems. 2018. *Regionalisation in Poultry Development in Eastern Africa.* Report 1121. Wageningen, Netherlands: Wageningen Livestock Research.

Visser, J., and J. Wangu. 2021. "Women's Dual Centrality in Food Security Solutions: The Need for a Stronger Gender Lens in Food Systems' Transformation." *Current Research in Environmental Sustainability* 3: 100094.

Waithanji, E., D.H. Affognon, S. King'ori, G. Diiro, D. Nakimbugwe, and K.K.M. Fiaboe. 2020. "Insects as Feed: Gendered Knowledge Attitudes and Practices among Poultry and Pond Fish Farmers in Kenya." *NJAS: Wageningen Journal of Life Sciences* 92 (1): 100312.

Wong, J.T., J. de Bruyn, B. Bagnol, H. Grieve, M. Li, R. Pym, and R.G. Alders. 2017. "Small-Scale Poultry and Food Security in Resource-Poor Settings: A Review." *Global Food Security* 15: 43–52.

Zemelak, G., C. Luizinho, W. Cassio, and A.B. Gudrun. 2016. "Characterization of Village Chicken Production Systems and Challenges across Agro-climatic Zones in Ethiopia." *International Journal of Livestock Production* 7 (11): 94–105.

Chapter 14

YOUTH ENGAGEMENT IN AGRICULTURE AND FOOD SYSTEMS TRANSFORMATION IN KENYA

Victor Mugo and Ivy Kinyua

Food systems incorporate many actors at different intersecting levels and spaces. Young people[1] constitute one of the most significant groups of these actors and contribute significantly to food systems in a variety of ways, from agricultural production and processing to food-related retail services, through formal and informal employment, paid and unpaid labor, and self-employment. In addition to engaging through work and livelihoods, young people are involved in research, conservation, and knowledge acquisition and transmission. They also participate in consumer pressure groups and social movements raising awareness on the need for food system transformation and demanding climate change action. Through all these contributions, young people support achievement of the Sustainable Development Goal targets such as those on food security, economic growth, poverty reduction, and environmental sustainability (HLPE 2021; FAO and AUC 2022).

Africa's delayed demographic transition means that the continent will soon become the main source of growth for the world's workforce. By 2100, Africa's youth will be equivalent to twice Europe's entire population and almost one-half of the world's youth (Rocca and Schultes 2020). However, youth employment stands at the crossroads of economic and demographic constraints (Mueller and Thurlow 2019). Despite rapid urbanization in recent years, urban employment growth has been concentrated in the relatively low-value services sector, rather than in higher-value manufacturing jobs (Lukalo and Kiminyei 2019). Meanwhile, the burgeoning youth population is located primarily in rural areas, which means that job creation should occur in rural areas to generate livelihoods for Kenya's youth and reduce the pressures on increasingly crowded urban areas.

1 Article 260 of Kenya's Constitution defines a youth as a person between ages 18 and 34 .

In particular, off-farm jobs that add value to agricultural produce (such as processing) are an opportunity to kick-start manufacturing, create jobs in rural areas, and spur structural transformation, as suggested in the Economic Transformation Agenda 2022–2027 of the current Kenyan administration. Further, with an estimated 20 percent absorption into the formal job sector, the informal sector (which includes many agricultural jobs) represents an immediate opportunity to catalyze economic growth and employment for some of the 20 million young people joining the labor market in Africa each year (Mueller and Thurlow 2019).

In Kenya, agriculture employs more than half of the workforce and, crucially, supports the livelihoods of 70 percent of rural households. The sector represents one of the most important platforms for employment, income generation, and poverty reduction in East Africa's largest economy (Afande, Maina, and Maina 2015; World Bank 2019). As such, Kenya's food system could present a large, untapped reservoir for employment and an opportunity to curb the runaway rate of youth unemployment, which stands at 16.9 percent (KNBS 2021).

However, despite youth making up 35 percent of the population, youth engagement and employment in agriculture has been disappointingly low: the sector directly engages less than 10 percent of the youth labor force (Ministry of Agriculture, Livestock and Fisheries 2018). Even worse, the elimination of agriculture as an exam subject in primary schools and the reduction of agricultural classes for secondary school students have exacerbated the perception that the sector does not provide intellectually stimulating, economically rewarding, and meaningful careers. The carryover effects can be felt at the tertiary level: a dramatic decline has been logged in the number of students enrolling in agricultural courses, from 24,221 students in 2017 to 18,165 in 2018—a 25 percent drop (KNBS 2018). This neglect of the sector is influencing youth cohorts to avoid it, taking away the opportunity to use youth's unbridled dynamism, energy, innovation, and drive to stimulate agricultural transformation and revitalize Kenya's agrarian systems. By strengthening the appeal of agriculture and food systems among Kenya's youth, the nation could harness the abundant returns from their effective participation and secure the future of its agriculture (Shujaaz 2021).

Nevertheless, even with the obstacles mentioned above, a large number of young people are actively engaged in Kenya's food value chains, as producers and traders of food, as workers, innovators, and entrepreneurs, and as policy actors. Unfortunately, despite their large numbers, their economic, social, and political contribution to food systems is frequently underestimated because of a lack of disaggregated data (FAO 2019). While some sources indicate that the mean age of farmers in Kenya is 60 years, this perception has yet to be empirically verified (Birch 2018; Yeboah and Jayne 2020). Moreover, it is statistically improbable as only 2.5 percent of the Kenyan population is 65 years and above, and less than half of this age group is economically active or even engaged in farming (Yeboah and Jayne 2020; World Bank 2022). In fact, emerging evidence suggests that, in sub-Saharan Africa, the average age of household heads who farm is 49, and when all individuals who spend some time on their own/family farm are considered, this average age falls to 32 (IFAD 2019). Within these figures, it is estimated that rural youth spend 50 percent or more of their time in farming.

Given this knowledge, farming still represents a significant portion of the jobs held by rural youth, albeit in the informal sector. In 2020, out of 17.4 million individuals employed in Kenya, the informal sector accounted for roughly 14.5 million. Youth made up over 63 percent of such employees, with nearly equal proportions of men and women (Faria 2022; Statista nd). The informality of youth employment in food systems often relegates them to participating as casual laborers on farms; as providers of unpaid labor on family farms; and work in the supply of inputs and other raw materials; in food processing and transportation, including as *boda boda* drivers (motorcycle taxis); in storage as warehouse workers; in preparing and serving meals; in selling food produce and products in formal and informal markets as butchers or grocers, or at kiosks; in disposing of food leftovers; in refuse handling; and in recycling. In all these jobs, education, age, and gender play an important role in determining their level of engagement and employment (Samuel Hall and DFID 2017; Glover and Sumberg 2020). For example, gender stereotypes and cultural perceptions mean that food handling is seen as the role of women; thus, out of necessity, young women make up most of the informal workers in food production, harvesting, processing, and selling of fresh vegetables and fruits, commonly referred to as *mama mboga* (Samuel Hall and DFID 2017).

These young people already working in the agriculture sector are underserved, underfunded, and under-engaged. Their meaningful involvement in the sector is crippled by lack of access to productive resources and assets, including land, knowledge and information, stable and structured markets, credit, supporting institutions, and representation in farmer organizations and unions (YEDF 2020). For the few in farmer-based organizations, including youth-led community organizations, these networks are fragmented, work in silos with no connection to each other, and are largely ineffective because of leadership and capacity gaps, limited access to funding, and inadequate coordination and access to support networks. As such, youth engaged in the agriculture sector do not have a powerful common voice to articulate their aspirations, challenges, and issues and, as a consequence, their representation during policy formulation and engagement processes is often tokenized and not representative of their participation in Kenya's food systems (Afande, Maina, and Maina 2015; Muiderman 2016; Birch 2018; FAO and AUC 2022).

These combined challenges are exacerbated by the dynamics of a changing climate and the impacts of this on food systems. This is especially important as evidence pits the success and sustainability of efforts to create jobs for young people in the agriculture sector as strongly dependent on the future climate. Moreover, rural youth bear significant risks from climate change. When measuring vulnerability to climate change through the dimensions of exposure, sensitivity, and adaptive capacity, rural youth are found to be especially disadvantaged (IFAD 2019). The negative effects of climate change that cause income and livelihood losses serve to reinforce youth's perception of agriculture as a sector unable to provide decent, meaningful jobs and resilient livelihoods.

Again, though, despite the extensive economic, social, and cultural barriers that have essentially dis-incentivized this demographic from seeking opportunities in the agriculture sector, youth are still engaged at different levels within food value chains. There are numerous examples of young people developing innovative and unconventional ideas, approaches, and initiatives that are transforming how the continent produces, adds value, stores, sells, and consumes food (Generation Africa 2019). Some of these youth pioneers reveal an emerging distinct approach to youth participation in agriculture. Understanding their nuanced attitudes, perceptions, and aspirations provides useful pointers on youth-transformative pathways for young men and women's effective engagement and productive employment in food systems. Additionally, knowledge of the current state of food systems and their relationships with young people is required in order to be able to seize opportunities for and better understand youth's various roles in food system transformation.

Using a food systems lens, this chapter highlights the critical roles youth have played and continue to play in food system transformation. It also demonstrates the diversity of youth engagement in Kenyan food systems, and their aspirations and drivers for participation, as well as the enabling policy environment, institutional frameworks, and support ecosystems. Finally, given the complexities of youth engagement in food systems, the chapter highlights the constraints limiting their participation and provides recommendations for improving youth participation in Kenya's food systems.

The changing narrative of youth engagement and employment in food systems in Kenya

As youth engaged in food systems are not homogenous, it is crucial to understand their perspectives, identities, and practices to enable policymakers and other stakeholders to formulate policies, interventions, and programs that meet their needs. Understanding their different points of view provides evidence for youth-adapted and responsive solutions that address the changing context and the nature of their engagement in food value chains.

To map the engagement of Kenya's young men and women in food systems—including who, in what form, and where—Mercy Corps (2019) attempts to classify young men and women along four youth "personas" based on demographic, behavioral, and attitudinal criteria. Consequently, youth are categorized as determined builders, opportunistic movers, static planners, or rootless climbers.

To start with, determined builders reveal high levels of digital capability and experimentation with new ideas and technologies. They have an aspiration for larger-scale farming and agribusiness with bigger returns and make intentional and progressive steps toward agribusiness growth and accumulation of assets. They wish to attain the status of community focal point on agribusiness matters and use ICT tools including social media to publicize their agribusiness activities. This means they are most likely to participate in policy processes and in the representation of youth interests in different forums. Their engagement in food systems is motivated by self-reliance and the freedom of being self-employed.

Opportunistic movers have many characteristics similar to determined builders but are differentiated from them based on their greater appetite for risk. Opportunistic movers engage in high-risk agribusiness ventures and seek opportunistic and quick wins from low-period investments. They have a good appreciation for agricultural technologies and use ICTs to build their information and knowledge. Occasionally, their high-risk ventures fail but, based on their high-risk tolerance, they will pivot to other agribusiness ventures to build back to generate new on-farm income streams.

Static planners encounter barriers to effective engagement, shaped by gender inequities, limited digital capabilities, and lack of financial independence to own productive assets. As such, they are reliant on their spouses and their engagement in the agriculture sector is a means to meet their family's dietary needs and to aid them in attaining their ambitions of financial independence. Although they are mostly risk-averse and do not want to put their productive assets at risk, they use village savings and loans associations and *chamas* (informal social groups) to save their limited funds and to create a social safety net for themselves and their families. As they have a limited financial arsenal to expand their farming enterprises, they tend to be conventional in their agricultural practices.

Rootless climbers, on the other hand, have low levels of education and very limited exposure to different value chains, and as a result adopt conventional agricultural practices. They often see limited employment prospects in agriculture and partake in other off-farm activities to supplement their income. Their ambition in engaging in agriculture is to obtain a means to venture into other income-generating activities. Often, they struggle to attain their growth ambition in agriculture as a result of limited access to finance.

Yet, within these four personas, there is a changing narrative of youth aspirations that reveals new and emerging approaches to youth engagement in food systems in Kenya. Understanding these alternative psychological models and distinct approaches will be crucial in mapping a route that builds youth agency into optimally contributing to the ambitious transformation of Kenya's food system.

These distinct approaches are described below.

Agriculture vs. agribusiness: The lure of agricultural entrepreneurship

Driven by a business mindset and an entrepreneurial spirit, there is an emerging preference among Kenyan youth to engage in agricultural entrepreneurship ("agripreneurship") rather than conventional agriculture. While agriculture typically evokes the image of an old man standing under the scorching sun next to a grass-thatched hut, holding a garden hoe, agripreneurship is viewed as an aspirational undertaking that enables the formation, growth, and scaling of livelihood-sustaining enterprises and the exploration of business opportunities along the food value chain.

The Ministry of Agriculture, Livestock, Fisheries, and Irrigation defined agripreneurship as "the application of entrepreneurial principles to identify, develop, and manage viable agricultural enterprises for profit and improved livelihoods" (Kenya, Ministry of Agriculture, Livestock, Fisheries and Irrigation 2018).

In fact, many young people's prevailing perceptions of agriculture are that it is the preserve of a older generation, which is menial and backbreaking and with little opportunity for the application of technology, and a means of transfer of generational poverty. Meanwhile, agripreneurship feeds youth's aspirations of becoming business owners, sometimes in a tech-enabled environment and following an avenue that has the inherent potential for them to apply their talents, imagination, and business acumen as innovators, disruptors, and entrepreneurs. In this, they prefer shorter-season, high-value farm enterprises such as horticulture, poultry, beekeeping, and rabbit rearing (Mercy Corps 2019).

Yet, within agripreneurship, there are various regimes of participation, based on the size and formality of the agribusiness, digital skills, risk tolerance, gender inequalities, and access to capital and support networks. The cadres include:

- Livelihood-sustaining enterprises: These are opportunity-driven, highly local businesses designed to maintain a source of income for an individual family.
- Dynamic enterprises: These are agribusinesses operating in more basic industries, deploying existing products through proven business models, seeking incremental growth.
- Niche ventures: These are agripreneurship ventures creating innovative products and services that target particular market segments, and that incorporate goals other than profit and scale.
- High-growth ventures: With disruptive business models and significant growth and scale potential, these agribusinesses are led by ambitious entrepreneurs with high-risk tolerance (Generation Africa 2019).

Most recently, one of the drivers of the emergence of agripreneurship has been the COVID-19 pandemic and the ensuing containment measures. In this, different incentives and motivations buoyed the upsurge of agri-based microbusinesses among youth. Some youth felt the wrenching effects of lost income during the period, whereas others were keen to milk the opportunity of increased time on their hands caused by movement restrictions. Others took advantage of the market gap caused by the high demand for some types of vegetables and fruits, spawned by the realization that optimal nutrition was crucial in boosting immunity against the virus (Mugo 2020). All in all, the greatest motivation was to obtain extra income, in a sector with low barriers to entry, that would allow these young farmers and their families to get by during a time of instability. In essence, working in the agribusiness economy provided a higher level of financial security for youth: evidence shows that young people who engaged in the agriculture sector in 2020 earned on average $12 per month more than their counterparts engaged in other sectors (Shujaaz 2021).

Nevertheless, these agripreneurs can thrive only where they find supportive ecosystems; as such, it will be crucial to complement the upsurge of agripreneurship with prioritized mentorship, financial access, and training and capacity building. Otherwise, the attraction of youth into agriculture, which blossomed during the COVID-19 pandemic, may be lost (Generation Africa 2019; Mugo 2020; Shujaaz 2021).

Beyond the farm: Youth clamor for opportunities along the value chain

Closely related to the lure of agripreneurship, Kenyan youth's approach to engagement in food systems demonstrates an urge to seek opportunities beyond the farm but along the agricultural value chain. While agripreneurship also entails transformation of the orientation of on-farm activities into a business, this distinct approach reveals the ambition of Kenya's young men and women to engage in other aspects of the value chain as agrodealers and extension agents, in agritech and digital services, in processing, in packaging and storage, in transportation and logistics, as trade and market facilitators, as food and produce retailers, as financial service providers, as food bloggers, and in agri-media, and so on.

This distinct approach is a move in the right direction, as the effective participation of youth in upstream and downstream nonfarm segments creates space for innovation, entrepreneurship, access to markets, and increased productivity and efficiency (International Youth Foundation 2014). Furthermore, youth engagement in value chains is estimated to have higher ripple effects in terms of creating more jobs and is predicted to have higher growth rates than the on-farm production segment (HLPE 2021). For example, in Tanzania, Nigeria, and Rwanda, the off-farm food system contributed 40, 16, and 11 percent of all new jobs, respectively (Yeboah and Thomas 2018).

A knack for digital agriculture: A new generation leveraging technology

Digital technologies are particularly impactful in revolutionizing Kenya's agriculture sector as they reduce the drudgery of agriculture and make value chains more productive, efficient, profitable, and resilient to climate change. Meanwhile, there has never been a more digitally capable generation in Kenya's history than the current youth generation. This is reflected by the fact that 90 percent of young farmers ages 18–35 already have high levels of engagement with digital technologies and are frequent users of social media (Mercy Corps 2019). These technologies can help demonstrate to youth that agriculture can be a viable and profitable business opportunity, thereby increasing the desirability of agriculture-related career paths. Yet young rural farmers indicate that they would further embrace technology if these digital agriculture services were affordable, were better designed to fit local environments, and met their needs as value chain actors (Malabo Montpellier Panel 2019; Heifer International 2021). To reap the benefits of the digitalization of agri-value chains, rural youth, young women, and other marginalized youth cohorts also need to be better served, as Kenya faces a significant digital divide, with 44 percent of the urban population having access to the internet compared with 17 percent in rural areas (World Bank 2019).

Ambition for climate smartness and resilience: Youth-led climate action

The motivation for youth to explore climate resilience in their agricultural activities is tied to the immense benefits gained and the losses averted as a result of their adoption of climate-friendly solutions. With the impacts of climate change posing multiple constraints and dictating youth's ability to make a livelihood from their engagement in agriculture, the capacity of climate-smart agriculture to sustainably increase productivity and thereby improve livelihoods is particularly attractive for them. Meanwhile, with climate change impacts disproportionally affecting youth, given their higher vulnerability compared with the rest of the population, the ability of climate-smart agriculture to enhance their resilience and adaptive capacity strengthens their resolve to adopt sustainable agriculture practices (Bullock et al. 2020; Kosciulek 2020).

In essence, these motivations build upon some key youth behavioral characteristics and exhibited attitudes. For instance, global youth are considered fundamentally more environmentally and climate-conscious, having demonstrated key awareness and understanding of climate change, including how it affects them and how it represents an acute threat to their livelihoods and an existential threat to future generations (Mungai et al. 2018). Coupled with their adaptive mindset, which is a precept for the adoption of new ideas (new climate-smart innovations and management practices), and a knack for technologies (including the use of climate-smart technologies), these attitudes can form a basis for the wide-scale adoption of climate-smart and resilient practices.

However, this ambition to adopt climate-resilient technologies, innovations, and management practices in agricultural activities is hampered by low levels of awareness of green jobs and opportunities for climate-smart agriculture in the different food value chains. Moreover, youth (especially rural youth and young women) lack the resources, including the information and capital, to adopt climate-smart agriculture (Kenya, Ministry of Agriculture, Livestock, and Fisheries 2017).

To alleviate this, investments targeting rural youth need to incorporate a climate lens while financing for climate adaptation needs to integrate and intentionally target youth (IFAD 2019).

Catalyzing meaningful participation of youth in food systems

Drivers of youth participation and engagement in food systems

Globalization, rapid urbanization, rising incomes, shifting diets, a growing population, and digitalization are causing fundamental changes in Africa's food systems (AGRA 2019). In particular, population growth and urbanization across the continent have contributed not only to changed diets and consumption patterns but also to a widespread increase in the consumption of refined or highly processed foods. This has led to the emergence of food processing and the creation of more employment and growth opportunities for young people in agribusiness, farm services, value addition, retail, and food services. Similarly, rapid urbanization has altered the supply of and demand for food and contributed to the expansion of the food industry and the globalization of food trade. Higher food demands and nutritional and employment needs create an opportunity for youth employment in developing countries with their burgeoning youth populations (AGRA 2017; Abdelradi et al. 2021; Babu et al. 2021). With the increasing policy focus on youth roles in food system transformation, improving skills, knowledge, and capacities is becoming an increasingly significant driver of participation.

Youth interact within numerous activities and functions of a food system in a variety of ways. Therefore, it may be challenging to define their individual participation and motivation for engagement in terms of only a few factors (Glover and Sumberg 2020). However, the numerous competing definitions of youth based on their social, cultural, political, and economic environments may have an impact on their food environment, interactions, habits, and diets. Thus, the factors that influence youth participation and engagement in food systems are frequently interconnected and reinforce one another, while intersecting with other youth characteristics such as gender, ethnicity, education, and class. In addition, they may intersect with larger systemic and structural dynamics, including those that privilege particular types of knowledge. These disparities often lead to disadvantages, especially for marginalized youth cohorts such as rural youth and young women, in engagement in food systems.

Key challenges curtailing meaningful youth participation

Youth involvement in food systems is unique, yet in many respects their issues are similar to those facing non-youth, particularly because food production is inherently risky as a result of its sensitivity and susceptibility to external shocks. The monotony of on-farm tasks, variable input prices, delayed returns, and production inefficiencies, among other factors, contribute to youth apathy and low participation in food value chains (Kising'u 2016; Muthomi 2017). Although food systems are a viable livelihood option for many young people, this group is still disadvantaged by lack of access to land, education, training, information, financial services (credit and loans), markets, and participation in policy dialogue and decision-making processes.

According to the literature, sociocultural norms and practices strongly established in customary land tenure systems heavily discourage youth access to land. Practices such as the transfer of land to male descendants only when they marry or when their fathers die means that many young men will have to wait a long time to obtain their own land rights, while subsequent male generations will inherit smaller and less valued parcels of farmland (AGRA 2015). For young women, it is even more difficult to obtain land. In some cultures, they are not allowed to inherit land or still have to rely on their male relatives to obtain land, despite Kenyan legislation that grants women equal rights to land and prohibiting gender discrimination in land law, customs, and practices. Even though land laws protect women's inheritance, succession, and matrimonial property, women still hold only about 5 percent of all land deeds (AGRA 2015). As a result, they have limited control and decision-making authority over the land they use for agricultural and food production. This also prevents them from accessing financial services that require land as collateral.

Arguably, social norms and practices can shape and affect the roles of young men and women, as well as their access to and control over resources that drive output, income, and participation. With changing discourses on youth engagement in food systems and the emergence of new opportunities in agriculture and food systems, youth employment strategies must consider gender dynamics, particularly in relation to young women (Njeru and Mwangi 2017), in order to provide them with the tools they need to navigate an already complex agriculture and food system environment. Young women still lack basic market skills, which prevents them from accessing productive resources, economic possibilities, and participation in decision-making processes, among other things (AGRA 2015). As a result, young women find it particularly challenging to rely on food production and related activities as a source of income. Lack of access to financial resources locks them out from buying or leasing, and many therefore end up migrating to urban areas seeking alternative sources of income. Low levels of literacy and lack of knowledge of their land rights impede land tenure security (ibid.).

Broader education and training are critical to develop highly trained youth capable of driving agricultural growth and food system transformation (Amwata 2020). However, young people lack the core entrepreneurial and financial literacy skills required for employment and entrepreneurship and have limited access to technical and vocational education and training. Current education and training is frequently not adapted to meet changing labor market demands, leaving youth without the knowledge and skills they need to succeed in the workforce and provide new solutions to a struggling food system. (Afande, Maina, and Maina 2015; Kising'u 2016). This translates to insufficient knowledge, skills, and capacities to tap into ready markets, and therefore low revenues and incomes.

Beyond technical capacities, access to agricultural finance and insurance to enable youth to enhance their productivity is still limited. Young women face additional constraints owing to non-economic roles and responsibilities that limit their action and time. Conversely, broader analysis of the food sector market suggests that having access to markets could be more sustainable and critical than just having access to capital (Sudarkasa 2019). Poor access to markets has been linked with poor participation and exclusion of youth in food systems owing to poor profitability of products. For this reason, more and more stakeholders are enhancing youth participation in food systems by focusing on a market systems approach through agriculture sector incubation and acceleration programs.

Many challenges limiting youth participation in food systems are intertwined and mutually reinforcing, including individual conditions and vulnerabilities like physical and mental health, poverty, and disability.

Institutional and policy frameworks

EXPLORING THE POLICY ENVIRONMENT AND INSTITUTIONAL FRAMEWORKS THAT SUPPORT EFFECTIVE YOUTH PARTICIPATION IN KENYA'S FOOD SYSTEM

At the level of global goals, Kenya has committed to reducing the proportion of youth not in employment, education, or training by 2030 under Sustainable Development Goal 8 ("Promote sustained, inclusive, and sustainable economic growth, full and productive employment, and decent work for all"). Kenya has also committed to enhancing impact and expanding action to address the needs, build the agency, and advance the rights of young people through the United Nations Youth Strategy. At the regional level, the African Union's 2014 Malabo Declaration on Accelerated Agricultural Growth and Transformation for Shared Prosperity and Improved Livelihoods recognizes youth as key players in halving poverty by 2025 and commits Kenya to creating jobs in agricultural value chains for at least 30 percent of youth. At the national level, the Kenyan Constitution commits to take whatever measures necessary, including affirmative action programs, to ensure youth have access to various resources, including employment opportunities, and participate in development processes. Kenya's Vision 2030, the Big Four Agenda, and other development strategies and programs identify youth as critical to achieving the country's targeted annual economic growth of 10 percent. Correspondingly, they emphasize gender equality, women, and youth in the design and implementation of various interventions aimed at addressing emerging social, cultural, economic, and political issues affecting youth participation in development initiatives. Such interventions include the establishment of funds (for example, Uwezo Fund,[2] the Youth Enterprise Development Fund,[3] and the Women Enterprise Fund[4]) all of which aim to improve financial access and expand business and enterprise opportunities for youth.

Specific youth policies, plans, strategies, and legislative and institutional mechanisms formulated since independence include the National Youth Service Act 1964, National Youth Policy 2007, National Youth Council Act 2009, Kenya National Development Youth Policy 2018, and Kenya Youth Agribusiness Strategy 2018–2022. The last establishes a coordinated national response to obstacles to the effective participation of youth and provides new opportunities for youth in agriculture and its value chains. Its framework will allow for the more coordinated implementation of existing and future initiatives and unique opportunities for youth, under a long-term vision to achieve sustainable growth of the agricultural economy. However, its implementation mandate ends in 2022, which means it is crucial to formulate a new youth strategy that takes into consideration the new electoral and governance cycle and new evidence and data as well as changing youth aspirations and emerging approaches to their engagement in food systems.

Similarly, policy and legal frameworks for agricultural investment (the Agricultural Sector Transformation and Growth Strategy 2019–2029, Kenya Climate Smart Agriculture Strategy 2017–2026, and Kenya Climate Smart

2 www.uwezo.go.ke/

3 www.youthfund.go.ke/

4 https://vision2030.go.ke/project/women-enterprise-fund/

Implementation Framework 2018–2027) cover a broad range of areas and address youth as a cross-cutting issue. Each of them integrates, aligns, or links to either or all four of the components that enable young people to participate in food systems: youth integration, engagement, participation, and employment. Additional related policies, plans, and strategies include the Kenya Agri-Nutrition Strategy 2020–2024, Youth Enterprise Development Fund Strategy 2020–2024, Women Enterprise Development Fund 2013–2017, National Policy on Gender and Development 2019, National Employment Authority Act 2016, Medium and Small Enterprise Act 2012, Technical and Vocational Education and Training Act 2013, and National Government Affirmative Action Fund (Public Finance Management Act 2012). Policy coordination and implementation are overseen by the ministry in charge of youth affairs in collaboration with relevant government ministries, departments, and agencies; the private sector; youth networks, platforms, and institutions; and all other stakeholders (national, international) involved in youth affairs.

ANALYZING THE EFFICACY OF POLICY AND INSTITUTIONAL FRAMEWORKS AND IDENTIFYING GAPS

Youth look to policymakers and institutions, both public and private, to fulfill their aspirations for decent, stable jobs as well as the opportunity to meaningfully participate in and contribute to development initiatives. Yet, as a variety of the aforementioned policy frameworks and initiatives demonstrate, youth are often considered in a generalized and oversimplified way because of an absence of evidence on their heterogeneity. This results in broad, multisectoral national youth development policies and programs that serve as the foundation for inequitable and poorly coordinated distribution of resources and public investments. Worse still, the portrayal of youth as both objects and subjects of development, as innovators eager to develop new technologies or as development actors incapable of identifying and comprehending their own needs (Glover and Sumberg 2020) creates shifting youth narratives that contribute to youth exclusion from policymaking processes. As a result, their involvement in development activities becomes limited or passive (FAO 2014). Regrettably, rural and impoverished youth bear an outsized burden, as the majority of public consultations tend to take place in urban areas. Young women are also often excluded because of their non-economic activities and responsibilities, which restrict their mobility and time to participate in such policy processes (Huyer et al. 2021). This results in a lack of specific incentives to empower rural youth and women agripreneurs operating along the value chain, which limits the overall effect.

Not all the policies developed or adopted are implemented fully. Some of them, like the Kenya Youth Agribusiness Strategy, are on target but fall short of the mark, owing to the paucity of research and evidence to support the proposed strategies for youth integration or involvement in agriculture value chains. Others are overconfident in their ability to achieve immediate results in a short period. Lastly, several of the policies lack the institutional capacity and accountability, impact evaluations, resources, and coherence to achieve their stated objectives. Aside from these limitations, policy implementation is also affected by shifting enabling environments, such as evolving global policy debates and political factors.

Synthesis and outlook: Recommendations for improving youth participation in food systems

Enhance structures of support for youth in agribusiness

With the knowledge that young farmers and agripreneurs can thrive where they find supportive ecosystems, it is crucial that policymakers and other stakeholders enhance the ecosystem of support for Kenyan youth engaged in food systems. As the Kenya Youth Agribusiness Strategy identifies, it will be crucial to extend a lending hand to equip these youths with appropriate agribusiness skills, knowledge, and information and to enhance their access to affordable and youth-friendly growth capital that can enable them to scale promising agribusiness ventures.

In unlocking capital for youth-led farming and agripreneurship ventures, it will be especially necessary to develop and implement de-risking measures and the use of catalytic capital from public and philanthropic sources to mobilize private sector finance for onward lending to youth agripreneurs. This is critical since, across East Africa, risk in agribusiness lending is at least twice as high and the returns 4–5 percent lower than in lending to other sectors (ACELI Africa 2020). With this, expanding the scale and improving the efficiency of affirmative funds such as the Youth Fund and the Women Enterprise Fund is advised. Moreover, the development of easily accessible and affordable credit schemes such as the Hustler Fund[5] will provide additional opportunities for financial inclusion and resourcing for youth engaged in agriculture.

5 The Hustler Fund is a financial inclusion fund set up by the Government of Kenya in 2023 to address unemployment and the lack of opportunities for low-income earners through affordable credit.

Other structures of support that can be extended include innovation support, business development services, market and distribution links, mentorship, and connection to peer networks (Generation Africa 2019). With a critical mass of skilled young Kenyans with access to finance and the knowhow to drive productivity growth in farming and related value chains, the country can harness the power of its youth in catalyzing agricultural transformation (Yeboah and Thomas 2018).

Operationalize school-based agricultural education: 4K clubs

The relaunch of the 4K[6] clubs provides an opportune momentum to scale up school-based agricultural education. With the aim of inculcating a positive mindset regarding agriculture among young students and nurturing, preparing, and building future leaders of the agriculture sector, the 4K clubs use a graduated approach to cultivate the next generation of leaders who are passionate and invested in transforming food systems (Mcknight 2021). Prioritizing and investing in the 4K clubs program will support its integration, adoption, and operationalization within different contexts to the benefit of school-going youth and the country at large.

Build youth-led professional networks: Facilitate the formation of a national youth-in-agriculture association

As documented, youth-led and youth-serving associations and networks connect with young farmers and agripreneurs better than many other stakeholders, organizations, or institutions can. With the ability to build youth agency through learning by doing, a national youth in agriculture association would have the power to provide young farmers with a more powerful voice, to break silos and foster coordination of youth action in the sector, to facilitate inclusive representation of youth interests and perspectives in policymaking processes, and to attract additional public and private sector funding for significant investments in youth development and engagement in the sector.

Improve the evidence base for youth engagement in food systems

Despite extensive evidence that supports the idea of direct farm employment as a critical form of youth engagement in food systems, there is limited quantitative and qualitative data on young people's participation in food systems at various levels (for example, rural or urban) and stages (including production, processing,

---

6 4K stands for Kuungana, Kufanya, Kusaidia Kenya, which is loosely translated as "Coming together, to act, in order to help Kenya."

retailing, and consumption), or their needs and issues in engaging in food systems and the barriers to their participation. This knowledge gap makes it difficult to determine the scope of young people's involvement in food systems and to guide policies to increase youth involvement. Additional evidence on youth participation in food systems is needed in four key areas to strengthen youth engagement in agriculture and food systems:

- Labor monitoring and statistics on the employment and wage patterns of young people
- Various forms of youth participation in food systems
- Youth-targeted policies for food system education, engagement, and employment
- Share of agriculture and food systems related funds targeted for youth engagement

As a result, investment in research and development activities will inform new types of engagement that take into account the diversity of the youth demographic. This will make it possible to transcend youth stereotypes and establish equitable policies based on empirically supported methods that accommodate their specific constraints and preferences (Glover and Sumberg 2020).

Strengthen youth-led policymaking

Policymakers must involve youth in policymaking processes in order to align youth skills and interests with the demand trends of the food system. They also need to distinguish between long- and short-term approaches as well as between demand- and supply-side solutions, each tailored to the local contexts and stakeholders of youth's food systems. This will enable opportunities for greater youth inclusion and participation. In the short term, it will be crucial to develop a new youth-in-agribusiness strategy to update and scale up development priorities for youth engagement and employment in food systems in Kenya.

Additionally, interventions should be promoted in tandem with existing supportive programs and institutional and legislative frameworks to provide organized support and boost youth engagement in food systems. For example, Kenya's Vision 2030 plans to disburse KSh 2.27 billion by 2022 through the Youth Enterprise Development Fund and the Women Enterprise Fund to improve work possibilities for youth through entrepreneurship (YEDF 2020). While this is not all channeled to agriculture-related activities, it presents an opportunity to enhance youth engagement in different parts of the food system.

Finally, at all levels of development, remove geographical, economic, social, legal, policy, and programmatic barriers to youth engagement in decision-making, planning, and implementation. This is currently demonstrated through the Access to Government Procurement Opportunities program, which aims to make it easier for youth, women, and people with disabilities to engage in government procurement by ensuring that 30 percent of government procurement opportunities are set aside particularly for these businesses. Other strategies could include tax reductions and levies for youth-owned small and medium enterprises to give them a chance for growth and survival.

References

Abdelradi, F., A. Admassie, J.A. Adjaye, M. Ayieko, O. Badiane, K. Glatzel, S. Hendriks, et al. 2021. "Policy Options for Food Systems Transformation in Africa." UN Food Systems Summit Brief. Bonn: United Nations Food System Summit.

Aceli Africa. 2020. *Bridging the Financing Gap: Unlocking the Impact of Agriculture SMEs in Africa.* Washington, DC.

Afande, F.O., W.N. Maina, and P.M. Maina. 2015. "Youth Engagement in Agriculture in Kenya: Challenges and Prospects." *Journal of Culture, Society and Development* 7: 1–17.

AGRA (Alliance for a Green Revolution in Africa). 2015. *Africa Agriculture Status Report: Youth in Agriculture in Sub-Saharan Africa.* Nairobi.

AGRA. 2017. *Africa Agriculture Status Report: The Business of Smallholder Agriculture in Sub-Saharan Africa.* Nairobi.

AGRA. 2019. *Africa Agriculture Status Report: The Hidden Middle: A Quiet Revolution in the Private Sector Driving Agricultural Transformation.* Nairobi.

Amwata, D. 2020. "The Role of the Agricultural Technical Vocational Education and Training Institutions (ATVET)." Policy Brief. Nairobi: East Africa Institute.

Babu, S.C., S. Franzel, K.E. Davis, and N. Srivastava. 2021. "Drivers of Youth Engagement in Agriculture: Insights from Guatemala, Niger, Nigeria, Rwanda, and Uganda." IFPRI Discussion Paper 02010. International Food Policy Research Institute (IFPRI), Washington, DC.

Birch, I. 2018. *Agricultural Productivity in Kenya: Barriers and Opportunities.* K4D Helpdesk Report. Brighton, UK: Institute for Development Studies.

Bullock, R., S. Huyer, T. Shai, and M. Nyasimi. 2020. *The CCAFS Youth and Climate-Smart Agriculture (CSA) Strategy.* Wageningen, Netherlands: CGIAR Research Program on Climate Change, Agriculture and Food Security (CCAFS).

FAO (Food and Agriculture Organization of the United Nations), CTA (Technical Centre for Agricultural and Rural Cooperation) and IFAD (International Fund for Agricultural Development). 2014. *Youth and Agriculture: Key Challenges and Concrete Solutions.* Rome: FAO.

FAO. 2019. R*ural Youth Employment and Agri-food Systems in Kenya: A Rapid Context Analysis.* Rome.

FAO and AUC (African Union Commission). 2022. *Investment Guidelines for Youth in Agrifood Systems in Africa.* Rome: FAO.

Faria, J. "Distribution of Informal Sector Employees in Kenya in 2019, by Gender and Age Group." Statista. Accessed February 22, 2022.

Generation Africa. 2019. *A Landscape Study: Youth Enterprise in AgriFood Systems.* Nairobi.

Glover, D., and J. Sumberg. 2020. "Youth and Food Systems Transformation." *Frontiers in Sustainable Food Systems* 4: 101.

Heifer International. 2021. *The Future of Africa's Agriculture: An Assessment of the Role of Youth and Technology*. Little Rock, AR.

HLPE (High Level Panel of Experts). 2021. *Promoting Youth Engagement and Employment in Agriculture & Food Systems*. Rome: FAO.

Huyer, S., E. Simelton, N. Chanana, A. Mulema, and M. Edwige. 2021. "Expanding Opportunities: A Framework for Gender and Socially-Inclusive Climate Resilient Agriculture." *Frontiers in Climate* 3: 718240.

IFAD. 2019. *Rural Development Report*. Rome.

International Youth Foundation. 2014. *Promoting Agricultural Enterpreneurship Among Rural Youth*. Best Practice Note No. 2. Baltimore, MD.

Kising'u, J.M. 2016. *Factors Influencing Youth Participation in Agricultural Value Chain Projects in Kenya: A Case of Kathiani Subcounty, Machakos County, Kenya*. Master's thesis, University of Nairobi, Nairobi.

Kenya, Ministry of Agriculture, Livestock, and Fisheries. 2017. *Kenya Climate Smart Agriculture Strategy 2017-2026*. Nairobi.

Kenya, Ministry of Agriculture, Livestock, Fisheries and Irrigation. 2018. *Kenya Youth in Agribusiness Strategy 2018–2022: Positioning the Youth at the Forefront of Agricultural Growth and Transformation*. Nairobi.

KNBS (Kenya National Bureau of Statistics). 2018. Economic Survey by the Kenya National Bureau of Statistics. Nairobi.

KNBS. 2021. *Quarterly Labour Force Report*. Nairobi.

Kosciule, D. 2020. *Strengthening Youth Participation in Climate-Related Policymaking*. Policy Briefing 225. Johannesburg: South African Institute of International Affairs.

Lukalo, D., and F. Kiminyei. 2019. *Promoting Structural Transformation for High Productivity Jobs in Kenya*. KIPPRA Policy Brief No. 53. Nairobi: The Kenya Institute for Public Policy Research and Analysis (KIPPRA).

Malabo Montpellier Panel. 2019. *Byte by Byte: Policy Innovation for Transforming Africa's Food System with Digital Technologies*. Dakar: IFPRI, Center for Development Research University of Bonn (ZEF), and Imperial College London.

McKnight, T. 2021. "Empower Youth, Transform Agriculture: An Introductory Guide to School-Based Agricultural Education in Sub-Saharan Africa." Washington, DC: Movement for School-Based Agricultural Education and Agricorps.

MercyCorps and Mastercard Foundation. 2019. *AFA Case Study:Digital Pathways for Youth in Agriculture*. Toronto: Mastercard Foundation.

Mueller, V., and J. Thurlow. 2019. *Youth and Jobs in Rural Africa: Beyond Stylized Facts.* Washington, DC: IFPRI.

Mugo , V. 2020. "Hope or Hype: The Rise of Youth Engagement in Agriculture During COVID-19." *Future Africa Forum,* October 21.

Muiderman, K. 2016. *Engaging Youth in Food Systems.* Utretcht, Netherlands: Netherlands Food Partnership.

Mungai, C., T. Muchaba, L. Szilagyi, M. Radeny, V. Atakos, and D. Ntiokam. 2018. *Youth Engagement in Climate-Smart Agriculture in Africa; Challenges and Opportunities.* Wageningen: CCAFS.

Muthomi, E. 2017. *Challenges and Opportunities for Youth Engaged in Agribusiness in Kenya.* Nairobi: United States International University.

Njeru, L.K., and J.G. Mwangi. 2017. "Influence of Gender Differences on Youth Participation in Agriculture in Kajiado Borth Sub County, Kenya." *International Journal of Development and Sustainability* 6 (8): 851–861.

Rocca , C., and I. Schultes. 2020. *Africa's Youth: Action Needed Now to Support the Continent's Greatest Asset.* London: Mo Ibrahim Foundation.

Samuel Hall and DFID (United Kingdom Department for International Development). 2017. *Youth Employment in Kenya.* London: Samuel Hall and DFID.

Shujaaz, Inc. "Compass: How Young People Can Lead Kenya's Agribusiness Transformation." *Shujaaz, Inc. News & Publications,* May 26.

Statista. undated. "Statista: Empowering People with Data." Accessed March 2022. https://www.statista.com/

Sudarkasa, M. 2019. "Why Expanding Value Chain Access Is Important for Young Agripreneurs." CTA Blog, February 19.

World Bank. 2019. "Kenya Economic Update: Securing Future Growth." Kenya Economic Update No. 20. Washington, DC: World Bank.

World Bank. 2022. "Population Ages 65 and Above (% of Total Population) - Kenya." World Bank data. Accessed April 19, 2022.

Yeboah, F.K., and J.S. Thomas. 2018. "Africa's Evolving Employment Trends." *Journal of Development Studies* 54 (5): 803–832.

Yeboah, F., and T. Jayne. 2020. "Is African Agriculture Becoming the Preserve of the Elderly?" International Development Working Paper. Department of Agricultural, Food and Resource Economics, Michigan State University, East Lansing, Michigan.

YEDF (Youth Enterprise Development Fund). 2020. *Youth Enterprise Development Fund Strategic Plan: 2020/21–2023/24.* Nairobi.

Chapter 15

FRESH PRODUCE VALUE CHAINS IN KENYA: CHALLENGES AND PROSPECTS FOR ENHANCED MARKET ACCESS AND SMALLHOLDER INCLUSION

Mulubrhan Amare, Bekele Shiferaw, Dolapo Adeyanju, Kwaw Andam, and Jane Mariara

Production and marketing of fresh produce provides opportunities for smallholder farmers in low- and middle-income economies to engage and benefit from high-value markets and value chains. Fresh produce markets can be valuable for transforming the food sector in many low-income countries but perishability and difficulties in organizing supply chain logistics mean access remains challenging for smallholders.

Contract farming is one potential mechanism that smallholder farmers in developing countries can use to participate in and benefit from domestic and global value chains (Okello and Swinton 2007; Barrett et al. 2012; Minot and Sawyer 2016; Ruben 2017; Ton et al. 2017). Linking smallholder farmers more directly with national and global consumers should increase both the demand and producer prices for their fresh produce. Increased access to and participation in such value chains increases farm income earned by smallholders. Improvements in inclusion and efficiency of value chains are vital to enhance the effectiveness of contracting models, and to enhance the market access and integration of smallholders. Inclusion is important because large buyers, including processors, modern retailers, and exporters, are often hesitant to engage with small and marginal farmers and may prefer working through brokers, which reduces benefits to farmers. Also, market integration is becoming increasingly important for smallholders in order to avoid marginalization of the less organized sector in the more organized, growing global fresh produce market (Ruben 2017; Ton et al. 2017).

Different types of out-grower schemes have been implemented in Kenya to connect smallholder farmers with high-value markets for fresh produce. Smallholder farmers are typically contracted as out-growers through farmer groups, often organized under common interest groups or cooperatives. A literature survey by Minot and Sawyer (2016) seems to support the idea that farmers benefit more from contract farming, concluding that "successful contract farming schemes generally raise the incomes of farmers." However, neither participation in such high-value markets nor beneficial gains are guaranteed, particularly for smallholder producers in low-income countries (Minten, Randrianarison, and Swinnen 2009). This is because smallholder producers may not be able to meet the stringent standards to participate effectively in such markets, or large agribusiness firms may exploit smallholder farmers, since unequal bargaining power can result in unfair benefit-sharing outcomes (Little and Watts 1994; Warning and Key 2002).

Contract farming can also contribute to social differentiation, increased concentration of land ownership, and dominance of agribusiness firms in decision-making at the expense of small-scale farmers (Maertens and Swinnen 2007; Subervie and Vagneron 2013). Barrett and colleagues (2012) attribute this in part to firms' preference for larger and more experienced farmers and selective encouragement of participation of wealthier and better-educated producers. Both patterns can lead to limited participation and inclusion of smallholder producers. However, participation in cooperatives and common interest groups could help overcome some the limitations these patterns impose.

Different categories of farmers generally choose the selling mode that works best for them. For instance, very small, often part-time, sellers may be better off selling at local markets. However, many farmers prefer to sell individually by hiring casual workers to pick fruits when they are ready, with local retailers or exporters collecting the harvest from the farm. When the spot market prices are high, this also leads to side selling and breach of contracts. Also, some farmers sell directly to exporters, processors, or local retailers. Over time, though, individual selling has been discouraged because of issues relating to product quality and high bulking costs. For example, contracting in Kenya for avocado may occur through farmer groups rather than individual farmers (Mwambi et al. 2014; Amare et al. 2019), which can limit individual (smallholder) farmer's participation in high-value markets. Thus, contract farming participation has become increasingly important for smallholder fresh produce farmers as an alternative marketing strategy.

As these value chains have become increasingly important in Kenya, the avocado value chain stands out. Avocado now accounts for 17 percent of total horticultural export earnings and more than 50 percent of the export value in the fruit subsector (KEHPA 2015), thanks to strong and growing local and global demand. Increasing international competition, together with increasing demand for food safety and quality standards as well as traceability and reliability of supply, has made it almost mandatory for smallholders to organize and enter into avocado contract farming through farmers' groups and cooperatives.

This chapter documents contract farming practices and assesses challenges and opportunities in enhancing inclusion and reducing marketing costs for small-scale producers (including youth and women) to improve their access to fresh produce markets in developing countries. The chapter builds on a focused review of selected fresh produce value chains in developing countries and critically examines recent experiences from the avocado sector in Kenya—distilling relevant lessons and policy implications. As the world's third-largest exporter of avocado to Europe, Kenya is particularly suitable for a case study on this product (FAO 2017). The chapter also assesses whether participation of smallholders in modern value chains for fresh produce enhances inclusion and efficiency of contracting approaches and improves their income and livelihoods.

More specifically, the chapter addresses the following questions:

- To what extent and under what conditions do smallholder farmers participate in fresh produce value chains via contract farming?
- To what extent can contract farming be an effective mechanism to facilitate inclusion of small-scale farmers (including women and youth) in modern market channels for fresh produce?
- Does participation in the avocado value chain improve the livelihoods of participating farmers?

The chapter is structured as follows. The next section highlights the challenges to smallholder farmers' participation in value chains for high-value markets. The chapter then presents the methodology used for the case study, highlighting the data type and the analytical approach. Next comes an overview of contract farming for fresh produce in developing countries, with a subsection focused on Kenya. This is followed by the results of the Kenya avocado case study. The concluding section highlights key lessons and policy implications.

Market imperfections and challenges to smallholder farmer participation in value chains for fresh produce

The concept of agricultural value chains as a means to improve smallholders' livelihoods has attracted many scholars, policymakers, and development practitioners in recent times (Barrett et al. 2020). However, for smallholders to engage in value chains for high-value food products, they must comply with certain standards and market requirements, such as economies of scale, food quality and safety standards, and consistency of supply. Complying with such standards is quite challenging and has made it difficult for small-scale farmers to fully integrate into competitive and globalized food value chains (Eaton and Sheppard 2001).

The requirements generate intense competition for all parties in the value chain. The need to meet the demands of customers, processors, distributors, and retailers imposes ever-increasing obligations on suppliers in terms of quality, timing, handling, and other delivery arrangements. Smallholder farmers require both technical skills and financial investments to meet stringent standards, which are often out of reach for many of them. They also face resource constraints, such as a lack of irrigation water and land, which invariably limit their production and productivity. Likewise, their engagement in high-value markets is deterred by limited access to productive resources and services, including extension, credit, and market information (Coulter, Entwistle, and Gilbert 1998; Barrett et al. 2012).

The production and marketing of perishable fresh produce such as fruits and vegetables for high-value markets require good agricultural practices (GAP) and safe handling to meet phytosanitary standards. This, however, involves specialized production and packaging methods, and refrigerated transport or cold chains. Additionally, many smallholders have limited knowledge on optimal harvesting methods (for instance, for pest management, picking, and harvest). When quality cannot be standardized, smallholder farmers incur significant losses because processors, retailers, and exporters reject their produce for failing to meet private value chain standards. Inadequate exposure to or capacity to implement modern harvesting techniques and other global GAP standards limit the ability of small-scale growers to participate in and benefit from high-value fresh produce markets. Even when farmers have the required knowledge and information about export and quality requirements, they do not always adequately supervise those who pick their produce because of high supervision costs or difficulty in hiring skilled harvesters (PEP 2018).

Availability of market information is another critical factor in farmers' decision to invest in fresh produce production. When farmers are not part of organized value chains or out-grower schemes through contract farming, they rarely receive specific and adequate information in advance from exporters concerning product grading, pricing, and demand in domestic and export markets. Lack of adequate price and market information exposes producers supplying small quantities to informal and less-organized supply chains to seasonal gluts and price plunges. Other challenges include trust issues among exporters, brokers, and farmers; lack of incentives for upgrading; poor vertical and horizontal linkages within the value chain; and the imposition of strict new rules and market standards in response to rising concerns of consumers regarding food safety (Snodgrass and Sebstad 2005; USAID/Kenya 2008).

These overlapping technical, financial, and market-related constraints discourage smallholders from engaging in the production and marketing of fresh produce—limiting their ability to diversify income sources beyond traditional agricultural activities such as food staples and livestock (World Bank 2008; Amare et al. 2021). Therefore, smallholder farmers have incentive to opt for contract farming, which has potential to overcome some of the underlying market imperfections for perishable and bulky fresh produce, through access to specialized technical support, inputs, finance, and product handling, along with reliable market information that helps buffer producers against extreme price volatility. However, such contracting is available mainly in certain high-potential areas, with adequate growing conditions to meet desired production quantities to supply local or export markets, and where infrastructure is not a limiting factor in the procurement and handling of produce.

Methods

We reviewed the challenges and prospects for enhancing market access and inclusion in developing countries, drawing evidence from the avocado sector in Kenya. The research questions were addressed using both qualitative and quantitative analytical approaches. A key part of the methodology was a structured review of the contract farming literature in developing countries, looking at both the theoretical and the empirical evidence, to examine the prospects, challenges, and way forward in improving the market access and participation of smallholder producers, especially for fresh produce sectors. This broad analytical review was used to assess the evidence and draw lessons from the avocado sector in Kenya. It entailed a comparative review of contract farming models and experiences in developing countries, as well as a deep dive into the literature on the impact of contract farming on livelihood outcomes as related to smallholder fresh produce farmers.

Individual and combined searches were conducted on specific terms including contract farming, out-grower schemes, fresh produce, avocado, smallholders, participation, value chains, livelihood, horticultural production, and developing countries. Three major criteria were considered for inclusion in the review: (1) direct relevance of the articles to the participation of smallholder farmers in contract farming and high-value fresh produce markets in developing countries, (2) articles published in peer-reviewed journals, and (3) a focus on Africa. In addition to the literature review, the Kenya case study assessed avocado production and export trends using time series data from the Horticultural Crops Directorate (HCD) and Food and Agriculture Organization of the United Nations (FAO) databases. Data were analyzed using simple descriptive tools to assess and visualize trends and growth patterns in the avocado sector in Kenya, relative to other major exporters. Supplemental evidence from unpublished sources including working papers, policy briefs, and technical reports was used to fill evidence gaps and contextualize the findings.

Contract farming for fresh produce

Contracting models for fresh produce

The relevance of contract farming for overcoming market imperfections and incorporating smallholder farmers into high-value markets cannot be overemphasized. Contract farming is conceptualized as a system in which an interested buyer, often an exporter, processor, or retailer, purchases the harvests of independent producers, with the terms of the engagement predetermined through a negotiated contract. The literature on contract farming highlights the diversity of contractual arrangements between contractors and farmers (Bijman 2008; Eaton and Shepherd 2001; Simmons, Winters, and Patrick 2005). This diversity is attributed to the technical skills required for production and the associated transaction costs (Simmons, Winters, and Patrick 2005). Contract farming has been widely adopted in developing countries as a means of reducing the underlying risks for both the producers and the buyers and ensuring consistent throughput levels of known price and quality (Kirsten and Sartorius 2002).

Similarly, the use of various forms of contracts to ensure quality, coordination, and desirable quality is quickly rising in developing countries. However, the prevalence of contract farming and the models adopted differ notably across different categories of buyers, destination markets, and commodities. The variation across commodities could be attributed to differences in perishability, economies of scale in production and processing, quality sensitivity, and seasonality of supply, among other factors. Accordingly, Eaton and Shepherd (2001) highlight five broad models of contract farming: centralized, nucleus estates, intermediary, multipartite, and informal models. Even though these models involve a contract between an agribusiness and farmers, they have distinct structures and management patterns.

The **centralized model** involves large firms contracting a large number of smallholders with predetermined specifics on the quality and quantity of the products to be supplied (Bijman 2008). According to Eaton and Shepherd (2001), this model is suitable for commodities that require significant logistics and processing, such as coffee, milk, tea, and sugarcane. For example, dairy processing companies collect fresh milk of standardized quality from small-scale dairy farmers in many East African countries, including Ethiopia, Kenya, and Uganda, through long-term contractual arrangements.

The second model is the **nucleus estate model**, whereby companies not only contract independent producers but also enter the production node through an estate or plantation. The estate model is typically used to ensure throughput for the final buyer of the produce—but sometimes only for research or breeding purposes. This hybrid model is used mainly for perennial crops (such as oil palm, mango, and citrus), but there are instances where it has been applied in other crops. An example is the dairy sector in Indonesia, where the central estate is used for the rearing of "parent stock" (Eaton and Shepherd 2001).

Under the **intermediary model**, an agro-industrial firm subcontracts engagement with farmers to a third party—an intermediary—who could be a village leader or a farmer cooperative or group. This intermediary then serves as the primary point of contact for farmers (Prowse 2012). The model is common in Southeast Asia and has been applied in Kenya for avocado production (Mwambi et al. 2016). In some cases, buyers may engage in different contracting models. For example, major avocado exporting companies such as Kakuzi maintain their own estates (nucleus estate model) while also contracting avocado farmers (via intermediaries) as out-growers, as part of meeting growing demand in destination markets. An important limitation of the intermediary model is that contractors are at risk of losing control of production in terms of quality and quantity of supply as well as prices farmers receive (Eaton and Shepherd 2001). Given the lack of direct contact between farmers and the contractor, this model has several limitations with regard to vertical coordination and offering proper incentives (Bijman 2008).

The **multipartite model** may involve a joint venture between a firm and an NGO or a public entity and one or two private firms, which enter into a contractual agreement with farmers (Prowse 2012). A multipartite model can develop from a nucleus estate model, for instance through farmers forming cooperatives or the participation of a financial institution. Governments in many developing countries have actively participated in contract farming as part of the liberalization process through joint ventures with private enterprises; this helps improve trust and reduce risks related to the honoring of contracts (Little and Watts 1994). When public sector players are able to enhance the capacity of producers or mitigate the underlying risks beyond what private companies are willing or able to do, this approach has the potential to enhance the inclusion of small and marginal farmers in value chains.

Lastly, **informal models** are characterized by seasonal arbitrary production contracts with individual traders or small firms (Hung et al. 2019). Contracts are usually on a verbal basis, with a limited number of farmers. This often arises in less organized markets that do not require high standards or where markets are not differentiated based on quality. The greater risk of extra-contractual marketing involved means government support services such as extension and research are required (Eaton and Shepherd 2001).

In the case of fresh produce, contract farming is very common in vegetable and fruit production for export or sale to domestic supermarket chains, with evidence coming from Madagascar, Kenya, and Senegal (Prowse 2012; Ochieng, Veettil, and Qaim 2017; Ogutu, Ochieng, and Qaim 2020). According to Ochieng, Veettil, and Qaim (2017), in place of traditional wholesale markets, large buyers such as supermarkets, export or processing companies, and other agribusinesses serving high-value markets often source fresh and perishable products such as fruits and vegetables directly from farmers through contracts to ensure high-quality products that meet market standards. This rise in supermarket contracts is being driven by rapidly growing urban populations and demand for high-quality fresh produce. Using the case of small-scale vegetable farmers in Kenya, Ogutu, Ochieng, and Qaim (2020) found that such contractual arrangements contribute to sizeable income gains and poverty reduction among smallholders.

According to Prowse (2012), the informal models are best suited to categories of crops that require minimal processing, have limited variability in quality, and rely on standard production techniques. However, this may not be the case for other fresh produce that requires more technical capacity in terms of production and processing. An example is the horticultural sector, where formal models are believed to be more suitable. For instance, Mwambi and colleagues (2016) provide evidence on the use of the intermediary model in avocado production in Kenya. Minot (2011) suggests that informal models may be used for products sold through traditional channels that do not involve any contractual agreements, especially those that can be processed or consumed at home. Horticultural production for export, on the other hand, has specific requirements in terms of quantity, quality, production, timing, and handling methods to meet sanitary and phytosanitary standards, which are better met via a contract.

TABLE 15.1 Features of the different contracting models

Contracting model	Distinct features	Examples of value chain
Centralized	• Involves large firms and a large number of smallholders • Most suitable for commodities that require significant logistics and processing	Coffee, milk, tea, sugarcane
Nucleus estates	• Involves large firms and independent producers • Sometimes used only for research or breeding purposes • Mainly used for perennial crops	Oil palm, mango, citrus
Intermediary	• Contract between a firm and a broker, who represents the farmer • Possible risk of losing control over production in terms of quality, quantity, and prices	Avocado
Multipartite	• Contract between a firm and an NGO or a public entity and private firms	
Informal	• Characterized by seasonal arbitrary production contracts with individual traders or small firms • Contracts usually verbal • Contracts do not require high quality standards	Vegetable and fruits

Source: Authors based on Eaton and Shepherd (2021).

Contract farming for fresh produce in Kenya

The horticultural sector provides an ideal case study for investigating the issues and potential of contract farming in Kenya. The fresh produce-based food economy (including fruits and vegetables) forms a relatively important and dynamic subsector contributing around 36 percent to total agricultural activity, and its gross value is growing at between 15 and 20 percent per year. With growing demand, the Kenyan export market for fruits and vegetables has grown significantly. Notably, avocado production and export in Kenya has trended upward in recent years (HCD 2017), leading farmers in some regions of the country, such as Meru, to shift from traditional cash crops like coffee and tea to avocado.

However, despite the high growth potential, smallholders have not yet realized the full benefits of the sector, owing to various challenges. Kenya's avocado sector has become so lucrative that organized criminal gangs have begun to target growers, forcing them to hire security guards to protect the trees at night. Under the threat of organized cartels, farmers are being forced to harvest early in order to save their crop. Early harvesting, however, significantly reduces the quality and makes the avocado unacceptable for export. Because of quality issues and growing domestic market demand, only about 20 percent of the harvest is exported. In 2021, these avocado exports earned Kenyan farmers $132 million (Igunza 2022).

Smallholder avocado producers—who account for most of the production—also face challenges related to the traditional nature of the production, marketing, and supply chain. Many have limited access to production technologies, institutional credit, and knowledge, and face high marketing risks that diminish their incentive to participate in fresh produce value chains. Other challenges include poor infrastructure (Omosa 2006), lack of access to current and reliable market information, difficulty in accessing inputs and advisory services, and lack of postharvest handling facilities, including cold chains for the supply of quality produce (World Bank 2008). As result, Kenyan avocado and other fresh produce exporters are increasingly looking at contract farming as a means to address these challenges and improve the integration of the smallholder avocado producers into export value chains.

Several NGOs have made an effort in Kenya to promote contract farming for fresh produce. For example, in 2004, East African Growers (EAG) started providing extension services and training on topics such as manure application, pruning, grafting, spraying, and grading, and arranged for the collection and transport of produce from farmers who sold the avocados to EAG. Indu-Farm Ltd. had signed contracts with farmers to educate them on the Euro-Retailer Produce Working Group for Good Agricultural Practice (EurepGAP) certification, while farmers supplied avocados to the firm. In 2006, the United States Agency for International Development (USAID) funded a program, Kenya Business Development Services (KBDS), which sensitized farmers and developed a spraying regime that met market requirements with farmers.

However, most of these arrangements have not been sustainable in the avocado value chain. In most cases, it is the smallholder farmers who breach contracts, by selling directly to brokers at the farmgate in exchange for cash. In the absence of organized market information systems, and when small farmers need cash for small quantities sold, "spot market" transactions directly with brokers at farmgate reduce product marketing costs and provide liquidity to farmers to meet immediate needs. However, this also may reduce the advantages from aggregation and collective bargaining through farmers groups that contract farming offers. Compared with other developing countries, earlier literature reported a higher failure rate of contract farming schemes in Kenya owing to contract biases against producers and poor enforcement (Singh 2002; Minot and Ngigi 2010; Minot and Sawyer 2016).

Additionally, Ochieng, Veettil, and Qaim (2017) identify high transaction costs, less transparent quality grading, and unfair risk-sharing arrangements as factors contributing to low farmer participation and high rates of dropout in contractual arrangements. This implies that there is a need for all stakeholders to better understand the challenges and find optimal solutions to make contract farming an attractive and sustainable venture for smallholder farmers. Also, side-selling is a major impediment to contract farming, especially when sellers do not feel well rewarded for their produce or see the long-term benefits of staying in the arrangement. Side selling is a common problem for avocado and less-established buyers. For bigger buyers (for example, export value chains), the difficulty that smaller suppliers face in meeting food safety, quality, and sustainability standards can be a major challenge (Minot and Sawyer 2016; Amare et al. 2019).

Strong public support might also be justified in view of the full social benefits resulting from diversification of cereal-based production systems toward high-value tree crops like avocado. Although empirical evidence is limited, avocado farming as a tree-based, low external input economic activity is more climate-resilient, environmentally sustainable, and nutrition-smart than cereals production, and its higher returns are likely to offer a better path in land-scarce areas for marginal and smallholder farmers to lift themselves out of poverty. In land-scarce dryland regions, farmers growing basic staples under rainfed conditions are less likely to "farm themselves out of poverty" (Harris and Orr 2014). Additionally, the Government of Kenya needs to develop policies and legal frameworks that enable and foster contract farming approaches in avocado farming. Such frameworks should require that contracts be established in a transparent and mutually beneficial manner; smallholder producers are protected from any exploitative practices; and contracts are enforced fairly through the existing judicial system, including mechanisms for third-party arbitration and adjudication of disputes. This requires trust building through repeated engagements, offering competitive prices and timely payments, and standardization of the collection process to reduce rejection of quality produce.

Effect of participation in high-value produce on inclusion and livelihoods of farmers

Inclusion of smallholder farmers and producers in fresh produce value chains through contract farming is expected to generate income benefits and improve their well-being. There is, however, mixed evidence on how smallholders' participation in contract farming affects their livelihoods. Even though there is evidence that access to produce markets via contracting has positive impacts on participants' welfare (Ashraf, Gine, and Karlan 2008; Bolwig, Gibbon, and Jones 2009; Miyata, Minot, and Hu 2009; Bellemare 2012; Bolwig 2012), this has been inconclusive. While significant and positive impacts have been found on smallholders' incomes (Miyata, Minot, and Hu 2009; Olomola 2010; Bellemare 2012; Ogutu, Ochieng, and Qaim 2020), some authors have found it to be less effective in generating benefits for smallholders (Glover and Kusterer 1990). Some scholars argue that the impact of contract farming on smallholder farmers' inclusion and livelihoods varies depending on the nature of the contract and the category of the firm in question (Gow 2000). For instance, Mwambi and colleagues (2016) found that the intermediary model of avocado contracting in Kenya did not favor producers because the farmers did not know the detailed terms of the contract agreed between the intermediary and the final buyer. These terms were negotiated by the intermediary without the farmers' involvement.

Regardless of the model adopted, fresh produce contract farming in developing countries has generally seen varied results. Positive impacts have been reported in cases where contracts provide productive resources such as technical assistance, credit, improved technologies, and other farm inputs required to increase production and productivity of nontraditional high-value crops and to reduce marketing risks (Glover 1984; Williams and Karen 1985). Birthal, Joshi, and Gulati (2005) examined contract production of vegetables in India, where innovative arrangements were made for production and marketing. Their results showed that contract farmers received higher prices compared with farmers who were not under contracts. This is also consistent with Minten, Randrianarison, and Swinnen (2009), who found that vegetable contract farming in Madagascar yielded positive outcomes, including higher welfare, more income stability, and shorter lean periods for contract farmers as compared with their counterparts who were not under any form of contract.

However, such development impact is not ensured—the right incentives and contracting systems are often required to achieve inclusive and efficient results. In most developing countries, contract farming arrangements that succeeded in engaging and benefiting small-scale producers required the presence of enabling conditions and mutually beneficial agreements between the parties involved. For instance, the positive evidence, in terms of higher incomes for smallholder farmers, from contract farming in horticultural production in Senegal found by Maertens and colleagues (2007) was attributed to guaranteed market access and access to inputs. Similarly, the positive impact reported by Minten, Randrianarison, and Swinnen (2009) could reflect farmers' access to seeds, fertilizer, and pesticides through the contracting firms.

Conversely, other research finds that some contracts favor large-scale farmers, leaving poorer farmers and smallholders out of the development process (Runsten 1992; Little and Watts 1994). Using nationally representative survey data from six countries, Meemken and Bellemare (2020) find that contract farmers in most countries exhibit increased demand for hired labor, which suggests that contract farming could stimulate local employment, but contract farmers obtained higher incomes than their counterparts without contracts in only some countries. These results challenge the supposition that contract farming indisputably betters welfare. Similarly, Dubbert, Abdulai, and Muhammed (2021) assessed the participation of cashew farmers in contract farming in Ghana and find that participation had negative impacts on use of sustainable farm practices, which could reduce incomes and affect livelihoods. This evidence shows that contract farming experiences for the fresh produce sector vary by countries and commodities and implies that commodity- and country-specific contexts should be explored.

Overall, these findings indicate that contracts that have proved beneficial to the farmers and contractors—and hence promote inclusion—are those that are transparent and developed through a participatory process, managed effectively, and that allow both parties to cultivate trust as a basis for a long-term relationship (Eaton and Shepherd 2001). In such cases, decisions are made jointly with farmers, and the contractor covers only some part of the production and price risks (Key and Runsten 1999). This risk-sharing arrangement fosters active participation of all parties involved, which is an important factor in success. Ineffectiveness in fostering inclusion and farmer livelihoods, on the other hand, has been linked to contracts with unequal terms and a lack of capacity development and coordination among small producers that would enable them to engage and negotiate better bargains (da Silva 2005). For instance, farmers may lose bargaining leverage with the company if they invest in specialized assets and become overly reliant on their contract crops, compelling them to accept exploitative or less attractive contract terms (Warning and Key 2002). The next section summarizes evidence from a case study in Kenya.

Kenya case study: Avocado value chain

Avocado production and exports in Kenya

Kenya is one of the fastest-growing fresh produce exporters in the world—and avocado is the world's fastest-growing fruit in terms of popularity (Altendorf 2017; Kathula 2021). Avocados are grown in various agroecological zones in Kenya, primarily by smallholder farmers for subsistence and/or sale in local and export markets. The Central and Eastern regions account for about 70 percent of avocado production in the country, with the former being the leading producer (Johnny et al. 2019). Local varieties account for roughly 70 percent of overall avocado production in Kenya, while Fuerte and Hass, two improved avocado varieties appropriate for the export market, account for approximately 20 percent and 10 percent, respectively (HCD 2015).

The Hass variety has a dark green and brown skin that is not thick when mature and can easily be removed from the pulp. It has a small seed with a nonfibrous pulp and is often referred to as a "brown skin." It is vigorous and highly productive, with an oil content of over 21 percent (Gupta et al. 2018). The Fuerte variety, referred to as "green skin," has a smooth, green skin with medium thickness. It has a big seed and a buttery pulp. Its oil content is about 16–18 percent (Saenger et al. 2013). Compared with Fuerte, the Hass yields a higher price, attributed to its higher oil content, higher resistance to pests and diseases, and ability to hide bruises. This has led to an increasing shift in production from Fuerte toward Hass.

With increased commercialization, notable growth has been recorded in both total hectares harvested and overall production over the years (Figure 15.1). The total area under avocado in 2013 in Kenya was about 11,000 ha. Area and production more than doubled between 2011 and 2020—increasing by 145 percent and 116 percent, respectively (FAO 2022). Most Kenyan avocado farmers are found in Murang'a, Nyeri, Kiambu, Kisii, Meru, and the entire Mount Kenya region. But avocado is also grown in smaller quantities in other counties, such as Nandi, Bomet, Uasine Gishu, Trans-Nzoia, Bungoma, and

FIGURE 15.1 **Trends in avocado area harvested, production, and yield in Kenya**

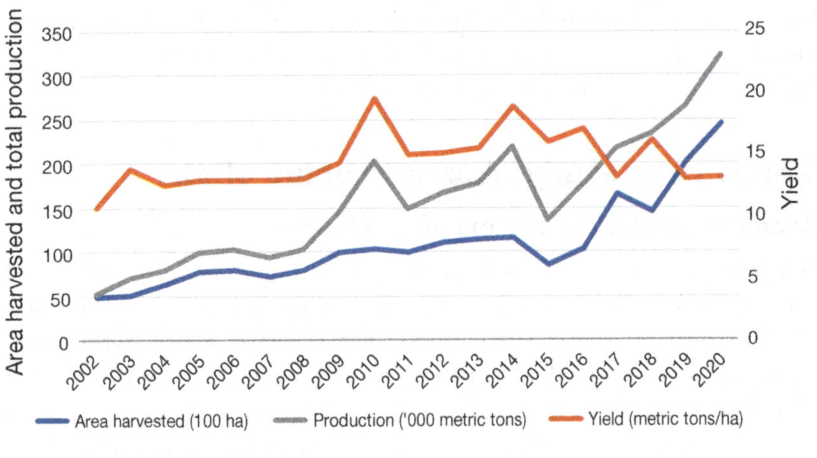

Source: FAO (2022); HCD (2015).

Siaya (Mwambi et al. 2016). Avocado exports to Europe nearly tripled in value between 2013 and 2017, according to the Dutch Centre for the Promotion of Imports from developing countries. This impact is being felt thousands of miles away, on farms in Kenya's highlands, where growers' fortunes are changing.

Avocado has emerged as the major fruit trade in the Kenyan export market in recent years, accounting for more than 7 percent of horticultural exports (HCDA 2015). Kenya is the third-largest producer (behind Mexico and Peru) and ranks eighth in the world on the list of largest exporters (Snel et al. 2021), with production reaching more than 300,000 metric tons in 2020 (Figure 15.2). This position clearly demonstrates the avocado crop's expanding potential and relevance in contributing to numerous aspects of economic development, such as raising rural household incomes, job creation, and export diversification.

Kenya's share of exports in total production by volume, however, ranks low compared with the other major avocado exporters. Chile and South Africa export about 60 percent and 55 percent of their total production, respectively; Kenya exports only about 20 percent. Moreover, the export share in total production has declined in Kenya over time, showing about a 20 percent decrease between 2015 and 2020 (Figure 15.3). The challenges in increasing the export share of production are often attributed to poor quality, in terms of both size and preferred varieties; lack of regulatory standards; weak institutional capacity of small-scale producers; and inadequate capacity and coordination of fruit

FIGURE 15.2 Trends in avocado production for major producing countries

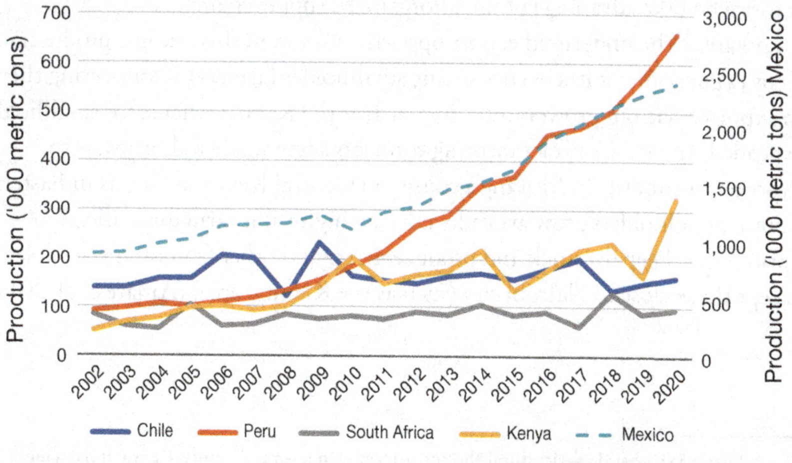

Source: FAO (2022).

FIGURE 15.3 Trends in the share of exports in total production of major avocado exporters

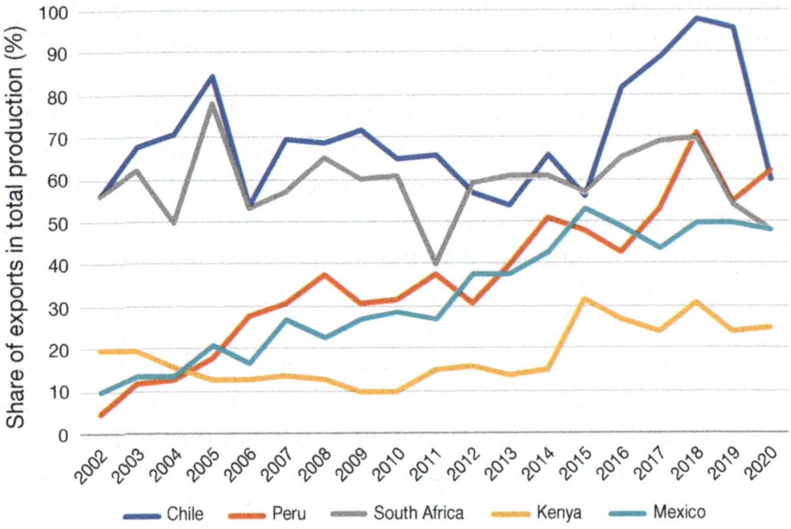

Source: HCD (2015); FAO (2022).

exports. Poor quality owes primarily to small-scale farmers' lack of knowledge of current production procedures, as they have traditionally produced the fruit for domestic markets or noncommercial purposes, as well as limited distribution and uptake of market-preferred varieties (Amare et al. 2019). Given the weak organization of avocado markets, most smallholder growers market their avocados through brokers, who may be legally certified agents or unofficial actors who take advantage of the unorganized supply chain.

Aware of the underused export opportunities from this sizeable production, Kenya's government has been assisting smallholder farmers by connecting them to exporters via out-grower initiatives such as the Kakuzi scheme for smallholder avocado farmers.[1] Contract farming connects a few small-scale growers to exporters, primarily in Murang'a county in Central Kenya, as well as in Eastern Kenya. Smallholders grow avocado as a new high-value farm diversification option; they keep livestock and produce maize and other fruits and vegetables, except those close to Nairobi city (such as the Kiambu area) (Amare et al. 2019).

1 Kakuzi is a large-scale agricultural enterprise located in Kiambu County, Kenya. It has a number of outgrower schemes, which are partnerships between the firm and smallholder farmers.

Impact of avocado contract farming in Kenya

Evidence from these avocado contract farming experiences in Kenya shows that participation in contracting is not sufficient to enhance the inclusion of smallholders in beneficial value chains and increase their incomes, especially where contract terms are unclear. According to Amare and colleagues (2019), farmers involved in export markets differ significantly from those who are not: their farms are relatively larger; they have had more training and own more Hass avocado trees, which have greater market value; and they tend to be older. Also, there is a positive link between participation in well-functioning avocado farmers' associations and participation in export markets. Participation in avocado export markets is associated with positive impacts on incomes, revenues, prices, and labor inputs.

Evidence shows that, while contract farming facilitates the inclusion of smallholder farmers in avocado value chains, this is skewed more toward farmers with larger endowments and planted areas and less effective in supporting the participation and inclusion of smaller and marginal farmers (Amare et al. 2019). These findings indicate that, when contract farming participating grower associations bring together farmers who own mature trees of market-preferred species, inclusion and beneficial gains are limited to those farmers who are group members. To enhance inclusion, there is a need for additional support and capacity development to organize and support non-benefiting growers to progressively improve their capacity and effectively engage in avocado value chains. This could also further increase the volume of production and expand export markets and, as demand increases, spur efforts to engage more farmers.

Other studies have found similar evidence on inclusion and livelihood impacts at the household level. Mwambi and colleagues (2016) found that, despite a positive and significant effect on avocado income, with a \$48–67 increase annually, this did not translate into higher household income, indicating that contract farming triggers some trade-offs between alternative income sources of smallholder avocado farmers. Also, the issues of side-selling and breach of contract were evident, partly because the fundamental aspects of the contract, such as the pricing and grading mechanisms, were not fully explained to the growers. Thus, to make contracts equally appealing and mutually beneficial for all parties involved, it is imperative to consider the conditions necessary to influence participation in contract framing, such as enhancing the knowledge and capacity of producers through education and advisory services, access to credit, and transparency about contract terms.

From a gendered perspective, a recent study by Muriithi and Kabubo-Mariara (2022) on the dynamics and role of gender in high-value avocado farming in Kenya found, contrary to past reports of men dominating cash crop farming, that gender was not a strong determinant of participation in avocado farming, with evidence of increased women's engagement in the suburbs of Nairobi city. However, the gender disaggregation showed that men-headed household participants were younger, had more years of schooling, and had more Hass trees than women-headed households. Conversely, women-headed households were better off in terms of access to finance, while no significant difference was found in terms of social capital and networking. The authors also highlighted the number of improved productive avocado trees as a prerequisite for participating in high-value avocado markets.

Many factors shape the positive impacts of contract farming. For instance, the null effect found among avocado farmers in Kenya has been attributed to the nature of the contract (intermediary model), which exposes farmers to price and production risks, among other factors (Mwambi et al. 2016). Where producers are not directly involved in negotiations, unclear contract terms are likely to negate the benefits of contracting, since incidents of side selling to other buyers offering lower but immediate payments may be common, leading to poor inclusion and lack of impact in terms of improving livelihoods. This indicates that the effects of contract farming on the inclusion of farmers in value chains and impacts on their livelihoods cannot be generalized and should instead be assessed and interpreted based on the specific context and contracting models, to identify and distill the underlying drivers and success factors.

Overall, most of the existing literature finds contract farming to be beneficial in increasing income and improving farmers' livelihoods. Using a gender lens, we see that in comparison with women, the men who participate in avocado contract farming are reportedly younger and more educated (Muriithi and Kabubo-Mariara 2022). Thus, they are likely to be more skillful and make better use of market information, consequently reducing market and other transaction costs.

Conclusions: Lessons and policy implications

Drawing evidence from the avocado sector in Kenya, this chapter has documented contract farming practices and challenges and opportunities in enhancing market access and inclusion for small-scale producers of perishable but high-value commodities in developing countries. To this end, it conducted a comprehensive review of relevant literature and analyzed data on avocado area and production, using descriptive but informative tools, to understand patterns in avocado production and exports. Results show that, even though participation in fresh produce value chains through contract farming could improve competitiveness and generate positive outcomes for smallholders, their level of participation is still low, explained by significant impediments including lack of technical knowhow, trust, innovations, access to market information, and training to enter and sustain engagements in globally competitive avocado export value chains. The mixed evidence on the impacts of participation in high-value produce on livelihoods is attributed to the varying contract farming models, the nature of firms, and the low level of producers' involvement in the implementation of contracts.

Given the knowledge-intensive nature of the avocado value chain and the potential social benefits to Kenya, public support and incentives to strengthen smallholder capacity and that of small and medium enterprises can play an important role in boosting global competitiveness, inclusion, and the creation of decent jobs in the avocado sector. For individual smallholder growers to access attractive markets and benefit from export opportunities, and make contract farming a successful strategy to transform the avocado economy in Kenya, it is particularly important to address the major impediments to smallholders' engagement and implement policies to strengthen both the incentives for and the loyalty of the contracting partners.

First, there is a need to invest in training on sustainable production, harvesting, and postharvest management techniques, and in the prevention of theft and illegal cartels that force farmers to harvest fruits early. Training on avocado production, harvesting, and handling methods could improve farmers' production techniques, thereby reducing the share of production rejected and increasing sales of high-quality avocados to the market. Also, training on the advantages of quality and advance contracting for accessing competitive markets may discourage side selling by raising farmers' knowledge about the benefits of contract farming and improving their marketing skills. Accordingly, when farmers become aware that they could obtain a higher value through contracts, they may be motivated to sell via agreed channels instead of side selling. Timely payment is critical and may need to be considered as part of contracting. Government may also play a role as part of creating the enabling environment.

Second, to build trust among contracting partners, there is a need to enforce contracts through a third-party such as the HCD, and to establish trust between farmers' groups and exporters. This can be done by disseminating relevant and up-to-date market information, such as on quality requirements and prices, as well as supporting avocado farmers—to improve productivity, quality, and consistency of supply. Also, there is a need to modernize the avocado sector through digital marketing platforms and e-commerce, to leverage Kenya's strong position in mobile banking and connectivity. Public–private partnerships (PPPs) can bring such innovation as well as best practices to improve the export demand for and competitiveness of Kenyan avocados. This is especially vital given the low proportion of smallholder produce that is exported and the lack of discipline among some exporters, which damages the credibility and reputation of Kenyan suppliers in global avocado markets. These are the noncompliant exporters who ship poor-quality avocados. Several exporters are currently reluctant to cooperate with researchers or provide information to develop and improve the avocado value chain, suggesting a need for PPPs to build trust and bring new ideas to modernize the avocado sector.

Creating an enabling environment for mainstreaming contract farming to strengthen agricultural value chains also requires appropriate legal frameworks, to facilitate contracting that entails clarity of terms, fairness, responsibility, and transparency to build trust in completing transactions between the different parties. Given the unequal market powers between the different parties, a legal framework would bring legal protection, uniformity, and consistency in engaging small-scale producers in the production and supply of products—including how the contracting parties will deal with desired quality, volume, timing, and prices through prior contracting. Legal frameworks can also facilitate access to finance and inputs for small producers when the buyer is willing to give technical advice and loans for contracted activities. This strengthens the bond and builds trust between the two parties. This can facilitate transformation of the fresh produce sector by improving access to productivity-enhancing inputs or upgrading essential infrastructure, such as irrigation; cold storage and transport; or equipment for harvesting, sorting, and grading of produce to meet required standards and reduce food loss for perishables.

Under clear legal frameworks, producer organizations can play a vital role in empowering small producers and in facilitating fair and transparent contracts, helping farmers sell their produce directly to food processors, retailers, or exporters. This has the potential to reduce transaction costs, improve producer prices, and enhance inclusion of small-scale producers, including women and youth. In place of internal economies of scale, reliance on external economies of scale through networking, associations, or other forms of grouping should be promoted. This could be among small enterprises or by linking small growers or enterprises with larger ones that no longer face major market entry barriers.

References

Altendorf, S. 2017. "Global Prospects for Major Tropical Fruits." In *The Food Outlook: Biannual Report on Global Food Markets*, 68–81. Rome: FAO.

Amare, M., J. Kabubo-Mariara, R. Oostendorp, and M. Pradhan. 2019. "The Impact of Smallholder Farmers' Participation in Avocado Export Markets on the Labor Market, Farm Yields, Sales Prices, and Incomes in Kenya." *Land Use Policy* 88: 104–168.

Amare, M., B. Shiferaw, H. Takeshima, and G. Mavrotas. 2021. "Variability in Agricultural Productivity and Rural Household Consumption Inequality: Evidence from Nigeria and Uganda." *Agricultural Economics* 52 (1): 19–36.

Ashraf, N., X. Gine, and D. Karlan. 2008. "Finding Missing Markets (and a Disturbing Epilogue): Evidence from an Export Crop Adoption and Marketing." Center Discussion Paper No. 967. Economic Growth Center, Yale University, New Haven, CT.

Barrett, C.B., M.E. Bachke, M.F. Bellemare, H.C. Michelson, S. Narayanan, and T.F. Walker. 2012. "Smallholder Participation in Contract Farming: Comparative Evidence from Five Countries." *World Development* 40 (4): 715–730.

Barrett, C.B., T. Reardon, J. Swinnen, and D. Zilberman. 2020. "Agri-food Value Chain Revolutions in Low- and Middle-Income Countries." *Journal of Economic Literature* 58: 1–67.

Bellemare, M.F. 2012. "As You Sow, So Shall You Reap: The Welfare Impacts of Contract Farming." *World Development* 40 (7): 1418–1434.

Bijman, J. 2008. *Contract Farming in Developing Countries: An Overview of the Literature.* Wageningen, Netherlands: Wageningen University.

Birthal, P.S., P.K. Joshi, and A. Gulati. 2005. "Vertical Coordination in High-Value Commodities: Implications for Smallholders." IFPRI Discussion Paper 85. International Food Policy Research Institute (IFPRI), Washington, DC.

Bolwig, S. 2012. "Poverty and Gender Effects of Smallholder Organic Contract Farming in Uganda." Uganda Strategy Support Program Working Paper 8. IFPRI, Washington, DC.

Bolwig, S., P. Gibbon, and S. Jones. 2009. "The Economics of Smallholder Organic Contract Farming in Tropical Africa." *World Development* 37 (6): 1094–1104.

Coulter, A., V. Entwistle, and D. Gilbert. 1998. *Informing Patients: An Assessment of the Quality of Patient Information Material.* London: King's Fund.

da Silva, C. 2005. "The Growing Role of Contract Farming in Agrifood Systems Development; Drivers, Theory and Practice." Working paper. FAO, Rome.

Dubbert, C., A. Abdulai, and S. Mohammed. 2021. "Contract Farming and the Adoption of Sustainable Farm Practices: Empirical Evidence from Cashew Farmers in Ghana." *Applied Economic Perspectives and Policy* 45 (1): 487–509.

Eaton, C., and A. Shepherd. 2001. *Contract Farming: Partnerships for Growth*. Rome: FAO.

FAO (Food and Agriculture Organization of the United Nations). 2017. *Major Tropical Fruits: Market Review*. Rome.

FAO. 2022. *Major Tropical Fruits: Market Review—Preliminary Results*. Rome.

Glover, D. 1984. "Contract Farming and Smallholder Outgrower Schemes in Less-Developed Countries." *World Development* 12 (11): 1143–1157.

Glover, D., and K. Kusterer. 1990. *Small Farmers, Big Business: Contract Farming and Rural Development*. New York: St Martin's Press.

Gow, H. 2000. "Restructuring the Agribusiness Sector and the Role of Foreign Direct Investment." In *Agricultural Policy and Enlargement of the European Union*, eds. A. Burrell and A. Oskam, 201–216. Wageningen, Netherlands: Wageningen Press.

Gupta, S.K., P. Singhal, A. Singh, R. Chauhan, and B. Kumar. 2018. "Nutritional and Pharmaceutical Benefits of Avocado Plant." *Journal of Advanced Scientific Research* 9 (2): 4–11.

Harris, D., and A. Orr. 2014. "Is Rainfed Agriculture Really a Pathway from Poverty?" *Agricultural Systems* 123: 84–96.

Hart, C. 2001. *Doing a Literature Search: A Comprehensive Guide for the Social Sciences*. London: SAGE Publications.

HCD (Horticultural Crops Directorate). 2015. *Horticultural Crops Production Report*. Nairobi: Government Printer.

HCD. 2017. *Horticulture Validated Report 2016-2017*. Nairobi: Government Printer.

Hung Anh, N., W. Bokelmann, N. Thi Thuan, T. Do Nga, and N. Van Minh. 2019. "Smallholders' Preferences for Different Contract Farming Models: Empirical Evidence from Sustainable Certified Coffee Production in Vietnam." *Sustainability* 11 (14): 3799.

Igunza, E. 2022. "Kenyan Vigilantes Taking on Avocado Gangs." BBC News, 16 January.

Johnny, E.G., J.K. Mariara, R. Mulwa, and G.M. Ruigu. 2019. "Smallholder Avocado Contract Farming in Kenya: Determinants and Differentials in Outcomes." *African Journal of Economic Review* 7 (2): 91–112.

Kathula, D.N. 2021. "Avocado Varieties and Export Markets for Sustainable Agriculture and Afforestation in Kenya." *Journal of Agriculture* 5 (1): 1–26.

KEHPA (Kenya Horticultural Export Promotion Authority). 2015. *The Kenya Avocado Industry: A Review of the Status and Opportunities*. Nairobi.

Key, N., and D. Runsten. 1999. "Contract Farming, Smallholders, and Rural Development in Latin America: The Organization of Agro Processing Firms and the Scale of Out Grower Production." *World Development* 27 (2): 381–401.

Kirsten, J., and J. Sartorius. 2002. "Linking Agribusiness and Small-Scale Farmers in Developing Countries: Is There a New Role for Contract Farming?" *Development Southern Africa* 19 (4): 505–529.

Little, P., and M. Watts. 1994. *Living under Contract: Contract Farming and Agrarian Transformation in Sub-Saharan Africa*. Madison, WI: University of Wisconsin Press.

Maertens, M., and J.F.M. Swinnen. 2007. "Standards as Barriers and Catalysts for Trade and Poverty Reduction." *Journal of International Agricultural Trade and Development* 4 (1): 47–61.

Maertens, M., L. Dries, F.A. Dedehouanou, and J.F.M. Swinnen. 2007. "High-Value Global Supply Chains, EU Food Safety Policy and Smallholders in Developing Countries: A Case-study from the Green Bean Sector in Senegal." In *Global Supply Chains, Standards and the Poor*, ed. J.F.M. Swinnen, 159–172. Oxon, UK: CABI Publishing.

Meemken, E., and F.M. Bellemare. 2020 "Smallholder Farmers and Contract Farming in Developing Countries." *Proceedings of the National Academy of Sciences* 117 (1): 259–264.

Minot, N. 2011. "Contract Farming in Sub-Saharan Africa: Opportunities and Challenges." Paper presented at a policy seminar, Smallholder-led Agricultural Commercialization and Poverty Reduction: How to Achieve It, April 18–22, 2011.

Minot, N., and M. Ngigi. 2004. "Are Horticultural Exports a Replicable Success Story? Evidence from Kenya and Côte d'Ivoire." IFPRI Discussion Paper 120. IFPRI, Washington, DC.

Minot, N., and B. Sawyer. 2016. "Contract Farming in Developing Countries: Theory and Practice, and Policy Implications." In *Innovation for Inclusive Value Chain Development: Successes and Challenges*, eds. A. Devaux, M. Torero, J. Donovan, and D. Horton, 127–158. Washington, DC: IFPRI.

Minten, B., L. Randrianarison, and J.F. M. Swinnen. 2009. "Global Retail Chains and Poor Farmers: Evidence from Madagascar." *World Development* 37 (11): 1728–1741.

Miyata, S., N. Minot, and D. Hu. 2009. "Impact of Contract Farming on Income: Linking Small Farmers, Packers, and Supermarkets in China." *World Development* 37 (11): 1781–1790.

Muriithi, B., and J. Kabubo-Mariara. 2022. "The Dynamics and Role of Gender in High-Value Avocado Farming in Kenya." *European Journal of Development Research* 34: 2272–2304.

Mwambi, M.M., J. Oduol, P. Mshenga, and M. Saidi. 2016. "Does Contract Farming Improve Smallholder Income? The Case of Avocado Farmers in Kenya." *Journal of Agribusiness in Developing and Emerging Economies* 6 (1): 2–20.

Ochieng, D.O., P.C. Veettil, and M. Qaim. 2017. "Farmers' Preferences for Supermarket Contracts in Kenya." *Food Policy* 68: 100–111.

Ogutu, S.O., D.O. Ochieng, and M. Qaim. 2020. "Supermarket Contracts and Smallholder Farmers: Implications for Income and Multidimensional Poverty." *Food Policy* 95: 101940.

Okello, J.J., and S.M. Swinton. 2007. "Compliance with International Food Safety Standards in Kenya's Green Bean Industry: Comparison of a Small- and a Large-Scale Farm Producing for Export." *Applied Economic Perspectives and Policy* 29 (2): 269–285.

Olomola, A.S. 2010. "Enhancing Productivity, Income and Market Access of Rural Producers in Africa: The Case of Contract Farming in Nigeria." Paper presented at the Inaugural National Science Foundation (NSF) Joint Workshop of the African Economic Research Consortium (AERC) and the International Growth Centre (IGC), Mombasa, Kenya.

Omosa, M. 2006. "Towards Food Security in Africa: Theories Images and Realities." *Eastern Africa Journal of Development Studies* 1 (1): 43–61.

PEP (Partnership for Economic Policy). 2018. *Avocado Contract Farming in Kenya: Does It Work?* NWO/WOTRO project Productive Employment in the Segmented Markets of Fresh Produce, 2014-17. Policy Brief No. 178.

Prowse, M. 2012. *Contract Farming in Developing Countries: A Review.* Vol. 12. Paris: Agence Française de Développement.

Ruben, R. 2017. "Impact Assessment of Commodity Standards: Towards Inclusive Value Chains." *Enterprise Development and Microfinance* 28 (1): 82–97.

Runsten, D. 1992. "Transaction Costs in Mexican Fruit and Vegetable Contracting: Implications for Association and Participation." Paper presented at the XVIII International Congress of the Latin American Studies Association, Atlanta, GA, March 10–12.

Saenger, C. , M., Qaim, M. Torero, and A. Viceisza. 2013. "Contract Farming and Smallholder Incentives to Produce High Quality: Experimental Evidence from the Vietnamese Dairy Sector." Agricultural Economics 44: 297–308.

Simmons, P., P. Winters, and I. Patrick. 2005. "An Analysis of Contract Farming in East Java, Bali, and Lombok, Indonesia." *Agricultural Economics* 33: 513–525.

Singh, S. 2002. "Contracting Out Solutions: Political Economy of Contract Farming in the Indian Punjab." *World Development* 30 (9): 1621–1638.

Snel, H.J., J. Broeze, F.S. Kremer, and E. Osena. 2021. *Opportunities to Reduce Post-harvest Losses in the Kenyan Avocado Sector through Agro-processing and Agro-logistics.* Wageningen, Netherlands: Wageningen Centre for Development Innovation.

Snodgrass, D., and J. Sebstad. 2005. *Assessing the Impact of Kenya BDS and Horticulture Development Center Projects in the Tree Fruit Value Chain in Kenya. Baseline Research Report.* Microreport No. 33. United States Agency for International Development (USAID): Washington, DC.

Subervie, J., and I. Vagneron. 2013. "A Drop of Water in the Indian Ocean? The Impact of GlobalGap Certification on Lychee Farmers in Madagascar." *World Development* 50: 57–73.

Ton, G., S. Desiere, W. Vellema, S. Weituschat, and M. D'Haese. 2017. "The Effectiveness of Contract Farming in Improving Smallholder Income and Food Security in Low- and Middle-Income Countries. A Mixed-Method Systematic Review." Washington, DC: International Initiative for Impact Evaluation (3ie).

USAID (United States Agency for International Development)/Kenya. 2008. *Impacts of the KBDS and KHDP Projects in the Tree Fruit Value Chain in Kenya*. Microreport No. 129. Nairobi.

Warning, M., and N. Key. 2002. "The Social Performance and Distributional Consequences of Contract Farming: An Equilibrium Analysis of the *Arachide de Bouche* Program in Senegal." *World Development* 30 (2): 255–263.

Williams, S., and R. Karen. 1985. *Agribusiness and the Small-Scale Farmer: A Dynamic Partnership for Development*. Boulder, CO: Westview Press.

World Bank. 2008. *World Development Report 2008: Agriculture for Development*. Washington, DC: World Bank.

PART 6

Toward More Sustainable Food Systems

Part 6

TOWARD MORE SUSTAINABLE FOOD SYSTEMS

Sustainability is a key pillar of food system transformation. Environmental sustainability is needed to ensure food systems can be more productive, resilient, healthy, and inclusive both today and for future generations. Further, programmatic sustainability can ensure that programs and interventions not only deliver impact during their lifetime but also are able to ensure outcomes can still be attained after their end. Part 6 discusses sustainability from both environmental and programmatic perspectives, giving insights into how transformed food systems can be sustained over time.

The arid and semiarid lands (ASALs) in Kenya are on the frontlines of the climate crisis. Part 2 (Productivity) highlighted the diversity of agroecological zones of Kenya yet its relatively homogenous, maize-dominated production systems. When crops are grown outside their optimal environment, heavy reliance on chemical fertilizers and pesticides may be needed to boost yields. In the long run, such production practices can damage soil quality and reduce productive capacity.

Chapter 16 presents an analysis of Kenya's flour blending policy, which aims to promote the blending of sorghum and millet into maize flour, creating a more nutritious flour for Kenya's staple food, *ugali*. Sorghum and millet are better suited than maize to ASALs and have a higher nutritional value than maize-only flour, making flour blending a potential win-win for producers and consumers. However, Chapter 16 points out that low production and demand for blended flour present a major constraint to its widespread adoption. Simultaneous and complementary measures are needed to ensure success: improved technology adoption, better access to markets for farmers, and demand-side campaigns to increase the acceptance of blended flour among Kenyan consumers.

While environmental sustainability in food systems is often seen from the perspective of on-farm practices, sustainability must also be built into off-farm components of the food system. Food loss and waste (FLW) presents a major challenge to achieving environmental sustainability within the food system, particularly in value chains with highly perishable produce, such as fresh fruits. Chapter 17 presents an analysis of FLW in the fresh mango value chain. Between 35 and 45 percent of mangoes are lost or wasted in Kenya. Reducing FLW can make the food system more sustainable by reducing pressure to increase yields as compensation for such FLW. Policies and programs aimed at reducing FLW should focus on improving harvesting practices and packaging, implementing cold chains, and developing small-scale processing for perishable foods. Investments can help overcome the challenge of poor transportation infrastructure, which increases transportation times and FLW. Further, regulations aimed at small and medium enterprises present a barrier to entry for small-scale processors, who are central to reducing FLW by taking perishable foods and processing them into products with longer shelf lives.

In terms of programmatic sustainability, the private sector is increasingly seen as a sustainable medium through which to deliver solutions to food system actors. This dynamic is particularly evident in the digital ecosystem, where private companies are driving innovation and potentially revolutionizing how services are delivered throughout the food system. However, Chapter 18 argues that digital solutions often struggle to scale up after their pilot phases, and may not have sufficiently transformative effects. This disconnect between pilots and scale-ups may be attributed to a lack of finance for digital innovators and competition with public service providers (for example, for extension services and input distribution). Further, many digital service providers do not have sustainable business models either because they are in the early stages of development or because it is difficult to balance donor funding and financial sustainability. Trust, digital literacy, and inclusion remain issues in scaling up digital services. Policymakers can help overcome these issues by setting clear regulations for digital services that enable trust and scaling, investing in digital literacy through innovation hubs and incubators, and continuing to invest in mobile and internet connectivity. While the private sector has the potential to deliver sustainable solutions, the public sector must ensure the enabling environment is in place for this to happen.

Part 6 addresses environmental and programmatic sustainability. The Kenyan government can play a large role in facilitating pathways to sustainability through systematic policy approaches, smart regulations, and investment in key infrastructure. Achieving sustainable food systems will ensure that food systems can be healthy, productive, resilient, and inclusive for future generations.

Chapter 16

AN ENABLING ENVIRONMENT FOR THE NATIONAL FLOUR BLENDING POLICY: A FOOD SYSTEMS ANALYSIS

Mequanint B. Melesse, Yohannis Mulu Tessema, Eric Manyasa, and Andrew Hall

A national flour blending policy is about to be implemented in Kenya. This requires maize flour (the country's main staple) to be blended with at least 10 percent of either one or a composite of traditional crops, such as sorghum and millet.[1] The blending ratio is expected to increase gradually, with the goal of ultimately reaching 30 percent. The policy envisages achieving several goals. The first is to improve the nutritional quality of maize flour: sorghum and millet (and other candidate blending crops) have micronutrient characteristics that are absent in maize. The second is to promote more climate-tolerant crops and technologies: sorghum and millet can be grown in less favorable arid and semiarid lands (ASALs), in the very conditions that many farmers face in Kenya. This is particularly important given that maize is more susceptible than other staple crops to climate change. The third is to reduce the country's overreliance on imported maize and concerns about its food sovereignty.

Early considerations of composite flours in developing countries were driven by economic arguments, primarily to reduce large-scale wheat imports that used up scarce foreign exchange resources (Fellers and Bean 1988; Abdelghafor 2011). For example, the Food and Agriculture Organization of the United Nations (FAO) initiated its Composite Flour Program in 1964 (Fellers and Bean 1988) and supported these early efforts by developing composite flour technologies and products in many countries, including Brazil, Bolivia, Colombia, Senegal, and Sudan.

Various cereals, legumes, and root crops, such as maize, rice, sorghum, millet, barley, sweet potato, amaranth, and cassava, have been used to replace wheat in composite flours (Fellers and Bean 1988; Nwanekezi 2013). Over time,

1 While the discussion in this chapter focuses on sorghum and millet as main candidates for maize flour blending, the policy also considers cassava and amaranth as target crops.

nutritional considerations have started to play a key role in many countries, identifying composite flours as supplementary foods to meet young children's daily micronutrient requirements and for the treatment of malnutrition in therapeutic and emergency feeding programs (Oduro-Obeng and Plahar 2017).

Studies have shown that composite flours are technically feasible (Nwanekezi 2013). However, composite flours to date have been limited primarily to the production of weaning, snack, confectionery, and other specialty foods that meet consumer demand in terms of preference, variety, nutrition, and low cost (Fellers and Bean 1988; Nwanekezi 2013). Specifically, composite flours for staple foods that involve large-scale flour processing have not evolved in most developing countries.

The flour blending policy is set to shake up the food system in Kenya significantly and is likely to increase the use of sorghum and millet by creating new market opportunities for the crops. However, two big issues need to be properly addressed for the policy to succeed: (1) lack of sufficient volume of sorghum and millet at the moment and (2) limited acceptance of and poor demand for these crops, as people have strong preference for white maize *ugali* (cornmeal). Using the food systems perspective, this chapter demonstrates how this policy change can best be leveraged and nudged toward achieving its stated goals. We emphasize that, if the flour blending policy is to succeed, it should not act alone. It needs to be supported by well-thought-out and articulated complementary interventions and measures at different stages of the food system, including targeting appropriate cultivars, agronomy, seed systems, market linkages, and demand. For example, farmers need access to the input and output markets for target crops. There is a clear need to promote consumer demand to provide incentives for their production.

The complex dynamics in food systems mean that apparently promising and well-meaning national flour blending policies could fail and could even have perverse or unexpected outcomes and impacts, such as soaring prices of target crops, which could undermine nutritional benefits for the poor. This complexity calls for recognizing the dynamic, ever-changing nature of food supply and consumption arrangements and the way they create winners and losers, and the powerful forces at play that do not necessarily support the agenda of maize flour blending—for example, potential opposition from market players for whom maize market shares would fall as a result of the policy.

In addition, while the private sector needs to respond to the new policy, there is a danger that potential benefits could be captured by market players and others who have the resources to exploit the opportunity more quickly and at greater scale—and this excludes smallholders (Lapar et al. 2003). This calls for a new form of intervention that couples research and technology development and delivery with capacity building, as well as a wide range of partnerships. This will make it possible to navigate and coordinate changes across the whole of the food system to simultaneously maximize expected benefits and mitigate potentially unexpected negative consequences.

The chapter next briefly outlines the essence of the food systems perspective as relevant to this topic. It then presents a food systems analysis of the Kenyan flour blending policy, followed by conclusions.

The food systems perspective

The food systems perspective is increasingly recognized as a useful analytical framework to enhance our understanding of systems thinking and the design of policies and strategic interventions toward achieving more desirable food system outcomes (HLPE 2017). Chapter 1 provides a schematic presentation of the food systems framework. The food systems approach describes the connections, feedback loops, and trade-offs among its elements and processes that potentially affect system outcomes through a multitude of mechanisms working simultaneously at various scales and levels (De Brauw et al. 2019; Melesse et al. 2020). At the core of a food systems analysis is understanding the relationships between system activities and outcomes, while navigating feedback loops linking these domains, managing trade-offs, and identifying drivers that shape the impact that the outputs of all activities have on system outcomes.

The food systems perspective has several things to offer. It provides a frame for thinking about the directionality of change at a broad scale within the food system, rather than at finer resolutions that may miss richer interactions across system components. It helps in articulating theories of change that draw on complex systems thinking, highlighting complementary measures needed to support core food system interventions. Importantly, it can guide mapping of the distribution of power among stakeholders, potentially revealing the role of dominant players in setting and maintaining the goals and dynamics of food system transformations.

A food systems analysis of the flour blending policy

We now turn our attention to understanding the points of leverage and resistance that the flour blending policy is likely to face and the potential trade-offs that need to be taken into account and mitigated. A natural starting point is

FIGURE 16.1 The food systems framework as applied to the flour blending policy analysis

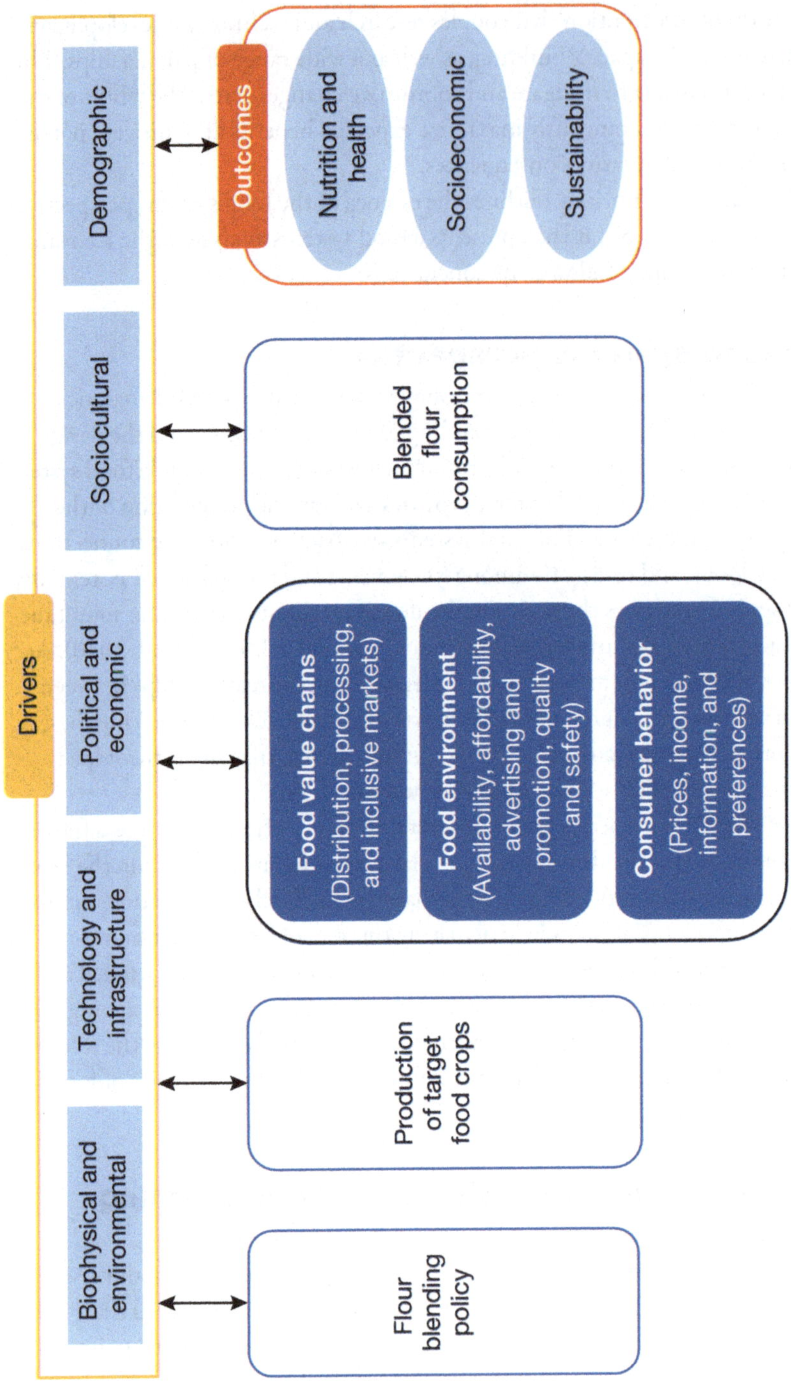

Source: Adapted from HLPE (2017); Mausch, Hall, and Hambloch (2020).

to recap the goals of the policy and key assumptions underpinning its impact pathway. We then explore some of their implications for responses.

Goals of the flour blending policy

The national flour blending policy is an action-driven initiative that is an outcome of a series of intensive multistakeholder consultative process with experts drawn from government ministries and agencies, research and development partners, academia, and the private sector. The policy requires that maize flour be blended with at least 10 percent of either one or a composite of underused high-nutrition crops (Kenya, Crops Act 2013). It also specifies that this ratio should increase gradually, reaching 30 percent as consumer behavior changes in favor of the product.

To coordinate and oversee the implementation of the policy, the government has established a Flour Blending Secretariat that consists of a panel of high-level experts coordinated by the Ministry of Agriculture and Livestock Development. The Secretariat has been in operation since 2018. In consultation with multidisciplinary experts, it has developed an implementation framework that outlines six key focus areas: (1) production systems and commercialization; (2) agri-nutrition, health, and consumer behavioral change; (3) private sector engagement, small and medium enterprises (SMEs), and investment; (4) standards, food safety, policy, and regulations; (5) national government, county, and partner implementation partnerships; and (6) resource mobilization, funding, and strategic partnerships.

Figure 16.1 provides a schematic representation of the proposed policy in light of the food systems perspective. It illustrates core areas in the food system for potential interventions and investments. The framework also reflects a complex set of drivers that influence core activities and potential system changes to shape the outcomes of the food policy. These drivers can broadly be considered socioeconomic drivers (for example, markets, policies, science and technology, social organizations, individual factors) and environmental drivers (for example, land, soils, water, climate, and biodiversity).

The policy seeks to achieve three broad sets of outcomes. First, it aims to improve nutrition and health, particularly of poor and vulnerable communities. Kenya is still grappling with various forms of malnutrition, and nationally about 26 percent of children under the age of five are stunted (Mbugua et al. 2014). Maize, the most important staple crop in the country, is deficient in essential micronutrients. It is also highly prone to unacceptably high levels of aflatoxin contamination that could lead to high rates of child stunting and a variety of noncommunicable diseases (Mutegi, Cotty, and Bandyopadhyay 2018).

TABLE 16.1 Nutrient composition of maize and target crops

	$\text{mg}100^{g-1}$			
Crop	Protein	Calcium	Iron	Zinc
Sorghum	10.4	25	5.4	4.4
Finger millet	7.7	350	9.9	1.5
Pearl millet	11.8	42	11.0	2.0
Maize	9.2	26	2.7	0.5

Source: National Research Council (1996).

Blending maize with millet and sorghum flours can improve the nutritional quality of maize flour since these crops (and other candidate blending crops) represent rich sources of energy, amino acids, and several essential micronutrients (Table 16.1), and also are slow energy release, thus beneficial to diabetics. Sorghum and millet are also less prone to mycotoxin contamination.

Second, the flour blending policy would contribute to several economic outcomes, including employment generation and import substitution in the country. It aims to stimulate domestic agriculture, which means more jobs and income along the value chain of the candidate crops for blending. Maize flour processing and consumption represents a sizable economic activity in Kenya (Khamila et al. 2019). Packaged maize flour produced by large-scale millers has a market value of about US$444 million a year (about 35 percent of the market), while *posho* maize flour produced by small and microenterprise mills, a critical part of the rural economy, is worth a staggering $840 million. Maize is the most important staple food crop for about 96 percent of Kenyans. Household surveys indicate that it accounts for 9–18 percent of total household expenditures in the country. Moreover, the country relies heavily on government-subsidized imported maize to meet its domestic demand. FAOSTAT data value maize imports at about $389 million for the year 2017 (see Figure 16.2), when a catastrophic drought caused shortages of maize and sent the domestic price soaring. Our calculations based on data obtained from several sources indicate that, if maize flour is blended with 10 percent of either millet or sorghum, Kenya's maize import volume could drop by up to 42 percent, thereby saving on average $44.8 million a year. Further, the policy would reduce the country's overreliance on maize and strengthen its food sovereignty—that is, its ability to feed its population without relying on imports.

Third, the flour blending policy would contribute to improved sustainability by building resilient farm and food systems through promoting climate-tolerant crops and technologies. Sorghum and millet production is more resilient than

FIGURE 16.2 Maize production and imports in Kenya, 2010–2020

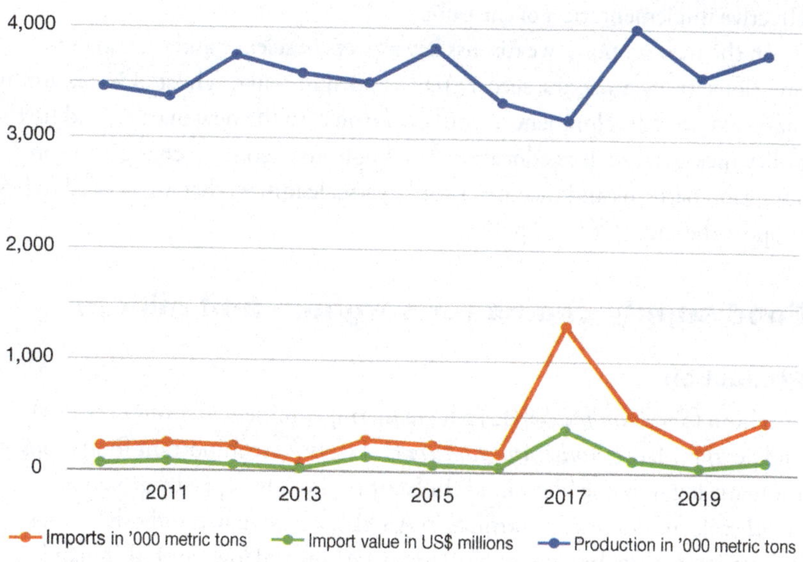

Source: FAOSTAT (2022).

maize production in the face of high climate change risks in drylands (Kilambya and Witwer 2019). Interest in these crops has grown recently as potential crops for a "New Green Revolution" (Goron and Raizada 2015). This is particularly pronounced in ASALs, where recurrent drought is a serious threat and climate change makes maize a riskier crop (Schipmann-Schwarze et al. 2015).

The proposed impact pathway of the policy is as follows: flour blending increases or creates demand for underused cereals, thus providing incentive to smallholder farmers and other value chain actors to work together and develop long-term business relationships. The efficacy of the impact pathway would essentially depend on the willingness and capacity of value chain actors to invest in the coordination, collaboration, and communication needed to respond to the new market opportunities. The success of the policy would also depend on capacity to manage potential trade-offs among expected outcomes for various actors (Mausch et al. 2020; Melesse et al. 2020). Maximizing returns of smallholder farmers through, for example, improved prices may undermine poor consumers' capacity to afford blended flours. Similarly, ecological sustainability considerations could also be at odds with the goal of maximizing farmers' returns. Balancing these competing goals in reality would not be easy. This calls for critical reflection on the assumptions, levers, and resistances that underpin the national flour blending policy, while meeting data and research needs for effective implementation of the policy.

In the next sections, we discuss key aspects of science and technology questions (traits, varieties, seed systems, and agronomy), a host of issues around market system development to connect farmers to the new market, and further policy measures, such as education campaigns to encourage changes in consumption patterns and incentives for large-scale millers that are needed to best support the success of this policy.

Food supply chains for sorghum and millets

Production

Sorghum (*Sorghum bicolor* (L.) Moench), finger millet (*Eleusine coracana*), and pearl millet (*Pennisetum glaucum* (L.) R. Br.) are important food crops for many low-income households in Kenya. They are typically grown by small-scale, resource-poor farmers in ASALs, where maize and other crops' cultivation is risky because of high temperatures and low rainfall (Kilambya and Witwer 2019). In addition, sorghum and millet require minimal external inputs and provide for increased soil cover when intercropped with legumes. While sorghum and millet can be grown in most parts of Kenya, their current production is mainly concentrated in the Eastern and Western regions. Despite the growing importance of sorghum and millet in the face of climate change, their production is characterized predominantly by low levels of input use and concomitantly low yield (Schipmann-Schwarze et al. 2015; see also Figure 16.3). Facing only their own demand with limited market access, households have weak incentives to adopt the new technologies necessary to increase productivity for more commercially oriented production.

The flour blending policy seeks to open up new and viable market prospects that provide incentives for farmers to produce beyond subsistence and commercialize these underused cereals. However, ensuring a regular supply of adequate volumes and suitable quality of sorghum and millet at relatively reasonable and stable prices will be important. This has proven the biggest deterrent to the development of a wider use of staple composite flours in the developing world (Fellers and Bean 1988; Nwanekezi 2013). Smallholders will likely need to improve their capacity to be able to adopt more productive and sustainable technologies, mitigate and manage different types of risks, and meet market quality standards (Karuho and Collins 2020). To achieve the immediate target of the policy—blending maize flour with 10 percent of either millet or sorghum,

FIGURE 16.3 Area allocation and crop productivity in Kenya, 2010–2020

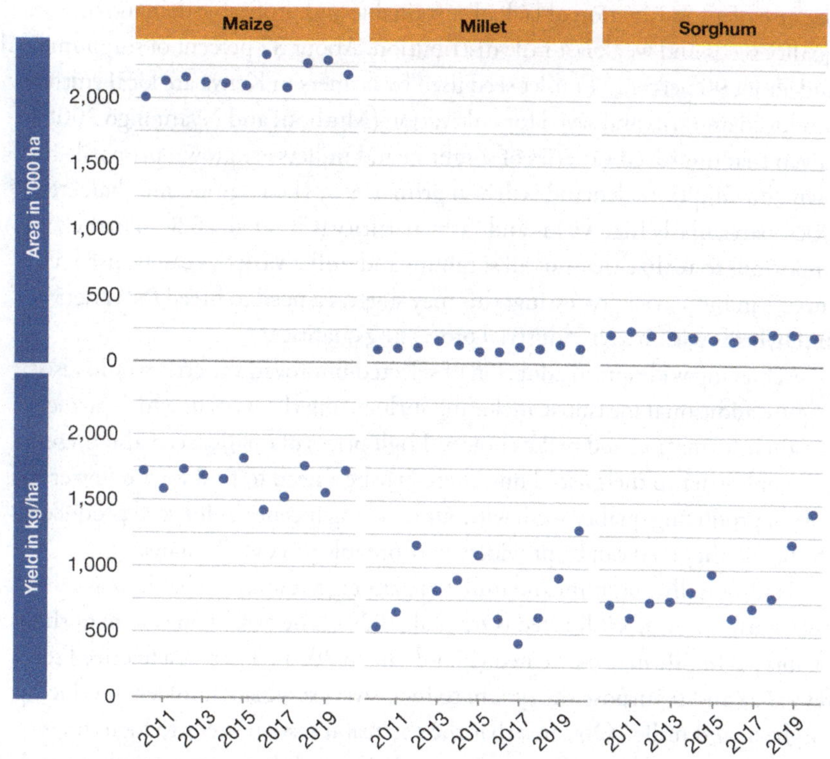

Source: FAOSTAT (2022).

production of millet and sorghum needs to increase by about 85 percent and 65 percent, respectively. Therefore, to produce the target crops at scale, smallholder farmers need to be leveraged through multiple interventions.

Developing effective seed systems and value chains is going to be critical for the timely and sustainable delivery of good-quality seed varieties that the flour market would demand. While more than 40 sorghum, more than 12 finger millet, and 3 pearl millet improved varieties have been developed and released to date through collaboration of the International Crops Research Institute for the Semi-Arid Tropics (ICRISAT), national agricultural research systems, especially the Kenya Agricultural and Livestock Research Organization, and universities in Kenya, seed systems for the crops are still undeveloped. Multiplication of sorghum and millet seeds has been minimal and is dominated by the public sector. Currently, there is no meaningful commercial seed delivery of these crops for smallholder farmers. Generally, seed is failing to reach smallholders because of a combination of lack of sustainable and profitable demand for quality seeds and weak last-mile distribution. About 87 percent of sorghum seed and about 90 percent of millet seed used by farmers in Kenya are local cultivars produced on-farm and saved for cultivation (Muthoni and Nyamongo 2008). Given that improved varieties of sorghum and millets are grown primarily for own-consumption, demand is driven primarily by their agronomic characteristics, particularly high yields and early maturity (Orr et al. 2020). Further, it is important to realize that not all sorghum and millet varieties can be used with success in industrial processing; this may suggest a need to breed for varieties and hybrids with better industrial processing qualities.

Achieving widespread adoption of selected improved varieties would also require additional measures, including influencing the ways in which farmers obtain information and make choices. High prices of quality seed also present a major barrier to their use. Thus, there may be a need to find ways to lower the cost of producing quality seed, without reducing incentives for seed producers, so that quality seed can be provided at affordable prices to farmers.

Traditionally, sorghum and millet are among a few crops that have received policy attention, albeit limited (Orr et al. 2020). The extension system of the country is highly maize-oriented (Handschuch 2014). There is a felt need to revitalize and repurpose the system to inclusively serve smallholders producing sorghum and millet. One area that the extension system needs to focus on is changing farmers' perceptions and attitudes toward these crops. Sorghum and millet are considered poor people's traditional food crops. This perception can keep farmers away from producing and consuming these crops (Bonke and Musshoff 2020). Sensitization, awareness campaigns, and training of farmers on the importance of farming and consumption of sorghum and millet may be needed to change farmers' preference for currently cultivated crops and expand the production of the targeted crops at scale. In addition, the extension system needs to transition farmer support services from "production-push" to climate-smart agronomic practices and business development services (Ferris et al. 2014).

Finally, while these are sporadic, sorghum and millet have competing uses. Market opportunities for sorghum and millet include as staple foods, raw material for the brewery industry, animal feed production, and export markets. But are smallholder farmers going to be able to supply sufficient quantities to satisfy these uses? This would likely require innovations beyond sustainably increasing crop productivity—for example, helping farmers access irrigation and reduce postharvest losses with improved logistics.

Grain market linkages

Commercial demand for millet is driven primarily by specialty markets (for example, food for children, expecting and nursing mothers, and healthy formulations for management of diabetics), while that for sorghum is limited to beer-making (Orr et al. 2020). The current commercial demand for the crops falls short of building significant capacity to absorb large grain volumes. The proposed flour blending policy is likely to lead to a significant increase in demand for sorghum and millet, which could create a lucrative market opportunity for smallholder farmers growing the crops.

However, many market constraints may preclude farmers from effectively participating in formal markets and hence from benefiting from this emerging market (Lapar, Holloway, and Ehui 2003; Barrett 2008). Further, business-minded actors may be strategizing how to derive maximum benefits from the policy. Individual smallholders may not be able to supply consistent and sufficient volumes of adequate quality to be attractive to market intermediaries. Semiarid drylands, where sorghum and millet are predominantly grown, are characterized by low population density, poor infrastructure, and limited access to markets, which can make transaction costs disproportionately high, increase seasonal price variations, and reduce incentives for both smallholder producers and buyers (Chamberlin and Jayne 2013; Melesse and Cecchi 2017). In such highly fragmented smallholder production systems, even if there is a surplus, it will be small and scattered, with ensuing problems of collection and increased cost of transport and of quality control. Thus establishing an efficient value chain would also require a very high level of planning and technological capability to ensure high and consistent quality. This is why estimates of food import requirements may not always be a function of production shortfalls in developing countries but may also of processors' inability to assemble adequate supplies from domestic sources to meet the needs of major urban centers. This is evident from the recent row when the international fast food chain, KFC, said it was unable to offer fries as it could not import its preferred pre-sliced potatoes, despite them being commonly grown locally and smallholder farmers having to sell them at low prices (Mwaura 2011).

Therefore, coordinated arrangements and marketing mechanisms are needed to better connect smallholder farmers into the supply chain. This may include reducing transaction costs, improving storage and market infrastructure, strengthening business services, and developing marketing mechanisms that distribute benefits equitably to farmers (for example, certification schemes) (Ferris et al. 2014). Market linkages need to be developed between sorghum and millet farmers and grain aggregators, assembly traders, large traders, and off-taker schemes. In most cases, local aggregators and traders are likely to play a vital role in buying sorghum and millet from markets in remote areas, which bigger traders and processors may not be able to access (Karuho and Collins 2020). Market intelligence and information systems can make markets more transparent and improve farmers' understanding of markets so they can engage more effectively in value chains. Digitizing market data and leveraging the potential of digital ICTs can enhance farmers' access to real-time market information across value chains by lowering transaction costs.

Furthermore, ensuring well-functioning markets requires building the institutional arrangements and support services necessary for competitive markets (Barrett 2008). Key entry points can include facilitating collective action models, warehouse receipt systems, and contract farming. Collective action models (for example, producer and marketing groups) can help smallholders achieve commercially viable volumes and strengthen their market power for better terms. Warehouse systems can reduce the pressure on farmers to sell immediately after harvest when prices are usually at their lowest levels, and potentially smooth seasonal price variations, while offering smallholders a market-based risk management instrument (Coulter and Onumah 2002). In Kenya, the National Cereals and Produce Board manages warehouse receipt systems serving maize marketing. These systems could be broadened to cover sorghum and millet. However, the seasonal variation in prices for sorghum and millet is currently much lower than for other crops (such as maize) (Orr et al. 2020); this may greatly decrease the potential benefits of a warehouse receipt system for these crops unless storage cost is reduced considerably.

Contract farming may also help foster certainty in market prices to assure farmers consistent and attractive financial benefits from market participation (Kaganzi et al. 2009) (see Chapter 15). Contract farming could possibly help ensure sufficient supplies of millet and sorghum grains, at least in the early stages of the policy cycle. This is because it can confer several benefits that could induce more farmers to produce these grains, including guaranteed income and reliable market access, risk sharing, and provision of inputs and appropriate technology.

The challenge for farmers is that price negotiation requires knowledge of their own costs of production and of prevailing market prices. A potential model to consider may be that millers contract directly with farmer groups, and also be used to provide extension and other input services. On the other hand, farmers must be able to supply the market in terms of the quality and reliability of supply required by buyers. Farmers often fail to honor contractual supply agreements, particularly in drought years, as growers prioritize household food security and are reluctant to sell (Orr, Mwema, and Mulinge 2014).

Finally, beyond traditional constraints to smallholder market participation, consideration should be given to smallholders' decision-making process and attitude to commercialization. Smallholders may not have commercial objectives and adequate technical competence to participate effectively in market development processes; gaining it may require a substantial change in farmers' attitudes and decision processes (FAO 2014).

Processing of blended flours

Flour processed for direct food consumption is currently the predominant value-added product from sorghum and millet in Kenya, accounting for 88 percent of use of the crops in the country (Schipmann-Schwarze et al. 2015; Orr et al. 2020). There is a large informal processing sector that consists mainly of traders who buy sorghum and millet grain, take it to a mill, and sell it at open air markets as flour. However, blending of flours from these crops with maize flour is currently not widely practiced in Kenya. Certain types of composite flours are marketed at supermarkets for baby foods, like *uji* flour.[2] Since standard recipes for *uji* and other blended flours do not exist, both composition and content vary from processor to processor.

Millers need to restructure and upgrade their mills and ramps to fit flour blending standards. For example, they may need to make capital investments in blending and metering equipment, as well as in storage and handling facilities. Small millers (SMEs), who account for about 70 percent of the overall maize flour production in the country (Khamila et al. 2019), may be unable to take advantage of new composite flour technology because they may lack modern equipment and technical expertise. However, smaller millers may have a unique opportunity to establish themselves close to local and regional collection centers in high-potential sorghum and millet production areas. Additionally, while large millers mainly target the formal market and urban consumers, SMEs working at local levels could benefit the poor and farm household consumption. Thus, they need to be supported in capacity development on blended flour standards, processing, and marketing. Other relevant support to them could include subsidies on equipment, low-cost financing options, technical assistance, and food safety and risk awareness trainings.

---

2 *Uji* flour is a mix of millet, sorghum, and several other crops, containing on average 30 percent millet.

Large millers are likely to be vital in the implementation process of the flour blending policy. Despite high potential gains from the new emerging market, millers are likely to be risk-averse and need to see value for their money before making investments to meet the demands of the new policy requirements. While the government pledges to provide subsidies and incentives to support the restructuring and upgrading of processing plants, millers are likely to be concerned about the long-term financial viability of blended flours. To mitigate this concern, a proof of concept is needed to demonstrate consumer acceptance and willingness to pay for blended flours, and whether premiums and market demand are large enough to induce millers to embed flour blending as a sustainable and profitable business.

Sorghum and millet appear to be among the better cereals for use in composite flour because of their predictable performance and compatibility with maize flour (Abdelghafor 2011), but availability may remain a concern for millers. Seasonal supply fluctuations and poor quality of grain may also pose challenges. Further, while sorghum and millet value chains are not well developed, smallholder value chains also behave in unpredictable ways (Orr 2018). There are some obvious reasons, like climatic events that lead to poor harvests, leaving smallholders with little to sell. For example, after poor rains in Eastern Kenya, only one-third of sorghum growers contracted by Smart Logistics for the sorghum beer value chain were willing to sell even part of their harvest (Orr 2018). Production of sorghum and millet for the new market could be scaled up by increasing their productivity through use of modern technologies and inputs, and allocation of more land for these crops, particularly in dryland regions of the country where other crops are less viable.

Demand for blended flours

While blending maize flour with sorghum and millet will boost the nutritional value of *ugali*, the blended flour is, however, likely to come with a change in color, taste, texture, and cooking procedure. Two considerations would play a key role in determining wider consumer acceptance of composite flours. First, most Kenyans have strong preference for pure white maize *ugali* (stiff porridge) as their staple food. Some would even say that they have not eaten food unless they have had maize *ugali*. Current consumer attitudes and habits may thus create inertia among consumers, inducing them to continue favoring consumption of pure maize over blended flours (Webb and Byrd-Bredbenner 2015). Such consumer preferences may also limit proportions of sorghum and millet to be used in blended flours (Orr, Gierend, and Choudhary 2017), implying that experimenting with blending formulation and pilot testing would be needed. Second, poor consumers may not be able to afford the relatively higher cost of blended flours.

Consumers make choices on what foods to eat based on their knowledge about nutrition and health benefits, their income, their conscious and unconscious preferences and motivation, and the price of food and availability (Sobal and Bisogni 2009). To increase consumption of blended flours, demand-side interventions may thus focus on changing relative prices or incomes or increasing consumer awareness of the nutrition and health benefits of target crops. Indeed, the flour blending strategy highlights the need for an elaborated plan for public awareness and image-building education to change consumer preferences, food habits, and perceptions to create enough demand for blended flours. However, this is easier said than done, especially for urban dwellers, the majority of whom have only known pure maize *ugali* and have never tasted *ugali* made of sorghum or millet flours. In addition, creating awareness to change consumer behavior at scale would require a huge investment.

Behavioral change communications can improve consumers' nutrition knowledge. However, acquiring nutrition information is only one step on the impact pathway of changing dietary behaviors. New knowledge alone is rarely enough to bring about desired changes in consumer behavior and eating habits—creating the well-known intention–behavior gap (Axelson, Federline, and Brinberg 1985). The recipient of the new information must have the motivation, ability, and opportunity to process the new information. Thus, it remains a challenge to shift consumers from the mode of passive recipients of external information or victims of external influences to the mode of active role players to change their dietary patterns and behavior. It is important to address the *why* aspect of recommended behaviors, by explaining the mechanisms through which the consumption of sorghum and millet can lead to better nutrition and health outcomes (Schneider and Masters 2019). This calls for well-laid strategies based on comprehensive market analysis and understanding of consumer behavior to enhance awareness about blended flours (Melesse and van den Berg 2021).

Commonly, dietary behavioral change interventions have found it difficult to impact behavior at a large scale. Also, they give very little support to the emergence of food movements that address the underpinning cultural attitudes and beliefs around food and consumption to embrace new consumption patterns. New perspectives and modalities are likely to be required to enable behavioral changes at large scale (community or country), including changes in underlying structures and mindsets at multiple levels.

Working through social structures, networks, and groups can facilitate quicker scaling of nutrition messages and behavioral change. In this approach, great potential lies in mobilizing consumer groups that can provide a platform where local consumer knowledge and insights can be combined with scientific knowledge to identify levers and barriers for consumer behavioral change. For example, schools could be leverage points to promote demand for and consumption of blended flours, because children's eating habits can be shaped as future consumers and schools have high potential to be effective hubs for promoting and accelerating the diffusion of information to reach adults through children-to-parent communication. Leveraging the power of procurement at large institutions, such as schools, hospitals, universities, prisons, and early childhood centers, and in food aid for emergency situations and community-based feeding programs, can play a key role in creating both market and consumer acceptance for blended flours. Targeting religious groups and harnessing neighboring effects can also be an important entry point. Importantly, such efforts need to target actors who can play reinforcing roles in promoting consumption of nutrient-dense blended flours, including nutrition and cooking clubs, community leaders, teachers, extension officers, nurses and health officers, community health workers, local administrations, and farmer organizations.

Further, consumer behavior and choices are shaped not only by individual factors but also by the complex food environment and system drivers (HLPE 2017). The key elements of the food environment that influence consumer food acceptability and choices are physical and economic access to food (proximity and affordability), food promotion, advertising and information, policy, and food quality and safety. As a result, interventions aiming at behavioral change need to be supported by everyday prompts, including physical proximity, affordability, labeling, advertising, quality, and safety signals.

As much as preferences, economic costs are likely to be a decisive factor, because they determine whether consumers are willing and able to pay a premium for blended flour. Sorghum and millet grains are more expensive than maize grains. Currently, sorghum sells at a 15–20 percent price premium over maize, and finger millet sells at a premium of 20 percent over sorghum and 40 percent over maize (Gierend and Orr 2015; Orr, Mwema, and Mulinge 2017) (see Figure 16.4). This could be attributed to low production, which means policies and investments are needed to support increased production of millet and sorghum. The cost of blended flour could also be affected by whether milling costs are the same as for maize, and whether sorghum and millet yield flour at the same extraction rate and by-products have the same value as maize.

FIGURE 16.4 Annual producer prices for crops in Kenya, 2010–2016

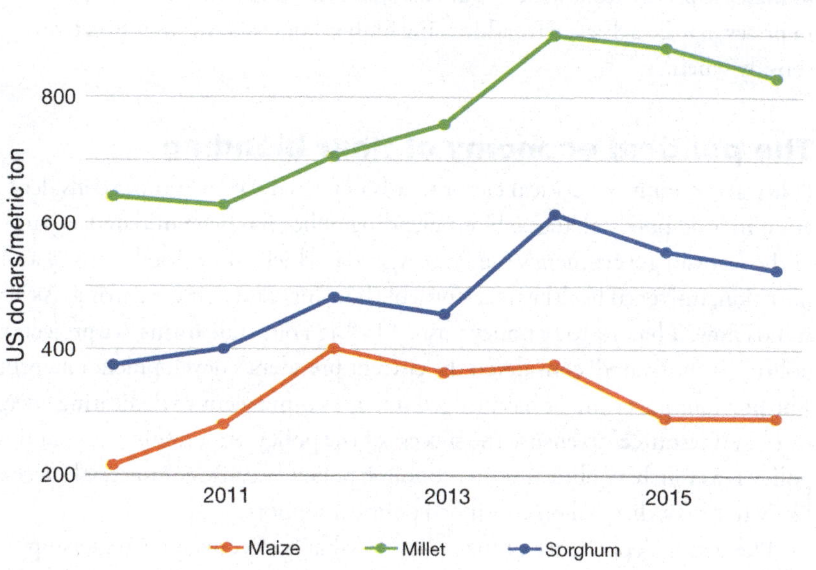

Source: FAOSTAT (2022).

Blended flours can be more expensive than the pure maize flour and hence receive little consumer support.

Further, market prices of blended flours may risk excluding poor urban households since they are dependent almost exclusively on the market for food and vulnerable to price increases. Thus, nutritional benefits may be felt mostly by high-income urban consumers who can buy commercially milled flours. On the other hand, it is important to recognize the unique position of farming households within the food value chain as both producers and consumers of food. While farmers may benefit through consumption of blended flours, the flour blending policy may raise grain and food prices in rural areas that traditionally have depended on sorghum and millet as staple foods. As a result, smallholder farmers may engage in selling the high-value crops (that is, sorghum and millet) and resort to lower-nutrition products for home consumption. This potentially creates trade-offs in rural vs. urban food security and nutrition goals. Thus, interventions that can benefit both producers and consumers are needed, such as increasing productivity and direct public investments in improving efficiency. Prices of sorghum and millet tend to remain relatively high because of scarcity. The policy could increase demand volumes to incentivize farmers to increase production, which could bring prices down. Further decreases in prices may be achieved by addressing high input costs through selective subsidy schemes.

The political economy of flour blending

Political economy is a critical factor that is not usually given proper consideration in food policy changes. Flour blending policy has been initiated as part of the Kenyan government's Big Four Agenda, which covers food security and nutrition, universal healthcare, affordable housing, and manufacturing. As it stands now, it has a strong policy push. The Big Four transformative projects are politically motivated, considered the current president's development blueprint. The previous government has pledged strong commitment to dedicating energy, time, and resources to ensure the success of the policy. As a result, the blending policy is as much a political as an economic policy, because commercial success is likely to be conditional on continued political support.

The strong dependence of the policy on what is effectively a partnership between government and business will make it vulnerable to policy changes. Additional complexity comes from the close links of business and politics in Kenya (Hornsby 2013) and the fact that priority-setting among policy objectives reflects the relative power of different ministries. The experience of the sorghum beer value chain in Kenya illustrates that conflicts between development and political objectives remain main sources of complexity and uncertainty in smallholder value chains (Orr 2018). In particular, political regime changes are a major source of uncertainty in developing countries.

Whether the blending policy will be supported and implemented as planned will depend on the level of support from the new administration. Signs are that agricultural policy may be leaning more toward maize production. In his inaugural speech, the president stopped consumption subsidies and announced maize production subsidies. Thus, the political economy around the blending policy is likely to be fraught with problems. Stakeholders may also remain skeptical of the stability of the current policy. As such, there is a need for appreciation on the part of policymakers of the complexity of the flour blending policy. The government needs to be realistic in its consideration of the policy so it can (1) integrate it into the national strategy and basic policies on agricultural production, agribusiness, incentives, regulations, and public communications and relations; (2) coordinate various ministries and agencies; and (3) allocate enough resources for the policy to succeed. Implementation of the flour blending policy needs robust support from county governments because agriculture is a devolved function in Kenya and resource allocation is pegged to county resources and budgets.

Existing political economy structures and factors may also align incentives of food system actors against target crops and the flour blending policy. For instance, sorghum and millet are largely "orphan" crops with regard to the country's dominantly maize-oriented extension system and public investment in agricultural development (Handschuch 2014), partly because drylands are less prioritized for maize production. Similarly, current arrangements and structures of grain markets are likely to work against sorghum and millet marketing since the prevailing market infrastructure and logistics in Kenya are set up primarily to accommodate the collection, processing, storage, and marketing of maize. Powerful, incumbent market actors (for example, maize importers and large millers) who are likely to lose some of their business may also lobby against the implementation of the policy and undermine political priorities and regulatory interventions. This may create lock-ins in the political economy. Engaging early with potential opposition or those who might lose from changes following the implementation of the flour blending policy may be a more effective strategy than simply pushing harder to scale the policy. This way, when the policy comes into play, it would not come as a shock and actors would not look for "work arounds." In sum, it is important to underscore the need for careful consideration of political economy factors, power dynamics, incentive structures, and social and cultural norms in existing systems, analyzing how they might enable or hinder the success of the flour blending policy.

Conclusions

Upcoming implementation of the national flour blending policy in Kenya is set to alter the food system landscape in the country and create a significant increase in demand for sorghum and millet. The policy seeks to create synergy between rural and urban development opportunities by linking urban consumption with rural agricultural production, prioritizing local markets and economic development. However, it remains to be seen who is going to benefit from the policy and which players along the value chain will adapt easily, and which others will be reluctant.

This chapter has applied a food systems perspective to analyze the flour blending policy, identifying knowledge gaps and challenges, as well as potential lock-ins and trade-offs, to inform strategic resource allocation and investments in the implementation of the policy. Our analysis reveals that a policy change of this sort deserves much more attention, and its success is likely to depend critically on a coordinated response across the different domains of the food system. Overall, the proposed flour blending policy needs to be approached with caution and holistically from a food systems perspective, but also from a human-centered perspective.

The complex dynamics in food systems imply that the flour blending policy cannot usefully be scaled in isolation (Hambloch et al. 2022). Many constraints within the food system may interact and mutually reinforce each other, creating systemic lock-ins and trade-offs between outcomes of the policy (Conti, Zanello, and Hall 2021). This means that the flour blending policy would need to be accompanied by systemic complementary interventions across the various domains of the food system. Such systemic interventions must extend beyond addressing technological and economic constraints to consider institutional, sociocultural, and political factors. Notably, proactive and coordinated interventions will be needed in the production, market, and policy arenas to continuously mitigate undesirable effects and potential trade-offs to ensure the success of the proposed flour blending policy.

Overall, achieving the goals of the proposed flour blending policy is likely to require a substantial shift in policy, and complementary changes and incentives in several thematic areas, including (1) technology choices, (2) institutions, policies, and incentives, (3) individual preferences, attitudes, and cultures, (4) power and political economy, (5) infrastructure, and (6) research and innovation priorities. This chapter has highlighted a range of potential interventions at various nodes of the food system, including addressing limited access to quality seeds of target crops; reorienting the current extension system to include and serve these crops; building capacity of aggregation systems and farmer organizations for collective action; locating processors near high-production areas of the target crops; and promoting the crops to create demand through consumer campaigns and targeting public procurement for blended flour (for example, for schools, hospitals, the military, and food aid).

Further, while the government is the main driving force, we should not overlook the roles of other stakeholders in navigating the complexity of this national flour blending policy. For example, ICRISAT, a global agricultural development research organization with sorghum and millet among its mandate crops, can help here to provide technology to help smallholders better respond to this opportunity. It can also provide systems research insights and tools to help better engage with and prepare for the complex, interrelated, and unpredictable nature of emerging opportunities in food systems.

References

Abdelghafor, R.F.M. 2011. *Quality of Bread From Composite Flour of Sorghum and Hard White Winter Wheat*. Khartoum: University of Khartoum.

Axelson, M.L., T.L. Federline, and D. Brinberg. 1985. "A Meta-Analysis of Food- and Nutrition-Related Research." *Journal of Nutrition Education* 17 (2): 51–54.

Barrett, C.B. 2008. "Smallholder Market Participation: Concepts and Evidence from Eastern and Southern Africa." *Food Policy* 33 (4): 299–317.

Bonke, V., and O. Musshoff. 2020. "Understanding German Farmer's Intention to Adopt Mixed Cropping Using the Theory of Planned Behavior." *Agronomy for Sustainable Development* 40 (6): 1–14.

Chamberlin, J., and T.S. Jayne. 2013. "Unpacking the Meaning of 'Market Access': Evidence from Rural Kenya." *World Development* 41: 245–264.

Conti, C., G. Zanello, and A. Hall. 2021. "Why Are Food Systems Resistant to New Directions of Change? A Systematic Review." *Global Food Security* 31: 100576.

Coulter, J., and G. Onumah. 2002. "The Role of Warehouse Receipt Systems in Enhanced Commodity Marketing and Rural Livelihoods in Africa." *Food Policy* 27 (4): 319–337.

De Brauw, A., I.D. Brouwer, H. Snoek, R. Vignola, M.B. Melesse, G. Lochetti, C. Van Wagenberg, M. Lundy, E. Maître d'Hôtel, and R. Ruben. 2019. "Food System Innovations for Healthier Diets in Low and Middle-Income Countries." IFPRI Discussion Paper 1816. International Food Policy Research Institute (IFPRI), Washington, DC.

FAO (Food and Agriculture Organization of the United Nations). 2014. *Understanding Smallholder Farmer Attitudes to Commercialization: The Case of Maize in Kenya*. Rome.

Fellers, D.A., and M.M. Bean. 1988. "Composite Flours." *Food Reviews International* 4 (2): 213–235.

Ferris, S., P. Robbins, R. Best, D. Seville, A. Buxton, J. Shriver, and E. Wei. 2014. *Linking Smallholder Farmers to Markets and the Implications for Extension and Advisory Services*. MEAS Brief No. 4. Urbana, Illinois: Monitoring Extension and Advisory Services Project.

Gierend, A., and A. Orr. 2015. "Consumer Demand for Sorghum and Millets in Eastern and Southern Africa: Priorities for the CGIAR Research Programme for Dryland Cereals." Socioeconomics Discussion Paper Series 35. International Crops Research Institute for the Semi-arid Tropics (ICRISAT), Patancheruvu, India.

Goron, T.L., and M.N. Raizada. 2015. "Genetic Diversity and Genomic Resources Available for the Small Millet Crops to Accelerate a New Green Revolution." *Frontiers in Plant Science* 6: 157.

Hambloch, C., K. Mausch, C. Conti, and A. Hall. 2022. "Simple Solutions for Complex Problems? What Is Missing in Agriculture for Nutrition Interventions." *Food Security* 15: 363–379.

Handschuch, C. 2014. "Traditional Food Crop Production and Marketing in Sub-Saharan Africa: The Case of Finger Millet in Western Kenya." PhD dissertation, Faculty of Agricultural Sciences, Georg-August-Universität Göttingen, Göttingen, Germany.

HLPE (High Level Panel of Experts). 2017. *Nutrition and Food Systems. A Report by the High Level Panel of Experts on Food Security and Nutrition*. Rome: FAO.

Hornsby, C. 2013. *Kenya: A History since Independence*. London: Bloomsbury Publishing.

Kaganzi, E., S. Ferris, J. Barham, A. Abenakyo, P. Sanginga, and J. Njuki. 2009. "Sustaining Linkages to High Value Markets through Collective Action in Uganda." *Food Policy* 34 (1): 23–30.

Karuho, O., and K. Collins. 2020. *Improving African Grain Markets for Smallholder Farmers*. Nairobi: Alliance for a Green Revolution in Africa (AGRA).

Kenya. 2013. Crops Act No. 16.

Khamila, S., D.S. Ndaka, A. Makokha, F. Kyallo, P.K. Kinyanjui, O.J. Kanensi, and J. Mwai. 2019. "Status of Commercial Maize Milling Industry and Flour Fortification in Kenya." *African Journal of Food Science* 13 (3): 65–82.

Kilambya, D., and M. Witwer. 2019. *Analysis of Incentives and Disincentives for Sorghum in Kenya*. Technical Note. Rome: FAO.

Lapar, M.L., G. Holloway, and S. Ehui. 2003. "Policy Options Promoting Market Participation among Smallholder Livestock Producers: A Case Study from the Phillipines." *Food Policy* 28 (3): 187–211.

Mausch, K., A. Hall, and C. Hambloch. 2020. "Colliding Paradigms and Trade-Offs: Agri-food Systems and Value Chain Interventions." *Global Food Security* 26: 100439.

Mbugua, S., E. Musikoyo, F. Ndungi, R. Sang, E. Kamau-Mbuthia, and D. Ngotho. 2014. "Determinants of Diarrhea among Young Children under the Age of Five in Kenya: Evidence from KDHS 2008-09." *African Population Studies* 28 (2): 1046–1056.

Melesse, M.B., and F. Cecchi. 2017. "Does Market Experience Attenuate Risk Aversion? Evidence from Landed Farm Households in Ethiopia." *World Development* 98: 447–466.

Melesse, M.B., and M. van den Berg. 2021. "Consumer Nutrition Knowledge and Dietary Behavior in Urban Ethiopia: A Comprehensive Study." *Ecology of Food and Nutrition* 60 (2): 244–256.

Melesse, M.B., M. van den Berg, C. Béné, A. de Brauw, and I.D. Brouwer. 2020. "Metrics to Analyze and Improve Diets through Food Systems in Low and Middle Income Countries." *Food Security* 12 (5): 1085–1105.

Mutegi, C.K., P.J. Cotty, and R. Bandyopadhyay. 2018. "Prevalence and Mitigation of Aflatoxins in Kenya (1960-to Date)." *World Mycotoxin Journal* 11 (3): 341–357.

Muthoni, J., and D.O. Nyamongo. 2008. "Seed Systems in Kenya and Their Relationship to On-farm Conservation of Food Crops." *Journal of New Seeds* 9 (4): 330–342.

Mwaura, W. 2022. "How KFC in Kenya Got Fried over Its Chip Shortage". BBC, Jan. 8.

National Research Council (NRC). 1996. *Lost Crops of Africa, Vol. I: Grains*. National Academy of Sciences, USA.

Nwanekezi, E.C. 2013. "Composite Flours for Baked Products and Possible Challenges: A Review." *Nigerian Food Journal* 31(2): 8–17.

Oduro-Obeng, H., and W.A. Plahar. 2017. "Development, Quality Evaluation and Estimated Contribution of Composite Flour Snack Foods to Nutrient Requirements of Young Children Aged 2 to 6 Years." *African Journal of Food Science* 11 (9): 318–329.

Orr, A. 2018. "Killing the Goose? The Value Chain for Sorghum Beer in Kenya." *Journal of Agribusiness in Developing and Emerging Economies* 8 (1) 34–53.

Orr, A., C. Mwema, and W. Mulinge. 2014. "The Value Chain for Sorghum Beer in Kenya." Socioeconomics Discussion Paper Series 16. International Crops Research Institute for the Semi-arid Tropics (ICRISAT), Patancheruvu, India.

Orr, A., A. Gierend, and D. Choudhary. 2017. "Value Chains for Sorghum and Millets in Eastern and Southern Africa: Priorities for the CGIAR Research Program for Dryland Cereals." Socioeconomics Discussion Paper Series 42. ICRISAT, Patancheruvu, India.

Orr, A., C. Schipmann-Schwarze, A. Gierend, S. Nedumaran, C. Mwema, E. Muange, E. Manyasa, and H. Ojulong. 2020. "Why Invest in Research and Development for Sorghum and Millets? The Business Case for East and Southern Africa." *Global Food Security* 26: 100458.

Schipmann-Schwarze, C., A. Orr, W. Mulinge, and J. Mafuru. 2015. "Sorghum and Finger Millet Flour Processing in Tanzania, Kenya, and Uganda." Socioeconomics Discussion Paper Series 32. ICRISAT, Patancheruvu, India.

Schneider, K., and W.A. Masters. 2019. "Orange Fanta versus Orange Fruit: A Novel Measure of Nutrition Knowledge in Malawi." *Maternal and Child Nutrition* 15 (1): e12656.

Sobal, J., and C.A. Bisogni. 2009. "Constructing Food Choice Decisions." *Annals of Behavioral Medicine* 38 (S1): s37–s46.

Webb, D., and C. Byrd-Bredbenner. 2015. "Overcoming Consumer Inertia to Dietary Guidance." *Advances in Nutrition* 6 (4): 391–396.

Chapter 17

TOWARD SUSTAINABLE TRANSFORMATION THROUGH POSTHARVEST MANAGEMENT: LESSONS FROM KENYA'S MANGO VALUE CHAIN

Jane Ambuko and Willis Owino

Management of postharvest food loss and waste (FLW) is an important strategy in efforts to sustainably meet the food and nutrition needs of the world's growing population. Sustainable food systems are critical to achieving food security and nutrition for all, now and in the future. Food systems cannot be sustainable when a large proportion of the food produced using limited resources is lost or wasted in the supply chain. At the global level, it is estimated that poor postharvest management means this is the case for 30 percent of the food produced for human consumption (FAO 2011, 2019).

The figure for Kenya is similar (Ministry of Agriculture, Livestock, Fisheries and Cooperatives 2018). The 2021 *Food Waste Index Report* (UNEP 2021) indicates that every Kenyan wastes about 100 kg of food every year, which adds up to 5.2 million metric tons[1] per year, excluding food loss that happens upstream, from production to retail. In monetary terms, wasteful consumption accounts for slightly over US$500 million annually (Mbatia 2021). FLW exacerbates food insecurity and has negative impacts on the environment through waste of precious land, water, farm inputs, and energy used in producing food that is not consumed. In addition, postharvest losses, caused by poor storage conditions, reduce income to farmers and contribute to higher food prices.

"Food security exists when all people, at all times, have physical, social and economic access to sufficient, safe and nutritious food which meets their dietary needs and food preferences for an active and healthy life" (World Food Summit 1996). Food and nutrition security for all remains an elusive global goal, and especially in sub-Saharan Africa, where one in five people suffer from some form of food insecurity. According to the Food and Agriculture Organization of the United Nations (FAO) (2021), about 26 percent of Kenya's population

1 Tons refers to metric tons throughout this volume.

is food-insecure, a situation that has been aggravated by events such the COVID-19 pandemic, locust plagues, and insufficient rainfall.

With an estimated growth rate of 2.3 percent per year, Kenya's current population of 53 million is set to rise to more than 100 million by the year 2050 (World Bank 2021). This calls for a paradigm shift in food production and consumption. Significant efforts have been made to increase production through expansion of agricultural land; increased inputs such as seed, water, and fertilizers; and overall intensification of production. However, increasing production of food that is ultimately not eaten, whether it is lost during the production and transformation processes or wasted at the consumption stage, entails a waste of economic and natural resources (HLPE 2014). Achieving food and nutrition security for the current population should not compromise the economic, social, and environmental bases for generating food security and nutrition for future generations. To create resilient and sustainable food systems, we must look beyond increasing production. Efforts must be made to ensure the food produced is used efficiently to reduce pressure on limited and inelastic production resources.

FLW reduction has become a subject of interest at the global, regional, and national level. At global level, it is enshrined in the Sustainable Development Goals (SDGs). Specifically, under SDG 12 on responsible consumption and production, target 12.3 calls for halving per capita global food waste at retail and consumer levels and reducing food loss along production and supply chains, including postharvest loss, by 2030. At the regional level, the African Union Heads of State and Government included in the 2014 Malabo Declaration a call to reduce postharvest losses by 50 percent by 2025. In Kenya, acknowledging the critical role of FLW reduction in efforts to address food and nutrition security, the Big Four agenda, under the Food and Nutrition Security pillar, sets a target of reducing FLW to 15 percent by 2022.

The benefits of FLW reduction in the food supply chain are subject to discussion, with opinions varying. In efforts to reduce FLW, there are both gainers and losers (HLPE 2014; FAO 2019). There is a cost to FLW reduction, and those who bear it may not necessarily enjoy the benefits of their efforts. The impact of FLW reduction on various actors in the supply chain (farmers, distributors, traders, processors, or consumers) depends on how the effect on food prices is distributed along the supply chain (FAO 2019). Therefore, in analyzing the impact of FLW reduction, optimal levels of FLW must be considered from both a private and societal perspective. Moreover, some level of FLW is unavoidable and tolerable and therefore acceptable as part of doing business (HLPE 2014).

Nevertheless, FLW represents needless use of limited resources to produce food that is not consumed and that ends up in landfills, with an even greater negative impact on food systems. Production, transportation, and handling of such food also has a significant negative impact on the environment. The total carbon footprint of food wastage is around 4.4 $GtCO_2$ eq per year, which is about 8 percent of total greenhouse gas emissions (WRI 2020). As such, FLW exacerbates the climate change crisis, thereby negatively affecting food production now and for future generations. Acknowledging the definition of "sustainable food systems" as ensuring food security and nutrition for all without compromising the economic, social, and environmental bases for generating the food security and nutrition of future generations (HLPE 2014), the critical role of FLW reduction is undeniable.

FLW is a complex food systems problem, which varies significantly with the context. Therefore, efforts to address FLW must be contextualized. Relevant factors include differences in region or location, including agroecological, socio-economic, sociocultural, and geopolitical variations. The causes and extent of FLW also vary significantly across food commodities, according to type, species, and even variety/breed within the same species. Food commodities have been categorized into five groups, namely cereals and pulses; fruits and vegetables; roots, tubers, and oil-bearing crops; animal products; and fish and fish products (FAO 2019).

In this chapter, we describe and discuss FLW in Kenya with a focus on the fruits and vegetables commodity group. We present a case study of mango because of its importance and contribution to Kenya's horticulture subsector. Over 80 percent of horticultural farmers in Kenya are smallholders who derive their livelihoods from 2–5 acres of land. Horticultural food crops produced in Kenya for the domestic and export market include fruits, vegetables, herbs, and spices. Among these, mango is the second most important fruit (by volume) produced in Kenya for the domestic and export markets (HCD 2018). Mango has great potential as a source of income and therefore economic empowerment for many smallholder farmers. The fruit is suited to different agroecological zones in Kenya (from sub-humid to semiarid) and therefore is grown in most of the 47 counties of Kenya as a cash crop. A steady increase in demand for mango, in both the domestic and the global markets, has led to expansion of the area under mango production from below 40,000 ha in 2015 to more than 50,000 ha in 2018 (HCD 2020).

However, as production volumes continue to increase in Kenya, high postharvest losses have been reported in the mango value chain. Such losses and other challenges in the mango value chain have hindered realization of the potential benefits of increased production volumes. Therefore, mango represents a good case study to highlight the importance of addressing FLW to complement efforts to increase production. The fruits and vegetables commodity group, to which mango belongs, sees high FLW (40–50 percent) (FAO 2011). The causes of and interventions to address FLW described in this chapter for mango are also relevant to other fruit and vegetable value chains.

Defining the problem of food loss and waste

To tackle the problem of FLW, there is a need for a common understanding of what it means—the extent of FLW and the causes or drivers, including the critical loss points. In addition, to trigger the necessary action, the impact of FLW on food and nutrition security and its environmental footprint must be demonstrated. Further, there is a need to highlight regional and national initiatives in place to reduce FLW.

Definitions and distinctions of terms used in postharvest management

FLW refers to the decrease in the mass of food (quantitative FLW) and the nutritional and/or economic quality of food (qualitative FLW) that was originally intended for human consumption. Food loss refers to food that is spilled, spoilt, or otherwise lost, or incurs a reduction of quality and value prior to the retail stage of the supply chain. Food loss typically takes place at the production, postharvest, processing, and distribution stages in the chain. It is a result of decisions and actions by food suppliers in the chain, excluding retailers, food service providers, and consumers (Figure 17.1). Food waste refers to food of good quality and that is fit for consumption that is not consumed because it is deliberately discarded. Food waste typically (but not exclusively) takes place at the retail

FIGURE 17.1 The distinction between food loss and food waste

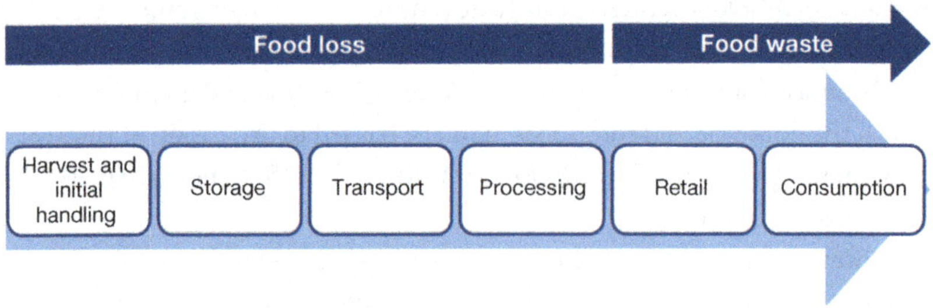

Source: Authors' illustration.

and consumption stages in the food supply chain (Figure 17.1). It results from decisions and actions by retailers, food service providers, and consumers.

A dimension of loss that is often ignored or sees little reporting is qualitative loss. Food quality loss or waste refers to the decrease of a quality attribute of food (nutritional, safety, or other aspect) that is linked to degradation at any stage of the food chain from harvest to consumption. Qualitative losses may occur without a decrease in the quantity of food and is therefore hardly reported. Decreases in nutritional value (for example, a decrease in vitamins) or economic value (for example, because of noncompliance with set standards) are examples of qualitative losses. Loss of quality can also lead to unsafe consumption that has a long-term effect on population health (HLPE 2014).

Extent of food loss and waste

Globally, an estimated 30 percent of food produced for human consumption is lost or wasted (FAO 2011, 2019. According to FAO (2011), FLW ranges between 26 and 36 percent globally but the distribution of food loss versus food waste along the supply chain differs across regions. For example, in sub-Saharan Africa, where FLW is estimated to be 36 percent, the highest losses (*food loss*) occur upstream at the production, postharvest handling, and storage stages. These stages alone account for 72 percent of total FLW, while the consumption stage accounts for only 5 percent. Developed economies in Europe and North America grapple more with downstream losses (*food waste*), with the retail and consumption stages accounting for most of the FLW (FAO 2011; HLPE 2014).

Certain value chains are more prone than others to FLW. For example, FLW in cereals and pulses is estimated to be about 8 percent, whereas 22 percent of fruits and vegetables are lost between production and the retail stage (FAO 2019). In Kenya, FLW in fruits and vegetables is even higher (40–50 percent or more) depending on the commodity. For mango in particular, FLW ranges between 35 and 45 percent (Gor et al. 2012; Snel et al. 2021). A major reason for high losses in fruits and vegetables is their perishability, which predisposes them to deterioration right from the point of harvesting up to consumption. The high losses in these nutritious food commodities have a negative effect on nutrition security, as Kenyan diets are generally deficient in fruits and vegetables.

Causes and drivers of food loss and waste

Identification of causes of FLW is critical in efforts to find context-appropriate solutions. Causes of FLW along the food supply chain are interrelated such that actions at one stage can affect all the rest. Immediate or direct causes of losses are linked to individuals' actions in dealing with the primary effects of a biological, microbial, chemical, biochemical, mechanical, physical, physiological, or psychological nature that lead to FLW. However, these immediate causes could be a result of other secondary reasons beyond the control of the individual actors.

Broadly, the causes of FLW can be organized into three levels, as micro-, meso-, and macro-level causes, based on the actors involved and the level of economic activity at which FLW is produced (HLPE 2014). Micro-level causes are primary causes of FLW that are attributed to actions or lack of action by individual actors at each stage of the supply chain from production to consumption. Meso-level causes are secondary or structural causes of FLW attributed to organizations or relationships of actors, the state of infrastructure, and other

FIGURE 17.2 Direct versus indirect causes of food loss and waste at different stages of the supply chain

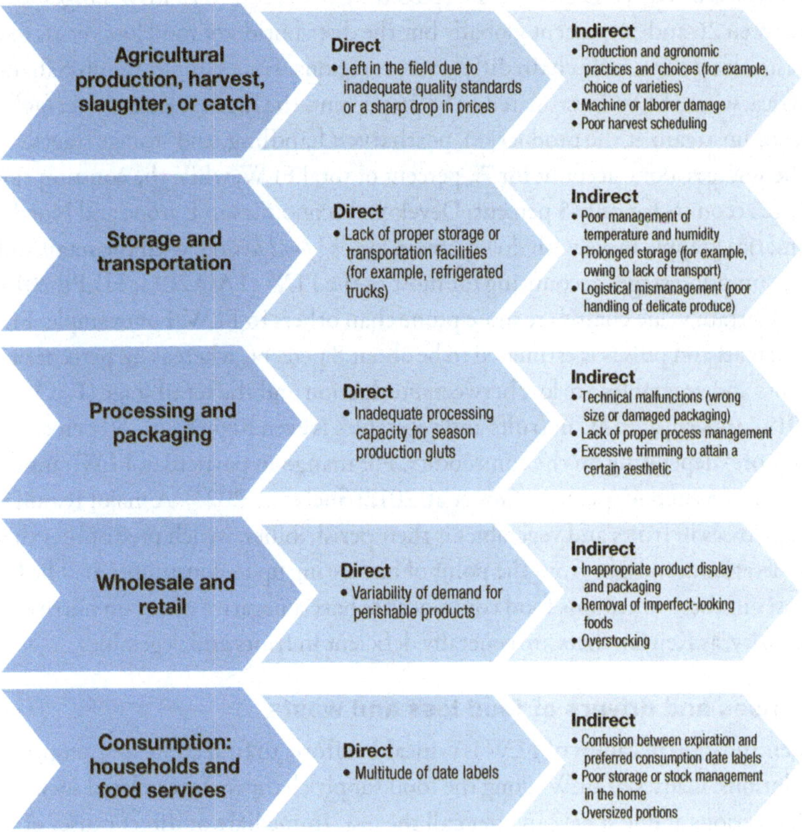

Source: Adapted by authors from FAO (2019).

factors beyond individual actions. They contribute to the occurrence and extent of micro-level losses. Macro-level causes are attributed to systemic issues such as an inadequate institutional or policy environment to enable proper functioning and coordination of food system actors. Macro-level causes point toward a food system malfunction.

Causes of FLW can also be categorized as direct and indirect (FAO 2019). The FAO describes direct causes as those attributed to actions (or lack of action) of individual actors that lead to FLW along the chain. Indirect causes are more systemic and concern the economic, cultural, and political environment of the food system in which the actors operate, and which may influence their decisions that lead to FLW. From a policy perspective, the indirect causes affect the decisions of individual actors that must be addressed so as to establish targeted interventions. It is noteworthy that, often, the losses observed at one stage of the supply chain could be a result of actions (or lack of action) at a different stage. Therefore, the supply chain should be viewed as a conveyor belt whereby the action of one actor at one stage could compromise the whole chain. Therefore, interventions to address FLW should be holistic and not isolated to the apparent causes at a single stage.

Causes of food loss and waste in the smallholder mango value chain

Mango is a climacteric fruit (like many other tropical fruits and vegetables), making it highly perishable, with a short shelf life after harvest. In postharvest handling, fruits such as mango are considered "living." This is because they still perform physiological functions such as respiration and transpiration, which are critical for living. As a climacteric fruit, mango can be harvested at physiological maturity and left to continue ripening after harvest. During ripening, the fruit undergoes compositional changes that transform the inedible form into the edible form with desirable eating characteristics. Biological/physiological processes that continue after harvest and that contribute to deterioration of mango fruits after harvest include respiration (which leads to depletion of stored food reserves), transpiration (which results in water loss), softening (which makes the produce prone to mechanical damage and subsequent rotting), and ethylene evolution (which triggers ripening and/or deteriorative changes). The rate of these biological processes depends on environmental factors including the temperature, humidity, and gas composition of the environment in which the fruit is stored or handled. Therefore, measures to preserve postharvest quality and slow deterioration and consequent FLW in mangoes are premised on the need to manage the biological and environmental factors that contribute to deterioration.

Poor harvest practices; poor handling of the harvested fruits; inappropriate storage practices, including poor cold chain management; and lack of capacity to transform the perishable fruit into shelf-stable products are some of the upstream micro-level causes of losses in mangoes and other fruits. At the meso level, poor public infrastructure, including roads, not only contributes to mechanical injuries during transport but also affects accessibility to markets. In addition, poor organization of actors in the value chain and stringent market standards are some of the meso-level drivers of FLW in mango value chains. At the macro level, FLW in mango is attributed to the impact of poor policies, laws, and regulations. For example, taxation regulations that deter the import of postharvest technologies have a negative impact on access to affordable technologies. Moreover, lack of investment in research and extension services by the government has negative impacts on efforts to develop and disseminate homegrown solutions to FLW. Figure 17.3 depicts the organization of common causes of FLW in the mango value chain (and those of other fruits and vegetables) at micro, meso, and macro levels.

FIGURE 17.3 Organization of common causes of losses in mango and other fruit and vegetable value chains

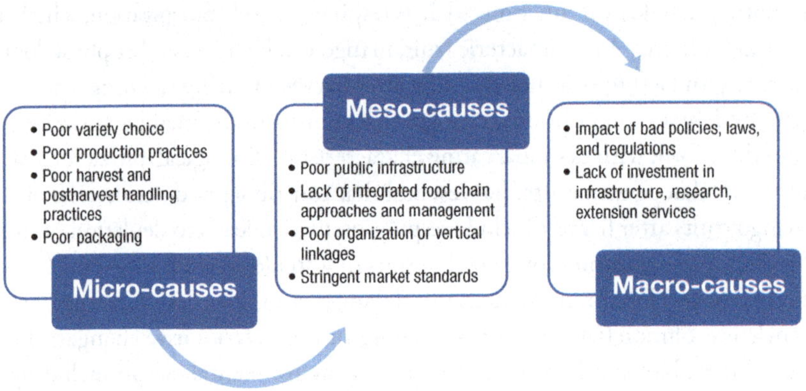

Source: Authors' illustration.

Addressing the causes of food loss and waste in the mango value chain

The solutions to FLW, like the causes, can also be organized at three levels (micro, meso, and macro) depending on the value chain. In the sections below, the interventions described focus on the mango value chain but are also applicable to most fruit and vegetable value chains in the Kenyan context.

Micro-level interventions to reduce food loss and waste

Interventions to address FLW must be designed to be context-appropriate, with attention to the characteristics of the geographic region, commodity, scale of operation, and stage of the supply chain, among other key considerations. There is no single intervention that can be recommended to holistically address the drivers/causes of postharvest losses. However, a set of technologies, practices, and other interventions may be combined to reduce FLW. Simple technologies and practices that individual actors (or actors organized in groups) can apply at different stages of the supply chain to preserve quality and reduce losses are described.

HARVEST AND IMMEDIATE HANDLING AFTER HARVEST

Poor harvest and immediate postharvest handing practices contribute significantly to postharvest losses in mango and other fruits and vegetables. Poor harvest practices result in mechanical injuries leading to immediate and/or later spoilage of commodities because they are predisposed to biological agents of deterioration. Although the injuries may not be evident at the time of harvest, they ultimately manifest at later stages of the supply chain, leading to disposal of the fruits (Snel et al. 2021). Therefore, harvesting practices should be aimed at minimizing mechanical injuries to the fruits. For example, in mango, simple harvesting tools such as fruit pickers can be used instead of the common practice of shaking tall trees to fell the fruits. After harvest, presorting the fruits to remove those that are infected or infested by pests and diseases is important to avoid contamination of the whole batch (Kader 2005). Further, sorting the fruits based on the size and stage of ripening can reduce on-farm losses because it reduces handling further down the supply chain.

PACKAGING AND TRANSPORT

Rough handling of produce during packaging for transport also results in mechanical injuries that affect the quality of the produce. Other handling practices such as using inappropriate containers may either cause injuries or predispose the fruits to faster deterioration as a result of the unfavorable environment in the packaging containers. Common materials used for packaging

FIGURE 17.4 Packing options for mango fruits

A. Standard bread crate B. Bread crates packed with mangoes C. Mangoes packed in a sack

Source: Authors.

fruits and vegetables by smallholders in developing countries like Kenya include wooden or plastic crates, nylon sacks, polythene bags, woven palm baskets, and jute sacks (Teutsch 2018). The aim of the packaging process is to protect the fruits from mechanical damage; to prevent physical, chemical, and biological contamination; and to avoid tampering with the fruits (Prasad and Kochhar 2014). Some of the packaging materials used by farmers and traders to package mango fruits, such as nylon sacks and polythene bags (Figure 17.4, plate C), accelerate produce deterioration leading to qualitative and quantitative losses.

Crates have been promoted as the preferred containers for handling fruits and vegetables. Traders prefer wooden crates because of the cost but their rough surfaces inflict injury on fruits. Packaging and handling of fruits and vegetables is shifting to plastic crates, which are clean, light, and durable. They deliver satisfactory protection against compression damage. Notably, unlike wooden types, they have a smooth inside finish, are easy to clean, and are reusable and stackable (Accorsi et al. 2014). A standard bread crate with a capacity of 20–30 kg (plates A and B) costs approximately $6. Special crates that are stackable, nestable, or collapsible cost a little more ($15–25) but have the advantage of saving space when empty. Nestable crates are especially recommended because they can be nested/stacked when packed with produce without causing any injuries to the produce.

COLD CHAIN MANAGEMENT IN MANGO VALUE CHAINS

Maintenance of low but safe temperatures during handling from harvest to end-user (the cold chain) is critical to preserve the quality of perishable produce such as mango (Ambuko et al. 2016). The cold chain for perishable produce ensures that cool temperatures are maintained as the fruit is handled during harvest, collection/aggregation, transport, storage, processing, and marketing until it reaches the final consumers (Kitinoja 2014). This includes minimizing delays between harvesting and cooling of the produce that could result in significant quantitative and qualitative losses (Kader 2002). Handling of perishable produce at suboptimal temperatures aggravates deterioration from biological agents. Research shows that an increase in temperature by 10°C above optimum increases the deterioration rate in perishable commodities two to threefold (Kader 2005).

A cold chain is often perceived as a complex and sophisticated system with high-tech facilities such as conventional cold rooms and refrigerated transport. However, a 2021 study of mango by Amwoka and colleagues revealed that an appreciable cold chain could be achieved through appropriate harvest and postharvest handling practices coupled with simple low-cost storage technologies. During harvest, the time of day at which harvesting is carried out has a significant effect on the postharvest longevity of the fruits (ibid.). For the majority of mango farmers, harvesting is a continuous process that can take place at any time of day as long as there is a buyer. However, harvesting produce at cooler times of day reduces the heat load on the produce that results from high temperatures and exposure to direct sunlight when fruits are harvested at hotter times of the day (Kiaya 2014; Amwoka et al. 2021). Harvested produce should be transported from the field to storage immediately. Delays in the field can expose the produce to more heat, leading again to a high heat load in harvested crops, which negatively affects shelf life and quality (Kiaya 2014). After harvest, mango fruits destined for long-term storage benefit greatly from precooling to remove the field heat. Precooling not only saves energy during cold storage but also ensures uniform produce temperature during storage (Amwoka et al. 2021).

Proper cold chain management practices complement storage technologies to preserve the quality of harvested produce. Conventional cold rooms provide the best temperature-controlled storage environment for fruits and vegetables. However, the cost of installation and maintenance of conventional cold rooms is beyond reach for most small-scale farmers in developing countries like Kenya. Unreliable access to electricity also presents a major constraint in adopting such technologies.

To overcome the challenge of access to conventional cooling for smallholders, there have been research efforts to find low-cost alternatives that are suited for rural areas in Kenya. These include off-grid evaporative cooling technologies, solar-powered cold storage, and affordable on-grid technologies. Off-grid evaporative cooling operates on the principle of evaporative heat exchange. When hot air from outside passes over a wetted pad/medium, the water in the wetted pad evaporates as it draws heat from the surrounding air, creating a cooling effect

(Lal Basediya, Samuel, and Beera 2013). The cooler and more humid conditions inside the evaporative cooling chamber preserve the quality and extend the shelf life of perishable horticultural produce. Research has shown that a temperature difference of 2–15°C between ambient air and inside an evaporative cooling chamber can be achieved depending on the time of day and season (Appendix 17.1). In addition, high relative humidity (≥99 percent) has been achieved in evaporative cooling chambers (Ambuko et al. 2017; Amwoka et al. 2021).

Various evaporative cooling technologies have been tested and proven effective at preserving quality in perishable produce. Examples of these include the evaporative charcoal cooler, the zero energy brick cooler, the pot-in-pot cooler, and the hessian sack cooling chamber. Appendix 17.1 describes the zero energy brick cooler and the evaporative charcoal cooler.

Other, off-grid solar-powered, cold storage technologies that have been promoted for application in mango and other perishable produce include Freshbox[2], Solar Freeze[3], and JuaBaridi, among others. Although these off-grid technologies have proven effective in preserving postharvest quality of mango and other fruits and vegetables, their adoption rate is still very low.

Low-tech on-grid solutions have also been proposed to overcome the cost constraints of conventional cold rooms. An example of this is the Coolbot™ cold room, which is a walk-in on-grid cold room that offers a low-cost alternative to conventional cold rooms. The Coolbot controller is an electronic gadget that uses multiple sensors and a programmed microcontroller that directs the air conditioner to operate at the desired temperature without freezing up (Dubey and Raman 2014). A 4 m by 4 m unit can cool up to 200 standard bread crates of stored fresh produce to temperatures as low as 4°C. The Coolbot technology is environmentally friendly, uses little energy, and has very low carbon emissions. The technology was introduced in Kenya on a pilot scale in 2015 and is available on order. On-station and on-farm studies have demonstrated its effectiveness in preserving quality and extending the shelf life of mango fruits (Karithi 2016; Ambuko et al. 2018a; Amwoka et al. 2021). Even though the Coolbot has significant cost advantages over conventional storage facilities, the costs are still out of reach for many smallholder farmers. A standard 4 m by 4 m unit can cost between $3,000 and $6,500 (compared with $10,000 for a conventional facility with similar capacity) depending on the level of sophistication and the availability of materials used in its fabrication (Kitinoja 2014; Karithi 2016; Ambuko

2 www.freshbox.co.ke

3 www.solarfreeze.co.ke

et al. 2018a). However, the Coolbot has the advantage of low installation and maintenance costs compared with conventional cold rooms.

COMPLEMENTARY TECHNOLOGIES FOR QUALITY PRESERVATION IN MANGO (AND OTHER FRUITS AND VEGETABLES)

In addition to improved practices and cold chains, there are complementary technologies that can be applied to enhance quality preservation in mango and other fruits. Examples of these are modified atmospheric packaging (MAP), the application of edible coatings and waxes, and natural plant hormones that can reduce losses. These technologies extend shelf life and preserve quality by reducing the rate of deteriorative processes such as respiration, transpiration, ethylene evolution, and pathological breakdown. For example, MAP using the right film has been shown to preserve the postharvest quality of mango fruits (Githiga et al. 2014). However, these beneficial effects of MAP can be realized only when the right film, whose permeability characteristics have been optimized to suit the physiological characteristics of the fruit, is used. In addition, the right storage temperature is important. A combination of the Coolbot cold room and MAP has been shown to extend the shelf of mango further compared with cold storage alone (Ambuko et al. 2018b).

The shelf life of mango fruits can also be extended through application of edible coatings or waxes. The thin film lowers the loss of water and slows gas diffusion resulting in reduced shriveling and respiration rates in the stored fruits. In addition, the thin film prevents fruit bruises during handling. Waxing effectiveness in mango has been demonstrated in the Apple mango variety, where waxing extended the fruits' shelf life by at least five days relative to unwaxed fruits (Maina et al. 2019).

The use of natural hormones can also improve the shelf life of fruits such as mango. An example of natural hormone-line compounds that have application in fruit quality preservation is 1-Methylcyclopropene (1-MCP). It is a competitive inhibitor of the ripening hormone, ethylene, which is known to trigger ripening and the related physiological processes that lead to spoilage of fruits and vegetables. The effectiveness of 1-MCP in extending the shelf life of mango fruit has been demonstrated in various mango varieties of commercial importance in Kenya, including Tommy Atkins and Apple (Ambuko et al. 2016). Although 1-MCP is widely used globally in fruit and vegetables, its adoption is limited in Kenya, and mainly in avocado fruits. Efforts are being made by the parent company, AgroFresh[4], to promote the use of 1-MCP in Kenya for other fruits and vegetables.

Meso-level interventions to reduce food loss and waste

Although the technologies described above have been shown to be effective and have the potential to reduce FLW in mango and other perishable commodities, their adoption is limited. Factors that limit adoption include high initial costs of installation (particularly for individual farmers), lack of scale to generate a positive return on investments, and absence of financial incentives to improve the quality of produce. This last issue arises because, especially in the local market, pricing is not guided by any quality standards. The organization of actors into groups can overcome these barriers and facilitate better vertical integration and market access. Operationalization of this can be achieved using different approaches. The sections below describe two approaches to the organization of horticultural farmers and linking them to markets (horizontal and vertical integration). The approaches represent meso-level interventions that have been tested and proven to work in Kenya's context.

SMALLHOLDER HORTICULTURE AGGREGATION CENTERS

Produce aggregation can help farmers achieve thc scale traders demand. In groups, farmers can collectively demand premiums for quality and share the costs of expensive technologies. In Kenya, the concept of produce aggregation has been pursued among smallholder mango farmers under the Rockefeller Foundation's YieldWise initiative. In this, smallholder mango farmers who are organized in groups gain access to cold storage facilities, allowing them to aggregate their individual small volumes over time to achieve the quantities traders demand. In addition, such centers set standards for the produce to be delivered by smallholder farmers, and thus accept only high-quality produce. This approach assures not only the quality but also the quantity and consistency of the produce aggregated. Box 17.1 describes the Karurumo smallholder horticultural self-help group (in Embu county in Kenya), which is one of the beneficiaries of the initiative's pilot. The farmers affiliated with the group have been able to aggregate their produce for targeted traders including exporters and local anchor buyers and traders.

---

4 www.agrofresh.com

BOX 17.1 Smallholder horticulture aggregation and processing centers: Mango case study

It is estimated that 40–50 percent of mango fruits produced in Kenya goes to waste, especially during the peak season between November and March. Because farmers lack storage facilities for the highly perishable fruit, they are at the mercy of brokers. A price survey conducted in 2017 and 2020 revealed that, while most traders buy mangoes at a paltry KSh 3–5 at the farmgate, the same fruits retail for as much as KSh 100 in Nairobi's retail outlets. In 2017, the University of Nairobi Postharvest Project team set out to change this with support from the Rockefeller Foundation's YieldWise initiative. The project sought to demonstrate the potential of smallholder aggregation and processing to reduce postharvest losses in fruits such as mango.

A smallholder horticultural aggregation and processing center was established for the Karurumo horticultural self-help group in Embu county. The center is a full-scale aggregation and processing center with cold storage facilities for aggregation complemented by simple equipment for small-scale wet and dry processing. The installed cold facilities include an evaporative charcoal cooler and two zero energy brick coolers as well as a Coolbot cold room. Based on best practices for horticultural produce handling and cold chain management, produce is sorted and graded based on the market destination. Thereafter, it is precooled in the evaporative coolers to remove the field heat prior to storage in the Coolbot cold room.

Installed small-scale processing facilities include a juice processing line and two solar tunnel dryers. These have enabled farmers to transform the unsold fruit into shelf-stable products. During the peak season, the fruits that are too ripe for the market or that have some defects that make then unsellable are pulped and pasteurized. The pulp is later used to make other products, such ready-to-drink juices, juice concentrates, and jam. With these facilities, farmers are no longer forced to sell the fruits to brokers at low prices. Meanwhile, with cold storage, they can aggregate produce that meets the requirements of traders, in quantity, quality, and consistency terms. This means they can collectively negotiate for better prices from traders. At this point, farmers have been able to negotiate KSh 10 per piece—more than twice the standard farmgate price paid by traders. And if traders are unwilling to pay better prices, farmers can transform the perishable fruits into shelf-stable products. With market access, these processed products have been shown to fetch even better returns than the fresh fruits. When mango fruits are out of season, farmers can use the storage and processing facilities for other fruits and vegetables.

Source: Ambuko (2020).

SHORT AND EFFICIENT SUPPLY CHAINS FOR FRUITS AND VEGETABLES

Production of fruits and vegetables must be linked to markets and consumers through stable value chains to ensure sustainability (FAO and CIRAD 2021). Long and inefficient supply chains contribute to high postharvest losses that affect supply. They also affect access to and affordability of fruits and vegetables, especially for low-income consumers (as seen in Chapter 4). Most urban consumers in Kenya depend on informal markets for their supply of fresh fruits and vegetables. The informal supply chain is highly inefficient, with very high postharvest losses reported at each stage. The cost of the losses is borne by smallholders and consumers, who end up paying for the inefficiencies in the supply chain.

Recognizing that 90 percent of the retail market in sub-Saharan Africa is informal and highly inefficient, Twiga Foods Ltd. (TFL)[5] saw a business opportunity that would address inefficiency by removing the many layers of intermediaries. This would in turn reduce postharvest losses and lower the cost of food, especially for urban consumers. Since the company entered the Kenya market space in 2014, it has revolutionized the retail trade that connects small-scale fruit and vegetable farmers in rural areas to traders in cities. TFL has addressed the market access challenge of smallholder farmers by replacing the unscrupulous brokers who take advantage of farmers by offering below-market prices for the produce or fail to buy produce during peak seasons. With the entry of TFL, farmers are assured of markets for their produce and fair pricing as TFL collects produce directly from them. TFL has registered traders who place orders through a sales representative or directly on TFL's app. The company then dispatches the order within 24 hours using its vehicles—free of charge. Payment to farmers is made within 24 hours of collection through the mobile money platform M-Pesa. This short and highly efficient chain has contributed to a reduction of postharvest losses from 30 percent to 4 percent for produce sold through the TFL platform. If this model could be replicated countrywide for various fruits and vegetables, it would not only reduce postharvest losses but also make fruits and vegetables accessible at affordable prices for many, while improving incomes for farmers.

Small-scale processing to reduce food loss and waste

Food processing minimizes undesirable biochemical changes that alter the nutritional and sensory composition or wholesomeness of food and thereby prolongs

5 https://twiga.com/

food shelf life. Food processing can be a game changer in sustainably reducing food loss and waste, boosting food security, contributing to livelihoods through gainful employment, and increasing national GDP. Processing perishable produce into shelf-stable products is especially important for perishable commodities such as fruits and vegetables where high postharvest losses are reported.

MANGO FRUIT PROCESSING—THE OPTIONS

Mango is a highly perishable seasonal fruit with significant postharvest losses reported during the peak season, which occurs between November and March in Kenya. To minimize such losses, the perishable fruit can be transformed to shelf-stable products through small-scale processing. Mangoes can be processed in a variety of ways, and each type of processing adds value to the final product, as shown in Appendix 17.2 (Owino and Ambuko 2021). For small-scale processors, some of these options are more viable than others. For example, preparing fresh cut mango is relatively low cost, and fresh cut mango is in demand in urban areas and can help improve the nutritional quality of street food. However, food safety remains an issue, and to assure fresh-like quality, minimize microbial contamination, and extend the shelf life of fresh cut mango, there is a need to use hygienic water with a combination of disinfectants, antimicrobials, anti-browning, and texture-maintaining preservatives (ibid.).

Mangoes can also be processed into pulp, to serve as a base for juice, wine, probiotic dairy drinks, or jellies (wet processing). The promotion of fruit-based beverages over soda can improve dietary quality. If mangoes are processed as dried fruits (for example, dehydrated), they can be added as supplements in formula or baked goods to increase micronutrient intake. In the case of drying, the type of mechanism used changes nutrient retention. For example, refractance window drying leads to better-quality and more nutritious mango leather than does solar drying (Owino and Ambuko 2021). The waste products of mango processing—peels and kernels—can be incorporated into other food products, cosmetics, and animal feed.

Kenya is well placed to take advantage of food processing as a way to reduce FLW, boost food security, generate value for small farmers and enterprises, and build food systems resilience. Both the demand and supply conditions are in place for transformative effects through processing. Data from the Kenya Association of Manufacturers (KAM) reveal that Kenya spends \$2.4 million on imports of food and beverages, pointing to local market demand for processed food products (KAM 2018). This demand is expected only to increase in coming years as a result of a growing middle class and urbanization.

The food processing subsector is already one of the largest in Kenya's manufacturing sector, contributing about 2.5 percent to national employment and 5.1 percent to Kenya's GDP (Chapter 2), while accounting for 15.3 percent of exports in 2021 (KIPPRA et al. 2023). The sector contains large-scale processors but is dominated by small and medium food processing enterprises (food processing SMEs) as well as informal businesses. In 2020, the Kenya National Bureau of Statistics economic survey (KNBS 2020) showed that the middle-class made up 44 percent of the population and was expected to continue expanding by an average annual rate of 5 percent. These rising numbers mean that more and more Kenyans have more disposable income and hence are demanding healthier diets (Chapter 4). Further, by 2030, 63 percent of the population is expected to reside in urban areas, where consumers are increasingly buying from supermarkets and county/municipal markets and turning to a more diversified diet, but also to easy-to-cook and highly palatable meals. These population trends point toward increased demand for processed, nutritious, and healthy food products, and consequently an opportunity for the expansion of food processing SMEs.

On the supply side, Kenya is the third-largest mango producer in Africa, with a production area of 50,550 ha, a production volume of 772,700 tons, and a value of KSh 11.72 billion in 2017 (TechnoServe 2021). Makueni, Machakos, Kilifi, and Kwale are the leading counties in terms of mango production, accounting for 28.2, 21.5, 15.0, and 7.7 percent shares, respectively. Only 3 percent of mangoes are exported, pointing toward a strong local market (SNV Netherlands and ProFound 2019; Mujuka et al. 2020). However, processed mango makes up a small portion of domestic sales: the domestic fresh market accounts for about 90 percent of mango produced while only about 5 percent is processed. With a short harvest season, more than 40 percent of production goes to waste as a result of processing and demand constraints. To take advantage of the potential of mango processing, several initiatives have been launched (Box 17.2).

INCREASED VALUE FROM MANGO PROCESSING

Although marketing of mangoes as fresh whole fruits is the most common practice among small-scale farmers in developing counties, processing the fruit into nutritious and safe products has the potential to accrue bigger profits for farmers and other actors (Owino and Ambuko 2021). As Figure 17.5 shows, any value addition to mango yields better returns compared with fresh mango sales. The most lucrative processed product from mango fruit is wine, with a net profit of $5,500 per ton of mango fruit. However, a sophisticated system is required to

BOX 17.2 Mango processing initiatives

A number of private and public sector initiatives for mango processing have been undertaken recently in Kenya. For instance, Makueni county government established the county-owned Makueni Fruit Processing Plant in 2017. Its objectives are to process mango pulp into purees and juices so as to stabilize fruit prices, reduce mango postharvest losses, provide local farmers with sustainable channels to generate an income, train farmers on new technology and processes, and create employment opportunities for community members.

Revival of the Tana River-based mango pulp manufacturer, Galole Fruit Processing Factory, in 2020 by the Coast Development Authority was intended to reduce mango postharvest losses, improve the living standards of over 30,000 coastal smallholder mango farmers, and create employment for youth in Tana River, Lamu, and Kilifi counties. The mango processing plant has the capacity to crush over 1,200 tons of mangoes per year.

Kitui Enterprise Promotion Company is a private business based in Kitui county that has taken advantage of the processing potential of mango and is involved in the production and distribution of mango juice, mango flakes, mango powder, and fortified flour, targeting mainly the local market.

Source: Makueni County Food Development and Marketing Authority, accessed June 2022. https://mcfdma.co.ke

FIGURE 17.5 Net profits for different processed products from mango fruit

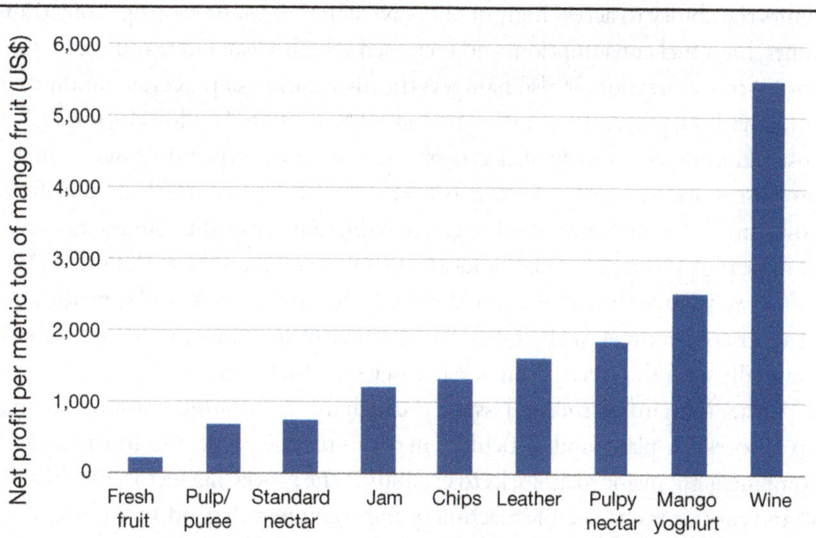

Source: Owino and Ambuko (2021).

produce the quality and quantity needed to compete favorably with imported wines in the Kenyan market. Mango puree requires only pulping and pasteurizing capacity, and is a common product for many small-scale processors but also the product with the lowest returns. The net profit on pulp from 1 ton of fruit is $700 (Owino and Ambuko 2021).

One of the simplest processing options for smallholder farmers/processors is drying (dehydration) because it does not require sophisticated equipment or facilities. The dried products include mango chips and mango leather (rolls). Mango chips and leather from 1 ton of mango fruits can fetch a net profit of $1,300 and $1,600, respectively. If mango drying is conducted using recommended good manufacturing practices and high hygiene standards, which ensure preservation of quality (nutritional and aesthetic) and safety, the dried products are highly recommended for small-scale farmers/processors in developing countries.

CHALLENGES TO SMALL-SCALE PROCESSING OF MANGO (AND OTHER FRUITS AND VEGETABLES)

Processing of mango (and other fruits and vegetables) in Kenya faces a number of challenges, particularly for small-scale processors. These challenges have hindered the contribution of small-scale processing to FLW reduction among smallholder farmers in mango (and other fruit and vegetable value chains).

First, lack of an all-season access road network, especially in rural areas, limits the ability to access high-quality raw materials for processing. Long transit times, high fuel consumption, and increased vehicle wear and tear increase the cost of transportation. It also hampers the distribution of processed products in rural markets, particularly during periods of heavy rain. In addition, the high cost and unreliable supply of electricity increases the production costs of small processors. Kenyan electricity tariffs are the fourth-highest in Africa, but the government has announced a 15 percent reduction across the country as a way of reducing production costs for locally manufactured products (KPLC 2021). However, voltage fluctuations and blackouts remain an issue, and substitution of alternative sources, such as diesel, is insufficient to reduce production costs, especially with the recent dramatic rise in global fuel prices.

Other than infrastructural issues, the high initial investment costs of setting up a processing plant, and difficulties in obtaining the proper machinery and equipment are major roadblocks to expanding the processing sector. Availability of and easy access to suitable machinery and equipment; a good and reliable supply of spares; equipment maintenance and other after-sale services; technical skills to operate the machinery; and efficient technology upgrading and advisory services are essential in the production of competitive food products. Currently, most food processing machinery and spare parts are imported from European countries, China, India, or Brazil, among others. This entails high import declaration fees, among other levies. Inability to accurately determine if foreign-manufactured machinery will suit local conditions can lead to the importation of inappropriate items. Availability of spares or service repairs may also be costly for imported machinery and equipment (Diao, Silver, and Takeshima 2016). Meanwhile, local manufacturers of processing machinery face a myriad of challenges, including high import duty on raw materials, poor aesthetics of locally fabricated equipment, high electricity costs, and proliferation of comparatively cheaper and aesthetically better food processing equipment (Ampah et al. 2021). The result is that many producers use obsolete equipment, which drives up their costs and makes their products less competitive on the market.

Access to finance has been one of the key challenges in expanding the activities of food processing SMEs, especially in their early growth and start-up phase, when they need to procure the prerequisite equipment and have sufficient operational capital. The perception of formal finance providers that there is a higher risk in lending to food processing SMEs leads to higher interest rates and an excessive collateral requirement, which, in turn, raises the cost of borrowing and limits access to finance (Were 2016).

The unpredictable supply of mango also constrains processers. Climate change impacts include adverse and erratic weather conditions, making supply fluctuations more common. High costs of production inputs (seed, fertilizer, etc.) can also result in a decline in levels of production, thus increasing the cost of raw materials for food processing SMEs. Maintaining the quality of the food raw material after harvest is another major constraint. This is partly the result of inadequate infrastructure for transporting or storing raw food materials, especially during periods of glut (Mujuka et al. 2020; George et al. 2021; Musyoka, Isaboke, and Ndirangu 2021; Snel et al. 2021).

In terms of marketing, food processing SMEs tend to produce and sell similar products to those of their competitors, with very few innovations to vary the composition, aesthetics/packaging, or even price. This lack of diversity weakens their positioning in the market since customers end up with limited variety (Chikez, Maier, and Sonka 2021). Further, counterfeit food products are displacing legitimate products in the market through informal channels. These are generally (although not always) retailed at lower prices than their legitimate equivalents, and with time they can squeeze the latter out of the market, reducing revenues for law-abiding companies. Despite the existence of quality inspection of imported food products by government agencies, counterfeit products still find their way into and distort local markets, affecting the profitability of food processing SMEs. Occasionally, counterfeit food products are seized and destroyed. However, it appears that government agencies lack capacity or willingness to deal with violations of regulations on the importation of counterfeit processed food.

Lack of expertise in processing constrains the sector too. Effective food processing depends on the availability of technical specialists. In general, most food processing SMEs are owned by local entrepreneurs, who generally have access to some capital to start the business but few to no technical skills in processing. Those food processing SMEs that engage experts with high levels of processing knowledge and management skills are more predisposed to adopt technologies and expertise that enable their products to penetrate markets and survive competition. Lack of adequate knowledge and management skills is one of the major causes of smallholder processing enterprise failure.

Academia is a source of knowledge creation, innovation, and technological advance, and ideally should generate the knowledge and technologies demanded by food processing SMEs. However, R&D in the food processing sector is largely governed by universities and research institutions, with very little involvement of the food industry. Universities are still regarded as ivory towers, generating knowledge without solving the challenges that would result in economic advancement for food processing SMEs.

Finally, the current regulatory framework poses a challenge to the sector. Kenya's national food safety system comprises 22 pieces of food safety and quality legislation that have been passed through various acts of parliament, and is managed by various agencies under different ministries and laws. The food processing SME business registration process and regulatory requirements are quite stringent and can be time-consuming. For instance, a business needs a Kenya Bureau of Standards (KEBS) certificate to operate. To obtain this, it needs processing facility approval by the public health authority, a hazard analysis and critical control point plan, compliance with labeling requirements, a National Environment Management Authority certificate, a public health certificate, and a medical certificate for each staff member. KEBS has 20 regional offices at which application for certification can be carried out. However, food products have to be sampled and taken for analysis at food laboratories in Nairobi, so the certification process can drag on for quite some time. Then there are taxes and levies, including municipal and county taxes and distribution levies, which can be prohibitive and drive food processing SMEs to informal operations.

Unless these challenges are addressed, the contribution of small-scale processing to FLW reduction efforts may not be realized in full.

Macro-level interventions to address food loss and waste

Although micro- and meso-level interventions have a direct effect on FLW reduction, an enabling policy environment is key to their success. Macro-level interventions are linked to the policy and regulatory environments that will affect actions (or lack thereof) by actors at the micro and meso levels.

POSTHARVEST MANAGEMENT POLICIES AND STRATEGIES

The African Union Commission's Continental Postharvest Loss Reduction Strategy developed in 2018 recognizes lack of relevant policies and coordination as one of the macro-level causes of FLW in most African countries. For example, in Kenya, although several national programs and strategies contain components of postharvest management, there is no specific policy to guide FLW reduction initiatives. A draft national strategy for postharvest management 2018–2025, cascaded from the continental strategy, is anchored on four pillars identified as drivers of postharvest loss reduction in Kenya: policies, institutions, reduction practices, and reduction services. Under the policy pillar, the strategy acknowledges that there is no policy focus on FLW reduction, along with no specific legislation and regulations on postharvest losses in Kenya. The overall framework on food losses is provided for in various laws. These include the Constitution of Kenya (2010), the Food, Drugs and Chemical Substances Act (Cap 254), the Crops Act (No. 16 of 2013), the Agriculture and Food Authority Act (No. 13 of 2013 revised 2015), the Meat Control Act (Cap 356), the Fisheries Act (Cap 378), the Dairy Industry Act (Cap 336), and the Standards Act (Cap 496), among others.

In addition, over the years, the Kenyan government has put in place several programs and strategies that have components aimed at addressing the drivers of FLW. Although these are not designed specifically to address postharvest loss reduction, there are initiatives therein aimed at FLW reduction. Examples include the National Agribusiness Strategy 2012, Kenya Youth Agribusiness Strategy 2018–2022, the Ministry of Agriculture, Livestock, and Fisheries Strategic Plan 2013–2017, the National Food and Nutrition Policy 2017–2022, the Agricultural Sector Transformation and Growth Strategy 2019–2029, and the food and nutrition pillar of the Big Four Agenda 2018–2022. All these strategies/programs allude to the importance of postharvest loss reduction through technology adoption, value addition, capacity building, and market access/linkages. The food and nutrition pillar of the Big Four Agenda has a set target to reduce overall postharvest losses from 30 percent in 2018 to 15 percent in 2022 and to increase agro-processing from 16 percent in 2018 to 50 percent in 2022.

These documents reflect the government's acknowledgement that FLW reduction is critical to the goal of attaining food and nutrition security. However, action on and support to FLW reduction initiatives remain limited. It is for this reason that Kenya falls short in the African Union's biennial review on progress toward realization of the 2014 Malabo targets for FLW reduction. For example, in the 2019 review, on the commitment to end hunger by 2025, Kenya scored 4.04 out of 10 against a target of 5.04 . On the commitment to postharvest loss reduction, Kenya scored a paltry 0.02 against a target of 3.00 out of 10. This indicates that Kenya is not on track to halve postharvest losses by 2025. Although this dismal performance is attributed in part to lack of data, it may also reflect a lack of commitment to assigning the resources to address the challenges identified. The national (draft) and continental strategies reveal that existing subsector policies focus more on boosting production and promoting markets, with less emphasis on FLW reduction along food supply chains. The strategies recommend a review of and update to existing FLW reduction policies and the development of policies that directly address FLW reduction.

STRENGTHENING INSTITUTIONAL CAPACITIES FOR FOOD LOSS AND WASTE REDUCTION

National institutions engaged directly or indirectly in FLW reduction activities require adequate capacity to collaborate with county governments, other public institutions, and the private sector in FLW reduction efforts. There is a need to strengthen existing institutional capacities toward effective implementation of FLW reduction interventions at national and county levels. This will require assessment of the existing institutional setting for FLW reduction and then strengthening technical capabilities, interaction and partnerships, reduction information management, and human capital and skills development.

INVESTMENT IN FOOD LOSS AND WASTE REDUCTION RESEARCH, CAPACITY BUILDING, AND EXTENSION

Globally, disproportionately small amounts of agricultural resources have been invested in the preservation of food (5 percent, compared with 95 percent invested in food production). Likewise, little research funding is allocated to postharvest management (Kitinoja et al. 2010). Education also puts more emphasis on production-inclined disciplines. It is noteworthy that most of the research in Kenya on postharvest management, including FLW reduction solutions, is supported by development partners. Therefore, research to find homegrown and context-appropriate solutions to FLW reduction is urgent. To address the knowledge and skills gap in postharvest management, capacity building at all levels is recommended. Very few tertiary institutions in Kenya offer diploma or degree programs in postharvest management (or anything closely related). This means that graduates lack hands-on skills in this domain. Curricula and short courses that target practitioners in food systems would help bridge the knowledge and skills gap among practitioners and extension agents.

In Kenya, access to extension services by farmers has continued on a downward trend over the years as the extension workforce ages and leaves the service. The devolution of agriculture and consequently extension services has aggravated the situation. Strengthening extension capacity is therefore critical to ensure extension agents are well equipped to reach out to farmers (and other practitioners) with the most current knowledge and skills on best practices and technologies for postharvest management.

Examples of food loss and waste reduction efforts that incorporate micro-, meso-, and macro-level solutions

FLW reduction requires multifaceted, multistakeholder, and complementary approaches that are context-appropriate. Such an approach is envisaged to include the application of appropriate technologies and practices, research to find sustainable solutions, education and training of food supply chain actors, and enabling policies. This section highlights two examples of multifaceted strategies that have been tested and proven effective to address FLW among smallholder fruit and vegetable farmers.

The YieldWise initiative

As described above, one approach to FLW reduction entails the smallholder aggregation and processing centers piloted under the YieldWise initiative supported by the Rockefeller Foundation. This has proven effective in addressing FLW among smallholder mango farmers.

The YieldWise initiative recognizes five barriers to addressing FLW reduction:

1. Limited knowledge of food loss and solutions
2. Broken distribution channels for loss-reducing technology
3. Limited capacity of farmers
4. Limited credit/financing
5. Difficulties in efficiently linking supply and demand

FIGURE 17.6 The four intervention areas of the YieldWise initiative

Market demand and linkages

- Linking large anchor buyers' demand for fresh and processed crops to smallholder supply, and local alternative markets to excess crops

Smallholder farmers' training and aggregation

- Aggregation of smallholder farmers into farmer groups to meet the quantity, quality, and consistency requirements of buyers
- Capacity building and other adoption measures to ensure smallholder farmers take up technology and other interventions, including adoption of best practices in the field (preharvest) to ensure high-quality produce

Technology

- Distribution and use of loss-reducing technologies to improve handling, storage, and processing of crops
- Supporting targeted breakthrough innovative technologies in specific value chains (for example, cold storage)

Financing

- Required to facilitate manufacturing, distribution, acquisition, and adoption of technologies, for example, loans and leasing models
- Investment capital to fund the scale-up of promising technologies and innovative distribution models

Source: TechnoServe (2021).

FIGURE 17.7 The Target-Measure-Act approach

Target

Countries/companies should set targets for FLW reduction that are aligned with the SDG 12.3 target of halving FLW by 2030 or the Malabo 2014 target of halving FLW by 2025. The hypothesis is that such targets will create ambition that will motivate action toward FLW reduction.

Measure

Each entity (government/company) should measure its own FLW. This is premised on the assumption that quantifying FLW within borders, operations, or supply chains can help decision-makers better understand how much food is lost/wasted, and where and why food is being lost or wasted. This evidence base provides a solid foundation for targeted and prioritized FLW reduction interventions. It is also key to monitoring progress toward achieving the set targets.

Action

Urgent action toward FLW reduction is needed to deliver results and progress on the set targets. There is no one solution to FLW that fits all, no silver bullet. All actions/interventions should be context-specific, taking into consideration the socioeconomic dynamics for each situation. There is a call to action by all actors in the food supply chain.

Source: Adapted from Flanagan (2019).

To address these barriers, the YieldWise strategy, which has been piloted in mango (Kenya), maize (Tanzania), and tomato and cassava (Nigeria), has focused on four intervention areas (Figure 17.6).

Although the pilot among smallholder mango farmers in Kenya faced some challenges that may have hindered full realization of the intended benefits, this is a promising approach that has been scaled to other commodities.

The Target-Measure-Act approach

The World Resources Institute proposes a multisectoral and multidisciplinary strategy for FLW reduction (Flanagan 2019). The strategy is anchored on three interventions—Target-Measure-Act, as Figure 17.7 shows. The generic model can be customized in the Kenyan context to target prioritized commodity value chains at the national or subnational levels or by individual companies.

Conclusion

The need to address FLW in our food supply chain is urgent, not only to realize food and nutrition security in sustainable food systems but also to ensure that the carbon footprint and negative impacts on the environment are reduced. The food system is complex, with diverse commodities and contexts, and requires solutions that are tailormade to each scenario. There is no single solution to FLW that fits all.

This chapter has used mango as a case study to represent the fruits and vegetables commodity group, which in Kenya and globally reports the highest losses. The causes of FLW in mango at the micro, meso, and macro levels and the corresponding solutions, as highlighted in this chapter, can be contextualized to other fruits and vegetables. Simple solutions, including low-tech postharvest handling practices and technologies for FLW reduction, have been described. These must be considered in context to achieve the intended impact.

Key to FLW reduction efforts is continued research to find homegrown and context-appropriate solutions, as well as capacity building of food supply chain practitioners on proper postharvest management. Similarly, there is a need to strengthen outreach and extension programs to ensure target users adopt research outputs. In addition, better coordination in supply chains in an enabling policy environment is a key ingredient to complement best practices and technologies adopted by individual actors in the supply chain.

Appendix 17.1 Off-grid evaporative cooling solutions

Evaporative cooling is a natural and physical phenomenon that operates on the principle of evaporative heat exchange. When hot air from outside passes over a wetted pad/medium, the water in the wetted pad evaporates as it draws heat from the surrounding air, creating a cooling effect (Lal Basediya, Samuel, and Beera 2013). The evaporative cooling decreases temperatures while increasing humidity inside the storage chambers. The cooler and high humidity conditions preserve the quality and extend the shelf life of perishable horticultural produce. Research has shown that evaporative coolers can achieve a temperature difference of as much as 15°C below ambient temperatures and increased humidity of ≥99 percent (Ambuko et al. 2017; Amwoka et al. 2021). Various evaporative cooling technologies have been tested and proven effective. Examples of these are the evaporative charcoal cooler, the zero energy brick cooler (zero energy cooling chamber), the pot-in-pot cooler, and the hessian sack cooling chamber. The zero energy brick cooler and evaporative charcoal cooler are described below.

Zero energy brick cooler (ZEBC): This is a double-walled structure made of bricks and covered with a moisture absorbing material. In between the double-walled bricks is sand that retains added water and keeps the inside of the ZEBC cool under the principle of evaporative cooling (Ambuko et al. 2016, 2017). As water evaporates from the wetted sand, it takes heat away from the produce and surrounding environment, leading to cooler and humid conditions inside the chamber. According to Roy (2011), the standard size of a ZEBC is 165 cm long, 115 cm wide, and 67.5 cm high, with the space between the doubled brick walls estimated to be 7.7 cm. The original design has some limitations with

FIGURE A17.1 Zero energy brick cooler designs

A. Original design ZEBC with a double brick wall and no reinforcement

B. Improved ZEBC using an improved single wall and cavitied bricks with reinforced steel rods

Source: Authors.

respect to capacity, stability, and longevity. There have been efforts to improve this through adaptive research. The improved ZEBC is larger and reinforced with steel rods to ensure stability of the structure, since the bricks are simply interlocked and not cemented. Figure A17.1 shows two versions of a ZEBC designed and fabricated by biosystems engineers from the University of Nairobi.

The ZEBC can achieve a temperature difference of between 2° and 15°C when compared with the ambient temperature and a relative humidity difference of up to 50 percent relative to the ambient room humidity. The relatively cool temperature and high humidity have been shown to extend the shelf life of mango fruits by 5–10 days in comparison with ambient room conditions (Amwoka et al. 2021).

Evaporative charcoal cooler (ECC): The evaporative charcoal cooler is a larger, walk-in, structure wherein the medium that holds water is charcoal. The charcoal is sandwiched between a double wall, usually made from chicken wire. The cooling efficacy of the ECC is similar to that of the ZEBC (Ambuko et al. 2017). There are various designs, with the choice depending on available resources and the prevailing conditions. Figure A17.2 shows the traditional charcoal cooler and an improved version, designed by biosystems engineers from the University of Nairobi. The improved version has been reinforced with an external fiberglass wall, which makes it stronger and able to withstand hard environmental conditions. The charcoal cooler can extend the shelf life of mango fruit by four days to two weeks depending on the harvest maturity of the fruits and the prevailing weather conditions.

Figure A17.3 presents a comparison of the cooling efficiency of the ZEBC and the EEC relative to ambient air conditions. Figure A17.4 shows differences in relative humidity for the evaporative cooling technologies and ambient air.

FIGURE A17.2 Evaporative charcoal cooler designs

A. Traditional charcoal cooler with improved and galvanized aluminum frames.

B. Improved evaporative charcoal cooler (right) reinforced with fiberglass walls.

Source: Authors.

FIGURE A17.3 Realtime changes in temperature in a ZEBC, EEC, and ambient room during a nine-day observation period

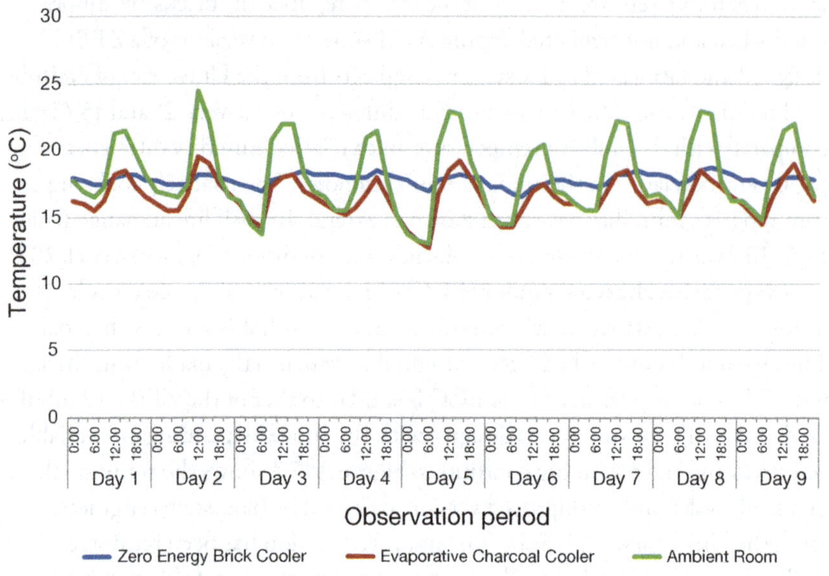

Source: Ambuko et al. (2018b).

FIGURE A17.4 Realtime changes in relative humidity in a ZEBC, EEC, and ambient room during a nine-day observation period

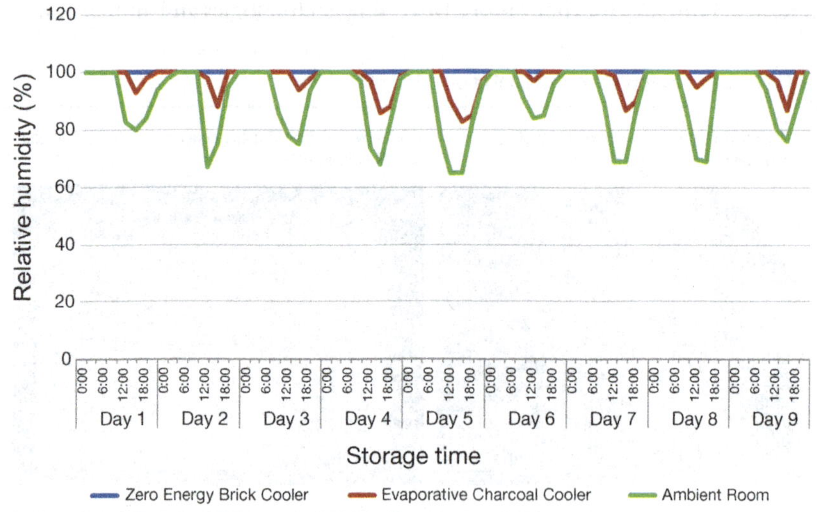

Source: Ambuko et al. (2018b).

Appendix 17.2 Processing of mango products

FIGURE A17.5 Processing of different products derived from mango

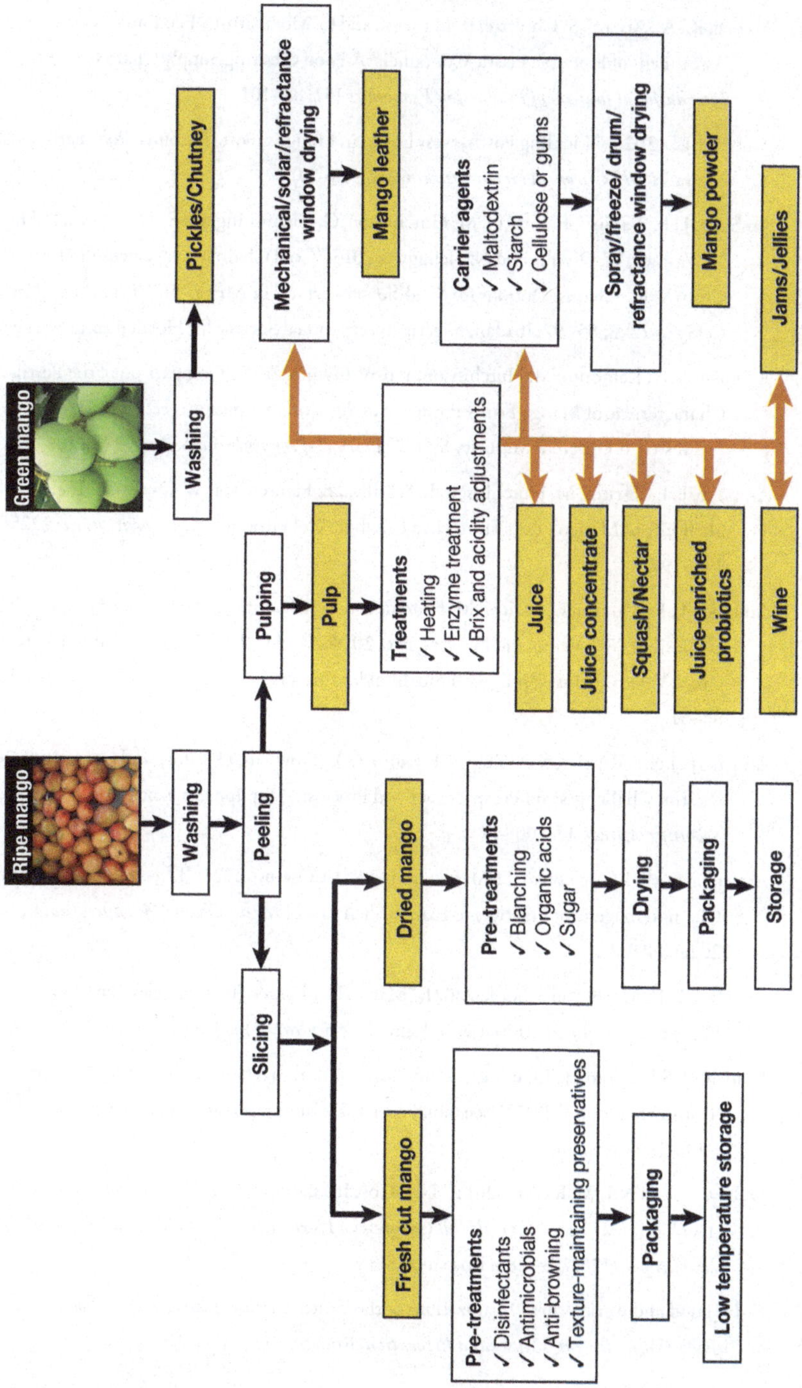

Source: Owino and Ambuko (2021).

References

Accorsi, R., A. Cascini, S. Cholette, R. Manzini, and C. Mora. 2014. "Economic and Environmental Assessment of Reusable Plastic Containers: A Food Catering Supply Chain Case Study." *International Journal of Production Economics* 152: 88–101.

Ambuko, J.L. 2020. "Tackling Postharvest Losses in Mango among Resource Poor Farmers in Kenya." *ISHS Chronica Horticulturae* 60 (2): 28–29).

Ambuko, J., F. Wanjiru, E. Karithi, M. Hutchinson,'G. Chemining'wa, E. Mwachoni, B. Hansen, L. Wasilwa, W. Owino, and N. Nenguwo. 2016. "Cold Chain Management in Horticultural Crops Value Chains: Options for Smallholder Farmers in Africa." In *III All Africa Horticultural Congress 1225,* 85–92. Ibadan, Nigeria: International Society for Horticultural Science (ISHS).

Ambuko, J., N. Kemunto, M. Hutchinson, and W. Owino. 2017. "Comparison of the Postharvest Characteristic of Mango Fruits Produced under Contrasting Agro-ecological Conditions and Harvested at Different Maturity Stages." *Journal of Agricultural Sciences* 9 (8): 181–192.

Ambuko, J., E. Karithi, M. Hutchinson, L. Wasilwa, B. Hansen, and W. Owino. 2018a. "Postharvest Shelf Life of Mango Fruits Stored in a Coolbot™ Cold Room." *Acta Horticulturae* 1225: 193–197.

Ambuko, J., F. Wanjiru, E. Karithi, M. Hutchinson, G. Chemining'wa, E. Mwachoni, B. Hansen, L. Wasilwa, W. Owino, and N. Nenguwo. 2018b. "Cold Chain Management in Horticultural Crops Value Chains: Options for Smallholder Farmers in Africa." *Acta Horticulturae* 1225: 85–91.

Ampah, J., J.X.F. Ribeiro, K.A. Bugyei, F. Kumi, J.O. Akowuah, H. Ofori, and C. Otchere. 2021. "Status, Challenges and Prospects of Food Processing Equipment Fabricators in Ghana." *Scientific African* 12: e00843

Amwoka, E.M., J.L. Ambuko, H.M. Jesang, and W.O. Owino. 2021. "Effectiveness of Selected Cold Chain Management Practices to Extend Shelf Life of Mango Fruit." *Advances in Agriculture* 2021: 8859144.

Chikez, H., D. Maier, and S. Sonka. 2021. "Mango Postharvest Technologies: An Observational Study of the Yieldwise Initiative in Kenya." *Agriculture* 11 (7): 623.

Diao, X., J. Silver, and H. Takeshima. 2016. "Agricultural Mechanization and Agricultural Transformation." IFPRI Discussion Paper 1527. International Food Policy Research Institute (IFPRI), Washington, DC.

Dubey, N., and N.L.M. Raman. 2014. "Low-Cost Insulation for Cool-Rooms Using 'CoolBot'." In *XXIX International Horticultural Congress on Horticulture: Sustaining Lives, Livelihoods and Landscapes 1120,* 279–284. Brisbane: ISHS.

FAO (Food and Agriculture Organizations of the United Nations). 2011. *Global Food Losses and Food Waste- Extent, Causes and Prevention.* Rome.

FAO. 2019. *The State of Food and Agriculture 2019: Moving Forward on Food Loss and Waste Reduction.* Rome.

Flanagan, K. 2019. "Is the World on Track to Cut Food Loss and Waste in Half by 2030?" *WRI Insights,* September 23.

Flanagan, K., K. Robertson, and C. Hanson. 2019. *Reducing Food Loss and Waste. Setting the Global Action Agenda.* Washington, DC: World Resources Institute.

George, A., L.M. Obiero, O.M. Wandayi, D.O. Gekonge, E.O. Geoffrey, and M. Villacampa. 2021. "Current Practices Concerning the Environmental Management Systems among Horticultural Processing MSMES in Kenya." *East African Journal of Science, Technology and Innovation* 2 (Special Issue): 1–16.

Githiga, R., J. Ambuko, M. Hutchinson, and W. Owino. 2014. "Effect of Activebag Modified Atmosphere Packaging on the Postharvest Characteristics of Mango Fruits, *Mangifera indica* L, Cultivar Tommy Atkins." *Journal of Applied Biosciences* 83: 7535–7544.

Gor, C., S. Agong, L. Acheng, W. Akuno, and D. Andika. 2012. The Interface between Mango Value Chain Analysis and the Socioeconomic Determinants." *African Journal of Horticulture Science* 6 (1): 1–16.

HCD (Horticultural Crops Directorate). 2018. *Horticulture, Validated Report 2016–2017.* Nairobi.

HCD. 2020. *Horticulture Data 2019–2020 Validation Report.* Nairobi: Ministry of Agriculture.

HLPE (High Level Panel of Experts). 2014. *Food Losses and Waste in the Context of Sustainable Food Systems.* Rome: FAO.

Kader, A.A. 2002. *Postharvest Technology of Horticultural Crops, Third Edition.* Oakland, CA: University of California.

Kader, A.A. 2005. "Postharvest Handling." In *The Biology of Horticulture, 2nd Ed*, eds. J.E. Preace and P.E. Reed, 379–407. Hoboken, NJ: John Wiley and Sons.

KAM (Kenya Association of Manufacturers). 2018. *Policy and Sustainability Report 2018.* Nairobi.

KAM. 2020. *Counterfeit Goods and Illicit Trade: Kenya Association of Manufacturers (KAM) Report 2020.* Nairobi.

KAM. 2020. *Policy and Sustainability Report 2020.* Nairobi.

Karithi, E.M. 2016. "Evaluation of the Efficacy of Coolbot™ Cold Storage Technology to Preserve Quality and Extend Shelf Life of Mango Fruits." Master's thesis, Department of Plant Science and Crop Protection, University of Nairobi, Nairobi.

Kiaya, V. 2014. *Post-harvest Losses and Strategies to Reduce Them.* Paris: Action Contre la Faim (ACF).

KIPPRA (The Kenya Institute for Public Policy Research and Analysis), Kenya National Bureau of Statistics (KNBS), and IFPRI. 2023. 2021 Social Accounting Matrix for Kenya: A Nexus Project SAM.

Kitinoja, L. 2014. *Use of Cold Chains for Reducing Food Losses in Developing Countries*. PEF White Paper No. 13-03. La Pine, OR: Postharvest Education Foundation.

Kitinoja, L., H.A. AlHassan, S. Saran, and S.K. Roy. 2010. "Identification of Appropriate Postharvest Technologies for Improving Market Access and Incomes for Small Horticultural Farmers in sub-Saharan Africa and South Asia." *Acta Horticulturae* 934: 31–40.

KNBS (Kenya National Bureau of Statistics). 2020. *Economic Survey 2020*. Nairobi.

KPLC (Kenya Power and Lighting Company). 2021. "Implementation of Power Bill Reduction." Press Release, December 16.

Lal Basediya, A., D.V.K. Samuel, and V. Beera. 2013. "Evaporative Cooling System for Storage of Fruits and Vegetables: A Review." *Journal of Food Science and Technology* 50 (3): 429–442.

Maina, B., J. Ambuko, M.J. Hutchinson, and W.O. Owino. 2019. "The Effect of Waxing Options on Shelf Life and Postharvest Quality of "Ngowe" Mango Fruits under Different Storage Conditions." *Advances in Agriculture* 2019: 5085636.

Mbatia, A. 2021. "UNFSS Independent Dialogue on Food Loss and Waste in Kenya." *Farm Africa News*, August 27.

Ministry of Agriculture, Livestock, Fisheries and Cooperatives. 2018. *Kenya Postharvest Loss Strategy 2018–2025*. Nairobi.

Mujuka, E., J. Mburu, A. Ogutu, and J. Ambuko. 2020. "Returns to Investment in Postharvest Loss Reduction Technologies among Mango Farmers in Embu County, Kenya." *Food and Energy Security* 9 (1): e195.

Musyoka, J.K., H.N. Isaboke, and S.N. Ndirangu. 2021. "Determinants of Quantity of Mango Supply among Small-Scale Farmers in Machakos County, Kenya." Paper presented at the 3rd International Conference at Machakos University, June 2021, Machakos, Kenya.

Owino, W.O., and Ambuko, J.L. 2021. "Mango Fruit Processing: Options for Small-Scale Processors in Developing Countries." *Agriculture* 11 (11): 1105.

Prasad, P., and A. Kochhar. 2014. "Active Packaging in Food Industry: A Review." *Journal of Environmental Science, Toxicology and Food Technology* 8 (5): 1–7.

Snel, H., J. Broeze, F. Kremer, E. Osena, J. Muyela, J. Erick, E. Wanjuu, and A. van Spronsen. 2021. "A Food System Analysis of Kenya's Mango, Avocado and Poultry Sectors: Assessing Opportunities to Reduce Food Losses," No. WCDI-21-185.Wageningen Centre for Development Innovation, Wageningen, Netherlands.

SNV Netherlands and ProFound. 2019. *Potential for Kenya to Export Fresh and Dried Mango to Europe: A European Market Analysis of the Opportunities, Trends, Requirements, and Channels to Gain Access to the European Market*. The Hague, Netherlands: SNV Netherlands.

Teutsch, B. 2018. *100 under $100: Tools for Reducing Postharvest Losses.* La Pine, OR: Postharvest Education Foundation.

TechnoServe. 2021. *The Impact of YieldWise on Mango Production in Kenya.*

UNEP (United Nations Environment Programme). 2021. *Food Waste Index Report.* Nairobi.

UNEP DTU Partnership and UNEP. 2021. *Reducing Consumer Food Waste Using Green and Digital Technologies.* Copenhagen: UNEP DTU Partnership; Nairobi: UNEP.

Were, A. 2016. "Manufacturing in Kenya: Features, Challenges and Opportunities." *International Journal of Science, Management and Engineering* 4 (6): 15–26.

WRI (World Resources Institute). 2020. *Food Loss and Waste: A Global Review.* Washington, DC.

World Bank. 2021. "Kenya." https://data.worldbank.org/country/KE

Chapter 18

DIGITAL INNOVATIONS AND AGRICULTURAL TRANSFORMATION IN AFRICA: LESSONS FROM KENYA

Martin Paul Jr. Tabe-Ojong, Gashaw T. Abate, Kibrom A. Abay, and David J. Spielman

Digital innovation is a key feature in the global and national discourse on food systems transformation. Efforts to better integrate food systems—defined here as the constellation of actors and their activities originating from agriculture, livestock, forestry, or fisheries, as well as the broader economic, societal, and natural environments in which they operate, including the production, aggregation, processing, distribution, consumption, and disposal of food products (Dwivedi et al. 2017; FAO 2018; Njuki et al. 2021)—will depend partly on how digital technologies can be used to bolster engagement, coordination, and innovation among a wider and more inclusive set of actors, including marginalized and vulnerable groups (Benfica et al. 2021).

Small-scale and resource-poor farmers are one such marginalized and vulnerable group, particularly within emerging African food systems. Already, there is rapidly growing enthusiasm around the potential of digital tools to transform smallholder agriculture, with substantial attention being paid to the opportunities in public policy discourse throughout Africa. The rapid spread of mobile phones and growing internet penetration in low- and middle-income countries, as well as promising evidence on positive impacts, justify this enthusiasm. In the past few decades, many public and private sector initiatives have been launched with the objective of building digital tools to transform smallholder agriculture in developing countries. These initiatives have led to the proliferation of digital innovations that aim to address smallholders' information, skills, and market constraints. These digital innovations mostly aim to address alternative forms and sources of market failures as well as institutional delivery bottlenecks.

Several evaluations show that digital innovations have considerable potential to revolutionize smallholder agriculture in Africa and, in particular, in Kenya (Kikulwe, Fischer, and Qaim 2014; Baumuller 2018; Abay et al. 2021; Benfica et al. 2021). Kenya was an early beneficiary of the boom in digital innovations to support smallholder agriculture in Africa. Pilots aiming to experiment with various digital innovations in sub-Saharan Africa are disproportionally centered in Kenya (CTA 2019). As such, Kenya has important lessons to offer many African countries on both how to leverage digital solutions to support smallholder agriculture and how to attract investment in digital innovation.

Despite these signs of progress, many digital innovations targeting smallholder farmers in Kenya—and in Africa more generally—remain in the pilot stage, with limited success at scale. In spite of encouraging results from the evaluations of these pilots, realizing the potential of digital development for agriculture requires much more investment and learning from large and successful scaling-up efforts. Indeed, many digital innovations targeting smallholders in developing countries fail to scale up (World Bank 2016); those that do scale up fail to be sustainable. While the engagement of the private sector in building digital innovations continues to be encouraging, national agricultural agencies and related institutions in many African countries are simply not changing at the pace that the digital revolution demands. The critical disconnect between pilots and scale-ups of digital tools targeting smallholder agriculture, as well as the lack of integration of these innovations within national agricultural systems, is likely to define whether smallholders can benefit from the digital revolution. These limitations apply to Kenya as well as to many other countries in Africa, albeit to different extents.

This chapter reviews existing digital innovations targeting smallholders in Kenya with the objective of drawing lessons that can benefit future scale-ups of digital innovations in Kenya and the rest of Africa. Our review aims to identify encouraging developments as well as important pitfalls that inhibit realization of the promise of digital innovations to transform smallholder agriculture in Kenya and in the region.

The chapter first briefly presents the landscape of digital innovations in Kenya. It then documents Kenya's success in the digital space by highlighting unique features in its digital ecosystem, including policies and regulatory systems. What follows highlights the disconnect between pilots and scale-ups of digital innovations in Kenya by exploring potential challenges and enabling factors. The chapter concludes with some lessons that the rest of Africa can learn from Kenya in digitalizing agriculture.

The landscape of digital innovations in Kenya

Several types of digital innovations have so far been used to support Kenya's agricultural and food system transformation. Some of these innovations have been used to support agricultural extension, market access and linkages, and various farm coordination activities, like the distribution of farm inputs and equipment. In particular, many of these innovations have been instrumental in reducing information and search costs through market advisory and information systems. Beyond these production- and market-geared solutions, Kenya has also made significant strides in introducing and scaling up digital innovations that provide digital financial services (Suri and Jack 2016; Suri 2017). The functionalities of these innovations have evolved over time, and there is now momentum behind the innovation process to continuously respond to the dynamic nature of Kenya's agriculture and food sectors.

This section develops a typology of digital innovations to examine six specific categories that are relevant to the Kenyan experience. The typology is drawn from a repository of relevant digital innovations, presented in Appendix 18.1. While some of the categories overlap in terms of definition, and certain digital innovations may fall in more than one category, given their multiple functions, the typology itself is nonetheless useful in organizing and structuring both national and global experience in a rapidly changing field.

Agricultural extension and advisory services

The digital innovations in this category mainly offer agronomic advisory services to farmers and aim to accelerate technical changes in food production systems. They mimic the typical agricultural extension and advisory services prevalent in many rural settings, sometimes functioning as a substitute and sometimes as a complementary service or an integrated augmentation of conventional services. They often focus on increasing agricultural production and productivity by tackling key production constraints facing farmers. For instance, iCow aims to improve productivity and profitability by providing extension advice and training on proper ways of raising livestock (Daum et al. 2022). Similarly, DigiCow addresses key production issues in the livestock sector like poor disease management and inbreeding.

In terms of how they function, some of the digital innovations in this category, such as iShamba, operate as call centers: interested farmers can dial in to speak to experts on a range of issues affecting their farm and livestock production. In doing so, farmers also receive additional information such as market prices and weather updates. Additional examples of digital innovations that provide extension and advisory services for farmers in Kenya include Arifu, M-shamba, and Viazi Soko. Other applications, like Digital Green and Kuza Biashara, are leveraging video extension services. Given the inability of public extension systems to effectively meet farmers' demand, there is increasing interest in revamping extension provision through public–private partnerships. Currently, many county governments are establishing links with applications like DigiCow to enable public–private extension.[1]

Weather information and services

In the face of weather and climate vagaries, digital innovations in this category disseminate information to farmers on weather patterns and events, including forecasts and long-term predictions. For instance, using the Kenya Agricultural and Livestock Research Organization's (KALRO's) Kenya Agricultural Observation Platform (KAOP) and the Climate, Livestock, and Markets (CLIMARK) project's MyAnga application, farmers can monitor rainfall and weather patterns through their mobile phones (CTA 2020). Beyond weather and climate information, some of the digital solutions in this category provide support to farmers by addressing energy and irrigation needs. An example is SunCulture, which uses off-grid solar technology to provide farmers with reliable access to water, irrigation, lighting, and mobile charging. Some of these applications also have the ability to combine solar water pumping technology with high-efficiency drip irrigation to enable farmers to improve production and land productivity.

Market information systems and linkages

Digital innovations in this category disseminate information about market prices and link buyers to sellers as well as importers and exporters of agricultural commodities. Some of them also act as a marketplace where buyers can source food and agricultural produce. They form the bulk of the category of agricultural digital innovations in Kenya. Their main objective is to reduce information asymmetries associated with selling and buying agricultural commodities. Typically, they reduce some aspects of transaction costs, enabling parties involved in trading to realize optimal gains. An example is M-Kilimo, founded and deployed in 2009 to reduce the informational constraints facing farmers in horticultural value chains (Misaki et al. 2018). Other applications in this category include iCow, N'Kayo, iShamba, Kenya Agricultural Market Information System's Kilimo, SunCulture, iProcure, DigiFarm, E-Tinga, Farmers Pride, Farmshine, Herdy Fresh, Kitchen, Soko, Mifugo.trade, Selina Wamucii, Taimba, TruTrade, Tulaa, Twiga Foods, Jumia, and Kuza Biashara.

---

1 This is mostly carried out under the government's Kenya Climate Smart Agriculture Program and the National Agricultural and Rural Inclusive Growth Project. The World Bank is also a partner here and facilitates financial support and partnerships between counties and private companies (for example, SunCulture, DigiCow, Apollo, Hello Tractor, One Acre Fund, and Kuza Biashara).

Digital financial services (intermediation and payment systems)

Digital innovations in this category often aim at facilitating financial transactions between trading actors. This usually takes the form of mobile money transactions. These are cashless payments, reducing the uncertainties and risks associated with market exchanges. Some of the digital innovations under this category act as physical banking and microfinance institutions that offer savings, loans, and credit facilities to farmers. Relatedly, they also offer financial management services through transparent and traceable transactions. The applications in this category have the potential to revolutionize smallholder agriculture as they have been critical in spurring financial inclusion (Suri 2017; Bharadwaj and Suri 2020). A typical example and success story has been M-Pesa, a mobile financial service that has been widely scaled up in Kenya. However, these financial services are underused in agriculture and do not seem to be having a transformative impact on the sector as yet (Parlasca, Johnen, and Qaim 2022). M-Pesa also has linkages and facilitates the services of various agriculturally geared programs and activities. For instance, it is used to facilitate the enrolment and payment of claims of index insurance services provided by the Agriculture and Climate Risk Enterprise. Another example of an application rendering financial services is M-Shwari, Kenya's most popular digital banking platform. Other examples under this category include DigiCow, Apollo, Dodore, FarmDrive, Musoni, Virtual City, and Mastercard Farmers Network.

Supply chain coordination (agricultural inputs and services)

The digital innovations in this category aid in various farm planning and coordination tasks. They cover production system management, sales and inventory management, and bookkeeping in farm production. They also ease access to various inputs, such as seeds, fertilizers, and agricultural machinery. An example here is the National Potato Council of Kenya's Viazi Soko, a digital platform that provides an efficient way to access quality farm inputs and services such as certified seeds, fertilizer, and mechanization services, among others (Parker 2021). The web-based portal and mobile-based application enable farmers to book and place orders for farm inputs as well as to receive accurate agronomic information and other services geared to improving productivity and profitability.

Another example worth mentioning is Smart Cow, which operates as dairy management software. It records important biographic details about livestock, such as on birth and pedigree, insemination, breeding, deworming, vaccination, and deaths as well as postmortem records. Other examples in this category include MTela, Apollo, DigiCow, M-shamba, E-Tinga, FarmIT, Tulaa, and Budget Mkononi.

Data and crowdsourcing services

This category of digital innovations responds to the increasing need for quality and real-time data to answer policy-oriented questions with regard to agricultural transformation (Benfica et al. 2021). These tools collect data using crowdsourcing and remote data collection methods. They thus offer the possibility of multiple data usage options, such as predictions and forecasting, which are relevant in informing policy decisions. Compared with the above-mentioned categories, these innovations are still at the early phase of development and deployment. They provide the basis for evidence-based policies by developing opportunities to source data from farmers. An example in this category is KAZNET, which sources livestock-related information from pastoralists in Kenya (Graham et al. 2021). Akin to the first category (agricultural extension), they also offer peer-to-peer engagement and expert support with the possibility of feedback loops. Closely related examples here include Nuru, DigiCow, and KAOP, which crowdsource data and share these to input suppliers to help them meet farmers' demand effectively.

Explaining Kenya's success in the digital space

Unique features of Kenya's digital ecosystem

Kenya represents one of the few countries in Africa that has made significant strides in the deployment and use of digital innovations. The country is ranked third in Africa and fifth in the Middle East and Africa region in its startup ecosystem (StartupBlink 2022). The country's success in the digital space owes largely to its reliable internet/broadband connectivity and its high mobile/internet subscription rate, combined with well-positioned incubation and accelerator centers, a thriving innovative and entrepreneurial environment, a (relatively) robust digital marketplace, and a conducive business environment that encourages active private sector participation (Drouillard 2017; Osiakwan 2017).

Reliable internet connectivity and near universal mobile subscription. Kenya has among the fastest and most reliable internet connectivity in Africa, mainly because of strategic public and private investment in internet infrastructure such as the national fiber optic backbone infrastructure (Osiakwan 2017; UNDP 2022). As a result, Kenya's internet access and use subscription has been increasing significantly over the past decade. By December 2021,

FIGURE 18.1 Connectivity in Kenya and sub-Saharan African regions

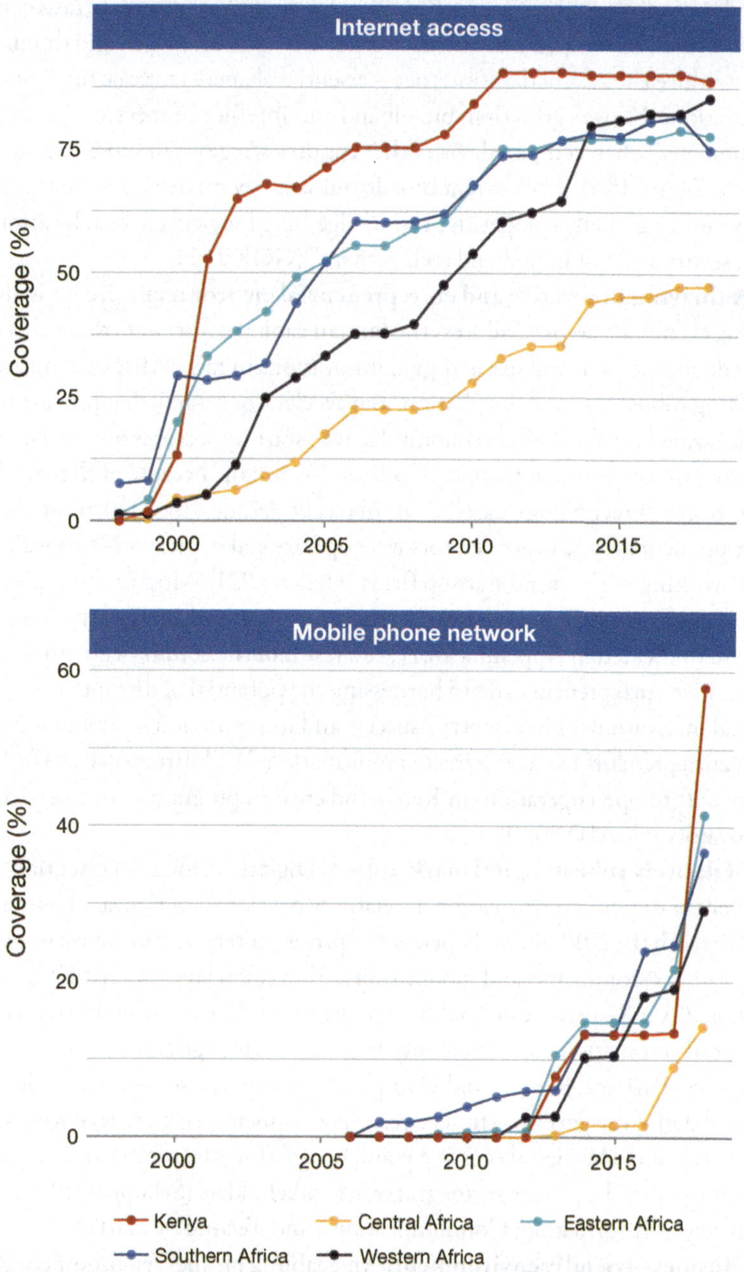

Source: Authors using GSMA (2022) data.

data/internet subscriptions stood at 46.3 million, of which 28.4 million were mobile broadband subscriptions (CA 2021). Kenya's broadband infrastructure, in particular, has helped increase network reach and bandwidth and dramatically reduce the cost of data/internet—essential elements to scale up digital innovations. Kenya's growth in broadband and internet connectivity remains phenomenal when compared with other countries/regions in sub-Saharan Africa (Figure 18.1). The current broadband strategy further aims "to transform Kenya into a globally competitive knowledge-based society enabled by affordable, secure and fast broadband connectivity" (GOK 2021).

A thriving innovative and entrepreneurial environment. Kenya is also taking exemplary steps to address the human capital constraints that otherwise impede the creation and use of digital innovations in most African countries, by building robust training, incubation, and accelerator centers that provide the skills needed for the digital economy. Kenya's start-up sector is one of the most vigorous on the continent, dubbed "Silicon Savannah" because of the increase in start-ups of technology-based firms in recent decades. By 2020, more than 30 start-up incubators and accelerators were operational in greater Nairobi alone, each working with several start-up firms (Hellen 2021). Most of the applications that are providing digital solutions in the areas of agriculture and the food system (indicated in Appendix 18.1) are a result of the country's openness to innovative entrepreneurs and to harnessing the potential of disruptive technological innovations. The country's success and its reputation as a vibrant hub for tech entrepreneurs has also attracted multinational tech firms such as IBM and Microsoft to open operations in Kenya and create a program to train aspiring innovators (USAID 2020).

Relatively robust digital marketplace. Digital business, an essential element of digital transformation, is relatively developed in Kenya, in both reach and breadth. In 2017, about 39 percent of private enterprises in Kenya were engaged in e-commerce, and mobile transactions alone represented KSh 3.2 trillion. Over 82 percent of smallholder farmers in Kenya use mobile-based financial transactions, although only 15 percent is for agriculture-related purposes (Parlasca, Johnen, and Qaim 2022). There is also a strong tendency toward digitizing government services: e-government plays a crucial role both as orchestrator of the digital ecosystem and in signaling the preferred direction of the country to the private sector and other stakeholders (Schuppan 2009; Kenya, Ministry of Information, Communications, and Technology 2019).

Business-friendly environment. An enabling business environment that supports active private sector participation is another crucial element driving digital innovation in Kenya. The government allows private firms and other stakeholders to test digital solutions under easy operation and registration procedures, and with guaranteed intellectual property rights. The business environment in Kenya also encourages a wide range of business models, including government funding, recognizing the public good characteristics of some of the digital innovations. For instance, KAOP, developed by KALRO, is a publicly funded digital innovation that provides information on weather as well as pest and disease forecasts. Meanwhile, iCow and DigiFarm are supported by a private mobile network operator (Safaricom) that is actively exploring/seeking more sustainable business models (Birner, Daum, and Pray 2021).

Policies and regulatory systems to support digital innovations

Kenya's success in digital innovation is also partly driven by adaptive and outcome-based policies and regulations. Policymakers and regulators in the country seem to have a relatively strong preference for responsive regulation that set rules ex-post as services and their providers evolve, as opposed to prescriptive ex-ante regulation. An example here is the "test and learn" approach that the country followed when M-Pesa was launched (Ndung'u 2019). Kenya's 2019 Digital Economy Blueprint also clearly states that regulations need to focus on the service delivered to consumers rather than on the type of company or technology that delivers the services—a regime that encourages companies to experiment and innovate in a multitude of products and services.

The need for strong coordination and collaboration among regulatory bodies is another key lesson other African countries can learn from Kenya. Key government bodies (the Central Bank, the Communications Authority, and the Competition Authority) have coordinated on regulatory issues related to digital innovations from the initial stages of digital transformation. This has allowed a coordinated set of regulatory actions to emerge across regulators over time (Ndung'u 2019; UNDP 2022).

Digital policies and regulation in Kenya also strive to support digital agriculture, and digital solutions are identified as a strong enabler in achieving the food security pillar of the country's Big Four Agenda (FAO 2021). Several policies and high-level commitments have also been put in place to ensure that poor smallholder households, and especially women, enjoy the benefits of digitalization. These policies and commitments have eased the growth and development of different business models that address the main constraints in meeting some of the key development outcomes of eradicating hunger and poverty.

Success stories

Some of the key successes in Kenya's digital space hail from applications that deliver financial services. One case is M-Pesa, which is arguably the most successful digital innovation in sub-Saharan Africa. Although its use is widespread, it is expected to be particularly beneficial for smallholder farmers, who usually lack access to financial services. Studies from Kenya have shown that these digital innovations are associated with increases in household welfare (Kikulwe, Fischer, and Qaim 2014; Suri and Jack 2016; Parlasca, Mußhoff, and Qaim 2020; Hartmann, Nduru, and Dannenberg 2021). These digital financial services have also been critical in protecting and cushioning households from income and health shocks through increased remittances (Jack and Suri 2014). These impacts have been shown to be more pronounced for women and women-headed households (Suri and Jack 2016). Related to M-Pesa is the use of M-Shwari, a mobile bank account innovation that offers credit and saving services in collaboration with M-Pesa (Bharadwaj and Suri 2020).

Despite the potential of digital financial services like M-Pesa and M-Shwari to revolutionize smallholder agriculture, some of these services are greatly underused by farmers for agricultural purposes, limiting their transformative impact. While the use of mobile money services is high (82 percent) among farmers in Kenya, only about 15 percent of farmers use such services for agriculture-related purposes and less than 1 percent engage in mobile banking and loans (Parlasca, Johnen, and Qaim 2022).

Beyond these digital financial innovations, success stories also emerge from more production-oriented digital innovations, such as DigiCow, iCow, and M-Kilimo. DigiCow and iCow offer extension and advisory services to pastoralists. iCow is a globally known digital and extension tool for livestock development (Daum at al. 2022). It offers both simple and smart services to pastoralists, including but not limited to feeding, gestation, disease control, and milking information, as well as other agronomic, forestry, and ecological information. This digital tool has been shown to induce better disease and hygiene management practices with positive implications for yields and animal health (Daum et al. 2022). Another success story from Kenya's digital space is KAZNET, a digital platform that relaxes some of the constraints associated with conventional data collection methods and assists in the collection of high-frequency data in remote pastoral regions (Chelanga et al. 2022). The KAZNET micro-tasking platform was used to crowdsource data from various livestock markets under COVID-19 to track the effect of the pandemic on livestock development (Graham et al. 2021).

Is Kenya's progress in digital innovation sufficiently transformative?

Although our review and assessment of Kenya's progress in digital agricultural innovation offer reasons to be optimistic about its reach and impact, we find that the achievements thus far are still largely unknown and not necessarily "transformative" in the way the term is conventionally understood in the agriculture and food system space. Many agriculture-focused digital innovations remain at a pilot stage, such that there seems to be a significant and persistent disconnect between piloting and scaling-up (see, for example, World Bank 2016; CTA 2019; Fabregas et al. 2019; Steinke et al. 2020).Thus, Kenya may not yet be reaping the full potential of digital innovation. This signals a need for additional investment to address critical infrastructural and human capital constraints that are impeding the creation and use of digital tools, particularly among Kenya's underserved smallholders. We examine these issues in greater detail below.

Disconnect between pilots and scale-ups

Despite the remarkable success Kenya has achieved in terms of nurturing digital innovations, there is a long way to go to achieve widespread scale-up and learning from these experiments. Although digital innovations lend themselves readily to rigorous evaluation, most studies to date have focused on identifying the impacts of just a few successful digital innovations. Most such evaluations have looked at the high-profile cases, such as M-Pesa and its associated services (for example, Suri and Jack 2016; Suri et al. 2021). Very few studies systematically and rigorously examine impacts on smallholder farmers and other market actors, leading to a lack of in-depth understanding on how digital innovations are improving the functioning and performance of input markets, value chains, or innovation networks in the agriculture and food system. This shortage of nuanced evidence and experience from both successful and unsuccessful scaling efforts impedes learning and makes it difficult for farmers and other actors to differentiate between what has truly worked and what has not.

Bottlenecks to widespread and meaningful scale-up

This limited evidence on widespread use and scale-up may point to a larger challenge: the simple fact that some of these innovations may not be scalable. However, there are also more complex supply- and demand-side constraints to scaling that may warrant greater attention. Below, we briefly describe these major constraints.

SUPPLY-SIDE CONSTRAINTS TO SCALING

Efforts to scale up digital innovations that can fundamentally transform Kenyan agriculture likely require considerable public investment in infrastructure and talent—investments that often have public goods attributes and thus can be implemented only by governments. At the same time, the scale-up of these innovations requires a certain degree of public sector withdrawal from the market so that input and commodity suppliers can compete on a level playing field and leverage digital innovations to reduce their costs and secure competitive advantages. The Kenyan digital innovation landscape faces *insufficient investment from both public and private sector sources*. Private investment in digital agriculture-related technologies lags because of the initial fixed costs required, the non-appropriability of added value from many information-based services, and the long lag times to profitability even with a credible business model (Fabregas et al. 2019). Yet public resources are scarce in Kenya, subject to fierce competition over their distribution and use, and tied to election cycles and other short-term outlooks. Overall, this means that the government is not always positioned to make the needed long-term investments in soft and hard infrastructure to support digital innovation. Meanwhile, only a few private entrepreneurs and investors have the financial capacity or appetite to invest in digital tools to support what are often perceived as high-risk, low-return activities in the agriculture sector. Although external donors—bilateral and multilateral funding agencies such as the World Bank—continue to support investments in digital technologies in Kenya, their impacts on smallholder agriculture remain marginal.

Second, some of the rapid progress in digital innovations in Kenya has not been accompanied by parallel progress in agricultural systems themselves. This relates to the *asynchronous pace of change*. In effect, the pace at which digital innovations bring solutions to agricultural activities and markets is not being matched by changes in the rate of institutional transformation that is needed to support and sustain these digital technologies (for example, CTA 2019). This is particularly the case in terms of delivering agricultural services and transactions, which continue to operate in environments characterized by isolation, limited scale, and risk (Benami and Carter 2021).

Third, many digital tools and innovations in Kenya and Africa fail to embed within existing institutional infrastructure. Digital innovations and technologies can thrive if they are integrated within a conducive business environment and regulatory framework. Kenya, like many countries in Africa, still *lacks comprehensive and dynamic digitalization policies* to support digital innovation in the agriculture sector.

Finally, most companies investing in agriculture-focused digital technologies in Kenya are still *struggling to develop sustainable business models,* partly because of the incipient nature of market development (that is, lower demand at early stages owing to some of the factors we discuss below). This limits the reach and scale-up of digital tools that can support smallholder agriculture. Several digital innovations developed through donor funding are struggling to ensure a healthy balance between generating impact and financially sustaining themselves.

DEMAND-SIDE CONSTRAINTS TO SCALING

The rapid emergence of digital tools in Kenya—and Africa—requires an appropriate response, including research on context-specific needs and cost-benefit analysis. However, many digital innovations enter the market without proper contextualized demand analysis. In the absence of such knowledge and the related infrastructure, new digital tools end up being alien to farmers and hence fail to attract demand. Indeed, some evaluations and willingness-to-pay analyses show that smallholders in low- and middle-income countries (including Kenya) do not want to pay the full cost for some of these innovations for several reasons, including lack of trust and the non-excludability of the information and services (shareability) most digital solutions provide (see, for example, Fabregas, Kremer, and Schilbach 2019; Cole and Fernando 2021).

Using digital tools and innovations usually requires some level of literacy and numeracy, skillsets smallholder farmers in Kenya often lack. Thus, the accessibility and usability of digital innovations in Kenya and Africa deserve critical attention. In addition, low digital literacy appears to be an important factor limiting wider usage and scale-up of digital innovations (CTA 2019). For example, 28 percent of digital technology enterprises report consumer-level digital literacy as an important barrier to the adoption and use of digital innovations in Africa (ibid.). Besides this, user confidence in digital tools is another important factor related to the scale-up of digital innovations. Smallholders are more likely to embrace digital innovations if they can access reliable and trustworthy services and information.

Finally, digital innovations in Kenya and many parts of Africa are not sufficiently inclusive. Entrepreneurs and companies are most likely to target easy-to-reach markets and customers (CTA 2019), thus some digital tools may not benefit marginalized smallholders. Indeed, many argue that differential access to digital innovations and associated complementary infrastructure is generating a digital divide in Africa. For example, less than 40 percent of smallholder households in Africa have access to the internet—while this increases with farm size (Mehrabi et al. 2021). Similarly, women have relatively lower access to the internet and mobile phones, triggering an important gender gap and digital divide.

Overall, although Kenya has made significant strides in the adoption of digital innovations, this progress has not been sufficiently transformative, especially in the agriculture sector. The most transformative digital innovations in Kenya have been those that facilitate financial transactions (for example, M-Pesa and M-Shwari), which have achieved significant scale. However, these innovations are still not widely used for agricultural purposes. As indicated above, the use of mobile financial services by farmers for agriculture-related purposes remains limited, and as a result such services have not yet had a transformative impact on smallholder farming (Parlasca et al. 2022). Smallholder farmers usually trade in small quantities and mostly on a face-to-face basis with input suppliers, market agents, and consumers. They prefer cash, given that mobile money payments entail a fee (Parlasca et al. 2022). This reality may not be unique to these digital platforms, as farmers rarely make use of credit services in sub-Saharan Africa (Adjognon, Liverpool-Tasie, and Reardon 2017).

What will it take to make Kenya's digital innovations transformative in agriculture?

Realizing the full potential of digital innovations in Kenya requires the scaling up of current efforts and momentum related to the development and use of wide-ranging digital solutions. An immediate step in this direction entails addressing the *human capital* constraints that usually impede both the creation and the use of digital tools (Birner, Daum, and Pray 2020; Kim et al. 2020; Jellason, Robinson, and Ogbaga 2021; Malabo Montpellier Panel 2019). Because of these human capital constraints, most existing digital agricultural innovations in Kenya offer a limited range of solutions. Efforts are evolving to support incubators and start-ups, which may help advance ICT skills. In particular, there is a need to scale up support to innovation hubs and IT incubators and other local tech networks to embrace the agriculture sector and target smallholders. These human capital investments should be accompanied by efforts to familiarize and train users of digital innovations, especially farmers, who may have limited digital literacy. While investment in the digital literacy of users is important, digital innovations should also be adapted to the realities on the ground (Trendov, Varas, and Zeng 2019; Daum et al. 2022).

Second, the digital infrastructural space in Kenya requires greater public and private investment in *complementary infrastructure,* which refers to the amenities essential to scale digital solutions. While infrastructure such as internet connectivity, mobile network coverage, and electricity coverage continue to increase, further investments and expansion to rural farm households can improve the usability and accessibility of digital innovations for smallholder farmers. Such infrastructure is a public good and hence this requires the government to be the first mover.

Third, realizing the full potential of digital innovations requires sustainable delivery of services, which in turn requires *sustainable business models* for private entrepreneurs. Although digital innovations providing financial services have sustainable operations, most digital solutions targeting smallholder farmers in Kenya rely heavily on public—particularly *donor*—funding. Although donor funding can be instrumental at the pilot stage, these funding agencies need to quickly push digital providers toward viable business models to ensure sustainability.

Finally, Kenya needs to maintain *dynamic policies and smart regulations* that can further spur digital innovations and ensure equitable distribution of digital dividends. This is partly because regular updating and adjustment of digital polices is a necessary condition to address the dynamic needs of digital ecosystems.

What can Africa learn from Kenya in digitalizing agriculture?

Kenya offers important lessons for other African countries that are aspiring to adopt and scale digital innovations to facilitate agricultural transformation. These lessons can be grouped into four major avenues. First, tapping the potential of digital innovations requires significant infrastructural investments. Kenya has made significant strides in the digital space in terms of improving mobile/internet infrastructure. Many other African countries have weak internet infrastructure and differential access to the internet and mobile phones across communities and households, and this is triggering variants of the digital divide (for example, Mehrabi et al. 2021). Such countries need to revitalize investments and policies to ensure the digital inclusion of marginalized households and communities.

Second, Kenya's digital infrastructure is supported by its relatively advanced digitalization policies, which aim "to transform Kenya into a globally competitive knowledge-based society enabled by affordable, secure and fast broadband connectivity" (GOK 2021). Many countries in Africa still lack comprehensive and dynamic digitalization policies and Kenya thus offers an important lesson here, especially with regard to its preference for more adaptive/responsive regulations to digital innovations as opposed to prescriptive regimes. Kenya's experience in ensuring and facilitating strong coordination and collaboration among regulatory bodies is also an important lesson for other African countries, some of which lack proper regulatory frameworks.

Third, Kenya's experience in addressing the human capital constraints that impede the creation and use of digital innovations, through building robust incubation and accelerator centers, is also an important lesson. Finally, Kenya's business environment also has lessons to pass on to other African countries in terms of supporting and attracting private sector participation in digital innovations. Kenya's business environment encourages experimentation in digital solutions, with its easy registration procedures and guaranteed intellectual property rights.

However, we note that Kenya's progress is still not sufficiently transformative, especially with respect to agriculture-focused digital innovations. Replicating some of the most transformative digital innovations in Kenya, mainly those that facilitate financial transactions (for example, M-Pesa and M-Shwari), in smallholder agriculture is an immediate next step. Future efforts should also be accompanied by systematic assessments of both successes and failures at different stages of the piloting and scaling up of digital solutions.

Appendix 18.1 Typology of agriculture-oriented digital tools

TABLE A18.1 Typology of agriculture-oriented digital tools in Kenya

Typology of digital tool	Name of tool	Description	Key functions
Agricultural extension and advisory services	iCow	iCow is an app that aims to reduce cow mortality rates and educate farmers on proper agricultural practices. It is a mobile-based agricultural platform designed to improve smallholder farmer productivity and profitability.	Agricultural extension Market information
	DigiCow	DigiCow is a mobile app designed to use data and feedback production, financial reports, and breeding and health reports. In addition, it has a digital loan facility meant to offer credit to farmers.	Agricultural extension Data-driven decisions
	Digital Green	Digital Green seeks to help the poor lift themselves out of poverty by empowering them through digital technology and grassroots partnerships.	Video extension services
	Kuza Biashara	Kuza Biashara operates a digital micro learning and community platform and aims to improve food security and revolutionize resilience through empowering communities to learn, connect, and grow to scale.	Video extension services
	iShamba	iShamba is a call center of agricultural experts; farmers can send an SMS or call to ask an expert for instant help. iShamba helps farmers improve their farms and get better yields.	Agricultural extension Farming information
	Viazi Soko	Viazi Soko is a platform to facilitate online marketing of potato-related products and services. It was developed with the aim of helping address various challenges facing potato farmers and other stakeholders.	Assistance to market potatoes
	Budget Mkononi	Budget Mkononi is an interactive web-based budgeting tool designed to help young and inexperienced farmers identify the basic costs and elements required to set up and run their farming enterprise. It also has a farming guide on each commodity listed, with further information on how to plant, manage, harvest, and finance an agribusiness.	Information to help manage finances
Weather information and services	KALRO KAOP	The KAOP app helps farmers monitor weather through their mobile phones. KAOP generates real-time and location-specific agro-advisories for farmers and other stakeholders to enable them to make more informed farming decisions.	Information on weather
	CLIMARK MyAnga	MyAnga is part of a wider project called CLIMARK whose aim is to disseminate actionable weather advisories to residents of Marsabit and Isiolo counties. The project was set up to design and deploy a blended weather information management system.	Weather advisories
	SunCulture	SunCulture aims to solve the energy challenges of smallholder farmers, using off-grid solar technology to provide farmers with reliable access to water, irrigation, lighting, and mobile charging, all with a single system.	Solar solutions

Continued

Typology of digital tool	Name of tool	Description	Key functions
Market information systems and linkages	Kenya Agricultural Commodity Exchange (KACE)	KACE is primarily an information service to enhance price discovery as well as a spot exchange. It facilitates linkage between buyers and sellers, exporters and importers of agricultural commodities in trade. It provides farmers and market intermediaries relevant information about markets, and other services to enhance their bargaining power and competitiveness in the marketplace.	Information on exchange rates
	M-Kilimo	M-Kilimo is a mobile platform that help farmers, livestock keepers, and fishers to obtain market information using mobile phones.	Disseminates market information
	Farmshine	Farmshine connects farmers with information, suppliers, and service providers.	Linking local farmers to global markets
	Herdy Fresh	Herdy Fresh works with local farmers, upcoming brands, and stores to bring consumers a wide selection of high-quality products and services at great value to ease the burden of city living.	Linking local farmers to consumers, especially in urban areas
	Kitchen Soko	Kitchen Soko links consumers directly to the source by letting them know their farmer or baker or butcher to ensure accountability. With Kitchen Soko, one can shop directly from the farmer and producer.	Direct link from famer to consumer enhances traceability
	Mifugo.trade	Mifugo.trade facilitates livestock trade through an online livestock and livestock products exchange that directly connects livestock producers to buyers.	Linking sellers and buyers of livestock
	Selina Wamucii	Selina Wamucii is a global sourcing platform for food and agricultural produce, incorporating the world's producers, cooperatives, processors, and farmers. It is a market access solution for farmers and integrates them with cooperatives, producer organizations, agro-processors, and other organizations.	Linking global farmers to global customers
	Taimba	Taimba is a mobile-based cashless platform that connects farmers to retailers such as small-scale traders "mama mbogas," mini-markets, and restaurants.	Connecting farmers to retailers
	TruTrade	TruTrade is a social enterprise providing smallholder farmers with a reliable route to market and fair prices for their produce.	Linking farmers to markets
	Tulaa	Tulaa connects farmers with suppliers of fertilizer, seeds, and finance.	Linking farmers to farm inputs
	Twiga Foods	Twiga simplifies the supply chain between fresh food producers, manufacturers, and retailers through a business-to-business e-commerce platform.	Linking food value chain supplies and consumers
	Farm to Market Alliance	The Farm to Market Alliance empowers smallholders to become reliable market players through access to four integrated pathways: predictable markets, affordable finance, technologies and quality inputs, and handling and storage solutions.	Linking farmers to markets
	iProcure	iProcure is a mobile app specifically designed to support wholesale suppliers. Its main purpose is to make it quick and easy for customers to place their orders.	Linking wholesale suppliers and consumers
	Jumia	Jumia is an online retailer. It mostly links sellers and buyers. This includes the agricultural market.	Linking buyers and sellers
	Farmers Pride	Farmers Pride is a mobile app that connects organic farmers to customers. This app provides customers with 100 percent organic products.	Connecting organic farmers to consumers

Continued

Typology of digital tool	Name of tool	Description	Key functions
Digital financial services (intermediation and payment systems)	M-Pesa	M-Pesa is a mobile phone-based money transfer service, payments, and microfinancing service, launched in 2007 by Vodafone and Safaricom, the largest mobile network operator in Kenya.	Transactions
	M-Shwari	M-Shwari is a savings and loan service that enables M-PESA customers to save and access credit.	Saving and overdrafts
	Dodore	Dodore provides a fintech solution that ensures financial flow between farmers, markets, agrovets/dealers, and suppliers. The agri-wallet also offers loans and savings options for farmers.	Financial management
	DigiFarm	DigiFarm offers smallholder farmers access to a suite of information and financial services, including discounted products, customized information on farming best practices, and access to credit and other financial facilities.	Information on financial facilities
Supply chain coordination (agricultural inputs and services)	Smart Cow	Smart Cow is a dairy management software app. It helps keep records of livestock and monitor daily milk production for each individual cow. It also keeps a well-organized record of insemination and breeding and all health records for individual cows and groups.	Dairy management system
	Viazi Soko	Viazi Soko is a platform to facilitate online marketing of potato-related products and services. It was developed with the aim of helping address various challenges facing potato farmers and other stakeholders.	Assistance to market potatoes
	E-Tinga	Tinga Rental Store is a platform that helps in accessing superior agricultural machinery services through a rental platform for a predetermined time.	Information on farm machinery for rental
	FarmIT	FarmIT facilitates real-time tracking of harvests.	Tracking of harvests
	MTela	MTela is a shop management app for agricultural input retailers. It helps with their sales and inventory management.	Financial management
	M-shamba	M-shamba supports digital learning on agronomy, regenerative agriculture, and food safety for farmers through its Interactive Voice Response service, USSD, and interactive SMS.	Using data to inform on demand for and supply of output
	Apollo Agriculture	Apollo bundles everything a farmer needs: financing, inputs, advice, insurance, and market access, when possible. It uses satellite data and machine learning to enable better credit decisions. It also uses automated operations to keep costs low and processes scalable.	Farm management decisions and automated operations
Data and crowdsourcing services	KAZNET	KAZNET is an android app developed to solve the need for accurate and timely market information from participants in remote areas.	Collecting market information
	Nuru	Nuru deploys artificial intelligence and Google's TensorFlow technology to improve surveillance and management of crop diseases.	Identifying cassava diseases

Source: Authors using multiple sources.

Note: This is not an exhaustive list of digital tools operating in Kenya.

References

Abay, K.A., G.T. Abate, J. Chamberlin, Y. Kassim, and D.J. Spielman. 2021. "Digital Tools and Agricultural Market Transformation in Africa: Why Are They Not at Scale Yet, and What Will It Take to Get There?" IFPRI Discussion Paper 2092, International Food Policy Research Institute (IFPRI), Washington, DC.

Adjognon, S.G., L.S.O. Liverpool-Tasie, and T. Reardon. 2017. "Agricultural Input Credit in sub-Saharan Africa: Telling Myth from Facts." *Food Policy* 67: 93–105.

Baumüller, H. 2018. "The Little We Know: An Exploratory Literature Review on the Utility of Mobile Phone-Enabled Services for Smallholder Farmers." *Journal of International Development* 30 (1): 134–154.

Benami, E., and M.R. Carter. 2021. "Can Digital Technologies Reshape Rural Microfinance? Implications for Savings, Credit, & Insurance." *Applied Economic Perspectives and Policy* 43 (4): 1196–1220.

Benfica, R., J. Chambers, J. Koo, A. Nin-Pratt, J. Falck-Zepeda, G-J. Stads, and C. Arndt. 2021. *Food System Innovations and Digital Technologies to Foster Productivity Growth and Rural Transformation.* UN Food Systems Summit Brief. Bonn: UN Food Systems Summit.

Bharadwaj, P., and T. Suri. 2020. "Improving Financial Inclusion through Digital Savings and Credit." *AEA Papers and Proceedings* 110: 584–588.

Birner, R., T. Daum, and C. Pray. 2021. "Who Drives the Digital Revolution in Agriculture? A Review of Supply-Side Trends, Players and Challenges." *Applied Economic Perspectives and Policy* 43 (4): 1260–1285.

Chelanga, P., F. Fava, V. Alulu, R. Banerjee, O. Naibei, M. Taye, M. Ber, D. Galgallo, W. Gobu, W. Lapario, K. Muendo, and N. Jensen. 2022. "KAZNET: An Open-Source, Micro-Tasking Platform for Remote Locations." *Frontiers in Sustainable Food Systems* 6: 730836.

Cole, S.A., and A.N. Fernando. 2021. "'Mobile'izing Agricultural Advice Technology Adoption Diffusion and Sustainability." *The Economic Journal* 131(633): 192–219.

CA (Communications Authority of Kenya). 2021. *Second Quarter Sector Statistics Report for the Financial Year 2021/2022 (1st October–31st December 2021).* Nairobi.

CTA (Technical Centre for Agricultural and Rural Cooperation). 2019. *The Digitalization of African Agriculture Report 2018–2019.* Wageningen, Netherlands.

CTA. 2020: *CTA Project Completion Report CLIMARK.* Wageningen, Netherlands.

Daum, T., T. Ravichandran, J. Kariuki, M, Chagunda, and R. Birner. 2022. "Connected Cows and Cyber Chickens? Stocktaking and Case Studies of Digital Livestock Tools in Kenya and India." *Agricultural Systems* 196: 103353.

Drouillard, M. 2017. "Addressing Voids: How Digital Start-Ups in Kenya Create Market Infrastructure." In *Digital Kenya: An Entrepreneurial Revolution in the Making*, eds. B. Ndemo and T. Weiss, 97–122. London: Palgrave Macmillan.

Dwivedi, S.L., E.T. Lammerts van Bueren, S. Ceccarelli, S. Grando, H.D. Upadhyaya, and R. Ortiz. 2017. "Diversifying Food Systems in the Pursuit of Sustainable Food Production and Healthy Diets." *Trends in Plant Science* 22 (10): 842–856.

Fabregas, R., M. Kremer, and F. Schilbach. 2019. "Realizing the Potential of Digital Development: The Case of Agricultural Advice." *Science* 366 (6471): eayy3038.

Fabregas, R., M. Kremer, M. Lowes, R. On, and G. Zane. 2019. "SMS-Extension and Farmer Behavior: Lessons from Six RCTs in East Africa." Working Paper, Agricultural Technology Adoption Initiative (ATAI).

FAO (Food and Agriculture Organization of the United Nations). 2018. *Sustainable Food Systems: Concept and Framework*. Rome.

FAO. 2021. *Farm Data Management, Sharing and Services for Agriculture Development*. Rome.

GOK (Government of Kenya). 2021. *National Broadband Strategy 2021–2030*. Nairobi.

Graham, M.W., P. Chelanga, N.D. Jensen, S.M. Leitner, F. Fava, and L. Merbold. 2021. "A Framework for Assessing the Effects of Shock Events on Livestock and Environment in Sub-Saharan Africa: The COVID-19 Pandemic in Northern Kenya." *Agricultural Systems* 192: 103203.

GSMA. 2022. *The Mobile Economy: Sub-Saharan Africa 2022*. London, UK.

Hartmann, G., G. Nduru, and P. Dannenberg. 2021. "Digital Connectivity at the Upstream End of Value Chains: A Dynamic Perspective on Smartphone Adoption amongst Horticultural Smallholders in Kenya." *Competition and Change* 25 (2): 167–189.

Hellen, A. 2021. "The Rise of the Startup Ecosystem in Kenya." In *The Changing Face of Entrepreneurship in Africa*, ed. C.M. Mwangi , 109-126. London, UK: Routledge.

Jack, W., and T. Suri. 2014. "Risk Sharing and Transactions Costs: Evidence from Kenya's Mobile Money Revolution." *American Economic Review* 104 (1): 183–223.

Jellason, N.P, E.J.Z. Robinson, and C.C. Ogbaga. 2021. "Agriculture 4.0: Is Sub-Saharan Africa Ready?" *Applied Sciences* 11 (12): 5750.

Kenya, Ministry of Information, Communications, and Technology. 2019. *Digital Economy Blueprint: Powering Kenya's Transformation*. Nairobi.

Kikulwe, E.M., E. Fischer, and M. Qaim. 2014. "Mobile Money, Smallholder Farmers, and Household Welfare in Kenya." *PLoS ONE* 9 (10): e109804.

Kim, J., P. Shah, J.C. Gaskell, A. Prasann, and A. Luthra. 2020. *Scaling Up Disruptive Agricultural Technologies in Africa*. Washington, DC: World Bank.

Malabo Montpellier Panel. 2019. *Byte by Byte: Policy Innovation for Transforming Africa's Food System with Digital Technologies*. Dakar: IFPRI; ZEF (Center for Development Research University of Bonn); and Imperial College London.

Mehrabi, Z., M.J. McDowell, V. Ricciardi, C. Levers, C. Diego, N. Mehrabi, H. Wittman, N. Ramankutty, and A. Jarvis. 2021. "The Global Divide in Data-Driven Farming." *Nature Sustainability* 4: 54–160.

Misaki, E., M. Apiola, S. Gaiani, and M. Tedre. 2018. "Challenges Facing Sub-Saharan Small-Scale Farmers in Accessing Farming Information through Mobile Phones: A Systematic Literature Review." *Electronic Journal of Information Systems in Developing Countries* 84 (4): e12034.

Ndung'u, L. 2019. "The Rise of Digital Financial Services in Kenya: The Role of M-Pesa." *Journal of African Business* 20 (1): 1–18.

Njuki, J., J. Parkins, and A. Kaler. 2021. "A Review of Evidence on Gender Equality, Women's Empowerment, and Food Systems." *Global Food Security* 27.

Osiakwan, E.M. 2017. "The KINGS of Africa's Digital Economy." In *Digital Kenya: An Entrepreneurial Revolution in the Making*, eds. B. Ndemo and T. Weiss, 55–92. London: Palgrave Macmillan.

Parker, M. 2021. *Partner NPCK Report: Viazi Soko Digital Platform*. Lima: International Potato Center (CIP).

Parlasca, M.C., O. Mußhoff, and M. Qaim. 2020. "Can Mobile Phones Improve Nutrition among Pastoral Communities? Panel Data Evidence from Northern Kenya." *Agricultural Economics* 51 (3): 475–488.

Parlasca, M.C., C. Johnen, and M. Qaim. 2022. "Use of Mobile Financial Services among Farmers in Africa: Insights from Kenya." *Global Food Security* 32: 100590.

Schuppan, T. 2009. "E-Government in Developing Countries: Experiences from Sub-Saharan Africa." *Government Information Quarterly* 26 (1): 118–127.

StartupBlink. 2022. *Global Startup Ecosystem Index 2022*. Haifa, Israel.

Steinke, J., J. van Etten, A. Müller, B. Ortiz-Crespo, J. van de Gevel, S. Silvestri, and J. Priebe. 2020. "Tapping the Full Potential of the Digital Revolution for Agricultural Extension: An Emerging Innovation Agenda." *International Journal of Agricultural Sustainability* 19 (5–6): 549–565.

Suri, T. 2017. "Mobile Money." *Annual Review of Economics*. 9 (1): 497–520.

Suri, T., and W. Jack. 2016. "The Long-Run Poverty and Gender Impacts of Mobile Money." *Science* 354 (6317): 1288–1292.

Suri, T., W. Jack, and S. Ramani. 2021. "The Impact of Digital Financial Services: Evidence from a Randomized Control Trial in India." *Quarterly Journal of Economics* 136 (1): 1–54.

Trendov, N.M., S. Varas, and M. Zeng. 2019. *Digital Technologies in Agriculture and Rural Areas: Status Report.* Rome: FAO.

UNDP (United Nations Development Programme). 2022. *Mapping the Innovation Ecosystem.* Nairobi.

USAID (United States Agency for International Development). 2020. *Kenya Country Development Cooperation Strategy (CDCS) 2020–2025.* Washington, DC.

World Bank. 2016. *World Development Report 2016: Digital Dividends.* Washington, DC.

Chapter 19

A WAY FORWARD: POLICY-DRIVEN TRANSFORMATION

Clemens Breisinger, Michael Keenan, and Juneweenex Mbuthia

This book has adopted a food systems framework as a new way of conceptualizing and designing food policies and research. Looking beyond agriculture and value chains makes it possible not only to turn food systems into a driver of economic transformation but also to better include health, productivity, resilience, inclusivity, and sustainability as integral parts of system transformation.

Such a fresh approach is urgently needed in light of limited development progress over the past years in Kenya and other countries. The share of manufacturing—traditionally a driver of economic transformation—in total output remains low; maize yields have been stagnating for the past 20 years; and poverty and food insecurity are on the rise again (Nafula et al. 2020; FAOSTAT 2022). In addition to structural challenges, growing challenges and vulnerabilities such as the threat of pandemics, commodity price crises, climate change, and conflicts, call for a new development and food policy paradigm (Breisinger et al. 2022; UNICEF 2022). At the same time, such a fresh approach can also help in harnessing the new opportunities that come with digitalization and with (policy) lessons from other countries that can be adapted to the Kenyan context.

The remainder of this concluding chapter offers several broader lessons for food systems research and then provides five sets of policy recommendations (one for each of the five dimensions: productivity, resilience, sustainability, health, and inclusivity) for Kenya. It concludes with a call to revisit existing development paradigms in order to truly transform food systems and development.

Lessons for food system transformation

One of the broader lessons emerging from this book is that **countries should be more selective when designing development projects and focus more strongly on strengthening national institutions.** All contexts are different, and every country will have to find its own pathway to food system transformation. One-size-fit-all policies overlook the complexity and nuance of national food systems, whereas, as a general principle, coordinated development and implementation of sound policy are at the core of food system transformation.

Kenya has in place an institutional framework that provides an enabling environment for food system transformation, bringing together key actors in the food system. One key actor is the Ministry of Agriculture, Livestock, Fisheries, and Cooperatives and specifically the Agriculture Transformation Office, tasked with coordinating the Agricultural Sector Transformation and Growth Strategy through performance management, spearheading intergovernmental actions, and guiding data and digitalization efforts (Kenya, Ministry of Agriculture, Livestock, Fisheries, and Cooperatives 2022).

Meanwhile, to support Kenya's agriculture and food sectors, the international community is organized into the Development Partner Group for Agriculture and Rural Development. These and other platforms are important for more coordinated collaboration between the government and development partners, especially given that an estimated 83 percent of funding for agriculture comes from the latter and only 17 percent from the former (Kenya Parliamentary Budget Office 2021). Over time, funding a higher share of agricultural spending from domestic sources is likely to build more ownership and sustainability. A more selective role of the government when designing development projects and a stronger focus on strengthening governmental institutions is also likely to improve longer-term development outcomes.

Research can play an important role in food system transformation. In Kenya, more efforts need to be made to better link national research institutes and universities with international research entities. Too often, research efforts are conducted in parallel, and results are not well communicated to policymakers and decision-makers. This can lead to missed opportunities and even confusion. For example, two different panel surveys exist for maize, one conducted by the Kenya Agricultural and Livestock Research Organization (KALRO) and the other by the International Maize and Wheat Improvement Center (CIMMYT). The data from KALRO suggest some improvements in maize yields in Kenya; results from the other survey suggest stagnation (see Chapters 7 and 8). More efforts are needed to increase the coordination and coherence among international research institutes and between international and national research institutes.

Collaboration between national and international researchers can lead to more relevant, rigorous, and coherent research. Peer-reviewed journal publications influence the direction of academic work and help shape policy objectives and design. Authors based in Northern countries write an estimated 75 percent of publications in peer-reviewed development journals (Amarante and Zurbrigg 2020). Not all, but some, of the Northern-based research in top development journals is characterized by rigorous methodology and/or theory but lacks context and relevance to study areas and national policies. Collaboration with in-country researchers is needed to overcome these barriers. Linking national and international researchers should take the form not of capacity building but rather of *capacity sharing* with national researchers, to enhance the contextual relevance and depth of research and contribute expertise in publishing in top development journals. This book is an example of bringing researchers together to create rigorous and relevant research in a cohesive form for national policymakers.

A second broader lesson is the importance of **encouraging a stronger focus on coherent policy design and effective implementation.** Sound policy development requires coherence in policy within food systems, across related areas of the economy and in conjunction with international policy, as called for in the 2014 Malabo Declaration (AUC 2014). Coherence of policies relies on addressing the complex web of interactions within food systems. Because changes in one area of a food system affect other areas, each specific policy must build synergies with other policies to reach the overall food system objectives. For example, policies affecting food supply chains must take into account the resulting effects on the food environment and consumer behavior. Further, policies in other sectors should align with food sector development, such that related policies do not undercut food system goals. Policies across international organizations and the national government should also be coherent, with national governments taking the lead in setting objectives (Chevallier 2022).

Without effective implementation, even well-designed policies will fail to meet their goals. Effective implementation requires government bodies to have the capacity to meet policy objectives and the needs of food system actors. This will require increases in efficiency at the national level and quality control of services offered by counties. Further, private–public partnerships can be leveraged to assist public institutions to reach the scale needed for food system transformation. For example, the draft bill of the Kenyan Agricultural Sector Extension Policy places private–public partnerships at its core to revive the extension system (Kenya, Ministry of Agriculture, Livestock, Fisheries, and Cooperatives 2022). However, concerns about inclusion and trust should be noted, as policy implementation must reach actors across different value chains—not just "priority value chains"—and all groups, particularly women and other marginalized actors. Especially when leveraging private–public partnerships, building trust within the food system is necessary to ensure actors buy into policy implementation.

A third broader lesson is the need to **harness the power of knowledge by strengthening the science–policy interface (SPI).** Policy development and implementation should rely on evidence shared through a robust SPI. An effective SPI must meet at least three criteria. First, it must leverage research to support the development and implementation of coherent, data-driven policies to achieve food system outcomes. Second, such research should be transparent, independent, and rigorous. Third, research agendas must be aligned with policy to produce relevant findings (Singh et al. 2021). Drawing on experience from previous agricultural value chain transformations, which relied heavily on strong SPIs, an additional criterion must be met: research must be on the ground and in touch with food system actors' needs. This means opening flexible communication channels between researchers, food system actors, and policymakers (Roseboom and Rutten 1998). Incentives for researchers may also need to be based on the practical, rather than academic, relevance of their research (Abraham et al. 2019). The Kenyan SPI can leverage national research institutes and universities and international research institutes by aligning research objectives to national policy and developing synergies through collaboration between the various institutes. Bridging the science–policy gap will also require more timely provision of research-based evidence when decision-makers need it and in a form that is digestible for them.

As an example of how to address these issues, the CGIAR Research Initiative on National Policies and Strategies[1] is currently in the process of co-creating an SPI in the form of a Community of Policy Practice in Kenya and other countries. These Communities of Policy Practice can also play an important role in bringing national and international policy researchers together.

---

1 www.cgiar.org/initiative/27-national-policies-and-strategies-for-food-land-and-water-systems-transformation/

Policy recommendations for food system transformation in Kenya

Kenya is well placed to foster transformation in the food system. At the UN Food Systems Summit in 2021, Kenya provided a blueprint for transformation of food systems, with a strong focus on climate resilience, investments in key sectors, and enhancing the role of women and youth in food systems. Kenya is already the regional leader in digital agriculture innovation, with 30 percent of all disruptive agricultural technologies in sub-Saharan Africa based in Kenya (Kim et al. 2020). It has implemented enabling policy frameworks and business environments in recent years (Kenya, Ministry of East African Community and Regional Development 2020) and is endowed with immense agricultural potential and diversity. Further, a youthful and enterprising population—if enabled—can drive innovation and help secure nutritious, productive, resilient, sustainable, and inclusive food systems (FAO 2019). As in other countries that have undergone radical transformation (albeit in agricultural value chains rather than entire food systems), enlightened, aligned, and ambitious public policy is the key ingredient for success (Abraham et al. 2019).

In the following, we summarize the key policy recommendations emerging from this book, organized along the main food system dimensions of health, productivity, resilience, inclusivity, and sustainability.

Health: Invest more in nutrition education and create smart regulation for food safety without overburdening businesses

Transforming Kenya's current food system for better nutrition and health will require a paradigm shift that puts consumer diets at the center of policymaking. As agriculture is by far the dominant sector in Kenya's food system, such a shift will entail striking a balance between traditional objectives like agricultural productivity growth, export stimulation, and farmer support, on the one hand, and the new responsibility for better nutrition and health for all Kenyans, on the other.

As Chapter 4 analyzes, malnutrition in Kenya is primarily a poverty problem. In addition to policies that lead to rising incomes, reducing relative food prices through demand-side policies (for example, targeted consumer subsidies) or supply-side policies (for example, targeted farm input subsidies or development of improved farm technologies) may be effective at narrowing the consumption gaps for nutritious foods. Weak consumer preferences and high price sensitivity may be an indication of a lack of consumer knowledge of the nutritional value of pulses, nuts, and seeds. Nutrition education, for example, in schools and through public information campaigns, may aid in changing consumer behavior.

As Chapter 3 shows, there is broad scope for improving animal health, feed standards, and breeds. For example, the government needs to strengthen the veterinary laboratory system to provide technical support for disease surveillance, diagnosis, and quality control. There is also a need for joint engagement of the State Department of Livestock and the Zoonotic Disease Unit under the Ministry of Health in controlling zoonotic diseases within the One Health Concept. It is also important to establish mechanisms for public and private partnership for controlling cross-county and transboundary infectious diseases and to coordinate with the Kenya Wildlife Service for the control of diseases at the livestock–wildlife interface.

Food safety is an important area for policy action. Chapter 5 outlines important actions to improve food safety, such as monitoring of water sources used for irrigation, and to remediate problems; providing water, sanitation, and hygiene infrastructure at markets and abattoirs; building capacity and incentivizing food safety among small-scale, informal businesses; implementing regular and comprehensive surveillance of high-risk foods; leveraging private sector capacity for self-monitoring under a coregulatory approach; and including food safety in infant and young child feeding recommendations for caregivers.

Productivity: Play a supporting role for the private sector to accelerate the transformation of input markets, food processing and service sectors, and mechanization efforts

As Chapter 2 shows, Kenya's food system is, as a whole, still in an early stage of transformation, with most growth and job creation occurring close to the farm. Successful transformation in Kenya requires even larger contributions from agro-processing and food services, with more value addition and jobs in the food system eventually generated off the farm. The value chains that are found to be the most effective in reducing poverty, creating jobs, and improving diets are also the ones that make up a large share of Kenya's current agriculture sector. This includes value chains producing animal products and traditional export crops. The value chains that are found to be least effective, such as cereals and root crops, often dominate agricultural landholding and account for a large share of public investments. Acceleration of structural changes within the food system through reorientation of the government's investment portfolio could enhance the contribution of the food system to broad development outcomes.

Chapters 2 and 8 argue that there is also a need for nonagricultural growth and economic transformation to absorb farmers in nonagricultural labor markets. For this, a set of policies is needed that includes measures to support the emerging commercial farmers who are expected to foster labor productivity growth, wage labor income, and integration in retail value chains toward domestic and export markets. Also relevant are policies and investments to shape the development of the industrial structure of the food and agriculture sector and the links at different levels of the value chain (Neven et al. 2009 on horticulture and supermarkets in Kenya; Lowder, Skoet, and Raney 2016). A third relevant policy area covers policies and institutions to facilitate the movement of labor out of agriculture and into nonagricultural sectors in this process. This further requires the private sector-led creation of rural and urban jobs in industry and services to move along hand-in-hand with agriculture and other forms of social protection such as social safety nets.

As an example of the discussion around supporting industrial development, Chapter 9 on mechanization argues that the development of local industry to manufacture machinery, implements, and equipment remains a feasible option—and it is happening, as in the case of Ndume Ltd., located in Gilgil, Kenya. A promising starting point is support to the development of a spare parts industry. Outside Kenya, while experiences in promoting local assembly have been mixed, some plants, like Ethiopia's Nazareth Tractor Assembly Plant, have remained operational for many years, providing almost half of tractors entering the country between 2005 and 2010 (Takeshima, Diao, and Aboagye 2020). Applied research in mechanization needs to enable the fast-tracking of progress made, through establishing more stable macroeconomic environments, liberalized markets, tighter fiscal regimes, and stronger institutional frameworks.

In addition to promoting industries, improving marketing systems is important in driving transformation. Chapter 3 highlights that unstructured marketing systems have a negative impact on industry, leading to its underperformance, using livestock as an example. There is a need to strengthen the capacities of producers and marketing groups in the production, processing, and storage of livestock products. As stated in Kenya's livestock policy, it is important to facilitate the dissemination of livestock marketing information to all value chain actors and to establish mechanisms for strengthening and harmonizing market information systems and developing linkages with local and international markets. Chapter 8 shows that, while agricultural input markets in Kenya are largely liberalized, the supply of the inputs is sometimes unreliable and the distribution networks are inefficient—hence supply is not in sync with demand, either temporally or spatially. Digitalization may play an important role in improving market information and efficiency, as discussed in Chapter 18.

Increasing agricultural productivity sustainably remains a priority. Chapter 9 argues that the government can strengthen support to farmers' training on machine operation, maintenance, and repair, to enhance efficiency and reduce the costs of service provision, as has been experienced in Ghana. These trainings can be provided in technical and vocational education and training institutes, universities, and research institutes, in particular the Agricultural Mechanization Training Institute.

In doing so, it is important to incorporate the knowledge of existing private sector hiring service providers, who often have experiences and expertise specific to their local business environment. In addition, machine hiring services should be promoted, including through value chain financing. Property rights and law need to be reviewed to enable more land rental activity, which has been shown in other countries to improve resource allocation and productivity. Integrating formal and informal seed systems, promoting better input management practices, and reducing input costs are critical areas for fostering agricultural growth. Reducing input costs will require investments in road infrastructure and nontariff barriers (for example, roadblocks).

Resilience: Develop policies to foster agricultural diversification and de-risk credit and insurance

An important element in building resilience to crisis is diversification. As Chapter 10 discusses, building a sustainable, resilient food system in a country such as Kenya requires a fundamentally different model of agriculture based on diversifying farms and farming landscapes, optimizing biodiversity, and stimulating interactions between different sectors for a sustainable healthy diet for all. Together, a varied and balanced diet, a wide range of crops and foodstuffs, and a diverse system of production and distribution make for a more resilient, stable, and healthier food system (EC 2020).

The ongoing global commodity price crisis highlights how Kenya relies on imports to meet its food, fertilizer, and energy needs—and there is considerable instability in fertilizer world prices. Therefore, Chapter 8 (on inputs) argues for expanding and diversifying private sector trade. A diversification of diets with a higher content of domestically grown crops and livestock could not only reduce imports but also improve diet quality and increase domestic producers' income (Chapter 4). Likewise a stronger focus on domestic energy sources (for example, expansion of geothermal energy) can provide more energy independence and a boost to domestic industrialization.

Diversification and greater use of domestically grown cereals can be also promoted through flour blending, argues Chapter 16. For this to happen, a range of potential interventions exist, at various nodes of the food system. These include addressing limited access to quality seeds of target crops, reorienting the current extension system to include and serve these crops, building capacity of aggregation systems and farmer producer organizations for collective action, locating processors near high-production areas of target crops, and promoting the crops to create demand through consumer campaigns and targeting public procurement for blended flour (for example, in schools and hospitals, with the military, and through food aid).

On building resilience at the farmer's level, Chapter 11 (on insurance) highlights how climate insurance is an increasingly important financial instrument to improve agricultural risk management for smallholder farmers, herders, and other value chains actors in the face of the present climate crisis.

Inclusion: Invest in capacity building for youth (particularly in agribusiness), for producers (that is, extension systems), and for women along the value chain, and set up clear contract enforcement mechanisms that protect small farmers

Providing better opportunities for women and youth in food system transformation will make food systems not only more inclusive but also likely more productive, argue Chapters 13 and 14 (on gender and youth). Chapter 13 (gender) argues that *women* play a critical role as both primary food producers and primary household caretakers, and are hence key stakeholders in sustainable and resilient food systems. To strengthen the role of women in food systems, existing challenges such as low land ownership and minimal participation in decision-making and food governance, existing disadvantages in owning/acquiring food production resources, and weaker networks all need to be addressed.

To give *youth* a better chance, structures of support for youth in agribusiness need to be improved to take advantage of the knowledge young farmers and agripreneurs have (Chapter 14). It will be crucial to lend a hand to equip these youth with appropriate agribusiness skills, knowledge, and information and to enhance their access to affordable and youth-friendly growth capital that can enable them to scale up promising agribusiness ventures. The relaunch of the 4K clubs will provide momentum to scale school-based agricultural education in Kenya to inculcate a positive mindset regarding agriculture among young students and to nurture, prepare, and build future leaders in the agriculture sector. In addition, the chapter argues that building youth-led professional networks, facilitating the formation of a national youth in agriculture association, improving the evidence base for youth engagement in food systems, and supporting youth engagement in the policymaking process are all important elements of successful food system transformation.

Better inclusion of resource-poor farmers in the food system is also important, argues Chapter 15. The authors provide a rationale for subsidies for resource-poor farmers. However, they also highlight the existing challenges that come with fertilizer subsidies in Kenya. While some county governments provide fertilizer subsidies, information about targeting, quantity, pricing, types, and mode of delivery is not publicly available. Lack of access to such information may hinder proper planning by fertilizer market players. This suggests room for improving the targeting of fertilizer subsidies and establishing a better monitoring and evaluation system for doing so.

Good targeting also plays a key role in insurance. For this, it is important to differentiate between different types of farmers and herders, distinguishing at the very least between the poorest farmers, typically landless laborers; the vulnerable non-poor, typically subsistence-oriented farmers; and more commercially oriented smallholder farmers. Each of these groups will require different insurance or social protection solutions (Chapter 11). This will also involve putting social and gender equity at the forefront of impact assessments, to ascertain that, among the most vulnerable segments of the population, insurance programs reach, benefit, and empower women and men alike.

Better access to credit is also a key factor in broadening access to food system activities. Chapter 12 shows that developing policies to hedge smallholders against systemic shocks, such as drought, is one way of enhancing access to credit. The use of formal insurance markets is a viable policy since it transfers the risk outside the household and hence protects its collateral. Bundling insurance with credit also minimizes the risk of default by smallholder borrowers, which reduces lenders' financial risks that threaten their business stability—a common phenomenon when rural agricultural production systems experience systemic shocks such as drought.

To better include smallholders in markets, Chapter 15 finds that contract farming can be a good strategy for individual smallholder growers to access attractive markets and benefit from export opportunities. For this to be successful, there is a need to invest in training on sustainable production, harvesting, and other postharvest management techniques, and on the prevention of theft and illegal cartels that force farmers to harvest fruits early. Creating an enabling environment for mainstreaming contract farming to strengthen agricultural value chains also requires appropriate legal frameworks to facilitate contracting that entails clarity of terms, fairness, responsibility, and transparency. This will build trust in completing transactions between the different parties.

Sustainability: Invest in targeted transportation infrastructure and digital networks and create smart regulations to support off-farm service providers and also adequately protect actors (for example, smart data regulations for the digital ecosystem)

Chapter 18 shows that Kenya has made significant strides in the digital space in terms of improving mobile/internet infrastructure and supportive policies. Kenya's exemplary efforts in addressing the human capital constraints that impede the creation and use of digital innovations in most African countries, through building robust incubation and accelerator centers, are also an important lesson. Promoting digital solutions can be transformative across the whole food system. For example, digital mechanisms can verify the quality of seeds and fertilizers, facilitate the provision of services (for example, the renting of machinery), and enable the marketing of food. However, the authors note that Kenya's progress is still not sufficiently transformative and sustainable, especially in agriculture-focused digital innovations. Replicating some of Kenya's most transformative digital innovations, mainly those that facilitate financial transactions (for example, M-Pesa and M-Shwari), in smallholder agriculture would be an immediate next step. Future efforts should also be accompanied by systematic assessments of both successes and failures at different stages of piloting and scaling of digital solutions.

Open questions and future research directions

Transforming food systems will also require revisiting some of the longstanding development paradigms and debates about agricultural and spatial development. Viewing these paradigms through the lens of the food system framework is useful because the framework can help in assessing the trade-offs between the different outcomes (health, productivity, resilience, inclusivity, and sustainability) (De Brauw et al. 2019). This section discusses some of the most relevant questions for Kenya based on the insights from this book and beyond, before we conclude the book with future research directions.

Should policies and investments be targeted more toward smallholder farmers or promote more larger-scale farming? On the one hand, targeting smallholders promotes inclusion. On the other hand, targeting larger farmers is more likely to improve productivity and even the financial sustainability of development programs. On this question for Kenya, results from Chapter 7 provide inconclusive evidence. The chapter also shows that the policy goal has often been to close the productivity gap between efficient and inefficient sub-counties, suggesting a policy bias toward smallholders. This policy model was imported to Africa because of the critical acclaim given to the Asian Green Revolution driven by small-scale farms—which is most farms in Africa at present (Hazell 2009). But Chapter 7 argues that, based on significant evidence from the most recent literature, farms between 20 and 70 ha are substantially more productive than farms under 5 ha (Muyanga and Jayne 2019). Results like these are a major challenge to the hypothesis that efficient smallholders are agents of change. Rather than an indication of efficiency, the small size of farms in Kenya could be part of a poverty trap whereby frictions in land markets prevent households from exiting agriculture to the extent that would be efficient (see, for example, Chen 2017; Gottlieb and Grobovšek 2019; and discussion in Gollin, Hansen, and Wingender 2021).

Are some farms becoming too small and should farmers be incentivized to move up or out of agriculture? (Fan 2014). Gollin and colleagues (2021) argue that, with well-functioning markets, it would be expected that the least effective farmers would move out of agriculture into other occupations, either selling or renting their land to farmers who are more skillful. Chapter 7 argues that the fact that this is not happening could imply that Kenya has institutional frictions or rigidities that prevent unproductive farmers from exiting the market (see discussion and references in Gollin, Hansen, and Wingender 2021). The outcome is aggregate inefficiency resulting from the misallocation of labor, land, capital, and managerial effort that creates a consequential drag on aggregate productivity.

Should the government and international partners invest more in Kenya's arid and semiarid lands (ASALs) or more in the more prosperous and fertile regions? Policy involving the ASALs touches on all the food system outcomes. Kenya's ASALs are on the frontline of the climate crisis and need heavy investment to promote climate-smart food systems and build resilience to increasingly volatile conditions. Further, ASALs are often marginalized in national and international development programs in favor of the more prosperous fertile regions, bringing up issues of inclusion. As Chapter 16 argues, strengthening production in the ASALs could also lessen the food system's dependence on maize and provide resilience to maize production shocks. However, investments here carry higher costs because of remoteness, conflict, and harsh conditions. Returns on investments may also be lower because of lack of market access and less favorable production conditions. Policy needs to adequately address these challenges, and research on ASALs should be increased to better inform policymakers.

Should the government simultaneously promote climate-smart agriculture and the increased use of chemical inputs? Chemical fertilizers vastly improve production, and the national government has been promoting their use for decades. But chemical fertilizers can also contaminate local water sources, and are subject to international price swings. At the heart of this question is a trade-off between increased production and sustainability and resilience. The government runs programs subsidizing chemical fertilizers alongside climate-smart agriculture programs (for example, the Kenya Climate Smart Agriculture Project). The recent drastic increase in natural gas and fertilizer prices is exposing the vulnerability of agricultural input markets to global market swings and making climate-smart agriculture a more appealing long-term alternative. However, policymakers and researchers need to consider how to incorporate climate-smart agriculture into existing policy frameworks to understand how to realize its benefits as a long-term alternative.

Other open questions that policymakers should consider and that require more research are: How can sustainability be built into food system development policies? How can public–private partnerships be leveraged while maintaining trust among food system actors and promoting sustainable business models? What regulations and institutions are needed to ensure that digitization efforts can be carried out while maintaining data privacy? How can national and international research agendas be aligned with evidence gaps in key policy areas? How can county governments be better empowered to foster food system transformation at the local level?

References

Abraham, D., T. Ngoga, J. Said, and M. Yachin. 2019. *How Israel Became a World Leader in Agriculture and Water: Insights for Today's Developing Countries*. London: Tony Blair Institute for Global Change.

Amarante, V., and J. Zurbrigg. 2020. "Research on Development Issues from the Southern World." Partnership for Economic Policy Working Paper No. 2020–24. Paternership for Economic Policy, Nairobi.

AUC (African Union Commission). 2014. "Malabo Declaration on Accelerated Agricultural Growth and Transformation for Shared Prosperity and Improved Livelihoods." Paper presented at the 23rd Ordinary Session of the African Union, June 30, 2014, Malabo, Equitorial Guinea.

Breisinger, C., X. Diao, P. Dorosh, J. Mbuthia, L. Omune, E.O. Oseko, A. Pradesha, J. Smart, and J. Thurlow. 2022. *Kenya : Impacts of the Ukraine and Global Crises on Poverty and Food Security*. Global Crisis Country Series Brief 1. Washington, DC: International Food Policy Research Institute (IFPRI).

Chen, C. 2017. "Untitled Land, Occupational Choice, and Agricultural Productivity." *American Economic Journal: Macroeconomics* 9 (4): 91–121.

Chevallier, R. 2022. "Policy Coherence Analysis in Climate-Smart Agriculture (CSA) in Africa." Working Paper No. 002. Accelerating Impacts of CGIAR Climate Research for Africa (AICCRA).

De Brauw, A., M. van den Berg, I.D. Brouwer, H. Snoek, R. Vignola, M. Melesse, G. Lochetti, C. van Wagenberg, M. Lundy, M.E. d'Hotel, and R. Ruben. 2019. *Food System Innovations for Healthier Diets in Low and Middle-Income Countries*. UN Food Systems Summit Brief. Bonn: UN Food Systems Summit.

EC (European Commission). 2020. *Resilience and Transformation: Report of the 5th SCAR Foresight Exercise Expert Group*. Luxembourg: Publications Office of the European Union.

Fan, S. 2014. "Moving Family Farms from Subsistence to Prosperity," *IFPRI Blog*, October 8.

FAO (Food and Agriculture Organization of the United Nations). 2019. *Rural Youth Employment and Agri-Food Systems in Kenya*. Rome.

FAO. 2022. FAOSTAT database. Accessed December 10, 2022. https://faostat.fao.org/

Gollin, D., C.W. Hansen, and A.M. Wingender. 2021. "Two Blades of Grass: The Impact of the Green Revolution." *Journal of Political Economy* 129 (8): 2344–2384.

Gottlieb, C., and J. Grobovšek. 2019. "Communal Land and Agricultural Productivity." *Journal of Development Economics* 138: 135–152.

Hazell, P.B.R. 2009. "The Asian Green Revolution." IFPRI Discussion Paper 911. IFPRI, Washington, DC.

Kenya, Ministry of Agriculture, Livestock, Fisheries, and Cooperatives . 2022. *Kenya Agricultural Sector Extension Policy: March 2022*. Nairobi.

Kenya, Ministry of East African Community and Regional Development. 2020. *Ease of Doing Business*. Nairobi.

Kenya, Ministry of Agriculture, Livestock, Fisheries, and Cooperatives. 2021. *National Agricultural Mechanization Policy*. Nairobi.

Kenya Parliamentary Budget Office. 2021. *Unpacking the Estimates of Revenue and Expenditure for 2020/2021 and the Medium Term*. Nairobi.

Kim, J., P. Shah, J.C. Gaskell, A. Prasann, and A. Luthra. 2020. *Scaling Up Disruptive Agricultural Technologies in Africa*. Washington, DC: World Bank.

Lowder, S.K., J. Skoet, and T. Raney. 2016. "The Number, Size, and Distribution of Farms, Smallholder Farms, and Family Farms Worldwide." *World Development* 87: 16–29.

Muyanga, M., and T.S. Jayne. 2019. "Revisiting the Farm Size-Productivity Relationship Based on a Relatively Wide Range of Farm Sizes: Evidence from Kenya." *American Journal of Agricultural Economics* 101 (4): 1140–1163.

Nafula, N., D. Kyalo, B. Munga, and R. Ngugi. 2020. *Poverty and Distributional Effects of COVID-19 on Households in Kenya*. Nairobi: African Economic Research Consortium.

Neven, D., M.M. Odera, T. Reardon, and H. Wang. 2009. "Kenyan Supermarkets, Emerging Middle-Class Horticultural Farmers, and Employment Impacts on the Rural Poor." *World Development* 37 (11): 1802–1811.

Roseboom, J., and H. Rutten. 1998. "The Transformation of the Dutch Agricultural Research System: An Unfinished Agenda." *World Development 26* (6): 1113–1126.

Singh, B.K., T. Arnold, P. Biermayr-Jenzano, J. Broerse, G. Brunori, P. Caron, O. De Schutter, et al. 2021. "Enhancing Science–Policy Interfaces for Food Systems Transformation." *Nature Food* 2 (11): 838–842.

Takeshima, H., X. Diao, and P.O. Aboagye. 2020. "Policies for Competitive and Sustainable Agricultural Production Systems: A Case Study of Ghana's Recent Mechanization Interventions." In *2020 Annual Trends and Outlook Report: Sustaining Africa's Agrifood System Transformation: The Role of Public Policies,* eds. D. Resnick, X. Diao, and G. Tadesse, 45–56. Washington, DC: IFPRI; Kigali: AKADEMIYA2063.

UNICEF. 2022. *Kenya Drought Situation* Issue 1. New York.

Contributors

Clemens Breisinger, Program Leader and Senior Research Fellow, International Food Policy Research Institute (IFPRI), Nairobi, Kenya.

Michael Keenan, Associate Research Fellow, IFPRI, Nairobi, Kenya.

Juneweenex Mbuthia, Research Officer, IFPRI, Nairobi, Kenya.

Jemimah Njuki, Chief, Economic Empowerment, UN Women, New York, USA.

Gashaw T. Abate, Research Fellow, IFPRI, Washington, DC, USA.

Kibrom A. Abay, Country Program Leader and Research Fellow, IFPRI, Cairo, Egypt.

Dolapo Adeyanju, Research Analyst, IFPRI, Abuja, Nigeria.

Silvia Alonso, Senior Scientist, International Livestock Research Institute (ILRI), Addis Ababa, Ethiopia.

Mulubrhan Amare, Research Fellow, IFPRI, Washington, DC, USA.

Jane Ambuko, Associate Professor, Department of Plant Science and Crop Protection, University of Nairobi, Nairobi, Kenya.

Kwaw Andam, Country Program Leader and Senior Research Fellow, IFPRI, Abuja, Nigeria.

Sirak Bahta, Senior Agricultural Economist, ILRI, Nairobi, Kenya.

Isabelle Baltenweck, Program Leader, ILRI, Nairobi, Kenya.

Salome A. Bukachi, Associate Professor, Institute of Anthropology, Gender and African Studies (IAGAS), University of Nairobi, Nairobi, Kenya.

Judith K. Chemuliti, Research Officer, Biotechnology Research Institute, Kenya Agricultural Research Institute, Kikuyu, Kenya.

Andrew R. Comstock, Senior Research Analyst, IFPRI, Washington, DC, USA.

Hugo De Groote, Principal Economist, International Maize and Wheat Improvement Center (CIMMYT), Nairobi, Kenya.

Xinshen Diao, Deputy Division Director and Senior Research Fellow, Foresight Policy Modeling Unit, IFPRI, Washington, DC, USA.

Olivier Ecker, Senior Research Fellow, IFPRI, Washington, DC, USA.

Dolapo Enahoro, Senior Agricultural Economist, ILRI, Nairobi, Kenya.

Andrew Hall, Senior Principle Scientist, Commonwealth Scientific and Industrial Research Organization (CSIRO), Canberra, Australia.

Vivian Hoffmann, Senior Research Fellow, IFPRI, Ontario, Canada.

Mercy Kamau, Senior Research Fellow and Director, Measurement, Learning and Evaluation Unit, Tegemeo Institute, Egerton University, Nairobi, Kenya.

Erastus Kang'ethe, Professor, Department of Public Health, Pharmacology and Toxicology, University of Nairobi, Nairobi, Kenya.

Joseph Karugia, Principal Scientist, ILRI, Nairobi, Kenya.

Pascal Kaumbutho, Founder and Managing Director, Agrimech Africa Ltd., Nairobi, Kenya.

Ivy Kinyua, Research Associate, Alliance of Bioversity International and CIAT, Nairobi, Kenya.

Lilian Kirimi, Senior Research Fellow, Tegemeo Institute of Agricultural Policy and Development, Egerton University, Nairobi, Kenya.

Leonard Kirui, Research Officer, ILRI, Nairobi, Kenya.

Berber Kramer, Senior Research Fellow, IFPRI, Nairobi, Kenya.

Jackson Langat, Measurement Learning and Evaluation Officer, Tegemeo Institute, Egerton University, Nairobi, Kenya.

Eric Manyasa, Product Placement Lead, International Crops Research Institute for the Semi-Arid Tropics (ICRISAT), Nairobi, Kenya.

Jane Mariara, Executive Director, Partnership for Economic Policy (PEP), Nairobi, Kenya.

Mequanint B. Melesse, Cluster Leader, ICRISAT, Nairobi, Kenya.

Charles Mensah, Research Officer, International Livestock Research Institute (ILRI), Accra, Ghana.

Victor Mugo, Head, Local Youth Action, World Food Forum, Food and Agriculture Organization of the United Nations (FAO), Rome.

Mercy M. Musyoka, Graduate Research Assistant, IAGAS, University of Nairobi, Nairobi, Kenya.

Michael Ndegwa, Associate Scientist, CIMMYT, Nairobi, Kenya.

Mariah Ngutu, Research Affiliate and Part-time Lecturer, IAGAS, University of Nairobi, Nairobi, Kenya.

Timothy Njagi, Development Economist and Research Fellow, Tegemeo Institute of Agricultural Policy and Development, Egerton University, Nairobi, Kenya.

Alejandro Nin-Pratt, Senior Research Fellow, IFPRI, Washington, DC, USA.

Isaac K. Nyamongo, Professor of Anthropology, Co-operative University of Kenya, Nairobi, Kenya.

Gideon Obare, Professor of Agricultural Economics, Egerton University, Nairobi, Kenya.

John Olwande, Research Fellow, Tegemeo Institute of Agricultural Policy and Development, Egerton University, Nairobi, Kenya.

Dalmas O. Omia, Research Fellow, IAGAS, University of Nairobi, Nairobi, Kenya,

Lensa Omune, Research Officer, IFPRI, Nairobi, Kenya.

Willis Owino, Associate Professor, Department of Food Science and Technology, Jomo Kenyatta University of Agriculture and Technology, Nairobi, Kenya.

Karl Pauw, Senior Research Fellow, IFPRI, Washington, DC, USA.

Apurba Shee, Associate Professor of Applied Economics, Natural Resources Institute, University of Greenwich, UK.

Bekele Shiferaw, Senior Evaluation Specialist, World Bank, Washington, DC, USA.

Jenny Smart, Senior Program Manager, IFPRI, Washington, DC, USA.

David J. Spielman, Director, Innovation Policy and Scaling Unit, IFPRI, Washington, DC, USA.

Martin Paul Jr. Tabe-Ojong, Associate Research Fellow, IFPRI, Cairo, Egypt.

Hiroyuki Takeshima, Senior Research Fellow, IFPRI, Washington, DC, USA.

Yohannis Mulu Tessema, Independent Researcher, Addis Ababa, Ethiopia.

James Thurlow, Director, Foresight Policy and Modeling Unit, IFPRI, Washington, DC, USA.

Calum G. Turvey, W.I. Myers Professor of Agricultural Finance, Cornell University, Ithaca, NY, USA.

John M. Ulimwengu, Senior Research Fellow, IFPRI, Washington, DC, USA.

Francis Wanyoike, Research Associate, ILRI, Nairobi, Kenya.

Liangzhi You, Senior Research Fellow, IFPRI, Washington, DC, USA.

Made in United States
Orlando, FL
04 November 2026

90207781R00295